Applied Mathematical Sciences
Volume 171

T0237270

For further volumes:
http://www.springer.com/series/34

Laurent Younes

Shapes and Diffeomorphisms

 Springer

Laurent Younes
Center for Imaging Science
Clark 324C
The Johns Hopkins University
3400 N.Charles Street
Baltimore MD 21218-2686
USA

Editors:

S.S. Antman
Department of Mathematics
and
Institute for Physical
    Science and Technology
University of Maryland
College Park, MD 20742-4015
USA
ssa@math.umd.edu

J.E. Marsden
Control and Dynamical
    Systems, 107-81
California Institute of
    Technology
Pasadena, CA 91125
USA
marsden@cds.caltech.edu

L. Sirovich
Laboratory of Applied
    Mathematics
Department of
    Biomathematical Sciences
Mount Sinai School
    of Medicine
New York, NY 10029-6574
isirovich@rockefeller.edu

ISSN 0066-5452
ISBN 978-3-642-26348-4   ISBN 978-3-642-12055-8(eBook)
DOI 10.1007/978-3-642-12055-8
Springer Heidelberg Dordrecht London New York

Mathematics Subject Classification (2010): 93B05, 1502, 57R27, 22E99, 37C10

Cover design: deblik, Berlin

Printed on acid-free paper

Springer is part of Springer Science+Business Media (www.springer.com)

*To Geneviève,
Hannah, Salomé, Simon*

Introduction

Shape is a fascinating object of study. Understanding how a single shape can incur a complex range of transformations, while defining the same perceptually obvious figure, entails a rich and enticing collection of problems, at the interface between applied mathematics, statistics and computer science. Various applications in computer vision, object recognition and medical imaging bring additional motivation for researchers to develop adequate theoretical background and methodology for solving these problems.

This book is an attempt at providing a description of the large range of methods that have been invented to represent, detect, or compare shapes (or more generally, deformable objects), together with the necessary mathematical background that they require. While being certainly a book on applied mathematics, it is also written in a way that will also be of interest to an engineering- or computer-science-oriented reader, including in several places concrete algorithms and applicable methods, including experimental illustrations.

The book starts with a discussion of shape representation methods (Chapters 1–4), including classical aspects of the differential geometry of curves and surfaces, but borrowing also from other fields that have positively impacted the analysis of shape in practical applications, like invariant moments or medial axes. Discretization issues, involved in the representation of these geometric objects in a digital world, are also discussed.

The second part (Chapters 5–7) studies curve and surface evolution algorithms and how they relate to segmentation methods that can be used to extract shapes from images, using active contours or deformable templates. A reader with enough background in differential geometry may start reading the book at Chapter 6, or at Chapter 8 if the main focus of interest is on diffeomorphic registration and comparison methods.

In Chapters 8 and 9, basic concepts related to diffeomorphisms are introduced, discussing in particular how using ordinary differential equations associated to vector fields belonging to reproducing kernel Hilbert space provide a computationally convenient framework to handle them. Chapters 10 and 11 then focus on the registration of deformable objects using diffeomorphisms; in Chapter 10, we catalog a large spectrum of deformable objects and discuss matching functionals that can be used to compare them. Chapter 11 addresses diffeomorphic matching, and focuses in particular on methods that optimize a matching functional combined with a regularization term that penalizes the distance of a diffeomorphism to the identity within the group.

The last two chapters (12 and 13) discuss metric aspects of shape analysis, with a special focus on the relation between distances and group actions. Both the global and infinitesimal points of view are presented. The classic Kendall's metric over configurations of labeled points is included, as well as a short discussion of Riemannian metrics on plane curves. Chapter 13 provides a presentation of the theory of metamorphosis.

In the Appendices are provided fundamental concepts that are needed in order to understand the rest of the book. The main items are some elements of Hilbert space theory (Appendix A), of differential and Riemannian geometry (Appendix B) and of ordinary differential equations (Appendix C). In all cases, the appendices do not provide a comprehensive presentation of these theories, but simply what is needed in the particular context of the book. Appendices D to F are short presentations of optimization algorithms, principal component analysis and dynamic programming.

Chapters 1 to 5, which are (with a few exceptions) rather elementary, provide an introduction to applied differential geometry that can be suitable for an advanced undergraduate class. They can be combined with Chapter 6 to form a graduate-level class on the same subject. Chapters 8 to 13 represent a specialized, advanced graduate topic.

I would like to thank my students and collaborators, who have helped make the ideas that are developed in these notes reach their current state of maturation. I would like, in particular, to express my gratitude to Alain Trouvé and Michael Miller, with whom the collaboration over the last decade has been invaluable. Special thanks also go to Darryl Holm, David Mumford and Peter Michor. This book was written while the author was partially supported by the National Science Fundation, the National Institute of Health and the Office for Naval Research.

Contents

List of Figures

1

Parametrized Plane Curves

1.1 Definitions

We start with some definitions.

Definition 1.1. *A (parametrized plane) curve is a continuous mapping $m :$ $I \to \mathbb{R}^2$, where $I = [a, b]$ is an interval.*

The curve m is closed if $m(a) = m(b)$.

A curve m is a Jordan curve if it is closed and m has no self-intersection: $m(x) = m(y)$ only for $x = y$ or $\{x, y\} = \{a, b\}$.

The curve is piecewise C^1 if m has everywhere left and right derivatives, which coincide except at a finite number of points.

The range of a curve m is the set $m([a, b])$. It will be denoted $\mathcal{R}_m$.

Notice that we have defined curves as functions over bounded intervals. Their range must therefore be a compact subset of $\mathbb{R}^2$ (this forbids, in particular, curves with unbounded branches).

A Jordan curve is what we can generally accept as a definition of the outline of a shape. An important theorem states that the range of a Jordan curve partitions the plane $\mathbb{R}^2$ into two connected regions: a bounded one, which is the interior of the curve, and an unbounded one (the exterior).

However, requiring only continuity for curves is a somewhat loose requirement, and allows for highly irregular curves. This is why we will always restrict ourselves to piecewise C^1, generally Jordan, curves. We will in fact often ask for more, and consider curves which are regular (or piecewise regular) in $\mathbb{R}^2$. Let $\dot{m}_u$ denote the derivative of a curve m with respect to its parameter u.

Definition 1.2. *A C^1 curve $m : I \mapsto \mathbb{R}^2$ is a regular curve if $\dot{m}_u \neq 0$ for all $u \in I$. If m is only piecewise C^1, we extend the definition by requiring that all left and right derivatives are non-vanishing.*

The previous definition is fundamental. This avoids, in particular, curves which are smooth functions (C^∞, for example) but with a range having geometric singularities. Consider the following example: let

$$m(t) = \begin{cases} (\varphi(t), 0), & t \in [0, 1/2] \\ (1, \varphi(t - 1/2)), & t \in [1/2, 1] \end{cases}$$

with $\varphi(t) = 16t^2(1-t)^2$, $t \in [0,1]$. It is easy to check that m is continuously differentiable, whereas the range of m is the corner $[0,1] \times \{0\} \cup \{1\} \times [0,1]$.

1.2 Geometric Equivalence

1.2.1 Open Curves

Definition 1.3. *Let* $m : I \to \mathbb{R}^2$ *be a plane curve. A change of parameter for* m *is a continuous function* $\psi : I' \to I$ *such that:*
(i) I' *is a bounded interval;*
(ii) ψ *is continuous, increasing (strictly) and onto (a homeomorphism).*
The new curve $\tilde{m} = m \circ \psi$ *is called a reparametrization of* m. *The ranges* $\mathcal{R}_m$ *and* $\mathcal{R}_{\tilde{m}}$ *coincide.*

When m belongs to a specific smoothness class, the same properties will be implicitly required for the change of parameter. For example, if m is (piecewise) C^1, ψ will also be assumed to be (piecewise) C^1 (in addition to the previous properties).

It is easy to see that the property for two curves of being related by a change of parameter is an equivalence relation, which is called "parametric equivalence". We will denote the parametric equivalence class of m by $[m]$. A property, or a quantity, which only depends on $[m]$ will be called geometric. For example, the range of a curve is a geometric notion.

Note that the converse is not true. If two curves have the same range, they are not necessarily parametrically equivalent: the piecewise C^1 curve defined on $I = [0,1]$ by $m(t) = (2t, 0), t \in [0, 1/2]$ and $m(t) = (2 - 2t, 0)$, $t \in [1/2, 1]$ has the same range as $\tilde{m}(t) = (t, 0), t \in [0, 1]$ but they are not equivalent. Also, if m is a curve defined on $I = [0,1]$, then $\tilde{m}(t) = m(1 - t)$ has the same range, but is not equivalent to m, since we have required the change of parameter to be increasing (changes of orientation are not allowed).

1.2.2 Closed Curves

Changes of parameters for closed curves must be slightly more general, because the starting point of the parametrization is not uniquely defined. For example, a circle can be parametrized as $m(t) = (r \cos t, r \sin t)$ for $t \in [0, 2\pi]$ and by the same equation for $t \in [-\pi, \pi]$. The relation between these parametrizations is not an increasing diffeomorphism, but the relation $\psi : [0, 2\pi] \to [-\pi, \pi]$ such that $\varphi(t) = t$ on $[0, \pi]$ and $\psi(t) = t - 2\pi$ on $[\pi, 2\pi]$. From a geometric point of view, this corresponds to changing the starting point of the parametrization from $(1, 0)$ to $(-1; 0)$ (considered as points in the unit circle).

Therefore, changes of parameter for closed curves must specify a new starting point for the parametrization (offset) as well as an increasing diffeomorphism. An additional restriction is that it should wrap smoothly: the right derivative(s) at the initial point must coincide with the left derivative(s) at the final point.

1.3 Unit Tangent and Normal

Let $m : I \to \mathbb{R}^2$ be a regular curve. The unit tangent at $u \in I$ is the vector

$$T_m(u) = \frac{\dot{m}_u}{|\dot{m}_u|}.$$

The unit normal is the unique vector $N_m(u)$ which extends T_m to a positively oriented orthonormal basis of $\mathbb{R}^2$: $(T_m(u), N_m(u))$ is orthonormal and $\det[T_m(u), N_m(u)] = 1$. The subscript m is generally dropped in the absence of ambiguity.

The frame (T, N) is a geometric notion: if $\varphi : I \to \tilde{I}$ is a C^1 change of parameter, and $m = \tilde{m} \circ \varphi$, then $T_{\tilde{m}}(\varphi(u)) = T_m(u)$ and similarly for the normal. It is therefore acceptable to consider T_m as attached to the point $m(u)$ and not to the parameter u. We will indifferently use the notation $T(u)$ or $T(m(u))$.

1.4 Arc Length and Curvature

Let $m : [a, b] \to \mathbb{R}^2$ be a parametrized curve. If $a = u_0 < u_1 < \cdots < u_n < u_{n+1} = b$ is a subdivision of $[a, b]$, the length of m can be approximated by $\sum_{i=1}^{n+1} |m(u_i) - m(u_{i-1})|$. Since this can be written

$$\sum_{i=1}^{n+1} \frac{|m(u_i) - m(u_{i-1})|}{u_i - u_{i-1}} (u_i - u_{i-1})$$

this converges, when the curve is C^1 (and also piecewise C^1) to the integral

$$\int_a^b |\dot{m}_u| du.$$

This leads to the definition:

Definition 1.4. *Let $m : [a, b] \to \mathbb{R}$ be a piecewise C^1 curve. Its length, L_m is defined by*

$$L_m = \int_a^b |\dot{m}_u| du.$$

The function $s_m : [a, b] \mapsto [0, L_m]$ defined by

$$s_m(u) = \int_a^u |\dot{m}_u| du$$

is called the arc length of the curve m (it is the length of the arc between $m(a)$ and $m(u)$).

We have the important result:

Proposition 1.5. *If m is a regular curve, then s_m is a change of parameter.*

Proof. Indeed, if m is regular, then $|\dot{m}_u| > 0$ for all u (or all but a finite number if m is piecewise C^1), Then s_m is strictly increasing, and onto by definition. Note also that s_m is differentiable with

$$\frac{ds_m}{du} = |\dot{m}_u|.$$

$\square$

The arc length is a geometric quantity: if $\tilde{m}$ is a curve, and $m = \tilde{m} \circ \psi$ a new parametrization of m, then

$$s_{\tilde{m}}(\psi(u)) = s_m(u).$$

This is a direct application of the change of variable formula:

$$s_m(u) = \int_{\psi^{-1}(a)}^u |\dot{m}_v| dv$$

$$= \int_{\psi^{-1}(a)}^u |\dot{\tilde{m}}_v \circ \psi(v)| \dot{\psi}_v dv$$

$$= \int_a^{\psi(u)} |\dot{\tilde{m}}_v| dv.$$

The curve $m \circ s_m^{-1}$ is called the arc length reparametrization of m. A curve m is said to be parametrized with arc length if $s_m = $ identity, or, equivalently, if $|\dot{m}_u| = 1$ for all t.

The curvature of a C^2 curve m is the speed of rotation of the tangent to m when m is parametrized with arc length. If $\theta(u)$ is the angle between $T(u)$ and the horizontal axis, then the curvature $\kappa_m(u)$ is defined by

$$\kappa_m(u) = \lim_{\varepsilon \to 0} \frac{\theta(u+\varepsilon) - \theta(u)}{s_m(u+\varepsilon) - s_m(u)} = \dot{\theta}_u/(\dot{s}_m)_u \tag{1.1}$$

and the usual notation is

$$\kappa_m = \frac{d\theta}{ds_m}.$$

In (1.1), the infinitesimal difference between angles $(\theta(u+\varepsilon) - \theta(u))$ is assumed to be the smallest possible choice in absolute value, modulo 2π.

Writing $T = (\cos\theta, \sin\theta)$ and $N = (-\sin\theta, \cos\theta)$ it is easy to prove the relation

$$\frac{dT}{ds_m} = \kappa N$$

which is often taken as the definition of the curvature.

Note also that the same kind of easy computation yields

$$\frac{dN}{ds_m} = -\kappa T.$$

The curvature is a geometric quantity: if $m = \tilde{m} \circ \psi$, then $\kappa_{\tilde{m}} \circ \psi = \kappa_m$; this can be deduced directly from (1.1) and the fact that both θ and s_m are geometric quantities.

1.5 Expression in Coordinates

1.5.1 Cartesian Coordinates

To provide explicit formulae for the quantities that have been defined so far, we introduce the space coordinates (x, y) and write, for a curve m: $m(u) = (x(u), y(u))$. The first, second and higher derivatives of x will be denoted $\dot{x}_u, \ddot{x}_{uu}, x_{uuu}^{(3)}, \ldots$ and similarly for y. The tangent and the normal vectors are

$$T = \frac{1}{\sqrt{\dot{x}_u^2 + \dot{y}_u^2}} \begin{pmatrix} \dot{x}_u \\ \dot{y}_u \end{pmatrix}, \quad N = \frac{1}{\sqrt{\dot{x}_u^2 + \dot{y}_u^2}} \begin{pmatrix} -\dot{y}_u \\ \dot{x}_u \end{pmatrix}.$$

The arc length is $ds_m = \sqrt{\dot{x}_u + \dot{y}_u}\, du$ and the curvature is

$$\kappa = \frac{\dot{x}_u \ddot{y}_{uu} - \dot{y}_u \ddot{x}_{uu}}{(\dot{x}_u^2 + \dot{y}_u^2)^{3/2}}. \tag{1.2}$$

The last formula is proved as follows. Since $\kappa N = dT/ds_m$, we have

$$\kappa = N^T \frac{dT}{ds_m}.$$

Since $T = dm/ds_m = \dot{m}_u/(\partial_u s_m)$, we have

$$N^T \frac{dT}{ds_m} = (\partial_u s_m)^{-2} \langle \ddot{m}_{uu}, N \rangle + (\partial_u s_m)^{-1} \partial_u \left((\partial_u s_m)^{-1} \right) \langle \dot{m}_u, N \rangle.$$

The last term vanishes, and the first one gives (1.2) after introducing the coordinates.

1.5.2 Polar Coordinates

Let (Oxy) be a fixed frame. A point m in the plane can be characterized by its distance, r, to the origin, O, and by θ, the angle between the horizontal axis (Ox) and the half-line Om. (Notice that this is different from the angle of the tangent with the horizontal for which we also used θ. These are unfortunately standard notation for both cases.) The relation between the Cartesian coordinates (x, y) of m and its polar coordinates (r, θ) is $x = r\cos\theta$ and $y = r\sin\theta$. This representation is unique, except for $m = O$ for which θ is undetermined.

A polar parametrization of a curve $t \mapsto m(t)$ is a function $t \mapsto (r(t), \theta(t))$. Often, the parameter t coincides with the angle θ and and the parametrization boils down to a function $r = f(\theta)$. Some shapes have very simple polar coordinates, the simplest being a circle centered at O for which the equation is $r = \text{cst}$.

Let us compute the Euclidean curvature for such a parametrization. Denote $\tau = (\cos\theta, \sin\theta)$ and $\nu = (-\sin\theta, \cos\theta)$. We have $m = r\tau$, and

$$\dot{m}_u = \dot{r}_u.\tau + r\dot{\theta}_u\nu\,,$$

$$\ddot{m}_{uu} = (\ddot{r}_{uu} - r\dot{\theta}_u^2)\tau + (2\dot{r}_u\dot{\theta}_u + r\ddot{\theta}_{uu})\nu.$$

Therefore,

$$\kappa = \frac{\det[\dot{m}_u, \ddot{m}_{uu}]}{|\dot{m}_u|^3} = \frac{r^2(\dot{\theta}_u)^3 - r\ddot{r}_{uu}\dot{\theta}_u + 2\dot{r}_u^2\dot{\theta}_u + r\dot{r}_u\ddot{\theta}_{uu}}{(\dot{r}_u^2 + r^2\dot{\theta}_u^2)^{3/2}}.$$

When the curve is defined by $r = f(\theta)$, we have $\theta = u$, $\dot{\theta}_u = 1$ and $\ddot{\theta}_{uu} = 0$, so that

$$\kappa = \frac{r^2 - r\ddot{r}_{uu} + \dot{r}_u^2}{(\dot{r}_u^2 + r^2)^{3/2}}.$$

This representation does not have the same invariance properties as the arc length (see next section), but still has some interesting features. Scaling by a factor λ simply corresponds to multiplying r by λ. Making a rotation with center O and angle α simply means replacing θ by $\theta + \alpha$. However, there is no simple relation for a translation. This is why a curve is generally expressed in polar coordinates with respect to a well-chosen origin, like its center of gravity.

1.6 Euclidean Invariance

The arc length and the curvature have a fundamental invariance property. If a curve is transformed by a rotation and translation, both quantities are invariant. The rigorous statement of this is as follows. Let R be a planar rotation and b a vector in $\mathbb{R}^2$. Define the transformation $g : \mathbb{R}^2 \to \mathbb{R}^2$ by $g(p) = R\bar{p} + b$. Then, if $m : I \to \mathbb{R}^2$ is a plane curve, one can define $g \cdot m : I \to \mathbb{R}^2$ by $(g \cdot m)(t) = g(m(t)) = Rm(t) + b$. Then, the statements are:

(i) $s_{g \cdot m}(t) = s_m(t)$, and in particular $L_{g \cdot m} = L_m = L$.

(ii) The curvatures κ_m and $\kappa_{g \cdot m}$, defined on $[0, L]$ (as functions of the arc length), coincide.

Even more important is the converse statement of (ii).

Theorem 1.6. *If two C^2 plane curves m and $\tilde{m}$ have the same curvature as a function of the arc length, $\kappa : [0, L] \rightarrow \mathbb{R}$, then there exist R and b such that $\tilde{m} = Rm + b$.*

We now proceed to the proof of these results.

Proof. If $\tilde{m} = Rm + b$, we have $\dot{\tilde{m}}_t = R\dot{m}_t$ and $\ddot{\tilde{m}}_{tt} = R\ddot{m}_{tt}$. In particular, $ds_{\tilde{m}} = |\dot{\tilde{m}}_t|dt = |R\dot{m}_t|dt = |\dot{m}_t|dt = ds_m$ since norms are conserved by rotations. This implies that $s_{\tilde{m}} = s_m$.

We have seen that the curvature was given by $\kappa_{\tilde{m}} = \det[\dot{\tilde{m}}_t, \ddot{\tilde{m}}_{tt}]/|\dot{\tilde{m}}_t|^3$. The denominator is also equal to $|\dot{m}_t|^3$ as before and

$$\det[\dot{\tilde{m}}_t, \ddot{\tilde{m}}_{tt}] = \det[R\dot{m}_t, R\ddot{m}_{tt}] = \det(R[\dot{m}_t, \ddot{m}_{tt}]) = \det R \det[\dot{m}_t, \ddot{m}_{tt}]$$

which proves that $\kappa_{\tilde{m}} = \kappa_m$ since $\det R = 1$.

Now, let $\kappa : [0, L] \rightarrow \mathbb{R}$ be an integrable function. We build all possible curves m that have κ as curvature and prove that they all differ by a rotation and translation. The curve m must be parametrized by arc length over $[0, L]$. The angle θ_m, defined on $[0, L]$, must satisfy:

$$\partial_t \theta_m = \kappa \text{ and } \dot{m}_t = (\cos \theta_m, \sin \theta_m).$$

Let $\theta(s) = \int_0^s \kappa(u)du$. The first equality implies that, for some $\theta_0 \in [0, 2\pi)$, we have $\theta_m(s) = \theta(s) + \theta_0$ for all $s \in [0, L]$. The second implies that, for some $b \in \mathbb{R}^2$,

$$m(s) = \int_0^s (\cos(\theta(u) + \theta_0), \sin(\theta(u) + \theta_0))du + b.$$

Introduce the rotation $R = \begin{pmatrix} \cos \theta_0 & -\sin \theta_0 \\ sin\theta_0 & \cos \theta_0 \end{pmatrix}$. From standard trigonometric formulas, we have

$$R \begin{pmatrix} \cos \theta \\ \sin \theta \end{pmatrix} = \begin{pmatrix} \cos(\theta + \theta_0) \\ \sin(\theta + \theta_0) \end{pmatrix}$$

so that, letting $m^*(s) = \int_0^s (\cos \theta(u), \sin \theta(u))du$, we have $m = Rm^* + b$. Since m^* is uniquely defined when κ is given, we obtain the fact that m is defined up to a rotation and translation. $\square$

1.7 Enclosed Area and Green (Stokes) Formula

When a closed curve m is smooth and has no self-intersection, its enclosed area can be computed with a single integral instead of a double one. Let

Ω_m be the bounded connected component of $\mathbb{R}^2 \setminus \mathcal{R}_m$. We assume that m is defined on $I = [a, b]$, and that *the curve is oriented so that the normal N points inward*. Since this is a convention that will be used repeatedly, we state it as a definition.

Definition 1.7. *A closed regular curve oriented so that the normal points inward is said to be positively oriented.*

We have the following proposition:

Proposition 1.8. *Using the notation above, and assuming that m is positively oriented, we have*

$$Area(\Omega_m) = \int_{\Omega_m} dx dy = -\frac{1}{2} \int_a^b \langle m(u), N(u) \rangle |\dot{m}_u(u)| \, du. \qquad (1.3)$$

Note that the last integral can also be written $-(1/2) \int_0^L \langle m(s), N(s) \rangle ds$ where s is the arc length. We also have $\langle m(s), N(s) \rangle = -\det[m(s), T(s)]$ which provides an alternate expression. We have indeed

$$-(1/2) \int_0^L \langle m(s), N(s) \rangle ds = (1/2) \int_0^L \det[m(s), T(s)] ds$$

$$= (1/2) \int_a^b \det[m(u), T(u)] |\dot{m}_u(u)| du$$

so that, using $T(u) = \dot{m}_u(u)/|\dot{m}_u(u)|$,

$$Area(\Omega_m) = (1/2) \int_a^b \det[m(u), \dot{m}_u(u)] du. \qquad (1.4)$$

We will not prove Proposition 1.8, but simply remark that (1.3) is a particular case of the following important theorem.

Theorem 1.9. *Green's formula. If $f : \mathbb{R}^2 \to \mathbb{R}^2$ is a smooth function (a vector field), then*

$$\int_a^b \langle f(m(u)), N(u) \rangle |\dot{m}_u(u)| \, du = -\int_{\Omega_m} \mathrm{div} f \, dx dy \qquad (1.5)$$

where, letting $f(x, y) = (u(x, y), v(x, y))$, $\mathrm{div} f = \dot{u}_x + \dot{v}_y$.

To retrieve equation (1.3) from equation (1.5), take $f(x, y) = (x, y)$ for which $\mathrm{div} f = 2$. Note that Green's formula is sometimes given with a plus sign, N being chosen as the *outward* normal.

Formula (1.3) can also be nicely interpreted as the limit of an algebraic sum of triangle areas. For this, consider a polygonal discretization of m with vertices $p_1, \ldots, p_N$. Let O be an arbitrary point in $\mathbb{R}^2$.

First consider the simple case in which the segment Op_k is included in the region Ω_m for all k (the polygonal curve is said to be star shaped with respect to O). In this case, the area enclosed in the polygon is the sum of the areas of the triangles. The area of (O, p_k, p_{k+1}) is $|\det[p_k p_{k+1}, Op_k]|/2$.[1] Assuming that the discretization is counterclockwise, which is consistent with the fact that the normal points inward, the vectors Op_k and $p_k p_{k+1}$ make an angle between 0 and π, which implies that their determinant is positive. We therefore get

$$\text{Area}(\Omega_m) = \frac{1}{2} \sum_{k=1}^{N} \det[Op_k, p_k p_{k+1}]. \tag{1.6}$$

Since this can be written $\frac{1}{2} \sum_{k=1}^{N} \det[Op_k, p_k p_{k+1}/|p_k p_{k+1}|]|p_k p_{k+1}|$, this is consistent with the continuous formula

$$\frac{1}{2} \int_0^L \det[Om(s), T(s)]ds.$$

The interesting fact is that (1.6) is still valid for polygons which are not star shaped around the origin. In this case, the determinant may take negative signs, which provides a necessary correction because, for general polygons, some triangles can intersect $\mathbb{R}^2 \setminus \Omega_m$.

Finally, the area and the perimeter of a simple closed curve are related by a classic inequality.

Theorem 1.10 (Isoperimetric Inequality). *It m is a simple closed curve with perimeter L and area A, then*

$$4\pi A \le L^2 \tag{1.7}$$

with equality if and only if m is a circle.

1.8 Rotation Index and Winding Number

Let m be a closed, C^1, plane curve, defined on $I = [a, b]$. The unit tangent $T : [a, b] \to S^1$ (the unit circle) can be written as a function $t \mapsto (\cos\theta(t), \sin\theta(t))$ where θ is a *continuous* function. This is not a completely obvious statement, since the equation $T = (\cos\theta(t), \sin\theta(t))$ defines θ up to a multiple of 2π. The difficulty is to pick this constant for each t so that the associated map is continuous, which is always possible, and is a particular instance of what is called a *lifting* theorem.

Since m is closed, we must have $T(b) = T(a)$, which implies that, for some k, one has $\theta(b) = \theta(a) + 2k\pi$. The integer k in this last expression is called the *rotation index* of the curve, and denoted r_m.

---

[1] The general expression of the area of a triangle (A, B, C) is $|\det(AB, AC)|/2$, half the area of the parallelogram formed by the two vectors.

The rotation index is a geometric quantity, since its value does not depend on the parametrization. If the curve is regular and C^2, then, taking the arc length parametrization, we find, using $\kappa = d\theta/ds$,

$$\theta(L) - \theta(0) = \int_0^L \kappa(s)ds$$

or

$$r_m = \frac{1}{2\pi} \int_0^L \kappa(s)ds.$$

This provides an algebraic count of the number of loops in the curve: a loop is counted positively if it is parametrized counter-clockwise (normal inward), and negatively otherwise. An "8", for example, has a rotation number equal to 0. This also provides an alternate definition of a positively oriented curve: *a simple closed curve is positively oriented if and only if its rotation index is* +1.

A similar notion is the winding number of the curve. It depends on a reference point $p_0 \in \mathbb{R}^2$, and is based on the angle between $p_0 m(t)/|p_0 m(t)|$ and the horizontal axis, which is again assumed to be continuous in t. Denoting this angle $\alpha_{p_0}(t)$, the winding number of m around p_0 is

$$w_{p_0}(m) = (\alpha_{p_0}(b) - \alpha_{p_0}(a))/2\pi.$$

It provides the number of times the curve loops around p_0. Again, it depends on the curve orientation.

If a curve is *simple* (i.e., it has no self-intersection), then it is intuitively obvious that it can loop only once. This is the statement of the theorem of turning tangents, which says that *the rotation index of a simple closed curve is either 1 or −1.* Proving this statement is not so obvious anyway (even in the differentiable case we consider), and the reader may refer to [64] for a proof.

1.9 More on Curvature

There is an important relation between positive curvature (for positively oriented curves) and convexity. One says that a simple closed curve is convex if the bounded region it outlines is convex (it contains all line segments between any two of its points). Another characterization of convexity is that the curve lies on a single side of any of its tangent lines. The relation with the curvature is stated in the next theorem.

Theorem 1.11. *A positively oriented C^2 curve is convex if and only if its curvature is everywhere positive.*

We only provide a partial justification of the only if part. Assume that m is positively oriented and that its interior, Ω_m, is convex. For a fixed arc length, s and ε small enough, we have (since m is positively oriented): $m(s)+\varepsilon N(s) \in \Omega_m$ if $\varepsilon > 0$ and $\in \overline{\Omega}_m^c$ if $\varepsilon < 0$. Now, using a second order expansion around s, we get

$$\frac{1}{2}(m(s+h) + m(s-h)) = m(s) + \frac{h^2}{s}\kappa(s)N(s) + o(h^2)$$

and this point cannot be in Ω_m if h is small and $\kappa(s) < 0$.

The local extrema of the curvature are also of interest. They are called the vertices of the curve. The four-vertex theorem, which we also state without proof, is another classic result for plane curves [153, 44, 164].

Theorem 1.12. *Every simple closed C^2 curve has at least four vertices.*

1.10 Discrete Curves and Curvature

1.10.1 Least-Squares Approximation

Because it involves a ratio of derivatives, the computation of the curvature is numerically unstable (very sensitive to noise). We give here a brief account of how one can deal with this issue.

Assume that the curve is discretized in a finite sequence of points, say $m(1), \ldots, m(N)$. The usual finite-difference approximations of derivatives are:

$$m'(k) = (m(k+1) - m(k-1))/2;$$
$$m''(k) = m(k+1) - 2m(k) + m(k-1).$$

The simplest formula for the approximate curvature is then

$$\kappa(k) = \frac{\det[m'(k), m''(k)]}{|m'(k)|^3}.$$

This is very unstable. A small variation in the position of $m(k)$ can have large consequences on the value of the estimated curvature. To be robust, curvature estimation has to include some kind of smoothing. The simplest procedure is to fit a polynomial of order 2 at each point. This is what we describe now.

Fix an approximation scale $\Delta \geq 1$ (Δ is an integer). For each k, compute three two-dimensional vectors $a(k), b(k), c(k)$ in order to have, for $-\Delta \leq l \leq \Delta$:

$$m(k+l) \simeq a(k)\frac{l^2}{2} + b(k)l + c(k).$$

Once this is done, $b(k)$ will be our approximation of the first derivative of m and $a(k)$ our approximation of the second derivative. The curvature will then be approximated by

$$\kappa(k) = \frac{\det[b(k), a(k)]}{|b(k)|^3}.$$

We will use least-squares estimation to compute a, b, c. We first build the matrix

$$A = \begin{pmatrix} \sum_{l=-\Delta}^{\Delta} \frac{l^4}{4} & 0 & \sum_{l=-\Delta}^{\Delta} \frac{l^2}{2} \\ 0 & \sum_{l=-\Delta}^{\Delta} l^2 & 0 \\ \sum_{l=-\Delta}^{\Delta} \frac{l^2}{2} & 0 & 2\Delta + 1 \end{pmatrix}$$

which is the matrix of second moments for the "variables" $l^2/2, l$ and 1. They can be computed in closed form as a function of Δ, but their computation as sums takes negligible time in any computer implementation, so we leave them as such. The second computation is, for all k:

$$z_0(k) = \sum_{l=-\Delta}^{\Delta} m(k+l),$$

$$z_1(k) = \sum_{l=-\Delta}^{\Delta} l\, m(k+l),$$

$$z_2(k) = \sum_{l=-\Delta}^{\Delta} \frac{l^2}{2} m(k+l).$$

Given this, the vectors $a(k), b(k), c(k)$ are provided by the row vectors of the matrix

$$A^{-1} \begin{pmatrix} z_2(k) \\ z_1(k) \\ z_0(k) \end{pmatrix}$$

where z_0, z_1, z_2 are also written as row vectors. As shown in Figure 1.1, this method gives reasonable results for smooth curves. However, if the figure has sharp angles, the method will oversmooth and underestimate the curvature.

1.10.2 Curvature and Distance Maps

There is another way to describe curvature, which is related to the notion of distance map to the curve. We consider a C^2 regular curve $s \mapsto m(s)$, parametrized with arc length, over the interval $[0, L]$. We assume that m is closed and simple (no self-intersection). As before, we let Ω_m denote the interior region of the curve m.

Let S^1 be the unit circle in $\mathbb{R}^2$ and consider the band

$$V_m^{(\varepsilon)}(s) = \{m(s) + tN(s), s \in [0, L], t \in (0, \varepsilon), N \text{ normal to } m \text{ at } s\}.$$

One can show that, for t small enough, the region $V_m^{(\varepsilon)}$ coincides with the set of points p such that $d(p, \Omega_m) \in (0, t)$.

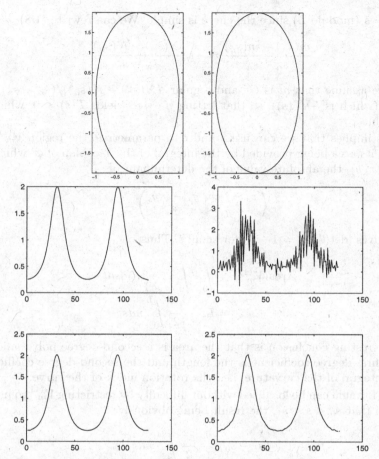

Fig. 1.1. Noise and curvature. The first curve on the left is an ellipse discretized over 125 points. The second on the right is the same ellipse, with an added Gaussian noise of standard deviation 10^{-3}. The difference is imperceptible. However, the second row shows the result of estimating the curvature without smoothing, on the first and the second ellipse, with a very strong noise effect. The last row shows the result of the second-order approximation with $\Delta = 5$, which is much better.

We want to compute the area of $V_m^{(\varepsilon)}$ when ε is small enough. The first remark is that, for small ε, the map $\varphi : (s, t) \mapsto m(s) + tN(s)$ is one-to-one on the set $(0, L) \times (0, \varepsilon)$. If this is not true, we can find sequences $(s_n), (\tilde{s}_n)$ and t_n such that $s_n \neq \tilde{s}_n$, $t_n \to 0$ and

$$m(s_n) + t_n N(s_n) = m(\tilde{s}_n) + t_n N(\tilde{s}_N). \tag{1.8}$$

Using a subsequence if needed, we can assume that s_n and $\tilde{s}_n$ both converge to limits s and $\tilde{s}$ in $[0, L]$. From (1.8), we must have $m(\tilde{s}) = m(s)$ which implies that $s = \tilde{s}$ (modulo L) since the curve is simple. We can rewrite (1.8)

$$\frac{m(\tilde{s}_n) - m(s_n)}{\tilde{s}_n - s_n} = -t_n \frac{N(\tilde{s}_n) - N(s_n)}{\tilde{s}_n - s_n}.$$

Since we assume that m is C^2 and regular, $(N(\tilde{s}_n) - N(s_n))/(\tilde{s}_n - s_n)$ has a limit (which is $-\kappa T(s)$), so that letting $n \to \infty$ yields $T(s) = 0$, which is impossible.

This implies that we can use s and t to parametrize the region $V_m^{(\varepsilon)}$ for small ε, its area being provided by the integral of the Jacobian of φ, which is, if $m = (x, y)$, the absolute value of the determinant of

$$\begin{pmatrix} \dot{x}_s - t\kappa\dot{x}_s & \dot{y}_s - t\kappa\dot{y}_s \\ -\dot{y}_s & \dot{x}_s \end{pmatrix}.$$

which gives $|\det(D\varphi)| = 1 - t\kappa$ for small t. Thus,

$$\text{Area}(V_m^{(\varepsilon)}) = \int_0^L \int_0^\varepsilon (1 - t\kappa) ds dt$$

$$= L\varepsilon - \frac{\varepsilon^2}{2} \int_0^L \kappa ds$$

The interesting conclusion is that the area is a second-degree polynomial in ε. The first-degree coefficient is the length and the second-degree coefficient is the integral of the curvature, i.e., the rotation index of the curve.

The formula can be localized without difficulty by restricting $V_m^{(\varepsilon)}$ to points s, t such that $s_0 < s < s_1$, the result being obviously

$$(s_1 - s_0)\varepsilon - \frac{\varepsilon^2}{2} \int_{s_0}^{s_1} \kappa ds,$$

with infinitesimal limit $\varepsilon(1 - \kappa(s)\varepsilon/2)ds$. This provides the infinitesimal area of the set of points that are close to the curve, and closer to some $m(s)$ for $s \in (s_0, s_1)$ than to any other point in the curve. This area is at first order given by the arc length times ε, with a corrective term involving the curvature.

This computation is a special case of a very general construction of what are called curvature measures [80]. They can be defined for a large variety of sets, in any dimension. We will see a two-dimensional description of them when we discuss discrete surfaces.

1.11 Invariance

Invariance is a fundamental concept when dealing with shapes. We already have encountered two types of invariance so far. The first one led to the definition of a geometric property and was the invariance with respect to change of parameter. The second was met in our discussion of arc length and curvature, and refers to the invariance of these quantities to rotations and translations. In this section, we want to give a presentation of this topic which will apply to more general transformations.

The first important additional transformation is scaling. This corresponds to replacing the curve m by $\tilde{m} = \lambda m$ where λ is a positive number. Visually, this corresponds to viewing the shape from a point that is closer or further away. Because of the renormalization, the unit tangent, normal and the angles θ_m are invariant. However, the length and arc length are multiplied by the constant factor λ. Finally, since the curvature is the rate of change of the angle as a function of arc length, it is divided by the same constant, $\kappa_{\tilde{m}} = \kappa_m / \lambda$.

It may also be interesting to consider invariance to affine transformations $m \mapsto Am + b$ where A is a general 2 by 2 invertible matrix (a general affine transformation). Arc length and curvature are not conserved by such transformations, and there is no simple formula to compute their new value. This section describes how new concepts, which will be called affine arc length and affine curvature, can be introduced to obtain the same type of invariance.

A comprehensive study of the theory of differential invariants of curves [161] lies, however, beyond the scope of this book. Here, we content ourselves with the computation in some particular cases. Although this repeats what we have already done with arc length and curvature, it will be easier to start with the simple case of rotation invariance. We know that s_m and κ_m are invariant by translation and rotation, and we now show how this can be obtained with a systematic approach.

1.11.1 Euclidean Invariance

The general approach to define generalized notions of length and arc length is to look for a function Q which depends only on the derivatives of a curve at a given point, such that $Q(\dot{m}_u, \ddot{m}_{uu}, \ldots)du$ provides the length of an element of curve between u and $u + du$.

The arc length is then defined by

$$\sigma_m(u) = \int_0^u Q(\dot{m}_v, \ddot{m}_{vv}, \ldots)dv.$$

(We will reserve the notation s_m uniquely for the Euclidean arc length.)

The function Q will be designed to meet invariance properties. The most important is probably its geometric nature (invariance by reparametrization); the requirement is: if $m = \tilde{m} \circ \varphi$, then $\sigma_m = \sigma_{\tilde{m}} \circ \varphi$. Computing the derivative of this identity yields, in terms of Q:

$$Q(\dot{m}_u, \ddot{m}_{uu}, \ldots) = \dot{\varphi}_u Q(\dot{\tilde{m}}_u \circ \varphi, \ddot{\tilde{m}}_{uu} \circ \varphi, \ldots). \tag{1.9}$$

Now, for $m = \tilde{m} \circ \varphi$, we have

$$\dot{m}_u = \dot{\varphi}_u \dot{\tilde{m}}_u \circ \varphi,$$
$$\ddot{m}_{uu} = \ddot{\varphi}_{uu} \dot{\tilde{m}}_u \circ \varphi + \dot{\varphi}_u^2 \ddot{\tilde{m}}_{uu} \circ \varphi,$$

and so on for higher derivatives.

As a consequence, if Q only depends on the first derivative, we must have

$$Q(\dot{\varphi}_u \dot{\tilde{m}}_u \circ \varphi) = \dot{\varphi}_u Q(\dot{\tilde{m}}_u \circ \varphi).$$

This is true in particular when, for all $z_1 \in \mathbb{R}^2$, $\lambda_1 > 0$:

$$Q(\lambda_1 z_1) = \lambda_1 Q(z_1).$$

This is the order 1 condition for Q. It is necessary by the discussion above, but one can show that it is also sufficient. Similarly, the order 2 condition is that, for all $z_1, z_2 \in \mathbb{R}^2$, for all $\lambda_1 > 0, \lambda_2 \in \mathbb{R}$:

$$Q(\lambda_1 z_1, \lambda_2 z_1 + \lambda_1^2 z_2) = \lambda_1 Q(z_1, z_2).$$

The argument can be applied to any number of derivatives. The general expression (based on the *Faa–Di Bruno formula*) is quite heavy, and we will not need it anyway, but the trick for deriving new terms is quite simple. Think in terms of derivatives: the derivative of λ_k is λ_{k+1} and the derivative of z_k is $\lambda_1 z_{k+1}$; then apply the product rule. For example, the second term was the derivative of the first term, $\lambda_1 z_1$, and therefore:

$$(\lambda_1 z_1)' = (\lambda_1)' z_1 + \lambda_1 (z_1)'$$
$$= \lambda_2 z_1 + \lambda_1^2 z_2$$

which is what we found by direct computation. The constraint with three derivatives would be

$$Q(\lambda_1 z_1, \lambda_2 z_1 + \lambda_1^2 z_2, \lambda_3 z_1 + 3\lambda_2 \lambda_1 z_2 + \lambda_1^3 z_3) = \lambda_1 Q(z_1, z_2, z_3).$$

The second type of constraint which is required for Q is the invariance by some class of transformations. If A is such a transformation, and $\tilde{m} = Am$, the requirement is $\sigma_{\tilde{m}} = \sigma_m$, or

$$Q(\dot{m}_u, \ddot{m}_{uu}, \ldots) = Q(\partial_u(Am), \partial_{uu}^2(Am), \ldots). \tag{1.10}$$

We consider affine transformations (the results will be extended to projective transformations at the end for this chapter). The equality is always true for translations $Am = m + b$, since Q only depends on the derivatives of m, and we therefore can assume that A is purely linear. Equality (1.10) therefore becomes: for all $z_1, z_2, \ldots \in \mathbb{R}^2$,

$$Q(z_1, z_2, \ldots) = Q(Az_1, Az_2, \ldots).$$

We now consider rotations. We will favor the lowest complexity for Q, and therefore first study whether a solution involving only one derivative exists. In this case, Q must satisfy: for all $\lambda_1 > 0$, for all $z_1 \in \mathbb{R}^2$ and for any rotation A,

$$Q(Az_1) = Q(z_1) \text{ and } Q(\lambda_1 z_1) = \lambda_1 Q(z_1).$$

Let $e_1 = (1,0)$ be the unit vector in the x-axis. Since one can always use a rotation to transform any vector z_1 into $|z_1|e_1$, the first condition implies that $Q(z_1) = Q(|z_1|e_1)$ which is equal to $|z_1|Q(e_1)$ from the second condition. We therefore find that $Q(z_1) = c|z_1|$ for some constant c, yielding $Q(\dot{m}_u) = c|\dot{m}_u| = c\sqrt{\dot{x}_u^2 + \dot{y}_u^2}$. We therefore retrieve the previously defined arc length up to a multiplicative constant c. The choice $c = 1$ is quite arbitrary, and corresponds to the condition that e_1 provides a unit speed: $Q(e_1) = 1$. We will refer to this σ_m as the Euclidean arc length, since we now consider other choices to obtain more invariance.

1.11.2 Scale Invariance

Let us now add scale to translation and rotation. Since it is always possible to transform any vector z_1 into e_1 with a rotation and scaling, considering only one derivative is not enough anymore.[2] We need at least two derivatives and therefore consider z_1 and z_2 with the constraints

$$Q(Az_1, Az_2) = Q(z_1, z_2) \text{ and } Q(\lambda_1 z_1, \lambda_2 z_1 + \lambda_1^2 z_2) = \lambda_1 Q(z_1, z_2).$$

Like for rotations, the first step is to use the first condition to place z_1 and z_2 into a canonical position. Consider the combination of rotation and scaling which maps e_1 to z_1. The first column of its matrix must therefore be z_1, but, because combinations of rotation and scaling have matrices of the form $S = \begin{pmatrix} a & -b \\ b & a \end{pmatrix}$, we see that, letting $z_1 = (x_1, y_1)$, the obtained matrix is

$$S_{z_1} = \begin{pmatrix} x_1 & -y_1 \\ y_1 & x_1 \end{pmatrix}.$$

Now take $A = S_{z_1}^{-1}$ to obtain, from the first condition:

$$Q(z_1, z_2) = Q(e_1, S_{z_1}^{-1} z_2).$$

A direct computation yields

$$S_{z_1}^{-1} z_2 = \frac{1}{x_1^2 + y_1^2} \begin{pmatrix} x_1 x_2 + y_1 y_2 \\ x_1 y_2 - x_2 y_1 \end{pmatrix}.$$

---

[2] This would give $Q(z_1) = Q(e_1) = \text{cst}$ and $Q(\lambda_1 z_1) = \lambda_1 Q(z_1) = Q(z_1)$ for all $\lambda_1 > 0$ yielding $Q = 0$.

So, we have so far obtained the fact that Q must be a function F of the quantities $a = \langle z_1, z_2 \rangle / |z_1|^2$ and $b = \det[z_1, z_2]/|z_1|^2$.

Now consider the second condition. The transformation $z_1 \to \lambda_1 z_1$ and $z_2 \to \lambda_1^2 z_2 + \lambda_2 z_1$ takes a to $\lambda_1 a + \lambda_2/\lambda_1$ and b to $\lambda_1 b$. Thus, if $Q(z_1, z_2) = F(a, b)$, we must have

$$F(\lambda_1 a + \lambda_2/\lambda_1, \lambda_1 b) = \lambda_1 F(a, b)$$

for all real numbers a, b, λ_2 and $\lambda_1 > 0$. Given a, b we can take $\lambda_2 = -\lambda_1^2 a$ and $\lambda_1 = 1/|b|$, at least when $b \neq 0$. This yields, for $b \neq 0$:

$$F(a, b) = |b| F(0, \operatorname{sign}(b)).$$

For $b = 0$, we can take the same value for λ_2 to obtain $F(0, 0) = \lambda_1 F(a, 0)$ for every λ_1 and a, which is only possible if $F(a, 0) = 0$ for all a. Thus, in full generality, the function Q must take the form

$$Q(z_1, z_2) = \begin{cases} c_+ |\det(z_1, z_2)|/|z_1|^2 & \text{if } \det(z_1, z_2) > 0, \\ 0 & \text{if } \det(z_1, z_2) = 0, \\ c_- |\det(z_1, z_2)|/|z_1|^2 & \text{if } \det(z_1, z_2) < 0, \end{cases}$$

where c_0, c_+, c_- are positive constants. To ensure invariance by a change of orientation, however, it is natural to choose $c_+ = c_-$. Taking this value equal to 1 yields

$$Q(z_1, z_2) = |\det(z_1, z_2)|/|z_1|^2.$$

We obtain the definition of the arc length for similitudes:[3]

$$d\sigma = \frac{|\dot{x}_u \ddot{y}_{uu} - \ddot{x}_{uu} \dot{y}_u|}{\dot{x}_u^2 + \dot{y}_u^2} du. \tag{1.11}$$

1.11.3 Special Affine Transformations

We now consider the case of area-preserving, or special affine transformations. These are affine transformations A such that $\det(A) = 1$. We need as before two derivatives, and the first step is again to normalize $[z_1, z_2]$ using a suitably chosen matrix A. Here, the choice is natural and simple, at least when z_1 and z_2 are independent: take A to be the inverse of $[z_1, z_2]$, normalized to have determinant 1, namely

$$A = \begin{cases} \sqrt{\det(z_1, z_2)}[z_1, z_2]^{-1} & \text{if } \det(z_1, z_2) > 0, \\ \sqrt{\det(z_2, z_1)}[z_2, z_1]^{-1} & \text{if } \det(z_1, z_2) < 0. \end{cases}$$

When $\det(z_1, z_2) > 0$, this yields

---

[3] To complete the argument, one needs to check that the required conditions are satisfied for the obtained Q; this is indeed the case although we skip the computation.

$$Q(z_1, z_2) = Q(\sqrt{\det(z_1, z_2)}e_1, \sqrt{\det(z_1, z_2)}e_2)$$

so that Q must be a function F of $\sqrt{\det(z_1, z_2)}$. Like for similitudes, we obtain, applying the parametrization invariance condition, that

$$F(\lambda_1^{3/2}\sqrt{\det(z_1, z_2)}) = \lambda_1 F(\sqrt{\det(z_1, z_2)})$$

which implies, taking $\lambda_1 = (\det(z_1, z_2))^{-1/3}$, that

$$Q(z_1, z_2) = F(1)(\det(z_1, z_2))^{1/3}.$$

The same result is true for $\det(z_1, z_2) < 0$, yielding

$$Q(z_1, z_2) = \tilde{F}(1)(\det(z_2, z_1))^{1/3}.$$

with a possibly different constant $\tilde{F}(1)$. Again, for orientation invariance, it is natural to define the area-preserving arc length by

$$d\sigma = |\ddot{x}_{uu}\dot{y}_u - \ddot{y}_{uu}\dot{x}_u|^{1/3}du.$$

We have left aside the case $\det(z_1, z_2) = 0$. In this case, assume that $z_2 = \alpha z_1$. The second condition implies, taking $\lambda_2 = -\lambda_1^2\alpha$:

$$\lambda_1 Q(z_1, \alpha z_1) = Q(\lambda_1 z_1, \lambda_1^2\alpha z_1 + \lambda_2 z_1) = Q(\lambda_1 z_1, 0)$$

but we can always find an area-preserving transformation which maps $\lambda_1 z_1$ to e_1 so that $\lambda_1 Q(z_1, \alpha z_1) = Q(e_1, 0)$ is true for every $\lambda_1 > 0$ only if $Q(z_1, \alpha z_1) = 0$. This is consistent with the formula obtained for $\det(z_1, z_2) \neq 0$.

Computations are also possible for the complete affine group and also for the projective group, but they require to deal with four and more derivatives and are quite lengthy. They will be provided at the end of this chapter.

1.11.4 Generalized Curvature

In addition to arc length, new definitions of curvature can be adapted to more invariance constraints. One way to understand the definition is to return to the rotation case, and our original definition of curvature.

We have interpreted the curvature as the speed of rotation of the tangent with respect to arc length. Consider the matrix $P_m = [T_m, N_m]$ associated to the tangent and normal to m. Because (T_m, N_m) is an orthonormal system, this matrix is a rotation, called a moving frame [40, 77, 82, 83], along the curve. The rate of variation of this matrix is defined by

$$W_m = P_m^{-1}\partial_s P_m.$$

In the present case, it is

$$W_m = \partial_s \theta_m \begin{pmatrix} \cos\theta_m & \sin\theta_m \\ -\sin\theta_m & \cos\theta_m \end{pmatrix} \begin{pmatrix} -\sin\theta_m & -\cos\theta_m \\ \cos\theta_m & -\sin\theta_m \end{pmatrix} = \kappa_m(s) \begin{pmatrix} 0 & -1 \\ 1 & 0 \end{pmatrix}.$$

This illustrates the moving frame method, which provides here the Euclidean curvature. It can be shown to always provide a function which is invariant by the considered transformations and change of parametrization. More precisely, we have the following definition. For a group G with associated arc length σ, we let $J_k(G)$ be the set of vectors $(z_0, z_1, \ldots, z_k) \in (\mathbb{R}^2)^{k+1}$ such that there exists a curve m parametrized with σ such that $z_k = d^k m / d\sigma^k$. That this condition induces restrictions on $z_1, \ldots, z_k$ is already clear in the case of rotations, for which one must have $|z_1| = 1$.

Definition 1.13. *Let G be a group acting on $\mathbb{R}^2$ (e.g., a subgroup of $GL_2(\mathbb{R})$). A G-moving frame of order k is a one-to-one function $P_0 : J_k(G) \to G$ with the following property. For all curves $m : I \to \mathbb{R}^2$, parametrized with G-arc length, define $P_m : I \to G$ by*

$$P_m(\sigma) = P_0(m, \dot{m}_u, \ddot{m}_{uu}, \ldots, m^{(k)}).$$

Then, one must have $P_{gm}(\sigma) = gP_m(\sigma)$.

We now consider affine transformations, with group $G = GL_2(\mathbb{R}) \ltimes \mathbb{R}^2$ (cf. Appendix B.4.3). An element of G is represented by a pair (A, b) for a linear map A and $b \in \mathbb{R}^2$. We will therefore write $P_0 = (A_0, b_0)$, $P_m = (A_m, b_m)$. We denote by G_0 the linear part of G, i.e., $(A, b) \in G \Rightarrow A \in G_0$. The invariance condition in Definition 1.13 yields, for all $U \in G_0, h \in \mathbb{R}^2$

$$A_0(Uz_0 + h, Uz_1, Uz_2, \ldots, Uz_k) = UA_0(z_0, z_1, z_2, \ldots, z_k), \qquad (1.12)$$
$$b_0(Uz_0 + h, Uz_1, Uz_2, \ldots, Uz_k) = Ub_0(z_0, z_1, z_2, \ldots, z_k) + h.$$

We have the following result which generalizes Theorem 1.6. We here use the same notation as in Appendix B.4.3.

Theorem 1.14 (Moving Frame: affine case). *Let $G = G_0 \ltimes \mathbb{R}^2$ be a subgroup of $GL_2(\mathbb{R}) \ltimes \mathbb{R}^2$. If $P_0 = (A_0, b_0)$ is a G-moving frame, then, for any plane curve m*

$$\bar{W}_m(\sigma) = A_m(\sigma)^{-1} \partial_\sigma P_m = (A_m(\sigma)^{-1} \partial_\sigma A_m, A_m(\sigma)^{-1} \partial_\sigma b_m)$$

is invariant by change of parametrization and by the action of G. It moreover characterizes the curve up to the action of G: if $\bar{W}_m = \bar{W}_{\tilde{m}}$, then $\tilde{m} = gm$ for some $g \in G$.

Proof. Invariance by change of parametrization is obvious, because the arc length is invariant. If $\tilde{m} = Um + h$, then $P_{\tilde{m}} = (UA_m, Ub_m + h)$ and

$$\bar{W}_{\tilde{m}} = A_m^{-1}U^{-1}(U\partial_\sigma A_m, U\partial_\sigma b_m) = P_m^{-1}\partial_\sigma P_m = \bar{W}_m$$

which proves G-invariance.

Conversely, assume that $\bar{W}_{\tilde{m}} = \bar{W}_m = W$. Let $g = (U, h) = P_{\tilde{m}}(0)P_m(0)^{-1}$. The proof that $\tilde{m} = gm$ for some g derives from the uniqueness theorem for ordinary differential equations (cf. Appendix C); $P_m = (A_m, b_m)$ and $P_{\tilde{m}} = (A_{\tilde{m}}, b_{\tilde{m}})$ are both solutions of the equation $\partial_\sigma(A, b) = AW$, and gP_m is another solution, as can be easily checked. Since $gP_m(0) = P_{\tilde{m}}(0)$ by definition of g, we have

$$P_0(\tilde{m}, \dot{\tilde{m}}, \ldots, \tilde{m}^{(k)}) = gP_0(\dot{\tilde{m}}, \ldots, m^{(k)}) = P_0(gm, U\dot{m}, \ldots, Um^{(k)}).$$

Because P_0 is assumed to be one-to-one, we have $\tilde{m} = gm$ which proves the theorem. $\qquad\square$

For affine groups, we select a moving frame P_0 of the form $P_0(z_0, z_1, \ldots, z_k) = (A_0(z_1, \ldots, z_k), z_0)$. This implies that

$$\bar{W}_m = \left(A_m^{-1}\partial_\sigma A_m, A_m^{-1}\dot{m}_u \right).$$

We will mainly focus on the first term, that we denote by

$$W_m = A_m^{-1}\partial_\sigma A_m.$$

The choice made for rotations corresponds to $A_0(z_1) = [z_1, Rz_1]$, R being the $(\pi/2)$-rotation. It is obviously one-to-one and satisfies the invariance requirements. The second term in $\bar{W}_m$ is constant, namely $A_m^{-1}\dot{m}_u = (1, 0)$.

It can be shown that W_m can lead to only one, "fundamental", scalar invariant. All other coefficients are either constant, or can be deduced from this fundamental invariant. This invariant will be called the curvature associated to the group.

Consider this approach applied to similitudes. Assume that the curve is parametrized by the related arc length, σ. The frame, here, must be a similitude A_m, and as above, we take

$$A_m = \begin{pmatrix} \dot{x}_\sigma & -\dot{y}_\sigma \\ \dot{y}_\sigma & \dot{x}_\sigma \end{pmatrix}.$$

Define $W_m = A_m^{-1}\partial_\sigma A_m$, so that

$$W_m = \frac{1}{\dot{x}_\sigma^2 + \dot{y}_\sigma^2} \begin{pmatrix} \dot{x}_\sigma & \dot{y}_\sigma \\ -\dot{y}_\sigma & \dot{x}_\sigma \end{pmatrix} \begin{pmatrix} \ddot{x}_{\sigma\sigma} & -\ddot{y}_{\sigma\sigma} \\ \ddot{y}_{\sigma\sigma} & \ddot{x}_{\sigma\sigma} \end{pmatrix}$$

$$= \frac{1}{\dot{x}_\sigma^2 + \dot{y}_\sigma^2} \begin{pmatrix} \ddot{x}_{\sigma\sigma}\dot{x}_\sigma + \ddot{y}_{\sigma\sigma}\dot{y}_\sigma & \ddot{x}_{\sigma\sigma}\dot{y}_\sigma - \ddot{y}_{\sigma\sigma}\dot{x}_\sigma \\ -\ddot{x}_{\sigma\sigma}\dot{y}_\sigma + \ddot{y}_{\sigma\sigma}\dot{x}_\sigma & \ddot{x}_{\sigma\sigma}\dot{x}_\sigma + \ddot{y}_{\sigma\sigma}\dot{y}_\sigma \end{pmatrix}.$$

When the curve is parametrized with arc length, we have

$$\frac{|\dot{x}_\sigma\ddot{y}_{\sigma\sigma} - \ddot{x}_{\sigma\sigma}\dot{y}_\sigma|}{\dot{x}_\sigma^2 + \dot{y}_\sigma^2} = 1$$

along the curve. Therefore

$$W_m(\sigma) = \begin{pmatrix} \frac{\ddot{x}_{\sigma\sigma}\dot{x}_\sigma + \ddot{y}_{\sigma\sigma}\dot{y}_\sigma}{\dot{x}_\sigma^2 + \dot{y}_\sigma^2} & \mp 1 \\ \pm 1 & \frac{\ddot{x}_{\sigma\sigma}\dot{x}_\sigma + \ddot{y}_{\sigma\sigma}\dot{y}_\sigma}{\dot{x}_\sigma^2 + \dot{y}_\sigma^2} \end{pmatrix}.$$

The computation exhibits a new quantity which is

$$K = \frac{\ddot{x}_{\sigma\sigma}\dot{x}_\sigma + \ddot{y}_{\sigma\sigma}\dot{y}_\sigma}{\dot{x}_\sigma^2 + \dot{y}_\sigma^2} \qquad (1.13)$$

(σ being the similitude arc length). This is the curvature for the group of similitudes: it is invariant by translation, rotation and scaling, and characterizes curves up to similitudes.

We now consider special affine transformations (affine with determinant 1). Assume that the curve is parametrized with the corresponding arc length, σ, i.e.,

$$|\ddot{x}_{\sigma\sigma}\dot{y}_\sigma - \ddot{y}_{\sigma\sigma}\dot{x}_\sigma|^{1/3} = 1.$$

One can choose $A_m = \begin{pmatrix} \dot{x}_\sigma & \ddot{x}_{\sigma\sigma} \\ \dot{y}_\sigma & \ddot{y}_{\sigma\sigma} \end{pmatrix}$ which has determinant 1. Since $A_m(1,0)^T = \dot{m}_\sigma$, the term $A_m^{-1}\dot{m}_\sigma$ is trivial. We have

$$A_m^{-1}\partial_\sigma A_m = \begin{pmatrix} \ddot{y}_{\sigma\sigma} & -\ddot{x}_{\sigma\sigma} \\ -\dot{y}_\sigma & \dot{x}_\sigma \end{pmatrix} \begin{pmatrix} \ddot{x}_{\sigma\sigma} & x_{\sigma\sigma\sigma}^{(3)} \\ \ddot{y}_{\sigma\sigma} & y_{\sigma\sigma\sigma}^{(3)} \end{pmatrix} = \begin{pmatrix} 0 & \ddot{y}_{\sigma\sigma}x_{\sigma\sigma\sigma}^{(3)} - \ddot{x}_{\sigma\sigma}y_{\sigma\sigma\sigma}^{(3)} \\ 1 & -\dot{y}_\sigma x_{\sigma\sigma\sigma}^{(3)} + \dot{x}_\sigma y_{\sigma\sigma\sigma}^{(3)} \end{pmatrix}.$$

Since $\partial_\sigma(\ddot{x}_{\sigma\sigma}\dot{y}_\sigma - \ddot{y}_{\sigma\sigma}\dot{x}_\sigma) = \dot{y}_\sigma x_{\sigma\sigma\sigma}^{(3)} - \dot{x}_\sigma y_{\sigma\sigma\sigma}^{(3)} = 0$, the only non-trivial coefficient is: $\ddot{y}_{\sigma\sigma}x_{\sigma\sigma\sigma}^{(3)} - \ddot{x}_{\sigma\sigma}y_{\sigma\sigma\sigma}^{(3)}$ which can be taken (up to a sign change) as a definition of the *special affine curvature*:

$$K = \det(\ddot{m}_{\sigma\sigma}, m_{\sigma\sigma\sigma}^{(3)}). \qquad (1.14)$$

Again, this is expressed as a function of the affine arc length and invariant by the action of special affine transformations.

1.12 Characterization of a Bounded Convex Set

We now describe a parametrization which is specific to boundaries of convex sets. We assume that that m is a closed non-intersecting curve and let Ω_m denote its interior and $R_m = \partial\Omega_m$ its range. We will also assume that m has enough derivatives to validate the computations to come. The results, however, can be extended to general convex sets (see, for example, [129]). We also let, without loss of generality, m be parametrized with arc length.

Recall that a set $\Omega \subset \mathbb{R}^2$ is convex if and only if, for all p, q in Ω, the line segment $[p, q]$ is included in Ω_m. Notice that, if Ω_m is convex and m is positively oriented, then κ (the curvature) is positive, by Theorem 1.11.

Proposition 1.15. *Ω_m is convex if and only if m lies everywhere on one side of its tangents.*

We first study the height function over Ω_m, defined by

$$h(\theta) = \sup \left\{ u(\theta)^T p, p \in \Omega_m \right\}$$

with $u(\theta) = (\cos\theta, \sin\theta)$.

We have the following proposition:

Proposition 1.16. *Assume that Ω_m is convex and that R_m has no linear part. There exists a unique function $\theta \mapsto \sigma(\theta)$ such that*

$$h(\theta) = u(\theta)^T m(\sigma(\theta)). \tag{1.15}$$

This function is one-to-one and onto from $[0, 2\pi)$ to $[0, L)$, and satisfies

$$\ddot{h}_{\theta\theta} + h = \dot{\sigma}_\theta. \tag{1.16}$$

Proof. Since Ω_m is convex, the maximum of the linear function $p \mapsto \langle u(\theta), p \rangle$ is attained on the boundary of Ω_m, which is the range, R_m, of the curve m. Still by convexity, if the maximum is attained at two distinct points, it is also attained at any point of the line segment joining the two points, and this line segment is also included in R_m. Since R_m does not contain any linear part, the maximum is unique, and, its arc length provides $\sigma(\theta)$.

This also implies that the function $s \mapsto u(\theta)^T m(s)$ is maximal at $s = \sigma(\theta)$, which implies, computing the first and second derivatives, that

$$u(\theta)^T T(\sigma(\theta)) = 0 \text{ and } \kappa(\sigma(\theta)) u(\theta)^T N(\sigma(\theta)) \leq 0.$$

Also, since h increases when passing from the interior to the exterior of Ω_m, we must have $u(\theta)^T N(\sigma(\theta)) \leq 0$ which implies that $u(\theta) = -N(\sigma(\theta))$ and incidentally that $\kappa(\sigma(\theta))$ is non-negative. This proves that θ can be uniquely recovered from $\sigma(\theta)$, so that σ is one-to-one.

Also, given s, define θ such that $u(\theta) = -N(s)$; let $\overline{s} = \sigma(\theta)$ and assume that $\overline{s} \neq s$. Then the tangents at s and $\overline{s}$ are equal, which implies that the orientation of the tangent must either be constant or pass through a strict extremum between s and $\overline{s}$. The first case implies that there is a linear segment on the curve between s and $\overline{s}$, which is contrary to the assumption, and the second case implies that the derivative of the angle must be negative at some point, which is also impossible since the derivative is the curvature and the curve is convex. So $s = \overline{s}$ and σ is onto.

We now compute the derivative of h with respect to θ. Letting $v(\theta) = (-\sin\theta, \cos\theta)$, we have, using (1.15),

$$\dot{h}_\theta = \langle v, m(\sigma) \rangle + \dot{\sigma}_\theta \langle u, T(\sigma) \rangle.$$

But, since $\langle u, T(\sigma) \rangle = 0$, this implies

$$\dot{h}_\theta = \langle v , m \circ \sigma \rangle . \tag{1.17}$$

Taking the second derivative, we have

$$\ddot{h}_{\theta\theta} = -\langle u , m \circ \sigma \rangle + \dot{\sigma}_\theta \langle v , T \circ \sigma \rangle .$$

This implies that h satisfies the equation

$$\ddot{h}_{\theta\theta} + h = \dot{\sigma}_\theta \langle v , T \circ \sigma \rangle .$$

Now, since $u = -N$, we have $v = T$ which yields (1.16). $\qquad\square$

We now can state the theorem:

Theorem 1.17. *The function $\theta \mapsto \dot{\sigma}_\theta$ characterizes Ω_m up to translation. Conversely, given a positive, 2π-periodic function $\theta \mapsto \lambda(\theta)$, such that*

$$\int_0^1 e^{i\theta} \lambda(\theta) d\theta = 0 , \tag{1.18}$$

there exists a closed curve m, unique up to translation, such that its interior is convex, and satisfies

$$\lambda(\theta) = \dot{\sigma}_\theta . \tag{1.19}$$

Proof. First note that the function h characterizes Ω_m. Indeed, we have

$$\Omega_m = \bigcap_{\theta \in \mathbb{R}} \{ p \in \mathbb{R}^2 , \langle u(\theta) , p \rangle \le h(\theta) \} \tag{1.20}$$

The fact that Ω_m is included in this intersection of half-planes comes directly from the definition of h. Conversely, if $p \notin \Omega_m$, there exists a line separating p and Ω_m (by the Hahn–Banach theorem). This implies that p is above the maximal height of Ω_m relatively to this line.

From the theory of second-order linear differential equations, the function $\dot{\sigma}_\theta$ characterizes h up to the addition of a solution of the homogeneous equation $\ddot{h}_{\theta\theta} + h = 0$. Since solutions of this equation take the form $\langle a , u(\theta) \rangle$ for some $a \in \mathbb{R}^2$, we see that, given $\dot{\sigma}_\theta$, two compatible functions h have to differ only by $\langle a , u(\theta) \rangle$. It is easy to check that such a variation in h exactly corresponds to translating Ω_m by a, so that the first part of the theorem is proved.

Assume now that λ is given, with the properties listed in the theorem. We can obtain h by solving (1.16), for which a solution is

$$h(\theta) = \int_0^\theta \sin(\theta - \alpha) \lambda(\alpha) d\alpha .$$

From (1.15) and (1.17) we see that we must set

$$m(\theta) = h(\theta) u(\theta) + \dot{h}_\theta(\theta) v(\theta)$$

for $\theta \in [0, 2\pi]$ (with $v(\theta) = (-\sin\theta, \cos\theta)$). Using (1.18), we see that m is a closed curve. Moreover $m_\theta = (h + \ddot{h}_{\theta\theta})v = \lambda(\theta)v$ so that $|\dot{m}_\theta| = \lambda(\theta)$.

Define

$$\Omega = \bigcap_{\theta \in \mathbb{R}} \{p \in \mathbb{R}^2, \langle u(\theta), p \rangle \leq h(\theta)\}.$$

Ω is convex, because it is defined as an intersection of convex sets. Define

$$\tilde{h}(\theta) = \sup\{\langle u(\theta), p \rangle, p \in \Omega\}.$$

If $p \in \Omega$, we have, for all θ, $\langle u(\theta), p \rangle \leq h(\theta)$ so that $\tilde{h}(\theta) \leq h(\theta)$. We now prove the converse inequality. Since $\langle u(\theta), m(\theta) \rangle = h(\theta)$, it suffices to show that $m(\theta) \in \Omega$, i.e., that, for all α, $h(\alpha) - \langle u(\alpha), m(\theta) \rangle \geq 0$. Fix α and denote $\rho(\xi) = \langle u(\alpha), m(\xi) \rangle$. Since m is continuous over $[0, 2\pi]$, ρ has a maximum which satisfies $\dot{\rho}_\xi = 0$. Since $\dot{\rho}_\xi = \lambda(\xi)\langle u(\alpha), v(\xi) \rangle$, we see that $v(\xi)$ is orthogonal to $u(\alpha)$, which is possible only if $\xi = \alpha$ modulo π. Since $\lambda(\xi) > 0$ and $\langle u(\alpha), v(\xi) \rangle \leq 0$ for $\xi \simeq \alpha$, and ≥ 0 for $\xi \simeq \alpha + \pi$, the only value that provides a maximum is for $\xi = \alpha$. This implies that $\rho(\theta) = \langle u(\alpha), m(\theta) \rangle \leq \rho(\alpha) = h(\alpha)$, and concludes the proof that $\tilde{h} = h$, and that λ is associated to the outline of a convex set. $\square$

The following formulas give the area and perimeter of Ω_m as a function of h: we have

$$L := \text{perimeter}(\Omega_m) = \int_0^{2\pi} h(\theta) d\theta,$$

$$\text{area}(\Omega_m) = \int_0^{2\pi} (h\ddot{h}_{\theta\theta} + h^2) d\theta.$$

The curvature is given by

$$\kappa = (\ddot{h}_{\theta\theta} + h)^{-1/2}.$$

1.13 Non-Local Representations

1.13.1 Semi-Local Invariants

The practical issue with the invariants which have been previously defined is that they depend on derivatives which can be difficult to estimate in the presence of noisy data. Semi-local invariants are an attempt to address this issue by replacing derivatives by estimates depending on nearby, but not coincident, points.

They end-up being what is commonly called a *signature* of the curve, which is a function which assigns, to each parameter t, a value which depends on some geometric properties of the curve near the point $m(t)$.

Let's consider a general approach: we fix an integer, k. To implement the method, we need:

1. An algorithm to select k points on the curve, relative to a single point $m(t)$.
2. A formula to compute the signature based on the k selected points.

Let's introduce some notation. First, let S_m represent the selection of k points along m. If $p = m(t)$ is a point on m, we let $S_m(p) = (p_1, \ldots, p_k)$. Second, let F be the signature function: it takes $p_1, \ldots, p_k$ as input and returns a real number.

We enforce invariance at both steps of the method. Note that geometric invariance is implicitly enforced by the assumption that S_m only depends on $p = m(t)$ (and not on t). Consider now the issue of invariance with respect to a class G of affine transformations. For A in this class, we want that:

1. The point selection process "commutes": if $S_m(p) = (p_1, \ldots, p_k)$, then $S_{Am}(Ap) = (Ap_1, \ldots Ap_k)$.
2. The function F is invariant: $F(Ap_1, \ldots, Ap_k) = F(p_1, \ldots, p_k)$.

Point 2 is quite easy to satisfy. The principle is to use a transformation A which places the first points in $S_m(p)$ is a generic position to normalize the function F. This will be clearer on examples. Assume that the class of transformations which are considered are translations and rotations. Then, there is a unique such transformations that displaces p_1 on O and p_2 on $|p_1 - p_2|e_1$ where e_1 is the unit vector of the horizontal axis. Denote this transformation by A_{p_1, p_2}. Then, we must have

$$F(p_1, p_2, \ldots, p_k) = F(A_{p_1, p_2} p_1, A_{p_1, p_2} p_2, \ldots, A_{p_1, p_2} p_k)$$
$$= F(0, |p_1 - p_2|e_1, A_{p_1, p_2} p_3, \ldots, A_{p_1, p_2} p_k).$$

Conversely, it is easy to prove that any function F of the form

$$F(p_1, p_2, \ldots, p_k) = \tilde{F}(|p_1 - p_2|, A_{p_1, p_2} p_3, \ldots, A_{p_1, p_2} p_k)$$

is invariant by rotation and translation. The transformation $A_{p_1 p_2}$ can be made explicit: skipping the computation, this yields ((x_i, y_i) being the coordinates of p_i)

$$A_{p_1, p_2} p_j = \frac{1}{|p_2 - p_1|} \begin{pmatrix} (x_2 - x_1)(x_j - x_1) + (y_2 - y_1)(y_j - y_1) \\ (x_2 - x_1)(y_j - y_1) - (y_2 - y_1)(x_j - x_1) \end{pmatrix}.$$

Thus, with three selected points, the general form of F is

$$F(p_1, p_2, p_3) = \tilde{F}\Big(|p_2 - p_1|, \frac{\langle p_2 - p_1, \, p_3 - p_1 \rangle}{|p_2 - p_1|},$$
$$\frac{\det[p_2 - p_1, p_3 - p_1]}{|p_2 - p_1|}\Big).$$

If scaling is added to the class of transformations, the same argument shows that the only choice with three points is:

$$F(p_1, p_2, p_3) = \tilde{F}\left(\frac{\langle p_2 - p_1, \, p_3 - p_1 \rangle}{|p_2 - p_1|^2}, \frac{\det[p_2 - p_1, p_3 - p_1]}{|p_2 - p_1|^2}\right).$$

Similar computations hold for larger classes of transformations.

There are several possible choices for point selection. The simplest one is to use the arc length (relative to the class of transformations) that we have defined in the previous sections, and choose $p_1, \ldots, p_k$ symmetrically around p, with fixed relative arc lengths $\sigma_m(p_1) - \sigma_m(p), \ldots, \sigma_m(p_k) - \sigma_m(p)$. For example, letting $\delta_i = \sigma_m(p_i) - \sigma_m(p)$, and if $k = 2l + 1$, one can take $\delta_1 = -l\varepsilon, \delta_2 = -(l-1)\varepsilon, \ldots, \delta_k = l\varepsilon$.

However, the arc length requires using curve derivatives, and this is precisely what we wanted to avoid. Some purely geometric constructions can be used instead. For rotations, for example, we can choose $p_1 = p$, and p_2 and p_3 to be the two intersections of the curve m and a circle of radius ε centered at p (taking the ones closest to p on the curves). For scale and rotation, consider again circles, but instead of fixing the radius in advance, adjust it so that $|p_2 - p_3|$ becomes smaller that $1 - \varepsilon$ times the radius of the circle. This is always possible, unless the curve is a straight line.

Considering the class of special affine transformations [33], one can choose p_1, p_2, p_3, p_4 such that the line segments (p_1, p_2) and (p_3, p_4) are parallel to the tangent at p, and the areas of the triangles (p_0, p_1, p_2) and (p_0, p_3, p_4) are respectively given by ε and 2ε.

1.13.2 The Shape Context

The shape context [22] represents a shape by a collection of histograms along its outline. We make here a presentation of this concept in the continuum and do not discuss discretization issues.

Let $s \mapsto m(s)$ be a parametrized curve, defined on some interval I. For $s, t \in I$, let $v(s, t) = m(t) - m(s)$. Fixing s, the function $t \mapsto v(s, t)$ takes values in $\mathbb{R}^2$. We now consider a density kernel, i.e, a function $K : \mathbb{R}^2 \to \mathbb{R}^2$ such that, for fixed x, $K(x, .)$ is a probability density on $\mathbb{R}^2$, usually symmetrical around x. The typical example is

$$K(x, y) = (1/(2\pi\sigma^2)) \exp(-|x - y|^2/(2\sigma^2)). \tag{1.21}$$

Using this kernel, we let, for $s \in I$

$$f^{(m)}(s,y) = \int_I K(y, v(s,t))dt.$$

The density $f^{(m)}(s,.)$ is the shape context of the curve at s and the function $f^{(m)}$ is the shape context of the whole curve. To discuss some invariance properties of this representation, we assume that the curve is parametrized with arc length (and therefore focus on translation and rotations), and that K is radial, i.e., $K(x,y)$ only depends on $|x - y|$, like in (1.21).

A translation applied to the curve has no effect on $v(s,t)$ and therefore leaves the shape context invariant. A rotation R transforms v in Rv, and we have $f^{(Rm)}(s, Ry) = f^{(m)}(s,y)$. The representation is not scale invariant, but can be made so with an additional normalization (e.g., by forcing the mean distance between different points in the shape to be equal to 1, cf. [22]).

The shape context is a robust global representation, since it depends for any point on the totality of the curve. To some extent, however, it shares the property of local representations that small variations of the contour will have a small influence on the shape context of other points, by only slightly modifying the density $f(s,.)$.

1.13.3 Conformal Welding

Conformal welding relies on a theorem in complex analysis that provides a representation of a curve by a diffeomorphism of the unit circle. This theorem requires more mathematical background than the rest of this book, but the resulting representation is interesting enough to justify the effort.

We will identify $\mathbb{R}^2$ to $\mathbb{C}$, via the usual relation $(x, y) \rightarrow x + iy$, and add to $\mathbb{C}$ a point at infinity that will confer a structure of a two-dimensional sphere to it. This can be done using the mapping

$$F(re^{i\theta}) = \left(\frac{2r \cos \theta}{r^2 + 1}, \frac{2r \sin \theta}{r^2 + 1}, \frac{r^2 - 1}{r^2 + 1} \right).$$

This mapping can be interpreted as identifying parallel circles on the sphere to zero-centered circles on the plane; zero is mapped to the south pole, the unit disc is mapped to the equator, and the representation tends to the north pole when $r \rightarrow \infty$. With this representation, the interior and the exterior of the unit disc are mapped to hemispheres and therefore play a symmetric role. We will let $\bar{\mathbb{C}}$ denote $\mathbb{C} \cup \infty$. The complex derivative of a function is defined as the limit of $(f(z + h) - f(z))/h$ when $h \rightarrow 0$ in $\mathbb{C}$.

Two domains $\Omega_1, \Omega_2 \subset \bar{\mathbb{C}}$ are said to be conformally equivalent if there exists a function $f : \Omega_1 \rightarrow \Omega_2$ such that f is onto and one-to-one and the complex derivative $f'(z)$ exists for all $z \in \Omega_1$, with $f'(z) \neq 0$. Such a function has the property to conserve angles, in the sense that, the angle made by two curves passing by z remain unchanged after a transformation by f.

The Riemann mapping theorem [176] states that any simply connected domain (i.e., any domain within which any simpled closed curve can be continuously deformed into a point) is conformally equivalent to the unit disc.

This domain may or may not include a point at infinity and therefore may or may not be bounded. For example, the transformation $z \mapsto 1/z$ maps the interior of the unit disc to its exterior and vice-versa. This conformal transformation is obviously unique up to any conformal mapping of the unit disc onto itself. It can be shown that the latter transformations must belong to a three-parameter family (a sub-class of the family of Möbius transformations of the plane), containing functions of the form

$$z \mapsto e^{i\alpha} \frac{z^{i\beta} + r}{rze^{i\beta} + 1} \tag{1.22}$$

with $r < 1$. We let M_1 be the set of such transformations (which forms a three-parameter group of diffeomorphisms of the unit disc). A transformation in M_1 can be decomposed into three steps: a rotation $z \mapsto ze^{i\beta}$ followed by the transformation $z \mapsto (z + r)/(zr + 1)$, followed again by a rotation $z \mapsto ze^{i\alpha}$.

The Riemann mapping theorem can be applied to the interior and to the exterior of any Jordan curve γ. Letting Ω_γ represent the interior, and $\overline{\Omega}_\gamma^c$ the exterior (the notation holding for the complementary of the closure of Ω_γ), and D being the open unit disc, we therefore have two conformal transformations $\Phi_- : \Omega_\gamma \to D$ and $\Phi_+ : \overline{\Omega}_\gamma^c \to D$. These two maps can be extended to the boundary of Ω_γ, i.e., the range R_γ of the curve γ, and the extension remains a homeomorphism. Restricting Φ^+ to R_γ yields a map $\varphi^+ : R_\gamma \to S^1$ (where S^1 is the unit circle) and similarly $\varphi^- : R_\gamma \to S^1$. In particular, the mapping $\varphi = \varphi^- \circ (\varphi^+)^{-1}$ is a homeomorphism of S^1 onto itself. It is almost uniquely defined by γ. In fact Φ^+ and Φ^- are both unique up to the composition (on the left) by a Möbius transformation, as given by (1.22), so that φ is unique up to a Möbius transformation applied on the left or on the right. The indeterminacy on the right can be removed by the following normalization; one can constrain Φ^+, which associates two unbounded domains, to transform the point at infinity into itself, and be such that its differential at this point has a positive real part and a vanishing imaginary part. Under this constraint, φ is unique up to the left action of Möbius transformations.

In mathematical terms, we obtain a representation of (smooth) Jordan plane curves by the set of diffeomorphisms of S^1 (denoted $\text{Diff}(S^1)$) modulo the Möbius transformations (denoted $PSL_2(S^1)$), writing

$$\text{2D shapes} \sim \text{Diff}(S^1)/PSL_2(S^1).$$

We now describe the two basic operations associated to this equivalence, namely computing this representation from the curve, and retrieving the curve from the representation. The first operation requires computing the trace of the conformal maps of the interior and exterior of the curve. Several algorithms are available to compute conformal maps. The plots provided in Figure 1.2 were obtained using the Schwarz–Christoffel toolbox developed by T. Driscoll.

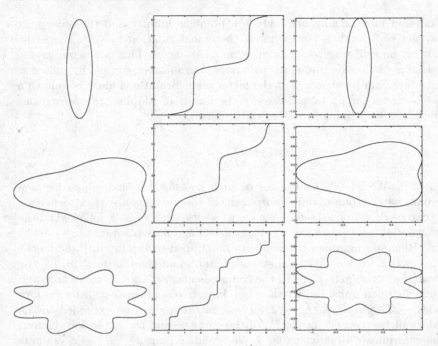

Fig. 1.2. Conformal disparity between the interior and exterior of three planar curves. *First column:* original curves; *second column:* signatures; *third column:* curves reconstructed from the signature.

The solution to the second problem (going from the representation to the curves) is described in [188, 189]. It is proved in [189] that, if φ is the mapping above, and $\psi = \varphi^{-1}$, the corresponding shape (defined up to translation, rotation and scaling) can be parametrized as $\theta \mapsto F(\theta) \in \mathbb{C}$, $\theta \in [0, 2\pi]$ where F is the solution of the integral equation

$$K(F)(\theta) + F(\theta) = e^{i\theta}$$

where $K(F)(\theta) = \int_0^{2\pi} K(\theta, \tilde{\theta}) F(\tilde{\theta}) d\tilde{\theta}$, and the kernel K is given by

$$K(\theta, \tilde{\theta}) = ctn\left(\frac{\theta - \tilde{\theta}}{2}\right) - \dot{\psi}(\tilde{\theta})\, ctn\left(\frac{\psi(\theta) - \psi(\tilde{\theta})}{2}\right)$$

which has limit $\ddot{\psi}(\theta)/2\dot{\psi}(\theta)$ when $\theta \to \tilde{\theta}$. The inverse representation can then be computed by solving, after discretization, a linear equation in F.

1.14 Implicit Representation

1.14.1 Introduction

Implicit representations can provide simple descriptions for relatively complex shapes and can in many cases be a good choice to design stable shape processing algorithms. If $f : \mathbb{R}^2 \to \mathbb{R}$ is a function that satisfies

$$f(p) = 0 \Rightarrow \nabla f(p) \neq 0, \tag{1.23}$$

its zero level set is the set C defined by the equation $f(p) = 0$ (cf. Figure 1.3). This set can have several connected components, each of them being the image of a curve (level sets can therefore be used to represent multiple curves). Our first goal is to show how local properties of curves can be computed directly from the function f. We will always assume, in this chapter, that the function f tends to infinity when p tends to infinity. This implies that the zero level sets are bounded and therefore represent closed curves.

In a neighborhood of any regular point of f (such that $\nabla f(m) \neq 0$), the set C can be locally parametrized, for example by expressing one of the coordinates (x, y) as a function of the other.[4]

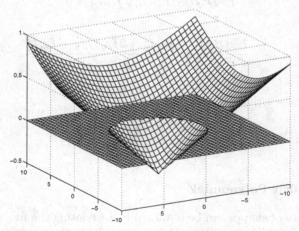

Fig. 1.3. Implicit curve from a function $f : \mathbb{R}^2 \mapsto \mathbb{R}$.

Assume that such a parametrization has been chosen. This results in a curve $m : I =] - \varepsilon, \varepsilon[\to \mathbb{R}^2$ such that $m(0) = m_0$ and $f(m(u)) = 0$ for $u \in I$ (m coincides with a subset of C). From the chain rule, we have:

---

[4] This is a consequence of the implicit function theorem.

$$\nabla f(m)^T \dot{m}_u = 0.$$

This implies that $\nabla f(m)$ is normal to m.

Orientation. *As a convention, we assume that $\nabla f(m)$ has a direction opposite to the normal N to m, so that $\nabla f(m) = -|\nabla f(m)|N$ (recall that (T, N) must have determinant 1, with $T = \dot{m}_u / |\dot{m}_u|$). If the curve is positively oriented (N points inwards), this implies that f is negative inside the curve and positive outside.*

If this is not true, it suffices either to replace f by $-f$ or to replace $m(u)$ by $m(-u)$. Assuming the convention, we obtain

$$T = \frac{1}{|\nabla f|}\left(-\dot{f}_y, \dot{f}_x\right).$$

From the second derivative of the equation $f(m(u)) = 0$, we have

$$\dot{m}_u^T D^2 f(m) \dot{m}_u + \nabla f(m)^T \ddot{m}_{uu} = 0.$$

(recall that the second derivative of f is a 2 by 2 matrix).

Since $\nabla f(m) = -|\nabla f(m)|N$ and $\ddot{m}_{uu}^T N = \kappa |\dot{m}_u|^2$, the previous equation yields (after division by $|\dot{m}_u|^2$),

$$T^T D^2 f(m) T - \kappa |\nabla f(m)| = 0.$$

so that

$$\kappa = \frac{T^T D^2 f T}{|\nabla f|} = \frac{\ddot{f}_{xx}\dot{f}_y^2 - 2\ddot{f}_{xy}\dot{f}_x\dot{f}_y + \ddot{f}_{yy}\dot{f}_x^2}{(\dot{f}_x^2 + \dot{f}_y^2)^{3/2}}.$$

This can also be written

$$\kappa = \mathrm{div}\,\frac{\nabla f}{|\nabla f|}. \tag{1.24}$$

1.14.2 Implicit Polynomials

A large variety of shapes can be obtained by restricting the function f to be a polynomial of small degree [125]. Doing so results in the need to only estimate a finite number of parameters. A polynomial in two variables and total degree less than n is given by the general formula

$$f(x, y) = \sum_{p+q \leq n} a_{pq} x^p y^q.$$

A shape represented as an implicit polynomial is simply given by the zero level set, C_f, of f: $C_f = \{z = (x, y), f(x, y) = 0\}$.

In full generality, C_f is what is called an algebraic curve (a curve defined by a polynomial equation), and may have branches at infinity, self-intersections, or multiple loops. There are, however, sufficient conditions for C_f to be a closed simple curve [53]. To avoid infinite branches, it suffices that the principal part of f has no real root, i.e.,

$$\sum_{k=0}^{n} a_{k,n-k}\lambda^k \neq 0$$

for $\lambda \in \mathbb{R}$. To avoid self-intersection, the condition is that $f = 0 \Rightarrow \nabla f \neq 0$. Figure 1.4 provides a few examples of zero level sets of implicit polynomials.

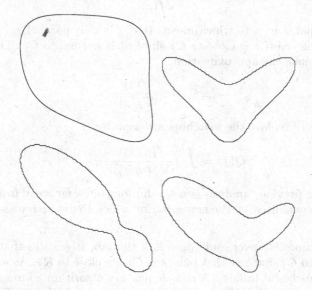

Fig. 1.4. Shapes generated by implicit polynomials of degree 4.

Let a shape be given and $\mathcal{R}$ denote its range. We consider the issue of approximating the set $\mathcal{R}$ by C_f for a suitable implicit polynomial.

Introduce the error functional

$$Q_0(f) = \int_0^L d(m(s), C_f)^2 ds\,,$$

with

$$d(z, C_f) = \inf_{z' \in C_f} |z - z'|\,.$$

The optimal f should minimize this expression. This can be made feasible with some approximations. If z is close to C_f, there exists $z' \simeq z$ such that $f(z') = 0$. We have the first-order approximation

$$0 = f(z') \simeq f(z) + \langle z' - z, \nabla f(z) \rangle$$

so that

$$\langle z' - z, \nabla f(z) \rangle \simeq -f(z).$$

Let $v_z = \nabla f(z) / |\nabla f(z)|$. Decompose z' in the form

$$z' = z + \langle z' - z, v_z \rangle v_z + w \simeq z + \frac{f(z)}{|\nabla f(z)|} v_z + w$$

with w orthogonal to v_z. This implies that

$$|z' - z|^2 \simeq \frac{f(z)^2}{|\nabla f(z)|^2} + |w|^2.$$

This is minimal if $w = 0$, which means that $z - z'$ is parallel v_z, which can always be achieved if z is close to C_f since v_z is normal to C_f. This implies that we can make the approximation

$$d(z, C_f) \simeq \frac{|f(z)|}{|\nabla f(z)|}.$$

We therefore replace the matching criterion by

$$Q(f) = \int_0^L \frac{f(m(s))^2}{|\nabla f(m(s))|^2} ds.$$

Of course, the previous analysis is not valid for points far away from C_f. But these are also penalized by the new criterion which therefore represent a valid alternative.

This criterion, however, only does half the job. It ensures that points in $\mathcal{R}_f$ are close to C_f, but not that points in C_f are close to $\mathcal{R}_f$. As a result, C_f may have branches at infinity. A complementary algorithm addressing this in the case when f has degree 4 is provided in [125], to which we refer for more details.

1.15 Invariance for Affine and Projective Transformations

The local invariants discussed in Section 1.11.4 (with respect to rotation, similitude and the special affine group) probably reach the limits of numerical feasibility, in the number of derivatives they require. Going further involves even higher derivatives, and has only theoretical interest. However, we include here, for completeness, the definition of the affine and projective arc lengths and curvatures. This section can be safely skipped. In discussing the projective arc lengths, we will use a few notions that are related to Lie groups and manifolds. The reader can refer to Appendix B for more details.

1.15.1 Affine Group

We use the same notation as in Section 1.11.4. We first introduce new parameters which depend on the sequence $z_1, \ldots, z_n$ that describe the first derivatives. We assume that $\det(z_1, z_2) \neq 0$ and let

$$\alpha_k = \alpha_k(z_1, \ldots, z_n) = \frac{\det(z_k, z_2)}{\det(z_1, z_2)}$$

$$\text{and } \beta_k = \beta_k(z_1, \ldots, z_n) = \frac{\det(z_1, z_k)}{\det(z_1, z_2)}.$$

These are defined so that

$$z_k = \alpha_k z_1 + \beta_k z_2 \tag{1.25}$$

which also yields

$$\begin{pmatrix} \alpha_k \\ \beta_k \end{pmatrix} = [z_1, z_2]^{-1} z_k.$$

In particular, we have $\alpha_1 = \beta_2 = 1$, $\alpha_2 = \beta_1 = 0$.

Assuming affine invariance, we must have

$$Q(z_1, \ldots, z_n) = Q([z_1, z_2]^{-1} z_1, \ldots, [z_1, z_2]^{-1} z_n)$$

which implies that Q must be a function of the α_k's and β_k's. We see also that we must have at least $n = 3$ to ensure a non-trivial solution. In fact, we need to go to $n = 4$, as will be shown by the following computation.

For $n = 4$, the parametric invariance constraint yields: for all $\lambda_1 > 0$, $\lambda_2, \lambda_3, \lambda_4$,

$$Q(\tilde{z}_1, \tilde{z}_2, \tilde{z}_3, \tilde{z}_4) = \lambda_1 Q(z_1, z_2, z_3, z_4)$$

with $\tilde{z}_1 = \lambda_1 z_1$, $\tilde{z}_2 = \lambda_2 z_1 + \lambda_1^2 z_2$, $\tilde{z}_3 = \lambda_3 z_1 + 3\lambda_2 \lambda_1 z_2 + \lambda_1^3 z_3$ and

$$\tilde{z}_4 = \lambda_4 z_1 + (3\lambda_2^2 + 4\lambda_3 \lambda_1) z_2 + 6\lambda_1^2 \lambda_2 z_3 + \lambda_1^4 z_4.$$

We now make specific choices for $\lambda_1, \lambda_2, \lambda_3$ and λ_4 to progressively reduce the functional form of Q. We will abuse the notation by keeping the letter Q to design the function at each step. Our starting point is $Q = Q(\alpha_3, \beta_3, \alpha_4, \beta_4)$.

We start by taking $\lambda_1 = 1$, $\lambda_2 = \lambda_3 = 0$, yielding $\tilde{z}_1 = z_1$, $\tilde{z}_2 = z_2$, $\tilde{z}_3 = z_3$ and $\tilde{z}_4 = z_4 + \lambda_4 z_1$. Define $\tilde{\alpha}_k, \tilde{\beta}_k$ for the α_k, β_k coefficients associated to the $\tilde{z}$'s. For the considered variation, the only coefficient that changes is α_4, which becomes $\tilde{\alpha}_4 = \alpha_4 + \lambda_4$. This implies that

$$Q(\alpha_3, \beta_3, \alpha_4, \beta_4) = Q(\alpha_3, \beta_3, \alpha_4 + \lambda_4, \beta_4).$$

Taking $\lambda_4 = -\alpha_4$, we see that Q does not depend on α_4, yielding the new functional form $Q = Q(\alpha_3, \beta_3, \beta_4)$.

Let's now consider $\lambda_1 = 1$, $\lambda_2 = \lambda_4 = 0$. In this case, z_1, z_2 remain unchanged, z_3 and z_4 become $\tilde{z}_3 = z_3 + \lambda_3 z_1$, $\tilde{z}_4 = z_4 + 4\lambda_3 z_2$. This implies

$\tilde{\alpha}_3 = \alpha_3 + \lambda_3$, $\tilde{\beta}_3 = \beta_3$ and $\tilde{\beta}_4 = \beta_4 + 4\lambda_3$. Taking $\lambda_3 = -\alpha_3$ yields the new functional form $Q = Q(\beta_3, \beta_4 - 4\alpha_3)$.

Now, take $\lambda_1 = 1$, $\lambda_2 = \lambda_4 = 0$, yielding $\tilde{z}_1 = z_1$, $\tilde{z}_2 = z_2 + \lambda_2 z_1$, $\tilde{z}_3 = z_3 + 3\lambda_2 z_2$ and $\tilde{z}_4 = z_4 + 6\lambda_2 z_3 + 3\lambda_2^2 z_2$, so that $\tilde{\beta}_3 = \beta_3 + 3\lambda_2$, $\tilde{\alpha}_3 = \alpha_3 - 3\lambda_2^2 - \lambda_2 \beta_3$ and $\tilde{\beta}_4 = \beta_4 + 3\lambda_2^2 + 6\lambda_2 \beta_3$. In particular,

$$\tilde{\beta}_4 - 3\tilde{\alpha}_3 = \beta_4 - 4\alpha_3 + 15\lambda_2^2 + 10\lambda_2 \beta_3.$$

Taking $\lambda_2 = -\beta_3/3$ yields $Q = Q(\beta_4 - 4\alpha_3 - 5\beta_3^2/3)$.

Finally, take $\lambda_2 = \lambda_2 = \lambda_4 = 0$ yielding $\tilde{\beta}_3 = \lambda_1 \beta_3$ $\tilde{\beta}_4 = \lambda_1^2 \beta_4$ and $\tilde{\alpha}_3 = \lambda_1^2 \alpha_3$. This gives

$$Q(\lambda_1^2(\beta_4 - 4\alpha_3 - 5\beta_3^2/3)) = \lambda_1 Q(\beta_4 - 4\alpha_3 - 5\beta_3^2/3).$$

Taking $\lambda_1 = 1/\sqrt{|\beta_4 - 4\alpha_3 - 5\beta_3/3|}$, assuming this expression does not vanish, yields

$$Q(\beta_4 - 4\alpha_3 - 5\beta_3^2/3) = \begin{cases} Q(1)\sqrt{|\beta_4 - 4\alpha_3 - 5\beta_3^2/3|} & \text{if } \beta_4 - 4\alpha_3 - 5\beta_3^2/3 > 0, \\ Q(-1)\sqrt{|\beta_4 - 4\alpha_3 - 5\beta_3^2/3|} & \text{if } \beta_4 - 4\alpha_3 - 5\beta_3^2/3 < 0. \end{cases}$$

Here again, it is natural to ensure an invariance by a change of orientation and let $Q(1) = Q(-1) = 1$ so that

$$Q(z_1, z_2, z_3, z_4) = \sqrt{|\beta_4 - 4\alpha_3 - 5\beta_3^2/3|}.$$

This provides the affine invariant arc length.

We can take the formal derivative in (1.25), yielding

$$z_{k+1} = \alpha_k' z_1 + \alpha_k z_2 + \beta_k' z_2 + \beta_k z_3 = (\alpha_k' + \beta_k \alpha_3)z_1 + (\alpha_k + \beta_k' + \beta_k \beta_2)z_2$$

so that $\alpha_{k+1} = \alpha_k' + \beta_k \alpha_3$ and $\beta_{k+1} = \beta_k' + \alpha_k + \beta_k \beta_3$. This implies that higher-order coefficients can always be expressed in terms of α_3, β_3 and their (formal) derivatives. In particular, using $\beta_4 = \beta_3' + \alpha_3 + \beta_3^2$, we get

$$Q(z_1, z_2, z_3, z_4) = \sqrt{|\beta_3' - 3\alpha_3 - 2\beta_3^2/3|}. \tag{1.26}$$

Returning to parametrized curves, let $\alpha_{m,k}$ and $\beta_{m,k}$ be the coefficients α_k, β_k in which $(z_1, z_2, \ldots)$ are replaced by their corresponding derivatives $(\dot{m}_u, \ddot{m}_{uu}, \ldots)$, so that

$$m^{(k)} = \alpha_{m,k} \dot{m}_u + \beta_{m,k} \ddot{m}_{uu}.$$

We want to express the affine arc length in terms of the Euclidean curvature. Assuming that m is parametrized with Euclidean arc length, we have $\ddot{m}_{ss} = \kappa R \dot{m}_s$ where R is the $\pi/2$ rotation. Taking one derivative yields (using $R^2 = -\text{Id}$)

$$m^{(3)} = \kappa R \ddot{m}_{ss} + \dot{\kappa}_s R \dot{m}_s = -\kappa^2 \dot{m}_s + (\dot{\kappa}_s/\kappa)\ddot{m}_{ss}.$$

This implies that $\alpha_{m,3} = -\kappa^2$ and $\beta_{m,3} = \dot{\kappa}_s/\kappa$; thus, (1.26) implies that the affine arc length, σ, and the Euclidean arc length are related by

$$d\sigma = \sqrt{|\partial_s(\dot{\kappa}_s/\kappa) + 3\kappa^2 - 2(\dot{\kappa}_s/\kappa)^2/3|}\,ds.$$

1.15.2 Projective Group

The problem is harder to address with the projective group (see Appendix B.4.3) because of the non-linearity of the transformations. We keep the same notation for α_k and β_k as in the affine case (since the projective group includes the affine group, we know that the function Q will have to depend on these reduced coordinates).

Before the computation, we need to express the effects that a projective transformation has on the derivative of the curve. We still let the symbol z_k hold for the kth derivative. A projective transformation applied to a point $z \in \mathbb{R}^2$ takes the form $g : z \mapsto (Uz+b)/(w^T z+1)$ with a 2 by 2 matrix U, and vectors $b, w \in \mathbb{R}^2$. Let's denote $\gamma_0 = (w^T z_0 + 1)^{-1}$ so that z_0 is transformed in $\tilde{z}_0 = \gamma_0(Uz_0 + b)$. We need to express the higher derivatives $\tilde{z}_1, \tilde{z}_2, \ldots$ as a function of the initial $z_1, z_2, \ldots$ and the parameters of the transformations. Letting γ_k represent the kth derivative of γ_0, the rule for the derivation of a product (Leibniz's formula) yields

$$\tilde{z}_k = \gamma_k(Uz_0 + b) + \sum_{q=1}^{k} \binom{k}{q} \gamma_{k-q} U z_q. \tag{1.27}$$

This provides a group action, which will be denoted $\tilde{z} = g \star z$. Our goal is to find a function Q such that $Q(z_1, z_2, \ldots, z_k) = Q(\tilde{z}_1, \tilde{z}_2, \ldots, \tilde{z}_k)$, and which is also invariant to the transformations induced by a change of variables. It will be necessary to go to $k = 5$ for the projective group.

We first focus on projective invariance, and make an analysis equivalent to the one that allowed us to remove z_0, z_1 and z_2 in the affine case. More precisely, we show that U, b, and w can be found such that $\tilde{z}_0 = 0$, $\tilde{z}_1 = e_1$, $\tilde{z}_2 = e_2$ and $\tilde{z}_3 = 0$, with $e_1 = (1, 0)$ and $e_2 = (0, 1)$.

First note that $\gamma_1 = -w^T z_1 \gamma_0^2$ and $\gamma_2 = -w^T z_2 \gamma_0^2 + 2(w^T z_1)^2 \gamma_0^3$. Take $b = -Uz_0$ to ensure $\tilde{z}_0 = 0$. We have $\tilde{z}_1 = \gamma_0 U z_1$, $\tilde{z}_2 = 2\gamma_1 U z_1 + \gamma_0 U z_2$ and

$$\tilde{z}_3 = 3\gamma_2 U z_1 + 3\gamma_1 U z_2 + \gamma_0 U z_3.$$

We therefore need

$$U z_1 = e_1/\gamma_0, U z_2 = e_2/\gamma_0 - (2\gamma_1/\gamma_0^2)e_1 = e_2/\gamma_0 + 2w^T z_1 e_1$$

and (after some algebra)

$$U z_3 = -3(\gamma_2/\gamma_0) A z_1 - 3(\gamma_1/\gamma_0) A z_2$$
$$= 3w^T z_2 e_1 + 3w^T z_1 e_2.$$

Using the decomposition $z_k = \alpha_k z_1 + \beta_k z_2$, we also have $A z_3 = \alpha_3(e_1/\gamma_0) + \beta_3(e_2/\gamma_0 - (2\gamma_1/\gamma_0^2)e_1)$ which yields the identification

$$w^T z_1 = \beta_3/(3\gamma_0) \text{ and } w^T z_2 = (3\alpha_3 + 2\beta_3^2)/9.$$

Using the definition of γ_0, this can be written

$$\begin{cases} w^T(z_1 - \beta_3/3 z_0) = \beta_3/3 \\ w^T(z_2 - (\alpha_3/3 + 2\beta_3/9)z_0) = (\alpha_3/3 + 2\beta_3/9) \end{cases}$$

which uniquely defines w, under the assumption (which we make here) that $z_0, (3/\beta_3)z_1, (9/(3\alpha_3 + 2\beta_3^2))z_2$ forms an affine frame. Given W, we can compute b and U. We have in particular, using the decomposition of z_k:

$$U z_k = (\alpha_k/\gamma_0 + 2\beta_k\beta_3/(3\gamma_0))e_1 + (\beta_k/\lambda)e_2.$$

Similarly, we have

$$w^T z_k = \alpha_k\beta_3/(3\gamma_0) + \beta_k(3\alpha_3 + 2\beta_3^2)/9.$$

With this choice for A, u and b, the resulting expressions of $\tilde{z}_3, \tilde{z}_4$ and $\tilde{z}_5$ can be obtained. This is quite a heavy computation for which the use of a mathematical software is helpful; the result is that the projective invariance implies that the function Q must be a function of the following four expressions:

$$A = \alpha_4 - \frac{8}{3}\alpha_3\beta_3 - \frac{8\beta_3^3}{9} + \frac{2}{3}\beta_3\beta_4$$

$$B = \alpha_5 - \frac{10}{3}\alpha_4\beta_3 + \frac{40}{9}\alpha_3\beta_3^3 + \frac{40\beta_3^4}{27} - \\ \frac{5}{3}\alpha_3\beta_4 - \frac{20}{9}\beta_3^2\beta_4 + \frac{2}{3}\beta_3\beta_5$$

$$C = -2\alpha_3 - \frac{4\beta_3^2}{3} + \beta_4$$

$$D = -\frac{10}{3}\alpha_3\beta_3 - \frac{5}{3}\beta_3\beta_4 + \beta_5.$$

Given this, it remains to operate the reductions associated to the invariance by change of parameter. This is done like in the affine case, progressively selecting the coefficients λ_i to eliminate one of the expressions and modify the others, at the difference that there is one extra constraint here associated to the fifth derivative. Note that with five constraints, we should normally be short of one expression, but one of the invariances is (magically) satisfied in the reduction process which would otherwise have required using six derivatives. We spare the reader the details, and directly provide the final expression for Q which is

$$Q = \left| \frac{40\beta_3^3}{9} + \beta_5 - 5\beta_3(\beta_4 - 2\alpha_3) - 5\alpha_4 \right|^{1/3}.$$

Like before, this can be expressed in terms of the formal derivatives of α_3 and β_3, yielding

$$Q = \left[\beta_3'' - 3\alpha_3' - 2\beta_3\beta_3' + 2\beta_3\alpha_3 + (4/9)\beta_3^3 \right]^{1/3}. \tag{1.28}$$

1.15.3 Affine Curvature

We can apply the moving frame method described in Section 1.11.4 to obtain the affine curvature of a curve m. We assume here that m is parametrized by affine arc length, σ. A moving frame on m is immediately provided by the matrix $A_m = [\dot{m}_\sigma, \ddot{m}_{\sigma\sigma}]$, or, with our z notation, $A_0 = [z_1, z_2]$. By definition of α_3 and β_3, the matrix $W_m = A_m^{-1} \partial_\sigma A_m$ is equal to

$$W_m = \begin{pmatrix} 0 & \alpha_{m,3} \\ 1 & \beta_{m,3} \end{pmatrix}.$$

Since the curve is parametrized with affine arc length, we have $Q = 1$ where Q is given by $\sqrt{|\dot{\beta}_3 - 3\alpha_3 - 2\beta_3^2/3|}$. This implies that $\alpha_{m,3}$ is a function of $\beta_{m,3}$ and $\dot{\beta}_{m,3}$ along the curve; the moving frame therefore only depends on $\beta_{m,3}$ and its derivatives, which indicates that $\beta_{m,3}$ is the affine curvature. Thus, when a curve is parametrized with affine arc-length, σ, its curvature is given by

$$\kappa_m(\sigma) = \frac{\det[\dot{m}_\sigma, m_{\sigma\sigma\sigma}^{(3)}]}{\det[\dot{m}_\sigma, \ddot{m}_{\sigma\sigma}]},$$

If the curve now has an arbitrary parametrization, the curvature is obtained by using $d\sigma = Q\,du$ where Q is given by (1.26). This yields the following expression:

$$\kappa_m(s) = \frac{1}{Q(u)} \frac{\det[\dot{m}_u, m_{uuu}^{(3)}]}{\det[\dot{m}_u, \ddot{m}_{uu}]} - 3\frac{\partial_u Q(u)}{Q(u)},$$

1.15.4 Projective Curvature

The moving frame method cannot be used exactly as described in section 1.11.4 in the projective case, because of the non-linearity of the transformations. The moving frame is still associated to a one-to-one function $P_0(z_0, \ldots, z_k) \in G = PGL_2(\mathbb{R})$. The invariance property in this case gives, with the definition of the action $z \mapsto g \star z$ given in (1.27), $P_0(g \star z) = g P_0(z)$. For Theorem 1.14 to make sense, we must use the differential of the left translation $L_g : h \mapsto hg$ on $PGL_2(\mathbb{R})$, and define

$$\bar{W}_m = DL_{P_m}(\mathrm{Id})^{-1} \partial_\sigma P_m$$

which belongs to the Lie algebra of $PGL_2(\mathbb{R})$. This is the general definition of a moving frame on a Lie group [82], and coincides with the definition that has been given for affine groups, for which we had $DL_g = A$ when $g = (A, b)$.

We first need to build the matrix A_0. For this, using as before the notation (e_1, e_2) for the canonical basis of $\mathbb{R}^2$, we define a projective transformation that takes the family $\omega = (0, e_1, e_2, 0)$ to the family $z = (z_0, z_1, z_2, z_3)$, i.e., we want to determine g such that $g \star \omega = z$ (we showed that its inverse exists in Section 1.15.2, but we need to compute it explicitly). Since this provides eight equations for eight dimensions, one can expect that a unique such transformation exists; this will be our $A_0(z)$.

Assuming that this existence and uniqueness property is satisfied, such a construction ensures the invariance of the moving frame by the group action. Indeed, letting z be associated to a curve m and $\tilde{z}$ to $\tilde{m} = g(m)$ for some $g \in PGL_2(\mathbb{R})$, we have $\tilde{z} = g \star z$. Since $A_0(z)$ is defined by $A_0(z) \star \omega = z$, the equality $A_0(\tilde{z}) \star \omega = \tilde{z}$ is achieved by $A_0(\tilde{z}) = gA_0(z)$, which is the required invariance. (Indeed, because $\star$ is a group action, we have $(gA_0(z)) \star \omega = g \star (A_0(z)\omega) = g \star z = \tilde{z}$.)

We now proceed to the computation. The first step is to obtain the expression of $g \star z$ for $z = (z_0, z_1, z_2, z_3)$. We do this in the special case in which g is given by:
$$g(m) = (Um + b)/(1 + w^T m)$$
w and b being two vectors in $\mathbb{R}^2$ and $U \in GL_2(\mathbb{R})$. Denote $g \star (\tilde{z}_0, \tilde{z}_1, \tilde{z}_2, \tilde{z}_3) = (z_0, z_1, z_2, z_3)$. From $(1 + \langle w, \tilde{z}_0 \rangle)z_0 = A\tilde{z}_0 + b$, we obtain

$$\begin{cases} (1 + w^T \tilde{z}_0)z_0 = U\tilde{z}_0 + b \\ (1 + w^T \tilde{z}_0)z_1 + w^T \tilde{z}_1 z_0 = U\tilde{z}_1 \\ (1 + w^T \tilde{z}_0)z_2 + 2w^T \tilde{z}_1 z_1 + w^T \tilde{z}_2 z_0 = U\tilde{z}_2 \\ (1 + w^T \tilde{z}_0)z_3 + 3w^T \tilde{z}_1 z_2 + 3w^T \tilde{z}_2 z_1 + w^T \tilde{z}_3 z_0 = U\tilde{z}_3. \end{cases} \quad (1.29)$$

Taking $\tilde{z} = \omega$, we get

$$\begin{cases} z_0 = b \\ z_1 + w_1 z_0 = u_1 \\ z_2 + 2w_1 z_1 + w_2 z_0 = u_2 \\ z_3 + 3w_1 z_2 + 3w_2 z_1 = 0. \end{cases} \quad (1.30)$$

where we have denoted $w = (w_1, w_2)$, $Ue_1 = u_1$ and $Ue_2 = u_2$. The third equation yields
$$z_3 = -3w_2 z_1 - 3w_1 z_2. \quad (1.31)$$

We will assume that z_1 and z_2 are linearly independent, so that $w = (w_1, w_2)$ is uniquely defined by this equation, and therefore $U = [u_1, u_2]$ by the middle equations of (1.30). Using again the notation $z_3 = \alpha_3 z_1 + \beta_3 z_2$, we get

$$\begin{cases} w_1 = -\beta_3/3 \\ w_2 = -\alpha_3/3. \end{cases}$$

This fully defines our moving frame $A_0(z)$.

Recall that the formal derivative of a quantity M that depends on $z_0, \ldots, z_3$ is given, with our notation, by $M' = \sum_{k=0}^{3}(\partial M/\partial z_k)z_{k+1}$. Since $b = z_0$, we have $b' = z_1$; from $u_1 = z_1 + w_1 z_0$, we get

$$u_1' = w_1 z_1 + z_2 + w_1' z_0,$$

and from (1.31) and $u_2 = z_2 + 2w_1 z_1 + w_2 z_0$,

$$u_2' = z_3 + 2w_1 z_2 + (2w_1' + w_2)z_1 + w_2' z_0$$
$$= (-2w_2 + 2w_1')z_1 - w_1 z_2 + w_2' z_0.$$

We have $w_1' = -\beta_3'/3$ and $w_2' = -\alpha_3'/3$, which are therefore directly computable along the curve.

By taking the representation of a projective transformation by the triplet (U, b, w), we have chosen a local chart on $PGL_2(\mathbb{R})$ which obviously contains the identity represented by $(\mathrm{Id}, 0, 0)$. To be able to compute the differential of the left translation $L_{A(z)}$, we need to express the product in this chart. One way to do this efficiently is to remark that, by definition of the projective group, products in $PGL_2(\mathbb{R})$ can be deduced from matrix products in $GL_3(\mathbb{R})$, up to a multiplicative constant. A function g with coordinates (U, b, w) in the chart is identified (up to multiplication by a scalar) to the matrix

$$\begin{pmatrix} U & b \\ w^T & 1 \end{pmatrix}$$

and the product of $\tilde{g} = (\tilde{U}, \tilde{b}, \tilde{w})$ and $\overline{g} = (\overline{U}, \overline{b}, \overline{w})$ is therefore identified to the product of the associated matrices which is

$$\begin{pmatrix} \overline{U} & \overline{b} \\ \overline{w}^T & 1 \end{pmatrix} \begin{pmatrix} \tilde{U} & \tilde{b} \\ \tilde{w}^T & 1 \end{pmatrix} = \begin{pmatrix} \overline{U}\tilde{U}' + \overline{b}\tilde{w}^T & \overline{U}\tilde{b} + \overline{b} \\ \overline{w}^T \tilde{U} + \tilde{w}^T & \overline{w}^T \tilde{b} + 1 \end{pmatrix}$$

which yields the chart representation for the product

$$\overline{g}\tilde{g} = \Big((\overline{U}\tilde{U} + \overline{b}\tilde{w}^T)/(1 + \overline{w}^T \tilde{b}),$$
$$(\overline{U}\tilde{b} + \overline{b})/(1 + \overline{w}^T \tilde{b}), (\tilde{U}^T \overline{w} + \tilde{w}')/(1 + \overline{w}^T \tilde{b})\Big).$$

To compute the differential of the left translation in local coordinates, it suffices to take $\tilde{U} = \mathrm{Id} + \varepsilon H$, $\tilde{b} = \varepsilon\beta$ and $\tilde{w} = \varepsilon\gamma$, and compute the first derivative of the product with respect to ε at $\varepsilon = 0$. This yields

$$d_{\mathrm{Id}} L_{\overline{g}}(H, \beta, \gamma) = (\overline{U}H + \overline{b}\gamma^T - \overline{w}^T \beta\overline{U}, \overline{U}\beta - \overline{w}^T \beta\overline{b}, \gamma + H^T \overline{w} - \overline{w}^T \beta\overline{w}).$$

We need to compute the inverse of this linear transformation, and therefore solve

$$\begin{cases} \overline{U}H + \overline{b}\gamma^T - \overline{w}^T \beta\overline{U} = \tilde{H} \\ \overline{U}\beta - \overline{w}^T \beta\overline{b} = \tilde{\beta} \\ \gamma + H^T \overline{w} - \overline{w}^T \beta\overline{w} = \tilde{\gamma}. \end{cases}$$

The second equation yields $\beta = (\overline{U} - \overline{b}\overline{w}^T)^{-1}\tilde{\beta}$. Substituting γ in the first by its expression in the third yields

$$\tilde{H} = (\overline{U} - \overline{b}\overline{w}^T)H + \overline{b}\tilde{\gamma}^T + (\overline{w}^T \beta)\overline{b}\overline{w}^T - \overline{w}^T \beta\overline{U}$$

so that
$$H = (\overline{U} - \overline{b}\overline{w}^T)^{-1}(\tilde{H} - \overline{b}\tilde{\gamma}^T) + (\overline{w}^T\beta)\mathrm{Id}.$$

Finally, we have
$$\gamma = \tilde{\gamma} - H^T\overline{w} + \overline{w}^T\beta\overline{w}.$$

$\overline{W}$ is obtained by applying these formulae to $\overline{g} = A(z) = (U, b, w)$ and $\tilde{H} = (\theta_1, \theta_2)$ with

$$\begin{cases} \theta_1 = u_1' = w_1 z_1 + z_2 + w_1' z_0 \\ h_2' = u_2' = (2w_1' - 2w_2)z_1 - w_1 z_2 + w_2' z_0 \\ \tilde{\beta} = z_1 \\ \tilde{\gamma} = w'. \end{cases}$$

Note that, since $(A - bw^T)h = Ah - w^T hb$, the identity $z_1 = u_1 - w_1 b$ implies
$$\beta = (U - bw^T)^{-1}z_1 = e_1.$$

Similarly, from $u_2 - w_2 b = z_2 + 2w_1 z_1$, we get
$$(U - bw^T)^{-1}z_2 = e_2 - 2w_1 e_1.$$

We have, using $b = z_0$ and $\tilde{\gamma} = w'$,
$$\tilde{H} - b\tilde{\gamma}^T = \big(w_1 z_1 + z_2, \ (w_1' - 2w_2)z_1 - w_1 z_2\big).$$

We therefore obtain
$$h_1 = (U - bw^T)^{-1}(w_1 z_1 + z_2) + w_1 e_1 = w_1 e_1 + e_2 - 2w_1 e_1 + w_1 e_1 = e_2,$$

$$h_2 = (U - bw^T)^{-1}((2w_1' - 2w_2)z_1 - w_1 z_2) + w_1 e_2$$
$$= (2w_1' - 2w_2)e_1 - w_1(e_2 - 2w_1 e_1) + w_1 e_2$$
$$= (2w_1' + 2w_1^2 - 2w_2)e_1.$$

Denoting $c = w_1' + w_1^2 - w_2$, we have obtained $W = \begin{pmatrix} 0 & 2c \\ 1 & 0 \end{pmatrix}$. Moreover, we have
$$\gamma = w' - H^T w + w_1 w = \begin{pmatrix} w_1' - w_2 + w_1^2 \\ w_2' - 2cw_1 + w_1 w_2 \end{pmatrix}.$$

Because we assume that $\left[\beta_3'' - 3\alpha_3' - 2\beta_3\beta_3' + 2\beta_3\alpha_3 + (4/9)\beta_3^3\right]^{1/3} = 1$, we see that $w_2' = -\alpha_3'/3$ can be expressed as a function of α_3 and the derivatives of β_3 (up to the second one), while c is equal to $-(\beta_3' - \beta_3^2/3 - \alpha_3)/3$. The invariant of smallest degree can therefore be taken to be $\beta_3' - \beta_3^2/3 - \alpha_3$ (in fact, $w_2' - 2cw_1 + w_1 w_2 = -c'/6$). The projective curvature can therefore be taken as

$$\kappa_m(\sigma) = \partial_\sigma\left(\frac{\det(\dot{m}_\sigma, m_{\sigma\sigma\sigma}^{(3)})}{\det(\dot{m}_\sigma, \ddot{m}_{\sigma\sigma})}\right) - \frac{\det(m_{\sigma\sigma\sigma}^{(3)}, \ddot{m}_{\sigma\sigma})}{\det(\dot{m}_\sigma, \ddot{m}_{\sigma\sigma})} + \frac{1}{3}\left(\frac{\det(\dot{m}_\sigma, m_{\sigma\sigma\sigma}^{(3)})}{\det(\dot{m}_\sigma, \ddot{m}_{\sigma\sigma})}\right)^2.$$

The computation of the expression of the curvature for an arbitrary parametrization is left to the reader. It involves the second derivative of the arc length, and therefore the seventh derivative of the curve.

Medial Axis

2.1 Introduction

The medial axis [24] transform associates a skeleton-like structure to a shape, which encodes its geometry. The medial axis itself (or skeleton) is the center of discs of maximal radii inscribed in the shape. The medial axis transform stores in addition the maximal radii.

More precisely, represent a shape by an open connected bounded set in the plane, denoted Ω. Let $B(p,r)$ denote the open disc of center $p \in \mathbb{R}^2$ and radius $r > 0$. One says that such a disc is maximal in Ω if and only if it is included in Ω, and no disc in which it is (strictly) contained is included in Ω. The skeleton of Ω, denoted $\Sigma(\Omega)$, is the set of all p such that $B(p,r)$ is maximal in Ω for some $r > 0$, i.e., $\Sigma(\Omega)$ is the set of loci of the centers of maximal discs. We shall also denote $\Sigma^*(\Omega)$ as the set of pairs (p,r) such that $B(p,r)$ is maximal. This is the medial axis transform (MAT). We have the following proposition.

Proposition 2.1. *The medial axis transform, $\Sigma^*(\Omega)$, uniquely characterizes Ω.*

Proof. Let

$$\tilde{\Omega} = \bigcup_{(p,r) \in \Sigma^*(\Omega)} B(p,r).$$

By definition of Σ^*, we have $\tilde{\Omega} \subset \Omega$ and we want to prove the reverse inclusion.

For $x \in \Omega$, let $r_x = \mathrm{dist}(x, \Omega^c)$ so that $b(x, r_x) \subset \Omega$ and define

$$G_x = \{y \in \Omega : B(y, r_y) \supset B(x, r_x)\}$$

and $r_x^* = \sup r_y, y \in G_x$. By definition, there exists a sequence (y_n) such that $r_{y_n} \to r_x^*$, and, since Ω is bounded, we can assume (replacing y_n by a subsequence is needed) that $y_n \to y^* \in \overline{\Omega}$. Obviously, y^* cannot belong to $\partial\Omega$ since this would imply $r_{y_n} \to 0$ and $r_x^* \geq r_x > 0$ (since $x \in G_x$). Also, since

$B(y_n, r_{y_n}) \subset \Omega$, we have at the limit $B(y^*, r_x^*) \subset \overline{\Omega}$ which implies $B(y^*, r_x^*) \subset \Omega$ because Ω is open. Similarly, passing to the limit in the inclusion $B(x, r_x) \subset B(y_n, r_{y_n})$ implies $x \in B(y^*, r_x^*)$.

We now show that $B(y^*, r_x^*)$ is maximal, which will prove that $\Omega \subset \tilde{\Omega}$, since we started with an arbitrary $x \in \Omega$. But if $B(y^*, r_x^*)$ is included in some ball $B(y, r) \in \Omega$, it will be a fortiori included in $B(y, r_y)$ and since $x \in B(y^*, r_x^*)$, we see that y must be in G_x with $r_y > . r_x^*$, which is a contradiction.

2.2 Structure of the Medial Axis

We assume that Ω is the interior of a piecewise smooth Jordan curve. Some structural properties of the skeleton can be obtained under some assumptions on the regularity of the curve [46]. The assumption is that the smooth arcs are analytic, which means infinitely differentiable, and such for each t, $m(t)$ is the limit of its Taylor series, except at a finite number of points; for these exceptional points, it is required that m has both left and right tangents. The simplest example of a curve satisfying this assumption is a polygon.

For such a curve, it can be shown that all but a finite number of points in the skeleton are such that the maximal disc $B(p, r)$ meets the curve m at exactly two points. Such points on the skeleton are called regular. Non-regular points separate into three categories.

The first type is when the maximal disc, $B(m, r)$, meets the curve in more than two connected regions. Such points are *bifurcation points* of the skeleton. The second possibility is when there is only one connected component; then, there are two possibilities: either m is the center of an osculating circle to the curve, or there exists a concave angle at the intersection of the curve and the maximal disc. The third possibility is when there are two connected components, but one of them is a sub-arc of the curve. This happens only when the curve has circular arcs.

The skeleton itself is connected, and it is composed of a finite number of smooth curves.

2.3 The Skeleton of a Polygon

There are at least two reasons for which the skeletons of polygons have practical interest. The first one is that digital shapes can always be considered as polygons (because they are described by a finite number of points), and numerical algorithms for skeleton computation rely on the description of the skeletons of polygons. The second one is that truly polygonal shapes (not only at the discretization level) are very common, because man-made objects often are polyhedrons.

Consider a closed polygon, without self-intersections. Denote its vertices $m_1, \ldots, m_N, m_{N+1} = m_1$. Let s_i denote the ith edge, represented by open line segments (m_i, m_{i+1}), for $i = 1, \ldots, N$. A maximal disc within the polygon has to meet the boundary at two points. We separate the cases depending on whether these points are on edges or vertices.

Let $B(m, r)$ be a maximal disc. Assume first that it is tangent to s_i at some point $p \in s_i$. Denote $T_i = (m_{i+1} - m_i)/|m_{i+1} - m_i|$ the unit tangent to s_i and N_i the unit normal. We assume that the orientation is such that N_i points inward. We must have

$$p = m - rN_i \text{ and } p = m_i + tT_i$$

for some $t \in (0, |m_{i+1} - m_i|)$. Taking the dot product of both equations with T_i and computing the difference yields

$$t = (m - m_i)^T T_i.$$

We therefore obtain the fact that $B(m, r)$ is tangent to s_i if and only if

$$m - rN_i = m_i + ((m - m_i)^T T_i)T_i$$
$$\text{with } 0 \leq (m - m_i)^T T_i \leq |m_{i+1} - m_i|.$$

We can distinguish three types of maximal discs: $B(m, r)$.

1. Bitangents: there exists $i \neq j$ with

$$m = m_i + ((m - m_i)^T T_i)T_i + rN_i = m_j + ((m - m_j)^T T_j)T_j + rN_j \text{ and }$$
$$0 \leq (m - m_i)^T T_i \leq |m_{i+1} - m_i|, 0 \leq (m - m_j)^T T_j \leq |m_{j+1} - m_j|.$$

2. Discs that meet the boundary at exactly one edge and one vertex: there exists $i \neq j$ such that

$$m = m_i + ((m - m_i)^T T_i)T_i + rN_i,$$
$$0 \leq ((m - m_i)^T)T_i \leq |m_{i+1} - m_i|$$
$$\text{and } |m - m_j| = r.$$

3. Discs that meet the boundary at two vertices: there exists $i \neq j$ such that $|m - m_i| = |m - m_j| = r$.

Note that a maximal ball can meet a vertex only if this vertex points inward (concave vertex). In particular, with convex polygons, only the first case can happen.

The interesting consequence of this result is that the skeleton of a polygon is the union of line segments and arcs of parabola. To see this, consider the equations for the three previous cases. For bitangents, we have

$$r = (m - m_i)^T N_i = (m - m_j)^T N_j$$

which implies

$$(m - m_i)^T (N_j - N_i) = (m_j - m_i)^T N_j.$$

If $N_i \neq N_j$, this is the equation of a line orthogonal to $N_i - N_j$. The case $N_i = N_j$ can never occur because the normals have to point to the interior of maximal balls and therefore coincide only if $s_i = s_j$.

For the second case, we have

$$m - m_i = ((m - m_i)^T) T_i T_i + |m - m_j| N_i$$

which yields

$$(m - m_i)^T N_i = |m - m_j|.$$

This is the equation of a parabola. To see why, express m as $m = m_i + \alpha T_i + \beta N_i$. The previous equations yield $\beta \geq 0$ and

$$\beta^2 = (\alpha - (m_j - m_i)^T T_i)^2 + (\beta - (m_j - m_i)^T N_i)^2$$

or

$$2(m_j - m_i)^T N_i \beta = (\alpha - (m_j - m_i)^T T_i)^2 + ((m_j - m_i)^T N_i)^2.$$

Finally, in the last case, the skeleton coincides with the line of points which are equidistant from the two vertices. We have therefore proved the following fact (which comes in addition to the properties discussed in Section 2.2).

Proposition 2.2. *The skeleton of a polygonal curve is a union of line segments and parabolas. For a convex polygon, the skeleton only contains line segments.*

2.4 Voronoï Diagrams

2.4.1 Voronoï Diagrams of Line Segments

The previous computation and the most efficient algorithms to compute skeletons are related by the theory of Voronoï diagrams. We start with their definition:

Definition 2.3. *Let $F_1, \ldots F_N$ be closed subsets of $\mathbb{R}^2$. The associated Voronoï cells are the sets $\Omega_1, \ldots, \Omega_N$ defined by*

$$x \in \Omega_i \Leftrightarrow d(x, F_i) < \min_{j \neq i} d(x, F_j).$$

The union of the boundaries, $\bigcup_{i=1}^{N} \partial \Omega_i$, forms the Voronoï diagram associated to $F_1, \ldots, F_N$.

In the case of a polygonal curve, the skeleton is included in the Voronoï diagram of the closed line segments that form the curve. A point of the skeleton has indeed to meet at least two segments (sometimes at their common vertices), and is at a strictly larger distance from the ones it does not intersect. It therefore belongs to the boundary of the cells. The converse is false: a point from the diagram is not necessarily in the skeleton (some points may correspond to external balls).

There exist very efficient algorithms to compute these diagrams. We shall not detail them here, but references can be found in [166, 159].

The notion of Voronoï diagrams for a polygon can be extended to a general curve. The question is to find sub-arcs $F_1, \ldots, F_N$ of the curve with the property that their diagram contains the curve's skeleton. What we have said concerning polygons applies, except in one case: when a maximal disc meets an arc at two distinct points. This could not happen with straight lines, and a condition ensuring that this does not happen for a given arc is as follows [126]. Recall that a vertex of a smooth curve m is a local extremum of the curvature.

Theorem 2.4. *A sub-arc of a C^2 closed curve which has two points belonging to a maximal disc necessarily contains a vertex.*

Therefore, it suffices to cut the curve at vertices to be sure that the obtained arcs cannot hold two contacts with maximal discs.

2.4.2 Voronoï Diagrams of the Vertices of Polygonal Curves

The medial axis can, in some sense, also be interpreted as the skeleton of the infinite family of points in the curve. This leads to the question of whether the Voronoï diagram of the sequence of points in a discrete curve is a good approximation of the skeleton when the discretization step goes to 0. The answer is: not at all, and almost yes.

The cells of the Voronoï diagram of a finite family of points are polygons, which makes them even simpler than the diagram of line segments. Moreover, there is an extensive literature on their computation, which is one of the basic algorithms in discrete and computational geometry [166, 162].

In the diagram of a polygonal curve, there are essentially two types of segments: those containing points which are equidistant to two consecutive vertices, and the rest. When the discretization is fine enough, the first type of segments cannot belong to the skeleton, and have to be pruned out of the Voronoï diagram. They are easily detectable, since they contain the midpoint of the two consecutive vertices, and therefore cross the curve. The computation of the skeleton of a discrete curve can therefore be done as follows: first compute the Voronoï diagram of the vertices. Then remove the segments that are not entirely in the interior of the curve. Figure 2.1 illustrates how discretization affects the skeleton in a simple example.

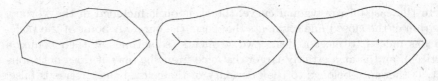

Fig. 2.1. Comparison of medial axes computed using Voronoï diagrams with different degrees of discretization.

2.5 Thinning

Thinning algorithms create their own kind of skeleton which does not necessarily correspond to the centers of maximal discs. They are, however, quite efficient and generally easy to implement. The principle is to progressively "peel" the boundary of the region until only a skeletal structure remains. One of the first methods is the Hilditch algorithm [111], in which a sequence of simple tests are performed to decide whether a pixel must be removed or not from the region. Another similar point of view uses the erosion operation in mathematical morphology [182]. We briefly describe the latter.

Define a structuring element B to be a symmetric subset of $\mathbb{R}^2$ (for example a small disc centered at 0). Using B, we define a sequence of operators that apply to a set X and creates a new set:

$$E_B(X) = \{x : x + B \subset X\} \text{ (erosion)},$$
$$D_B(X) = \{x : x + B \cap X \neq \emptyset\} \text{ (dilation)},$$
$$O_B(X) = D_B \circ E_B(X) \text{ (opening)},$$
$$L_B(X) = X \setminus O_B(X).$$

Erosion is like peeling X with a knife shaped like B. Dilation spreads matter around X, with a thickness once again provided by B. Opening is an erosion followed by a dilation, which essentially puts back what the erosion has removed, except the small structures which have completely been removed and cannot be recovered (since there is nothing left to spread on). The last operation, L_B, precisely collects these lost structures (called linear parts), and is the basic operator for the morphological skeleton which is defined by

$$S(X) = \cup_{n=1}^N L_B(E_{nB}(X)).$$

This is the union of the linear parts of X after successive erosions.

2.6 Sensitivity to Noise

One of the main issues with the medial axis transform is its lack of robustness to noise. Figure 2.2 provides an example of how small variations at the

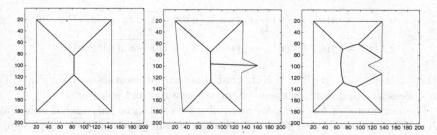

Fig. 2.2. Effect of a small shape change in the boundary on the skeleton of a rectangular shape.

boundary of a shape can result in dramatic changes in the skeleton. In fact, we have seen in our discussion of polygons that the addition of a convex vertex automatically results in a branch in the skeleton reaching to it.

Because of this, many skeletonization algorithms come with a way to prune the skeleton of spurious branches. There are two ways to do this.

- *Prior smoothing of the curve.* Curve smoothing algorithms have been described in Section 6. For polygons, smoothing can be done by removing small structures of flattening vague angle. It is interesting to note that smoothing curves does not always result in simplifying the skeleton (see [16] for a discussion).
- *Pruning.* Branches can be removed after the computation of the skeleton. This can be based on several principles, since branches resulting from small incidents at the boundary can be detected on the skeleton [93].

2.7 Recovering the Initial Curve

Given a parametrized sub-arc of the medial axis transform, one can explicitly reconstruct the part of the boundary $\partial\Omega$ which is associated to it (the contact points of the maximal balls with $\partial\Omega$). Assume that a C^1 function γ, from (a, b) to $\Sigma^*(\Omega)$ is given. Denote $\gamma(u) = (m(u), r(u))$.

Without loss of generality, assume that $u \mapsto m(u)$ is arc length ($|\dot{m}_u| = 1$). Assume also that $B(m(u), r(u))$ has exactly two contacts with $\partial\Omega$ (this is typically true on $\Sigma^*(\Omega)$ except at a finite number of points).

If $x \in \partial\Omega \cap B(m(u), r(u))$, then $|x - m(u)| = r(u)$ and, for all $\varepsilon \neq 0$, $|x - m(u + \varepsilon)| \geq r(u + \varepsilon)$ (because $B(m(u + \varepsilon), r(u + \varepsilon)) \subset \Omega$). Thus, letting $f(\varepsilon) = |x - m(u + \varepsilon)|^2 - r(u + \varepsilon)^2$, we have $f(0) = \partial_\varepsilon f(0) = 0$, with

$$\partial_\varepsilon f(0) = -2\langle x - m(u), \dot{m}_u(u) \rangle + 2r(u)\dot{r}_u(u).$$

Solving this equation, we obtain two solutions for x, given by

$$x_+(u) = m(u) + r(u)\left[-\dot{r}_u(u)\dot{m}_u(u) + \sqrt{1 - \dot{r}_u(u)^2}q(u)\right],$$

$$x_-(u) = m(u) + r(u)\left[-\dot{r}_u(u)\dot{m}_u(u) - \sqrt{1 - \dot{r}_u(u)^2}q(u)\right],$$

with $q(u) \perp \dot{m}_u(u)$, $|q(u)| = 1$. Note that this computation shows that $|\dot{r}_u| < 1$ is a necessary condition for the existence of two distinct solutions.

The curvature of the boundary can also be related to the medial axis via an explicit formula. Let ρ_+ (resp. ρ_-) be the vector $-\dot{r}_u\dot{m}_u + \sqrt{1 - \dot{r}_u^2}q$ (resp. $-\dot{r}_u\dot{m}_u - \sqrt{1 - \dot{r}_u^2}q$) so that $x_+ = m + r\rho_+$ and $x_- = m + r\rho_-$. The following discussion holds for both arcs and we temporarily drop the $+$ and $-$ indices in the notation.

We have $x = m + r\rho$; ρ is a unit vector, and since the maximum disc is tangent to the curve at x, ρ is normal to the curve. Since r is positive and ρ is a radial vector for a maximal disc, ρ points outward from the curve at point x and therefore is oriented in the opposite direction to the normal (assuming that the curve is positively oriented). Introduce the vector $h = \dot{m}_u + \dot{r}_u\rho$. We have $h^T\rho = -\dot{r}_u + \dot{r}_u = 0$ so that h is orthogonal to ρ. Since $|\rho| = 1$, $\dot{\rho}_u$ is also orthogonal to ρ and there exists a number c such that $\dot{\rho}_u = -ch$ (we have $|h|^2 = 1 - \dot{r}_u^2 > 0$ so that $h \neq 0$). Since $\rho = -N$, we also have

$$\dot{\rho}_u = \dot{\rho}_s\frac{ds}{du} = \kappa\frac{ds}{du}T$$

where κ is the curvature of the considered arc of curve. Likewise, $\dot{x}_u = (ds/du)T$ so that $\dot{\rho}_u = \kappa\dot{x}_u$. We now use this relation to compute κ: we have $\dot{x}_u = \dot{m}_u + \dot{r}_u\rho + r\dot{\rho}_u = (1 - cr)h$. This implies

$$\kappa = -c/(1 - cr)$$

which provides a very simple relation between c and the curvature.

To be complete, it remains to compute c. From $\dot{\rho}_u = -c(\dot{m}_u + \dot{r}_u\rho)$, we get

$$\dot{\rho}_u^T\dot{m}_u = -c(1 + \dot{r}_u\rho^T\dot{m}_u) = -c(1 - \dot{r}_u^2).$$

We also have

$$-\ddot{r}_{uu} = \partial_u(\rho^T\dot{m}_u) = \dot{\rho}_u^T\dot{m}_u + \rho^T\ddot{m}_{uu} = \dot{\rho}_u^T\dot{m}_u + K\rho^Tq$$

where K is the curvature of the skeleton. Writing $\rho^Tq = \varepsilon\sqrt{1 - \dot{r}_u^2}$ with $\varepsilon = \pm1$, we get the equation:

$$\dot{\rho}_u^T\dot{m}_u = -\ddot{r}_{uu} - \varepsilon K\sqrt{1 - \dot{r}_u^2}$$

which yields (reintroducing the $+$ and $-$ subscripts for each contact) $c_+ = \ddot{r}_{uu}/(1 - \dot{r}_u^2) + K/\sqrt{1 - \dot{r}_u^2}$ and $c_- = \ddot{r}_{uu}/(1 - \dot{r}_u^2) - K/\sqrt{1 - \dot{r}_u^2}$.

2.8 Generating Curves from Medial and Skeletal Structures

The previous section described how to retrieve a curve once its medial axis transform has been computed. We want here to discuss the issue of specifying a curve by providing the medial axis transform.

This is a more difficult problem, because not any combination of curves and radii is a valid medial axis. Even when the skeleton consists of only one curve, we have already seen conditions in the above section, like $|\dot{r}_u| < 1$ at all points in the interior of the medial curve, that are required in the skeletal representation. We must also ensure that the specified curve is regular on both sides of the axis, which, since $\dot{x}_u = (1 - cr)h$, must ensure that $1 - cr$ does not vanish along the curve. In fact, $1 - cr$ must be positive. To see this, note that we have proved that $1 - cr = (1 - r\kappa)^{-1}$. At a convex point $(\kappa > 0)$, r must be smaller than the radius of curvature $1/\kappa$ so that $1 - r\kappa > 0$. Since points of positive curvature always exist, we see that $1 - cr$ must remain positive along the curve in order to never be zero. Using the expression for c found in the previous section, this provides a rather complex condition:

$$1 - \frac{r\ddot{r}_{uu}}{1 - \dot{r}_u^2} > \frac{|K|r}{\sqrt{1 - \dot{r}_u^2}}. \tag{2.1}$$

To ensure continuity of the reconstructed curve when branches meet at the extremities of the axis, we need $|\dot{r}_u| = 1$ there. Also, if the medial axis has multiple branches, the corresponding parts of the curve must have the same limits on both sides. More conditions are needed to ensure that the contacts at these points are smooth. This provides a rather complicated set of constraints that must be satisfied by a generative medial axis model. This can be made feasible, however, in some simple cases, as shown in the following examples.

2.8.1 Skeleton with Linear Branches

Let's consider the situation in which each branch of the medial axis is a line segment, i.e., $K = 0$. The constraints on r are then $r > 0$, $\dot{r}_u^2 < 1$ and $r\ddot{r}_{uu} + \dot{r}_u^2 < 1$. The last inequality comes from the fact that $cr < 1 \Leftrightarrow r\ddot{r}_{uu} < 1 - \dot{r}_u^2$. Introducing $z = r^2/2$, this can also be written $\ddot{z}_{uu} < 1$.

Let's assume that $\ddot{z}_{uu} = -f$ with $f > -1$ and see what happens with the other conditions in some special cases.

Integrating twice, we find

$$\dot{z}_u(u) = \dot{z}_u(0) - \int_0^u f(t)dt$$

$$z(u) = z(0) + u\dot{z}_u(0) - \int_0^u (u - t)f(t)dt. \tag{2.2}$$

Shapes with a Single Linear Branch

Start with the simplest situation in which the medial axis is composed of a single segment, say $m(u) = (u, 0)$, $u \in [0, 1]$. Since $|\dot{r}_u| = 1$ at the extremities and the medial axis cannot cross the curve, we need $r_u(0) = 1$ and $\dot{r}_u(1) = -1$. Denote

$$M_0(u) = \int_0^u f(t)dt$$

$$M_1(u) = \int_0^u tf(t)dt.$$

Using the identities $\dot{z}_u(0) = r(0)$, $\dot{z}_u(1) = -r(1)$, $z(0) = r(0)^2/2$ and $z(1) = r(1)^2/2$, we can solve (2.2) with respect to $r(0)$ and $r(1)$ to obtain:

$$r(0) = \frac{M_0(1) + M_0(1)^2/2 - M_1(1)}{1 + M_0(1)}$$

$$r(1) = M_0(1) - r(0) = \frac{M_0(1)^2/2 + M_1(1)}{1 + M_0(1)}.$$

These quantities must be positive, and we will assume that f is chosen with this property (note that the denominator is always positive since $f > -1$). These equations imply that z, and therefore m, are uniquely determined by f. Of course, this does not imply that the remaining constraints, which are (in terms of z) $z > 0$ and $\dot{z}_u^2 < 2z$ are satisfied on $(0, 1)$. Since the latter implies the former, we can concentrate on it, and introduce the function $h(u) = 2z(u) - \dot{z}_u(u)^2$. We have $h(0) = r(0)^2$ and $h(1) = r(1)^2$. Moreover,

$$\dot{h}_u = 2\dot{z}_u(1 - \ddot{z}_{uu}) = 2\dot{z}_u(1 + f).$$

Since $1 + f > 0$, $\dot{h}_u$ vanishes for $z_u = 0$, or $M_0(u) = r(0)$. Note that $\dot{h}_u(0) = 2r(0)(1 + f) > 0$ and $\dot{h}_u(1) = -2r(1)(1 + f) < 0$ so $\dot{h}_u$ changes signs over $(0, 1)$.

Also, since the extrema of h only occur when $\dot{z}_u > 0$ (and $h = 2z$ at these points), h will be positive under any condition that ensures that $z > 0$ when $\dot{z}_u = 0$, which reduces to $r(0)^2/2 + M_1(u) > 0$ whenever $M_0(u) = r(0)$.

There is an easy case: if $f > 0$, then $M_1(u) > 0$ and the condition is satisfied. Moreover, if $f > 0$, then $M_1(1) \leq M_0(1)$ also so that $r(0)$ and $r(1)$ are positive. However, as Figure 2.3 shows, interesting shapes are obtained when $f < 0$ is allowed.

Shapes with Three Intersecting Linear Branches

Let's now consider a slightly more complex example with one multiple point and three linear branches. So we have three lines, ℓ_1, ℓ_2, ℓ_3, starting from a single point p_0. Let $\ell_i = \{p_0 + uw_i, u \in [0, s_i]\}$ where w_1, w_2, w_3 are unit vectors.

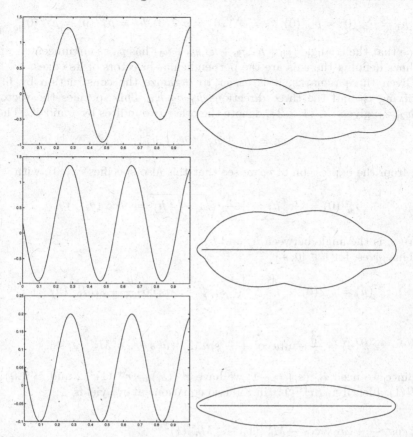

Fig. 2.3. Shapes with horizontal medial axes. The shapes are obtained with $\partial^2_{uu}(r^2) = -2f$; f is in the left column and the curves are in the right one.

Let q_i be a unit vector completing w_i in a positively oriented orthonormal frame. Finally, let $r^{(1)}, r^{(2)}$ and $r^{(3)}$ be the radii along each of these lines and $z^{(i)} = (r^{(i)})^2/2$. Assume that $\ddot{z}^{(i)}_{uu} = -f_i(u/s_i)$ for $u \in (0, s_i)$, where $f_i > -1$ as before, and is defined over $[0, 1]$.

We need to work out the compatibility conditions for the $r^{(i)}$ at the intersection point, $u = 0$. Assume that the branches are ordered so that $(w_1, w_2), (w_2, w_3)$ and (w_3, w_1) are positively oriented. The compatibility conditions are

$$x^{(1)}_+(0) = x^{(2)}_-(0), \; x^{(2)}_+(0) = x^{(3)}_-(0), \; x^{(3)}_+(0) = x^{(1)}_-(0).$$

Identifying the norms, we see that the radii must coincide: $r^{(1)}(0) = r^{(2)}(0) = r^{(3)}(0) := r_0$. So, defining h_1, h_2, h_3 by

$$h_1 = \rho_+^{(1)}(0) = \rho_-^{(2)}(0), h_3 = \rho_+^{(2)}(0) = \rho_-^{(3)}(0), h_2 = \rho_+^{(3)}(0) = \rho_-^{(1)}(0),$$

we see that the triangle $(p_0 + h_1, p_0 + h_2, p_0 + h_3)$ has p_0 as circumcenter, and the lines defining the axis are the perpendicular bisectors of its edges.

Given these remarks, it is easier to organize the construction by first specifying p_0 and the three directions h_1, h_2, h_3. This specifies the vectors w_1, w_2, w_3: given $i \in \{1, 2, 3\}$, denote the other two indices by j and j'. Then

$$w_i = (h_j + h_{j'})/|h_j + h_{j'}|$$

and, from the expression of ρ, we see that this also specifies $\dot{r}_u^{(i)}(0)$, with

$$\dot{r}_u^{(i)}(0) = -z_i^T h_j = -\frac{1}{\sqrt{2}}\sqrt{1 + h_j^T h_{j'}} = -\cos(\theta_i/2)$$

where θ_i is the angle between h_j and h_j'.

This gives, for $u \in [0, s_i]$

$$\dot{z}_u^{(i)}(u) = \dot{z}_u^{(i)}(0) - \int_0^u f^{(i)}(t/s_i)dt = -r_0 \cos\frac{\theta_i}{2} - s_i M_0^{(i)}(u/s_i)$$

and

$$z^{(i)}(u) = \frac{r_0^2}{2} - r_0 u \cos\frac{\theta_i}{2} - s_i u M_0^{(i)}(u/s_i) + s_i^2 M_1^{(i)}(u/s_i).$$

Since we need $\dot{r}_u^{(i)}(s_i) = -1$, we have $z^{(i)}(s_i) = r^{(i)}(1)^2/2$ and $\dot{z}_u^{(i)}(s_i) = -r^{(i)}(1)$. Identifying $r^{(i)}(1)^2$ in the two equations above yields

$$r_0^2 \cos^2\frac{\theta_i}{2} + 2r_0 s_i \cos\frac{\theta_i}{2} M_0^{(i)}(1) + s_i^2 M_0^{(i)}(1)^2$$

$$= r_0^2 - 2r_0 s_i \cos\frac{\theta_i}{2} - s_i^2 M_0^{(i)}(1) + s_i^2 M_1^{(i)}(1)$$

or

$$\left(M_0^{(i)}(1)^2 + 2M_0^{(i)}(1) - 2M_1^{(i)}(1)\right)\frac{s_i^2}{r_0^2}$$

$$+ 2\cos\frac{\theta_i}{2}\left(1 + M_0^{(i)}(1)\right)\frac{s_i}{r_0} - \left(1 - \cos^2\frac{\theta_i}{2}\right) = 0. \quad (2.3)$$

Assuming that $f^{(i)}$ satisfies

$$M_0^{(i)}(1)^2/2 + M_0^{(i)}(1) - M_1^{(i)}(1) > 0,$$

which a condition already encountered in the previous case, this equation has a unique solution, specifying s_i. The curve is then uniquely defined by $p_0, h_1, h_2, h_3, f^{(1)}, f^{(2)}, f^{(3)}$, with constraints on the $f^{(i)}$'s similar to those obtained in the one-branch case. Examples are provided in Figure 2.4.

Note that this construction does not really specify the medial axis, but only the orientation of its branches (since the s_i's are constrained by the rest of the parameters). One possibility to deal with this is to relax the specification of the $f_i's$ by adding a factor α_i, using

$$\ddot{z}_{uu}^{(i)} = -\alpha_i f^{(i)}.$$

This implies that $M_0^{(i)}$ and $M_1^{(i)}$ must be replaced by $\alpha_i M_0^{(i)}$ and $\alpha_i M_1^{(i)}$ in the computation above, and equation (2.3), with fixed s_i, becomes a second-degree equation in α_i. The consistency conditions (existence of a solution to this equation, requirement that $\alpha_i f^{(i)} > -1$, etc.) are, however, harder to work out in this case.

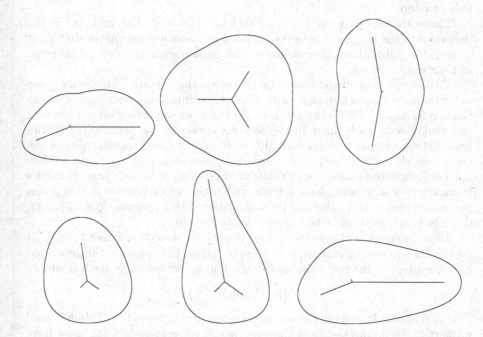

Fig. 2.4. Shapes generated from a medial axis with three linear branches.

Shapes with Generic Linear Branches

Conceptually, the above construction can be generalized to any skeleton with a ternary tree structure and linear branches. Indeed, the derivative $\dot{r}_u$ is uniquely specified at the extremities of any branch: it is -1 if the branch ends and $-\cos\theta/2$ at an intersection, where θ is specified by the branch geometry as above. Also, the radii at all branching points are uniquely specified as soon as one of them is (the constraint propagates along the tree). Of course, as before, the fact that the solution is uniquely defined does not guarantee consistency, which become harder to specify when the structure gets more complex. Finally, it is important to note that, for all the previous methods, even if the consistency conditions are satisfied, there is still a possibility for the shape to self-intersect non-locally (without singularity).

2.8.2 Skeletal Structures

One way to simplify the construction of a shape from a skeleton is to relax some of the conditions that are associated to medial axes. Skeletal structures, which we briefly describe now, have been introduced in [57, 58, 59] with this idea in mind.

There are two parts in Damon's skeletal structure. The first one is the skeletal set (the skeleton), which is a union of smooth open curves that meet at singular points (branching points or end-points) with well-defined tangents at their extremities.

The second part of the skeletal structure is the vectors which correspond to $r\rho$ in our previous notation, with some smoothness and consistency conditions; referring to [57] for details, here are the most important ones. Like with the medial axis, each point in the smooth curves of the skeletal set carries two of these vectors (one on each side of the curve), and singular points can carry one vector (end-points) or more than two at branching points. When one continuously follows one of these vectors along a smooth branch until a branching point, it must have a limit within the set of vectors at this point, and all vectors at this point can be obtained by such a process. At end-points, there is a unique vector which is tangent to the curve.

To summarize, a skeletal structure requires a skeletal set, say S, and, at each point p in the skeletal set, a set $U(p)$ of vectors that point to the generated curve, subject to the previous conditions. The generated curve itself is simply

$$C = \{p + U(p), p \in S\}.$$

The medial axis transform does induce a skeletal structure, but has additional properties, including the facts that, at each p, all vectors in $U(p)$ must have the same norm, and if p is on a smooth curve, the difference between the two vectors in $U(p)$ must be perpendicular to the curve. These properties are not required for skeletal structures.

Most of the analysis done in the previous section on the regularity of the generated curve with the medial axis transform can be carried over to skeletal structures. Along any smooth curve in the skeletal structure, one can follow a smooth portion of the generated curve, writing

$$x(u) = m(u) + r(u)\rho(u)$$

and assuming an arc-length parametrization in $m(u)$. Letting $c = -\dot{m}_u^T \dot{\rho}_u$, we can write, for some $\alpha \in \mathbb{R}$,

$$\dot{\rho}_u = -c\dot{m}_u + \alpha\rho$$

because ρ is assumed to be non-tangent to the skeletal set (except at its end-points). This definition of c generalizes the one given for the medial axis, in which we had $\dot{\rho}_u = -ch = -c\dot{m}_u + c\dot{\rho}_u\rho$. Since we have $\dot{x}_u = (1-cr)\dot{m}_u + (\alpha + \dot{r}_u)\rho$, we see that $cr < 1$ is here also a sufficient condition for the regularity of the curve.

We need to check that different pieces of curves connect smoothly at branching points. With the medial axis, a first-order contact (same tangents) was guaranteed by the fact that the generated curve was everywhere perpendicular to ρ. With skeletal structures, we have (since $\dot{\rho}_u^T\rho = 0$)

$$\dot{x}_u^T\rho = \dot{m}_u^T\rho + \dot{r}_u.$$

So, a sufficient condition for smooth contacts at branching points and at end-points is that $\dot{r}_u + \dot{m}_u^T\rho$ vanishes at the extremities of the smooth curves (while this quantity vanishes everywhere with the medial axis transform).

Obviously, these conditions are much less constraining than those associated to the medial axis transform. One can start fixing ρ, which defines c, then r such that $rc < 1$, with a few end-point conditions that must be satisfied. This simplification that is brought to curve generation, however, comes with a price, which is that a skeletal structure is not uniquely specified by a given curve, as the medial axis transform was. It cannot really be qualified as a "curve representation" like the ones we have considered so far.

3

Moment-Based Representation

3.1 Introduction

Moments are global descriptors which can be used to represent shapes [117, 199, 177]. They are defined as integrals, either along the boundary of the shape, or over its interior. To simplify the discussion, we introduce a common notation for both cases. Let m be a non-intersecting closed curve, and Ω_m its interior. Let f be a function defined on $\mathbb{R}^2$. We define the following moments.

Boundary moment: $I_{bdry}^m(f) = \int_0^{L_m} f(m(s))ds$ where s is the arc length on m and L_m the length of the curve.

Interior moment: $I_{int}^m(f) = \int_{\Omega_m} f(x,y)dxdy$.

(We will use the notation $I^m(f)$ for definitions or statements that apply to both moments.)

Let $\mathcal{F}$ be a finite or infinite set of real-valued functions, defined on $\mathbb{R}^2$. To a curve m, one can associate the family of numbers: $\mathcal{F}(m) = \{I^m(f), f \in \mathcal{F}\}$. This will be called the representation of m by $\mathcal{F}$-moments.

Of course, the interest is in finding a set $\mathcal{F}$ as small as possible that provides useful information on a shape. Another important constraint is the invariance with respect to rigid transformations: if the shape is translated and rotated, the moments should be left unchanged, or at least change in a predictable way.

3.2 Moments of Inertia and Registration

Low-degree polynomials provide the center and moments of inertia of the shape. When $f = 1$, $I_{bdry}^m(1)$ is the length of the curve and $I_{int}^m(1)$ is its area. The center of inertia is the point $c_m = (x_m, y_m)$ with $x_m = I^m(x)/I^m(1)$ and $y_m = I^m(y)/I^m(1)$ (with the abuse of notation of using x for the function $(x,y) \to x$). An affine transformation g acting on m transforms c_m into $g(c_m)$ (g being restricted to rotations and scaling in the boundary case).

The matrix of inertia of the shape is the symmetric matrix

$$J^m = \begin{pmatrix} J^m_{xx} & J^m_{xy} \\ J^m_{yx} & J^m_{yy} \end{pmatrix}$$

with

$$J^m_{xx} = I^m\big((x - x_m)^2\big)/I^m(1),$$
$$J^m_{xy} = I^m\big((x - x_m)(y - y_m)\big)/I^m(1) = J^m_{yx},$$
$$J^m_{yy} = I^m\big((y - y_m)^2\big)/I^m(1).$$

The matrix of inertia is invariant by translation, and the effect of a linear transformation (rotation and scaling in the boundary case) $m \to Am$ is $J \to AJA^T$. Note that, in general, translation-invariant moments can only be achieved by centering the shape around c_m, as we did. This corresponds to replacing moments $I^m(f(\cdot))$ by $I^m(f(\cdot - c_m))$.

This property leads to the simplest algorithm for shape registration. Since J^m is a symmetric matrix, it can be diagonalized, and assuming that it has distinct eigenvalues, $\lambda_1 > \lambda_2$, there is a unique rotation matrix R such that

$$J^m = R^T \begin{pmatrix} \lambda_1 & 0 \\ 0 & \lambda_2 \end{pmatrix} R.$$

The axes generated by the eigenvectors of J^m are called the axes of inertia of the shape. From the previous formula, the rotated shape Rm has a diagonal matrix of inertia.

Now, if a second shape $\tilde{m}$ is given, one can build its matrix of inertia $J^{\tilde{m}}$ and the associated rotation $\tilde{R}$. The shape $\tilde{m}$ can be aligned to m by aligning their axes of inertia, which is done by the transformation $\tilde{m} \mapsto R^T \tilde{R}\tilde{m}$. Similarly, the square root of the ratio between the traces of the matrices of inertia can be used for scale registration.

3.3 Algebraic Moments

The numbers $E^m_{kl} = I^m(x^k y^l)$ provide the algebraic moments associated to the curve m. Our goal in this section is to derive functions of these moments which are invariant to some family of linear transformations. This will require discussing how these linear transformations affect homogeneous polynomials, and make a brief incursion into the theory of algebraic invariants. We only consider linear transformations, since invariance by translation can be obtained by centering the shape at its center of inertia.

If f is a function defined on $\mathbb{R}^2$ and A a linear transformation, we can define a new function, denoted $A \cdot f$, by $(A \cdot f)(p) = f(A^{-1}p)$, $p \in \mathbb{R}^2$.

Let $\mathcal{P}$ be the set of homogeneous polynomials with two variables. An element of $\mathcal{P}$ takes the form

$$f(x,y) = \sum_{k=0}^{n} a_k x^k y^{n-k}.$$

If A is a linear transformation, and f a polynomial in $\mathcal{P}$, then $A \cdot f$ also belongs to $\mathcal{P}$, with new coefficients. The following arguments show how to relate invariant functions of algebraic moments to functions of the coefficients of homogeneous polynomials which are invariant by the transformations induced by linear maps.

The generating function of the shape is defined by $M^m(u,v) = I^m(e^{ux+vy})$. Expanding the exponential, we have

$$\begin{aligned}
M^m(u,v) &= \sum_{j \geq 0} I^m \left(\frac{(ux+vy)^j}{j!} \right) \\
&= \sum_{j \geq 0} \sum_{k=0}^{j} I^m \left(\frac{u^k x^k y^{j-k} v^{j-k}}{k!(j-k)!} \right) \\
&= \sum_{j \geq 0} \sum_{k=0}^{j} \frac{u^k v^{j-k}}{k!(j-k)!} E^m_{k,j-k} \\
&= \sum_{j \geq 0} f^m_j(u,v)
\end{aligned}$$

where $f^m_j(u,v)$ is the homogeneous polynomial of degree j with coefficients $E^m_{k,j-k}/k!(j-k)!$.

Given a linear transformation A, we now express M_{Am} as a function of M_m. We need to separate boundary and interior moments.

For boundary moments, we restrict A to be a similitude (rotation composed with scaling). Letting λ be the scaling parameter, we have, for an arbitrary function h, $I^{Am}_{bdry}(h) = \lambda I^m_{bdry}(A^{-1} \cdot h)$. Indeed, take a common parametrization for m and Am on a fixed interval, say $[0.1]$. Then

$$I^{Am}_{bdry}(h) = \int_0^1 h(Am(u)) |\partial_u(Am)| \, du = \lambda \int_0^1 h(Am(u)) |\partial_u m| \, du = \lambda I^m_{bdry}(A^{-1} \cdot h)$$

since, by definition, $A^{-1} \cdot h = h \circ A$. This can also we written $I^{Am}_{bdry}(h) = (\det A)^{1/2} I^m(A^{-1} \cdot h)$.

For interior moments, we have

$$I^{Am}_{int}(h) = \int_{A\Omega_m} h(p) \, dp = \int_{\Omega_m} h(Ap') |\det A| \, dp'$$

with the change of variables $p = Ap'$. This yields $I^{Am}_{int}(h) = |\det A| I^m_{int}(A^{-1}h)$. The computation is valid for any linear (invertible) A.

We now apply this to M^m. We particularize the discussion to interior moments (boundary moments are handled similarly, with the restriction to rigid transformations). For $p = (u,v)$, we have

$$M^{Am}(p) = I^{Am}(e^{\langle p, \cdot \rangle})$$
$$= |\det A| I^m_{int}(e^{\langle p, A(\cdot) \rangle})$$
$$= |\det A| I^m_{int}(e^{\langle A^T p, \cdot \rangle})$$
$$= |\det A| M^m(A^T p).$$

Using the expansion in terms of the f_j^m's, we obtain the fact that, for all j, $f_j^{Am}(p) = |\det A| f_j^m(A^T p)$, or

$$f_j^{Am} = |\det A| A^{-T} \cdot f_j^m$$

(where A^{-T} is the inverse of the transpose).

So, up to the scaling factor $|\det A|$, the polynomial f_j^m is transformed by a linear change of variable with the transformation $m \mapsto Am$. Any function of the coefficients of f_j^m which is invariant by a linear change of variables will therefore remain invariant via this transformation. Since the coefficients of f_j^m are the algebraic moments of m, we obtain the characterization of the invariants we are looking for. More precisely, if F is a function of the coefficients of homogeneous polynomial of degree n which is invariant (up to a power of the determinant) by linear changes of variables (F is called an *algebraic invariant*), then the same function applied to $E_{k,n-k}^m / k!(n-k)!$ will be invariant (up to a power of the determinant) to linear transformations applied to the curve m.

We will not discuss further the theory of algebraic invariants, except in the case of rotations, which will be handled in the next section. The simplest example is for polynomials of degree 2 (of the form $f(x, y) = ax^2 + bxy + cy^2$), for which one invariant is the discriminant, equal to $b^2 - 4ac$. This yields the invariant moment

$$\left(\frac{E_{11}}{1!1!}\right)^2 - 4\frac{E_{20}}{2}\frac{E_{02}}{2} = E_{11}^2 - E_{20}E_{02}$$

or

$$\int xy\, d\mu_m - \int x\, d\mu_m \int y\, d\mu_m$$

which is the covariance of the coordinates on the shape.

3.4 Rotation Invariance

The computation of algebraic invariants in the case of rotations is easier than with the general linear group. We will work with complex numbers, letting $z = x + iy$. We first notice that a polynomial $P(x, y)$ can be transformed into another polynomial Q via the equality $P(x, y) = Q(z, \overline{z})$. The coefficients of Q can be expressed as a function of those of P as follows. Let $P(x, y) = \sum_{k=0}^n a_k x^k y^{n-k}$ and $Q(z, \overline{z}) = \sum_{k=0}^n A_k z^k \overline{z}^{n-k}$. We have

$$Q(z, \overline{z}) = P(x, y) = \frac{1}{2^n} P(2x, 2y).$$

Using $2x = z + \overline{z}$, $2iy = z - \overline{z}$, we get

$$2^n Q(z, \overline{z}) = \sum_{k=0}^{n} a_k i^{k-n} (z + \overline{z})^k (z - \overline{z})^{n-k}$$

$$= \sum_{k=0}^{n} a_k i^{k-n} \left(\sum_{j=0}^{k} \binom{k}{j} z^j \overline{z}^{k-j} \right) \left(\sum_{q=0}^{n-k} (-1)^{n-k-q} \binom{n-k}{q} z^q \overline{z}^{n-k-q} \right)$$

$$= \sum_{k=0}^{n} \sum_{j=0}^{k} \sum_{q=0}^{n-k} \binom{k}{j} \binom{n-k}{q} (-1)^q a_k i^{n-k} z^{j+q} \overline{z}^{n-j-q}$$

$$= 2^n \sum_{l=0}^{n} A_l z^l \overline{z}^{n-l}$$

yielding

$$A_l = 2^{-n} \sum_{k=0}^{n} a_k i^{n-k} \sum_{j=\max(k+l-n,0)}^{\min(k,l)} \binom{k}{j} \binom{n-k}{l-j} (-1)^{l-j}.$$

Let us consider a transformation of x and y by a rotation with angle θ. The corresponding transformation on z is very simple: it is $z \mapsto e^{i\theta} z$, and on $\overline{z}$: $\overline{z} \mapsto e^{-i\theta} \overline{z}$. Consider the transformed polynomial

$$Q(e^{i\theta} z, e^{-i\theta} \overline{z}) = \sum_{l=0}^{n} B_l z^l \overline{z}^{n-l}.$$

Identifying the coefficients, we get $B_l = e^{(2l-n)i\theta} A_l$.

This implies in particular that

$$A_l^{2k-n} / A_k^{2l-n} = B_l^{2k-n} / B_k^{2l-n}$$

and if n is even, $A_{n/2} = B_{n/2}$.

To apply this to obtain invariant functions of the moments E_{kl}, define, for a given n

$$U_l^m = 2^{-n} \sum_{k=0}^{n} i^{n-k} \frac{E_{k,n-k}^m}{k!(n-k)!} \sum_{j=\max(k+l-n,0)}^{\min(k,l)} \binom{k}{j} \binom{n-k}{l-j} (-1)^{l-j}$$

then the ratios

$$(U_l^m)^{2k-n} / (U_k^m)^{2l-n}$$

are invariant by rotation of the curve m and can be used as rotation-invariant curve descriptors.

4

Local Properties of Surfaces

In this chapter, we start discussing representations that can be associated to three-dimensional shapes, where surfaces now replace curves. For this reason, we start with basic definitions and results on the theory of surfaces in $\mathbb{R}^3$. Although some parts are redundant with the abstract discussion of submanifolds that is provided in Appendix B, we have chosen to provide a more elementary presentation here, very close to [64], to ensure that this important section can be read independently.

4.1 Curves in Three Dimensions

Before addressing surfaces, we extend our developments on plane curves to the three-dimensional case. A three-dimensional parametrized curve is a function $\gamma : [a, b] \mapsto \mathbb{R}^3$. It is regular if it is C^1 and $|\dot{\gamma}_t| \neq 0$ for all $t \in [a, b]$. For regular curves, the unit tangent is defined like in two dimensions by $T = \dot{\gamma}_t / |\dot{\gamma}_t|$, and the arc length is $ds = |\dot{\gamma}_t| dt$.

Assume that the curve is C^2 and parametrized with arc length. One then defines the *curvature* of γ at s by $\kappa(s) = |\dot{T}_s|$. This differs from the planar case for which a sign was attributed to the curvature: here, the curvature is always positive.

One says that the curve is bi-regular if $\kappa(s) \neq 0$. In this case, one uniquely defines a unit vector N by the relation $\dot{T}_s = \kappa N$; N is perpendicular to T since T has unit norm. Finally, the binormal is the unique unit vector B which completes (T, N) into a positive orthonormal basis of [1] $\mathbb{R}^3$: $B = T \times N$. The

---

[1] If $h = (a, b, c)$ and $k = (a', b', c')$ are three-dimensional vectors, their cross product $h \times k$ is defined by

$$h \times k = (bc' - cb', a'c - ac', ab' - a'b).$$

It is orthogonal to both h and k and vanishes if and only if h and k are collinear. Moreover, for any third vector l: $\langle h \times k, l \rangle = \det[h, k, l]$.

frame (T, N, B) is called the Frénet frame, and the plane passing through $\gamma(t)$ and generated by T and N is called the osculating plane.

The variation of the normal is orthogonal to N and can therefore be written $\dot{N}_s = aT + bB$. We have $\langle N, T \rangle = 0$ so that $a = \langle \dot{N}_s, T \rangle = -\langle N, \dot{T}_s \rangle = -\kappa$. The *torsion* of the curve is given by $-b$ by definition and denoted τ so that $\dot{N}_s = -\kappa T - \tau B$. Using the fact that $\langle \dot{B}_s, T \rangle = -\langle B, \dot{T}_s \rangle = -\kappa \langle B, N \rangle = 0$ and $\langle \dot{B}_s, N \rangle = -\langle \dot{N}_s, B \rangle = \tau$, we have $\dot{B}_s = \tau N$, which provides the third equation of Frénet's formulae for three-dimensional curves:

$$\begin{cases} \dot{T}_s = \kappa N, \\ \dot{N}_s = -\kappa T - \tau B, \\ \dot{B}_s = \tau N. \end{cases} \tag{4.1}$$

There is a three-dimensional version of Theorem 1.6:

Theorem 4.1. *Two C^2 curves γ and $\tilde{\gamma}$ have the same curvature and torsion as functions of their arc length if and only if there exist a rotation R and a vector b such that $\tilde{\gamma} = R\gamma + b$.*

Proof. Let F be the 3 by 3 rotation matrix associated to the Frénet frame, i.e., $F = (T, N, B)$. Then, the Frénet formulae can be written as

$$\partial_s F = FA \tag{4.2}$$

where A is the skew-symmetric matrix

$$A = \begin{pmatrix} 0 & -\kappa & 0 \\ \kappa & 0 & \tau \\ 0 & -\tau & 0 \end{pmatrix}.$$

So let γ and $\tilde{\gamma}$ have the same curvature and torsion, which means that they have the same matrix A, and let F and $\tilde{F}$ be their respective Frénet frames. Then,

$$\partial_s(\tilde{F}F^T) = \tilde{F}AF^T + \tilde{F}A^T F^T = 0$$

since A is skew symmetric. Letting $R = \tilde{F}(0)F(0)^T$ (which is a rotation matrix), we therefore have $\tilde{F}(s) = RF(s)$ for all s. This implies in particular that $\tilde{T}(s) = RT(s)$ for all s, and since $T = \partial_s m$, we obtain $\tilde{\gamma}(s) = R\gamma(s) + b$ for some constant vector b.

The converse statement is obvious, since if $\tilde{\gamma}(s) = R\gamma(s) + b$, then $\tilde{F}(s) = RF(s)$ for all s, and $\partial_s \tilde{F}(s) = R\partial_s F(s)$, which implies (by (4.2)) that curvature and torsion coincide. $\square$

4.2 Regular Surfaces

Curves being represented by one parameter, one may think of surfaces as bi-parametrized objects, like functions $(u, v) = m(u, v)$ defined on some subset of $\mathbb{R}^2$.

Definition 4.2. *A C^p parametrized (regular) surface is a one-to-one, C^p map $m : U \mapsto \mathbb{R}^3$, where U is an open subset of $\mathbb{R}^2$ such that:*

1. *Its inverse, $m^{-1} : V = m(U) \to U$ is continuous (m is a homeomorphism between U and V).*
2. *For all $q \in U$, the differential $Dm(q)$ is one-to-one.*

The last statement is equivalent to the fact that the 3 by 2 matrix of partial derivatives $[\partial_u m, \partial_u m]$ has rank 2. It is the counterpart of regularity for curves. The fact that m is one-to-one prevents the surface from self-intersecting.

Finally, letting $S = m(U)$ be the *range* of m, we will often abuse the terminology by saying that S (the geometric object) is a parametrized surface. However, for many interesting surfaces, it is generally impossible (or at least not convenient) to find a parametrization which satisfies the previous requirement and covers the whole surface. This is a fundamental difference with the theory of plane curves. To be able to handle interesting cases, we need to limit our requirement for parametrizations to hold only on *patches* over the surface, with additional conditions ensuring that the surface is smooth and non-intersecting, and that patches fit well together.

Definition 4.3. *A subset $S \subset \mathbb{R}^3$ is a C^p regular surface, if, for each $p \in S$, there exists an open set V in $\mathbb{R}^3$, with $p \in V$ and a C^p parametrization of the surface patch $V \cap S$. The local parametrizations are also called local charts.*

This definition requires more than just M being covered with parametrized patches. These patches must be obtained from intersections of S with three-dimensional open sets. In particular, this prevents non-local self-intersection, since, along such an intersection, the surface would contain two local patches and would not be locally parametrizable. Figure 4.1 provides an illustration of how local parametrized patches can be combined to cover a surface.

If $m : U \to V \cap S$ is as in the definition, for any p in $V \cap S$, there exist parameters $(u(p), v(p))$ in U such that $m(u(p), v(p)) = p$. The functions $p \mapsto u(p)$ and $p \mapsto v(p)$ are called the local coordinates on $V \cap S$.

4.2.1 Examples

The simplest example of a parametrized surface is the graph of a C^1 function $f : U \subset \mathbb{R}^2 \to \mathbb{R}$. The parametrization is then $m(u, v) = (u, v, f(u, v))$. Since the inverse of (u, v, z) on the surface is (u, v), this is a homeomorphism, and the differential is

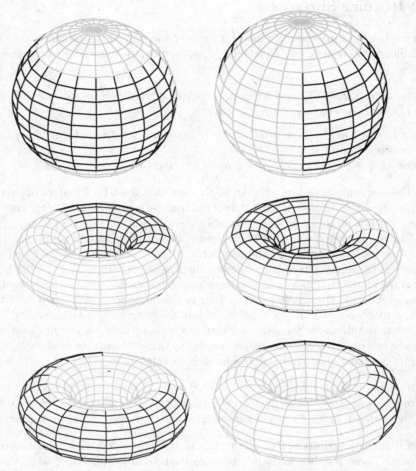

Fig. 4.1. Examples of decompositions of surfaces in local charts. Parametrizations are represented by grids over the surface, black inside local patches and gray outside.

$$(u, v) \mapsto \begin{pmatrix} 1 & 0 \\ 0 & 1 \\ \dot{f}_u & \dot{f}_v \end{pmatrix}$$

which has rank 2.

But a parametrized surface does not need to be a graph. An example is the helicoid, defined over $(0, a) \to \mathbb{R}$ by

$$m(u, v) = (u \cos(v), u \sin(v), v).$$

The simplest example of a surface which, according to our definition, cannot be globally parametrized is the cylinder, which can be defined by the set of $m(u, v) = (\cos u, \sin u, v)$, for $u \in [0, 2\pi)$ and $v \in \mathbb{R}$. This map is one-to-one and in fact a homeomorphism, and the only reason for which this in not a parametrization is that we have required parametrizations to be defined on open sets ($[0, 2\pi) \times \mathbb{R}$ is not open). The cylinder is a regular surface, by considering patches for the same map m, defined on $(0, 2\pi) \times \mathbb{R}$ and say $(-\pi, \pi) \times \mathbb{R}$.

Consider now the example of the unit sphere, which is denoted

$$S^2 = \left\{ p \in \mathbb{R}^3, |p| = 1 \right\}.$$

Like the cylinder, this surface cannot be globally parametrized. The simplest choice of local charts are the projections: $(u, v) \mapsto (u, v, \sqrt{1 - u^2 - v^2})$ and $(u, v) \mapsto (u, v, -\sqrt{1 - u^2 - v^2})$, both defined for $u^2 + v^2 < 1$, the open unit disc. The two maps cover the whole sphere, except the equator for which the third coordinate is 0. One can add other projections, like $(u, v) \mapsto (u, \pm\sqrt{1 - u^2 - v^2}, v)$, $(u, v) \mapsto (\pm\sqrt{1 - u^2 - v^2}, u, v)$ to cover everything, or use cylindrical-like charts close to the equator.

Another useful coordinate system for the sphere is the (properly named) spherical coordinates: $(u, v) \mapsto (\cos u \cos v, \sin u \cos v, \sin v)$. They cover the whole sphere when (u, v) varies in $[0, 2\pi) \times [-\pi/2, \pi/2]$ but this is not a local parametrization, since this set is not open (and the map is not one-to-one for $v = -\pi/2$ and $v = \pi/2$). Restricting to the open intervals requires using other charts to cover the meridian $u = 0$, for example the same coordinates on $(-\pi, \pi) \times (-\pi/2, \pi/2)$ which now only leave the poles uncovered. A neighborhood of the poles can be covered by the previous projection maps.

As a last example, consider the torus (a surface with a shape like a donut). It is the image of the $[0, 2\pi) \times [0, 2\pi)$ by the map

$$m(u, v) = ((R + r \cos v) \cos u, (R + r \cos v) \sin u, r \sin v),$$

where $0 < r < R$, which is one-to-one but once again not defined on an open set. The whole torus can be covered by considering this map restricted to open subsets of $[0, 2\pi) \times [0, 2\pi)$. Let us check that the rank of the differential of m is always 2. We have

$$Dm = \begin{pmatrix} -(R + r \cos v) \sin u & -r \sin v \cos u \\ (R + r \cos v) \cos u & -r \sin v \sin u \\ 0 & r \cos v \end{pmatrix}.$$

The determinant of the first two rows is $-r \sin v (R + r \cos v)$. Since $r < R$, it can only vanish when $\sin v = 0$. For the remaining two determinants, which are $r(R + r \cos v) \sin u \cos v$ and $r(R + r \cos v) \cos u \cos v$, to vanish together, one needs $\cos v = 0$, which is contradictory.

4.2.2 Changing Coordinates

As we have seen, several different valid parametrizations can be defined at a single point of a surface. Like for curves, "geometric" properties should not depend on the parametrization. We will define a few of them in the following: normals, curvature, length, area, etc.

It can be deduced from the requirements in Definition 4.2 that changes of coordinates are C^1 homeomorphisms. To be more specific, assume that in a neighborhood V of a point p on S, there exist two parametrizations $m : U \to V$ and $\tilde{m} : \tilde{U} \to V$. Then, because of the invertibility of the parametrization, one can go from U to V via m, then from V to $\tilde{U}$ via the inverse of $\tilde{m}$. The resulting map, $\varphi = \tilde{m}^{-1} \circ m : U \to \tilde{U}$, is called a change of coordinates, and is a *diffeomorphism* between U and $\tilde{U}$ (it is C^1, invertible, with a C^1 inverse). This consequence of Definition 4.2 can be proved using the inverse mapping theorem.

4.2.3 Implicit Surfaces

An implicit surface is defined by an equation of the form $f(p) = 0$ where $f : \mathbb{R}^3 \to \mathbb{R}$ is a scalar function which is such that $\nabla f(p) \neq 0$ if $f(p) = 0$. In this case, the set

$$S = \left\{ p \in \mathbb{R}^3, f(p) = 0 \right\}$$

is a regular surface. (This is a consequence of the implicit function theorem.)

4.3 Tangent Plane and Differentials

4.3.1 Tangent Plane

For a curve, we were able to define a unique unit tangent, but this is obviously not possible anymore for surfaces. Still, curves provide a simple way to define tangent vectors to surfaces.

A curve $m : I \to \mathbb{R}^3$ is supported by a surface S if and only if, for all $t \in I$, $m(t) \in S$. We have the following definition:

Definition 4.4. *Let S be a regular surface. A vector $T \in \mathbb{R}^3$ is tangent to S at a point $p \in S$ if and only if, for some $\varepsilon > 0$, there exists a curve $\gamma : (-\varepsilon, \varepsilon) \to S$ such that $\gamma(0) = p$ and $\dot{\gamma}(0) = T$.*

Assume, in the previous definition, that ε is chosen small enough so that the curve γ is completely inscribed in a parametrized patch of the surface S. Let $m : (u, v) \mapsto m(u, v)$ be the parametrization. Since m is one-to-one, one can express $\gamma(t) = m(u(t), v(t))$. The plane curve $t \mapsto (u(t), v(t))$ is the expression of γ in the local coordinates. From the chain rule, we have

$$\dot{\gamma}_t = \dot{u}_t \partial_u m + \dot{v}_t \partial_v m.$$

Thus, $\dot{\gamma}_t$ must be a linear combination of the two independent vectors $\partial_u m$ and $\partial_v m$. If $p = m(u_0, v_0)$, then, for any $\alpha, \beta \in \mathbb{R}$, the vector $\alpha \partial_u m + \beta \partial_v m$ is the derivative of $t \mapsto m(u_0 + \alpha t, v_0 + \beta t)$ and is therefore tangent to S at p. This proves the following proposition:

Proposition 4.5. *Let S be a regular surface, $p \in S$ and $m : U \to S$ a parametrization of S in a neighborhood of p. The set of tangent vectors to S at p is the plane generated by $\partial_u m$ and $\partial_v m$.*

The tangent plane to S at p will be denoted $T_p S$. Although the generating vectors $\partial_u m$ and $\partial_v m$ depend on the local parametrization m, the plane itself is a geometric quantity, since we gave a parametrization-independent definition of tangent vectors.

If S is defined implicitly by $f(p) = 0$, the tangent plane at p is characterized by the equation $\langle \nabla f(p), T \rangle = 0$. Indeed, if γ is a curve on S, then $f \circ \gamma(t) = 0$ for all t, and the chain rule implies: $\langle \nabla f(\gamma(0)), \dot{\gamma}(0) \rangle = 0$. This implies that $T_p S \subset (\nabla f(p))^\perp$, but since $T_p S$ is a plane, like $(\nabla f(p))^\perp$, they coincide.

4.3.2 Differentials

Differentials describe how measurements made on a surface vary locally. Start with a scalar function $f : S \to \mathbb{R}$. Take a local parametrization on S, $m : U \to V \cap S$. For $(u, v) \in U$, we can define the function $f_m(u, v) = f(m(u, v))$; this is a function from an open subset of $\mathbb{R}^2$ to $\mathbb{R}$, and provides the expression of f in the local system of coordinates: we have $f(p) = f_m(u(p), v(p))$. We have the following definition:

Definition 4.6. *Let S be a regular surface. A function $f : S \to \mathbb{R}$ is C^1 at $p \in S$ if and only if, for some local parametrization m on S around p, the function f_m is C^1 at $m^{-1}(p)$.*

We say that f is C^1 on S if it is C^1 at all $p \in S$.

(Because changes of coordinates are C^1, the definition does not depend on the choice of local parametrization at p.)

We now want to evaluate the effect that small variations in p have on the function f, i.e., we want to define the derivative of f. Usually, a small variation of $p \in \mathbb{R}^3$ in the direction h is represented by $p + \varepsilon h$, with small ε. This cannot be applied to S, since there is no reason for $p + \varepsilon h$ to belong to S if p does. It is reasonable, and rather intuitive, to define a small variation of p as a piece of curve on S containing p. This leads to:

Definition 4.7. *Let S be a regular surface and $p \in S$. A small variation of p in the direction $h \in \mathbb{R}^3$ is a C^1 curve $\gamma : (-\varepsilon, \varepsilon) \to S$ such that $\gamma(0) = p$ and $\dot{\gamma}_t(0) = h$.*

Note that, from this definition, small variations on S can only arise in directions which are tangent to S.

Now, we can define the differential of a scalar function f defined on S as the limit (if it exists) of the ratio $(f(\gamma(\delta)) - f(p))/\delta$ when δ tends to 0, γ being a small variation of p. This will be denoted $Df(p)h$, with $h = \gamma'(0)$. Implicit in this notation is the fact that this limit only depends on $\gamma'(0)$, which is true if f is C^1 as stated in the next proposition.

Proposition 4.8. *Let f be a C^1 scalar function on a regular surface S. Then, for any $p \in S$, and $h \in T_pS$, the differential of f at p in the direction h exists, and is equal to the limit of the ratio $(f(\gamma(\delta)) - f(p))/\delta$ for any C^1 curve γ on S with $\gamma(0) = p$ and $\dot\gamma_t(0) = h$.*

Proof. What we need to prove is that the limit of the ratio exists for any γ and only depends on h. Take a local parametrization m around p. We know that the function $f(m(u,v))$ is C^1, and letting $\gamma(t) = m(u(t), v(t))$, we have

$$\lim_{\delta \to 0} \frac{f(\gamma(\delta)) - f(p)}{\delta} = \lim_{\delta \to 0} \frac{f_m(u(\delta), v(\delta)) - f_m(u(0), v(0))}{\delta}$$
$$= \partial_u(f_m)\dot u_t(0) + \partial_v(f_m)\dot v_t(0).$$

This proves the existence of the limit. We have $h = \dot\gamma_t(0) = \dot u_t(0)\partial_u m + \dot v_t(0)\partial_v m$: since $(\partial_u m, \partial_v m)$ has rank 2, $\dot u_t(0)$ and $\dot v_t(0)$ are uniquely specified by h and thus the limit above only depends on h. The notation $Df(p)h$ is therefore valid.   □

Note that the expression provided in this proof shows that $Df(p)h$ is linear with respect to h. In other terms, $Df(p)$ is a linear form from T_pS to $\mathbb{R}$. Most of the time, the computation of $Df(p)$ is easy, because f can be expressed as the restriction to S of a differentiable function which is defined on $\mathbb{R}^3$. In this case, $Df(p)h$ coincides with the usual differential of f, but for our purposes, only its restriction to T_pS is relevant.

The proof above also provides a simple way to compute differentials in local charts: let $f : S \to \mathbb{R}$ be C^1, $p \in S$ and m be a local parametrization around p. Then, if $h = \alpha\partial_u m + \beta\partial_v m$, we have

$$Df(p)h = \alpha\partial_u(f_m) + \beta\partial_v(f_m). \tag{4.3}$$

When f is a vector-valued function ($f : S \to \mathbb{R}^d$), the differential $Df(p)$ is defined in the same way, and is also vector-valued. It is a map from T_pS to $\mathbb{R}^d$.

The simplest example of a differentiable map are the coordinates: if $m : U \to V \cap S$ is a local chart, the function $f = m^{-1}$ is such that $f_m(u,v) = (u,v)$ which is the identity map and therefore differentiable. In particular, the coordinates: $p \mapsto u(p)$ and $p \mapsto v(p)$ are scalar differentiable maps. If $T = \alpha\partial_u m + \beta\partial_v m$, we have $Du(p)T = \alpha$, $Dv(p)T = \beta$ and $Df(p)T = (\alpha, \beta)$.

Consider now the example of the sphere S^2. The tangent plane is easy to describe if one uses the fact that S^2 can be defined by the implicit equation $|p|^2 = 1$. If $\varphi(p) = |p|^2$, we have $\langle \nabla\varphi(p), h \rangle = 2\langle p, h \rangle$ so that h is tangent to S^2 at p if and only if $\langle p, h \rangle = 0$ (h is perpendicular to p). Fix a vector $p_0 \in S^2$ and consider the function $f(p) = \langle p, p_0 \rangle$. Then, since f is well-defined on $\mathbb{R}^3$, we can use its restriction which yields $Df(p)h = \langle h, p_0 \rangle$. This was an easy result, but for illustration purposes, let us retrieve it via local charts, which will require a little more computation.

Consider the parametrization $m(u, v) = (\cos u \cos v, \sin u \cos v, \sin v)$. Then,

$$\partial_u m = (-\sin u \cos v, \cos u \cos v, 0) \text{ and}$$

$$\partial_v m = (-\cos u \sin v, -\sin u \sin v, \cos v).$$

A straightforward computation shows that both $\partial_u m$ and $\partial_v m$ are orthogonal to $m(u, v)$. In the chart, letting $p_0 = (a, b, c)$, the function f_m is

$$f_m(u, v) = a \cos u \cos v + b \sin u \cos v + c \sin v.$$

Obviously, $\partial_u(f_m) = \langle p_0, \partial_u m \rangle$ and $\partial_v(f_m) = \langle p_0, \partial_v m \rangle$, so that, if $h = \alpha \partial_u m + \beta \partial_v m$, we get, by equation (4.3),

$$Df(p)h = \alpha \partial_u(f_m) + \beta \partial_v(f_m) = \langle p_0, h \rangle.$$

4.4 Orientation and Normals

Let S be a surface and m a local parametrization on S. The vector $\partial_u m \times \partial_v m$ is non-vanishing and orthogonal to both $\partial_u m$ and $\partial_v m$. Since $\partial_u m$ and $\partial_v m$ generate $T_p S$ at $p = m(u, v)$, $\partial_u m \times \partial_v m$ is normal to the tangent plane at p.

In particular, the vector $N = \partial_u m \times \partial_v m / |\partial_u m \times \partial_v m|$ is a unit normal to the tangent plane. It is called *a normal to the surface S*. Since unit normals to a plane are defined up to a sign change, the one obtained from another parametrization must be either N or $-N$. This leads to the following definition:

Definition 4.9. *Two local parametrizations, m and $\tilde{m}$, on a regular surface S have the same orientation at a given point at which they are both defined if*

$$\frac{\partial_u m \times \partial_v m}{|\partial_u m \times \partial_v m|} = \frac{\partial_y \tilde{m} \times \partial_v \tilde{m}}{|\partial_u \tilde{m} \times \partial_v \tilde{m}|}.$$

and have opposite orientation otherwise.

The surface S is said to be orientable if it can be covered by local parametrizations, which have the same orientation wherever they intersect.

A surface is therefore orientable if there is a consistent (continuous) definition of its normal all over it. All surfaces are not necessarily orientable (Figure 4.2). The typical example is a twisted ring (the Möbius band).

4.5 Integration on an Orientable Surface

Let S be an orientable surface and $f : S \to \mathbb{R}$ be a continuous function. We want to compute the integral of f over S. It is not possible to provide a global computational formula, but it can be done in a chart. So let $m : U \to V \cap S$

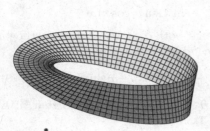

Fig. 4.2. Two examples of non-orientable surfaces. The left one is the Möbius band; the second one is similar, with an odd number of twists.

be a parametrized patch of the surface S. To motivate the definition, let U be divided into small rectangular cells (neglecting boundary issues). Consider a cell of the form $(u_0 - \varepsilon/2, u_0 + \varepsilon/2) \times (v_0 - \varepsilon/2, v_0 + \varepsilon/2)$. In this cell, we can make a first-order expansion of m in the form

$$m(u, v) = m(u_0, v_0) + (u - u_0)\partial_u m(u_0, v_0) + (v - v_0)\partial_v m(u_0, v_0) + o(\varepsilon)$$

so that, at first order, the image of the rectangular cell by m is a parallelogram in space, centered at $p_0 = m(u_0, v_0)$, namely

$$\sigma_0 = \{p_0 + \alpha \partial_u m + \beta \partial_v m, \alpha \in (-\varepsilon/2, \varepsilon/2), \beta \in (-\varepsilon/2, \varepsilon/2)\}.$$

Its area is given by $\varepsilon^2 |\partial_u m \times \partial_v m|$, and the integral of a function f over this parallelogram can legitimately be estimated by $\varepsilon^2 f(p_0)|\partial_u m \times \partial_v m|$. Summing over cells and letting ε tend to 0 leads to the following definition:

Definition 4.10. *Let f be a function defined on a regular surface S, and $m : U \to V \cap S$ a regular patch on S. The integral of f on $V \cap S$ is defined and denoted by*

$$\int_{V \cap S} f(p)d\sigma_S(p) = \int_U f_m(u, v) \, |\partial_u m \times \partial_v m| \, dudv. \tag{4.4}$$

   The integral of f over the whole surface S is defined as the sum of such integrals over non-overlapping local patches that cover S (maybe leaving out a finite number of curves or points on S). It is denoted

$$\int_S f(p)d\sigma_S(p).$$

This can be shown to be independent of the chosen family of patches.

Another (equivalent) way to globally define the integral is to use partitions of unity. Given a family $((U_i, m_i), i = 1, \ldots, n)$ of local parametrizations which cover the surface (so that $\bigcup_i m_i(U_i) = S$), but may overlap, one defines a partition of unity as a family of continuous functions $(\omega_i, i = 1, \ldots, n)$ where each ω_i is defined on S and takes values in $[0, 1]$, with $\omega_i(p) = 0$ if $p \notin m_i(U_i)$, and for all $p \in S$,

$$\sum_{i=1}^{n} \omega_i(p) = 1.$$

Such partitions of unity always exist, and one can define

$$\int_S f(p) d\sigma_S(p) = \sum_{i=1}^{N} \int_{U_i} \omega_i(m_i(u, v)) f_{m_i}(u, v) |\partial_u m_i \times \partial_v m_i| du dv$$

and here also, the result does not depend on the local parametrizations, or on the partition of unity.

The notation $d\sigma_S$ refers to the so-called area form on S, defined on a local chart by $d\sigma_S = |\partial_u m \times \partial_v m| du dv$.

The right-hand side of (4.4) does not depend on the chosen parametrization. This should be clear from the approximation process which led to the definition (which was purely geometric), and can be checked directly as follows. Let $\tilde{m} : \tilde{U} \to V \cap S$ be another parametrization of the same patch. For $p \in V \cap S$, the equation $p = m(u, v) = \tilde{m}(\tilde{u}, \tilde{v})$ provides a relation between homologous coordinates given by

$$\partial_u m = \frac{\partial \tilde{u}}{\partial u} \partial_{\tilde{u}} \tilde{m} + \frac{\partial \tilde{v}}{\partial u} \partial_{\tilde{v}} \tilde{m}$$

$$\text{and } \partial_v m = \frac{\partial \tilde{u}}{\partial v} \partial_{\tilde{u}} \tilde{m} + \frac{\partial \tilde{v}}{\partial v} \partial_{\tilde{v}} \tilde{m}.$$

The left-hand sides are computed at (u, v) and the right-hand sides at $(\tilde{u}, \tilde{v})$. This implies

$$\partial_u m \times \partial_v m = \left(\frac{\partial \tilde{u}}{\partial u} \frac{\partial \tilde{v}}{\partial v} - \frac{\partial \tilde{v}}{\partial u} \frac{\partial \tilde{u}}{\partial v} \right) \partial_{\tilde{u}} \tilde{m} \times \partial_{\tilde{v}} \tilde{m}.$$

Letting φ be the change of variables $(\varphi(u, v) = (\tilde{u}, \tilde{v}))$, this is $\partial_u m \times \partial_v m = (\det \varphi)(\partial_{\tilde{u}} \tilde{m} \times \partial_{\tilde{v}} \tilde{m}) \circ \varphi$. Therefore

$$\int_U f(m(u, v)) |\partial_u m \times \partial_v m| du dv$$

$$= \int_U f(\tilde{m} \circ \varphi(u, v)) |\partial_u m \times \partial_v m| \circ \varphi(u, v) |\det \varphi(u, v)| du dv$$

$$= \int_{\tilde{U}} f(\tilde{m}(\tilde{u}, \tilde{v})) |\partial_{\tilde{u}} \tilde{m} \times \partial_{\tilde{v}} \tilde{m}| d\tilde{u} d\tilde{v}.$$

The formula for integrals can be used to compute the area of the unit sphere, which can be parametrized (poles excepted) by

$$m(u, v) = (\cos u \cos v, \sin u \cos v, \sin v).$$

Then

$$\partial_u m = (-\sin u \cos v, \cos u \cos v, 0),$$
$$\partial_v m = (-\cos u \sin v, -\sin u \sin v, \cos v)$$

and $|\partial_u m \times \partial_v m|^2 = \cos^2 v$, so that

$$\int_{S^2} d\sigma = \int_0^{2\pi} \int_{-\pi/2}^{\pi/2} \cos v \, du dv = 2\pi [\sin v]_{-\pi/2}^{\pi/2} = 4\pi.$$

4.6 The First Fundamental Form

4.6.1 Definition and Properties

Let S be a regular surface. When h and k are two tangent vectors at $p \in S$, their dot product in $\mathbb{R}^3$ will be denoted $\langle h, k \rangle_p$. It is simply the usual dot product, the sum of products of coordinates, but gets a specific notation because it is restricted to $T_p S$. The associated quadratic form is called *the first fundamental form*, and denoted

$$I_p(h) := |h|_p^2. \qquad (4.5)$$

This form is the key instrument for metric measurements on surfaces. Although its definition is straightforward, one must remember that surfaces are mostly described by local charts, and the expression of the form in such charts is not the standard norm anymore. Indeed, let m be a local parametrization around p, and $h = \alpha \partial_u m + \beta \partial_v m \in T_p S$. Then

$$I_p(h) = \alpha^2 \langle \partial_u m, \partial_u m \rangle_p + 2\alpha\beta \langle \partial_u m, \partial_v m \rangle_p + \beta^2 \langle \partial_v m, \partial_v m \rangle_p$$
$$= \alpha^2 E + 2\alpha\beta F + \beta^2 G$$

with the notation

$$E = \langle \partial_u m, \partial_v m \rangle_p, \; F = \langle \partial_u m, \partial_v m \rangle_p, \; G = \langle \partial_v m, \partial_v m \rangle_p. \qquad (4.6)$$

E, F and G are the coefficients of the first fundamental form in the chart. They depend on the parameters u, v.

These coefficients provide all the required information to compute lengths of curves on S: let γ is such a curve; assuming that γ is contained in a parametrized patch and letting $\gamma(t) = m(u(t), v(t))$, we have

$$|\dot{\gamma}_t|^2 = |\dot{u}_t \partial_u m + \dot{v}_t \partial_v m|^2 = \dot{u}_t^2 E + 2\dot{u}_t \dot{v}_t F + \dot{v}_t^2 G$$

so that the length of the curve from its expression in local coordinates is provided by

$$\text{length}(\gamma) = \int_a^b \sqrt{\dot{u}_t^2 E(u,v) + 2\dot{u}_t \dot{v}_t F(u,v) + \dot{v}_t^2 G(u,v)} dt.$$

Similarly, one defines the energy of a curve γ by

$$\text{energy}(\gamma) = \frac{1}{2} \int_a^b |\dot{\gamma}_t|^2 dt = \frac{1}{2} \int_a^b (\dot{u}_t^2 E(u,v) + 2\dot{u}_t \dot{v}_t F(u,v) + \dot{v}_t^2 G(u,v)) dt.$$

Curves of minimal energy on a surface are called *geodesics*, as formalized by the following definition.

Definition 4.11. *Given two points p and p' on a surface M, a curve γ on M achieving the minimum energy among all curves on M linking p and p' is called a (minimizing) geodesic.*

In addition to minimizing the energy, it can be shown that geodesics are curves of minimal length between two points [64, 65]. If γ is a minimizing geodesic between p and p', and $h(t)$ is for all t a vector tangent to the surface at $\gamma(t)$, one can define, for small ε, a one-parameter family of curves $\gamma(t, \varepsilon)$ such that $\gamma(t, 0) = \gamma(t)$ and $\dot{\gamma}_\varepsilon(t, 0) = h(t)$. Since γ is minimizing, the function $\varepsilon \mapsto \text{energy}(\gamma(., \varepsilon))$ has a vanishing derivative at $\varepsilon = 0$. This derivative is given by

$$\int_a^b \dot{\gamma}_t^T \dot{h}_t dt = -\int_a^b \ddot{\gamma}_{tt}^T h dt$$

by integration by parts. The fact that this expression vanishes for any h tangent to the surface along γ implies that the "acceleration" $\ddot{\gamma}_{tt}$ is normal to the surface. By extension, curves satisfying this property are also called geodesics. They generalize the notion of straight lines in a plane.

Definition 4.12. *A C^2 regular curve γ on M is called a geodesic if its second derivative $\ddot{\gamma}_{tt}$ is always normal to M.*

Let's use this definition to compute the geodesics on the unit sphere. Such geodesics must satisfy $|\gamma(t)| = 1$ for all t and, in order to be normal

$$\ddot{\gamma}_{tt}(t) = \lambda(t)\gamma(t)$$

for some real-valued function λ. The derivative of $|\dot{\gamma}_t|^2$ being $2\langle \dot{\gamma}_t, \ddot{\gamma}_{tt} \rangle$, we obtain the fact that $|\dot{\gamma}_t| = cst$ (this is true for geodesics on any surface, according to Definition 4.12). For the sphere, we can also write, since $\langle \gamma, \dot{\gamma}_t \rangle = 0$,

$$0 = \partial_t \langle \gamma, \dot{\gamma}_t \rangle = |\dot{\gamma}_t|^2 + \lambda(t)|\gamma|^2$$

which implies that λ is constant, since $|\gamma| = 1$. So geodesics must satisfy the equation $\ddot{\gamma}_{tt} = \lambda\gamma$. By making a constant time change, we can assume that $|\dot{\gamma}_t| = -\lambda(t) = 1$, and that γ is parametrized with arc length. Since $\partial_t \ddot{\gamma}_{tt} = \dot{\gamma}_t$, we see that the curve has unit curvature and zero torsion and therefore coincides with a portion of unit circle. The only unit circles included in the sphere must be centered at 0, and constitute the great circles on the sphere. So we find that geodesic on the sphere are great circles parametrized at constant speed.

We conclude this section on the first fundamental form by stating without proof a proposition that allows us to use convenient local charts around a given point.

Proposition 4.13. *If S is a regular surface and $p \in S$, there exists a local parametrization $m : U \mapsto S$ around p such that $(\partial_u m, \partial_v m)$ is orthogonal on U.*

4.6.2 The Divergence Theorem on Surfaces

We finally give the Green formula for surfaces. A vector field on S is a function $h : S \to \mathbb{R}^3$ such that, for all p, $h(p) \in T_p S$. We start with a simple definition of the divergence of a C^1 vector field.

Definition 4.14. *Let h be a C^1 vector field on a regular surface S. The divergence of h on S is defined by*

$$\operatorname{div}_S h(p) = \langle e_1, Dh(p)e_1 \rangle + \langle e_2, Dh(p)e_2 \rangle \tag{4.7}$$

whenever e_1, e_2 is a positively oriented orthonormal basis of $T_p M$ (the result being independent of the choice made for e_1, e_2).

In this definition, $Dh(p)$ is a linear transformation between $T_p S$ and $\mathbb{R}^3$, the dot products being the $\mathbb{R}^3$ dot product (since $T_p S \subset \mathbb{R}^3$). If h is defined on S and takes values in $\mathbb{R}^3$ (not necessarily in TS), the definition remains meaningful. We will use the notation $\operatorname{div}'_S(h)$ for the left-hand side of (4.7) in that case. In fact, if h decomposes as $h = h_T + \mu N$ where h_T is a vector field on S, we have

$$\operatorname{div}'_S(h) = \operatorname{div}_S(h_T) + \mu \operatorname{div}'_S(N). \tag{4.8}$$

Another way of understanding the definition is by introducing the orthogonal projection on $T_p S$ (denoted $\pi_{T_p S}$) and the operator

$$\nabla h(p) = \pi_{T_p S} Dh(p) : T_p S \to T_p S. \tag{4.9}$$

This operator is the covariant derivative on S, as described in Appendix B, and Definition 4.14 simply says that

$$\operatorname{div}_S h(p) = \operatorname{trace}(\nabla h(p)) \tag{4.10}$$

Note that we have, for $\xi \in T_pS$

$$\nabla h(p)\xi = Dh(p)\xi - \langle Dh(p)\xi,\, N \rangle N.$$

This definition can be made explicit in a chart. This yields the following proposition (the proof, which is just computation, is left to the reader):

Proposition 4.15. *If m is a local chart on S and the C^1 vector field h decomposes as $h = \alpha\partial_u m + \beta\partial_v m$ in this chart, we have*

$$\mathrm{div}_S h = \dot{\alpha}_u + \dot{\beta}_v + (\alpha\dot{\rho}_u + \beta\dot{\rho}_v)/\rho \qquad (4.11)$$

where $\rho = |\partial_u m \times \partial_v m| = \sqrt{EG - F^2}$.

We also have the nice formula, still valid in a chart, that says that

$$\partial_u h \times \partial_v m + \partial_u m \times \partial_v h = \rho(\mathrm{div}_S h)N. \qquad (4.12)$$

This result is a direct consequence of the following simple computation in linear algebra, the proof of which is left to the reader.

Lemma 4.16. *Let A be a linear operator from M, an oriented two-dimensional linear subspace of $\mathbb{R}^3$, to $\mathbb{R}^3$. Let n be the normal to M. Define, for $e_1, e_2 \in M$*

$$\varphi_A(e_1, e_2) = \langle Ae_1,\, e_2 \times n \rangle + \langle Ae_2,\, n \times e_1 \rangle.$$

Then, there exists a real number $\rho(A)$ such that

$$\varphi_A(e_1, e_2) = \rho(A)\det(e_1, e_2, n),$$

which is also equal to $\rho(A)|e_1 \times e_2|$ if e_1, e_2 are positively oriented. Moreover, we have

$$\rho(A) = \mathrm{trace}((\mathrm{Id} - nn^T)A) \qquad (4.13)$$

where $(\mathrm{Id} - nn^T)A$ (which is A followed by the projection on M) is considered as an operator from M to itself.

Equation (4.12) just comes by applying Lemma 4.16 with $M = T_pM$, $A = Dh(p)$, $e_1 = \partial_u m$ and $e_2 = \partial_v m$.

We now give the divergence theorem on a surface, which is a direct generalization of the one we saw on $\mathbb{R}^2$ (Theorem 1.9):

Theorem 4.17. *Let S be an oriented regular surface, and h a smooth vector field on S. Then, if Σ is a subdomain of S with smooth boundary, we have*

$$\int_{\partial\Sigma} h^T n_\Sigma dl = -\int_\Sigma \mathrm{div}_S(h)d\sigma_S,$$

where the first integral is a line integral over the curve $\partial\Sigma$, and n_Σ is the inward normal to Σ (normal to $\partial\Sigma$ and tangent to S).

The proof (which we skip) is an application of the Green formula in $\mathbb{R}^2$ combined with a decomposition in local coordinates.

In addition to the divergence, one can define the gradient operator on a surface S, which applies to scalar-valued functions.

Definition 4.18. *Let $f : S \to \mathbb{R}$ be C^1. The gradient of f at $p \in S$ is denoted $\nabla_S f(p)$ and defined by $\nabla_S f(p) \in T_p S$ and*

$$\forall \xi \in T_p S, \quad \langle \nabla_S f(p), \xi \rangle_p = Df(p)\xi.$$

In a chart $(u, v) \mapsto m(u, v)$, we have

$$\nabla_S f = \frac{G f_u - F f_v}{EG - F^2} \partial_u m + \frac{E f_v - F f_u}{EG - F^2} \partial_v m. \tag{4.14}$$

The usual formula, $\operatorname{div}(fh) = \langle \nabla f, h \rangle + f \operatorname{div} h$ extends to surfaces with

$$\operatorname{div}_S(fh) = \langle \nabla_S f, h \rangle + f \operatorname{div}_S h \tag{4.15}$$

for a scalar function f and a vector field h on S.

The generalization of the Laplacian on $\mathbb{R}^2$ is the Laplace–Beltrami operator on S. It is defined as follows:

Definition 4.19. *The Laplace–Beltrami operator on a regular surface S associates to a scalar function f on S the scalar function $\Delta_S f$ defined by*

$$\Delta_S f = \operatorname{div}_S \nabla_S f. \tag{4.16}$$

The Laplace–Beltrami operator in a chart is therefore given by the combination of (4.14) and (4.11), which yields a formula notably more complex than the ordinary Laplacian.

Theorem 4.17 relates surface integrals to linear integrals over the surface. Surface integrals can also be related to three-dimensional integrals, if the surface is closed, via the three-dimensional divergence theorem.

Theorem 4.20. *Let Ω be a bounded domain in $\mathbb{R}^3$ and assume that $S = \partial\Omega$ is a regular surface. If v is a C^1 vector field on $\mathbb{R}^3$, we have*

$$\int_\Omega \operatorname{div} v = -\int_S v^T(m) N(m) d\sigma_S(m).$$

where $N(m)$ is the inward normal to S at m.

Like with curves, this theorem permits us to compute the volume of Ω as an integral over its boundary, namely (taking $v(x, y, z) = (x, y, z)$)

$$\operatorname{volume}(\Omega) = -\frac{1}{3} \int_S Om^T N(m) d\sigma_S(m). \tag{4.17}$$

4.7 Curvature and Second Fundamental Form

Let S be a C^2 orientable regular surface, and N be its normal. The function N can be seen as a map defined on S with values in $\mathbb{R}^3$ (in fact in the unit sphere S^2 since $|N| = 1$) which is called the Gauss map. It therefore has a differential, DN. For any $p \in S$, $DN(p)$ is a linear map from T_pS to $\mathbb{R}^3$. The fact that $|N|^2 = 1$ implies that $\langle DN(p)h, N(p) \rangle = 0$ for all $h \in T_pS$ so that the range of $DN(p)$ is orthogonal to $N(p)$ and therefore coincides to T_pS. We can therefore consider $DN(p)$ as an endomorphism

$$DN(p) : T_pS \to T_pS.$$

This endomorphism (also called the *shape operator*) is essential for describing the curvature of the surface, which measures how the surface bends in a neighborhood of a point p. It has the interesting property of being symmetric:

Proposition 4.21. *Let S be a regular surface and $p \in S$: for any $h, k \in T_pS$, we have*

$$\langle DN(p)h, k \rangle_p = \langle h, DN(p)k \rangle_p.$$

Proof. It suffices to show this for a basis of T_pS and let's take the one provided by a local parametrization around p: $h = \partial_u m$ and $k = \partial_v m$. Let $N_m = N \circ m$ be the expression of N as a function of the parameters, so that

$$DN(p)(\alpha \partial_u m + \beta \partial_v m) = \alpha \partial_u N_m + \beta \partial_v N_m.$$

In particular, $DN(p)\partial_u m = \partial_u N_m$ and $DN(p)\partial_v m = \partial_v N_m$, and what we need to show is

$$\langle \partial_u N_m, \partial_v m \rangle = \langle \partial_u m, \partial_v N_m \rangle.$$

But, from $\langle \partial_u m, N_m \rangle = 0$, we get $\langle \partial_u m, \partial_v N_m \rangle = -\langle \partial_v \partial_u m, N_m \rangle$. Similarly, $\langle \partial_v m, \partial_u N_m \rangle = -\langle \partial_u \partial_v m, N_m \rangle$. Since partial derivatives commute, the two quantities are equal, yielding the required identity. $\qquad\square$

Let γ be a curve on S, and assume that γ is parametrized by arc length. Let $T^{(\gamma)}$ be the unit tangent of γ, $\kappa^{(\gamma)}$ its curvature and $N^{(\gamma)}$ its unit normal, such that $\dot{T}_s^{(\gamma)} = \kappa^{(\gamma)} N^{(\gamma)}$. The normal $N^{(\gamma)}$ does not coincide with N in general, and we define the normal curvature of γ by the (algebraic) normal part of $\dot{T}_s^{(\gamma)}$ to the surface S. The interesting point is that it only depends on γ via $T^{(\gamma)}$.

Definition 4.22. *The normal curvature at p of an arc length parametrized curve γ on a regular surface S is $\kappa_N(s) = \langle \dot{T}_s^{(\gamma)}(s), N(s) \rangle$ where $T^{(\gamma)} = \dot{\gamma}_s$.*

The fact that the normal curvature only depends on $T^{(\gamma)}$ can be proved as follows: let γ be a curve on S such that $\dot{\gamma}_s(0) = T^{(\gamma)}$. For all s, we have $\langle T^{(\gamma)}, N \rangle = 0$ since $T^{(\gamma)}$ is tangent to S. Computing the derivative with respect to arc length and applying the chain rule yields

$$\langle \dot{T}_s^{(\gamma)}, N \rangle + \langle T^{(\gamma)}, DN(\gamma)T^{(\gamma)} \rangle = 0$$

so that

$$\kappa_N = -\langle T^{(\gamma)}, DN(\gamma)T^{(\gamma)} \rangle. \tag{4.18}$$

One also defines the *geodesic curvature* of γ at s_0 by the curvature (at s_0) of the projection of γ on the tangent plane to S at $\gamma(s_0)$, which is

$$\bar{\gamma}(s) = \gamma(s) - \langle \gamma(s) - \gamma(s_0), N(s_0) \rangle N(s_0).$$

Computing first and second derivatives in s and computing them at $s = s_0$ yields $\dot{\bar{\gamma}}_s(s_0) = \dot{\gamma}_s(s_0)$ and

$$\ddot{\bar{\gamma}}_{ss}(s_0) = \ddot{\gamma}_{ss}(s_0) - \kappa_N(s_0)N(s_0).$$

Denoting the geodesic curvature by $\kappa_g(s_0)$, we find (using the definition of the curvature for plane curves in the oriented tangent plane) that

$$\kappa_g = \det[\ddot{\gamma}_{ss}, \dot{\gamma}_s, N]$$

and that $\kappa_g^2 + \kappa_N^2 = \kappa^2$, the squared curvature of γ.

This expression in equation (4.18) involves another important quantity on S, its second fundamental form.

Definition 4.23. *Let S be a regular surface and $p \in S$. The second fundamental form at p is the quadratic form defined on T_pS by*

$$II_p(h) = -\langle h, DN(p)h \rangle.$$

In particular, we have the expression of the normal curvature of an arc length parametrized curve γ:

$$\kappa_N = II_\gamma(\dot{\gamma}_s).$$

Because $DN(p)$ is symmetric, it can be diagonalized in an orthonormal basis of T_pS: let (e_1, e_2) be such a basis, with corresponding eigenvalues $-\kappa_1$ and $-\kappa_2$ such that $\kappa_1 \geq \kappa_2$. The numbers κ_1 and κ_2 are called the *principal curvatures* of the surface at p. The reason for this terminology is that any unit vector in T_pS can be written, for some θ, in the form $h = \cos\theta e_1 + \sin\theta e_2$ and

$$II_p(h) = -\langle h, DN(p)h \rangle = \kappa_1 \cos^2\theta + \kappa_2 \sin^2\theta.$$

This implies that $\kappa_2 \leq II_p(h) \leq \kappa_1$, the lower bound being attained for $h = e_2$ and the upper bound for $h = e_1$: κ_1 and κ_2, respectively, are the maximum and minimum normal curvatures of curves passing through p.

Definition 4.24. *If κ_1 and κ_2 are the principal curvatures of a surface S at $p \in S$, one defines the mean curvature at p by $H(p) = (\kappa_1 + \kappa_2)/2$, and the Gauss curvature by $K(p) = \kappa_1\kappa_2$. They respectively coincide with the trace of $-DN(p)/2$ and the determinant of $DN(p)$.*

From this definition, we can also write

$$2H = -\text{div}'_S(N) \tag{4.19}$$

and rewrite (4.8) as

$$\text{div}'_S(h) = \text{div}_S(h_T) - 2H\mu. \tag{4.20}$$

4.8 Curvature in Local Coordinates

In this section, we give the expression of the curvature in local coordinates, as functions of the coefficients of the first and second fundamental forms. Recall the notation (4.6) for the first fundamental form and a local parametrization m. We introduce similar notation for the second form, letting

$$II_p(\alpha\partial_u m + \beta\partial_v m) = \alpha^2 e + 2\alpha\beta f + \beta^2 g$$

and

$$e = -\langle \partial_u m,\, \partial_u N \rangle = \langle \partial_u\partial_u m,\, N \rangle, f = -\langle \partial_u m,\, \partial_v N \rangle = \langle \partial_u\partial_v m,\, N \rangle,$$
$$g = -\langle \partial_v m,\, \partial_v N \rangle = \langle \partial_v\partial_v m,\, N \rangle. \tag{4.21}$$

Let $DN = \begin{pmatrix} a & c \\ b & d \end{pmatrix}$ in the basis $(\partial_u m, \partial_v m)$ (the matrix is not necessarily symmetric since the basis is not assumed to be orthonormal). We find:

$$-e = \langle \partial_u m,\, DN\partial_u m \rangle = aE + bF$$
$$-f = \langle \partial_v m,\, DN\partial_u m \rangle = aF + bG$$
$$-f = \langle \partial_u m,\, DN\partial_v m \rangle = cE + dF$$
$$-g = \langle \partial_v m,\, DN\partial_v m \rangle = cF + dG$$

which yields, in matrix form: $-\begin{pmatrix} e & f \\ f & g \end{pmatrix} = \begin{pmatrix} a & c \\ b & d \end{pmatrix}\begin{pmatrix} E & F \\ F & G \end{pmatrix}$. This implies that, in the basis $(\partial_u m, \partial_v m)$, DN is given by the matrix

$$-\begin{pmatrix} e & f \\ f & e \end{pmatrix}\begin{pmatrix} E & F \\ F & E \end{pmatrix}^{-1}.$$

From this, it can be deduced that

$$K = \frac{eg - f^2}{EG - F^2}$$

because it is just the ratio of the determinants. Also, after computation, one finds

$$H = \frac{eG - 2fF + gE}{2(EG - F^2)}.$$

The principal curvatures are then given by $\kappa = H \pm \sqrt{H^2 - K}$.

4.9 Implicit Surfaces

Like curves, surfaces can be defined by an implicit equation

$$S = \{p \in \mathbb{R}^3, f(p) = 0\}$$

where f is a C^2 function from $\mathbb{R}^3$ to $\mathbb{R}$ with $\nabla f \neq 0$ on S. We have already noticed that the tangent plane to S is orthogonal to ∇f, and therefore $N(p) = -\nabla f(p)/|\nabla f(p)|$ is a smooth unit normal to S which therefore is orientable (and we take the orientation provided by this choice of N).

The interesting feature in this representation is that, since f is defined on $\mathbb{R}^3$, the function N can be extended to $\mathbb{R}^3$ (denote the extension by $\hat{N}$) so that $DN(p)$ is simply the restriction to $T_p S$ of $D\hat{N}(p)$, In particular, the trace of $DN(p)$ is, by definition, $\langle e_1, DN(p)e_1 \rangle + \langle e_2, DN(p)e_2 \rangle$ for an arbitrary orthonormal basis of $T_p S$. It therefore suffices to add $\langle D\hat{N}N, N \rangle$ to obtain the trace of $D\hat{N}$, but this added quantity vanishes, since $\left|\hat{N}\right| = 1$. Thus, we have, for the mean curvature:

$$H = -\operatorname{trace}(D\hat{N})/2 = \frac{1}{2}\operatorname{div}\frac{\nabla f}{|\nabla f|}. \tag{4.22}$$

(This is the usual divergence on $\mathbb{R}^3$, not to be confused with the S-divergence in Definition 4.14.)

Let P_N be the projection on $\hat{N}^\perp$: $P_N = \operatorname{Id}_{\mathbb{R}^3} - \hat{N}\hat{N}^T$. The Gauss curvature can be computed after diagonalizing the matrix $P_N D\hat{N} P_N = D\hat{N} P_N$ which is symmetric and coincides with DN on $T_p S$. Using $\hat{N} = -\nabla f/|\nabla f|$, we get

$$\langle D\hat{N}P_N h, P_N k \rangle = -\frac{1}{|\nabla f|}\langle D^2 f P_N h, P_N k \rangle + \frac{1}{|\nabla f|}\langle D^2 f P_N h, \hat{N}\rangle\langle P_N k, N \rangle$$

$$= -\frac{1}{|\nabla f|}\langle D^2 f P_N h, P_N k \rangle$$

which is symmetric in h and k. The matrix $P_N D^2 f P_N/|\nabla f|$ has one vanishing eigenvalue since $P_N N = 0$, and the other two are the principal curvatures of S. Their product provides the Gauss curvature.

4.9.1 The Delta-Function Trick

When a surface or a curve is defined implicitly, integrals over its interior can be described in a straightforward way using the Heaviside function. Assume that S is the set $f(p) = 0$ for some smooth function f, and let Ω be its interior, defined by $f < 0$. Introduce the Heaviside function H_0 defined on $\mathbb{R}$ by $H_0(x) = 1$ if $x \geq 0$ and $H_0(x) = 0$ otherwise. Then, clearly, for any function V on $\mathbb{R}^3$, we have

$$\int_\Omega V(x)dx = \int_{\mathbb{R}^3} (1 - H_0(f(x)))V(x)dx. \tag{4.23}$$

More interesting is that contour or surface integrals can be defined via a level-set representation, albeit requiring passing to a limit. For this, we need to replace H_0 by a smooth approximation denoted H_ε, which must be an increasing function that tends to H_0 when ε tends to 0. Possible examples are (cf. [229, 43, 163]) $H_\varepsilon(x) = 0$ for $x < -\varepsilon$, $H_\varepsilon(x) = 1$ for $x > \varepsilon$ and, on $[-\varepsilon, \varepsilon]$:

$$H_\varepsilon(x) = \frac{1}{2}\left(1 + \frac{x}{\varepsilon} + \frac{1}{\pi}\sin\left(\frac{\pi x}{\varepsilon}\right)\right), \tag{4.24}$$

or, for all $x \in \mathbb{R}$:

$$H_\varepsilon(x) = \frac{1}{2}\left(1 + \frac{2}{\pi}\arctan\left(\frac{x}{\varepsilon}\right)\right). \tag{4.25}$$

This choice being made, let δ_ε denote the derivative of H_ε. The function δ_ε can be considered as a smooth approximation of the Dirac function δ_0, in the sense that, for any bounded function u on $\mathbb{R}$ which is continuous at $t = 0$, we have

$$\lim_{\varepsilon \to 0} \int_\mathbb{R} \delta_\varepsilon(t)u(t)dt = u(0). \tag{4.26}$$

We leave the easy proof to the reader (simply divide the integral over domains around 0 or away from 0).

We now describe how surface integrals for implicitly defined surfaces can be approximated using δ_ε.

Proposition 4.25. *Let $d = 2$ or 3 and $f : \mathbb{R}^d \to \mathbb{R}$ be a C^2 function with $\nabla f \neq 0$ if $f = 0$, and such that the implicit curve/surface $S = f^{-1}(\varepsilon)$ is bounded in a neighborhood of 0. Then, if $V : \mathbb{R}^d \to \mathbb{R}$ is continuous, we have*

$$\lim_{\varepsilon \to 0} \int_{\mathbb{R}^d} \delta_\varepsilon \circ f(x)V(x)|\nabla f(x)|dx = \int_S V(m)dm. \tag{4.27}$$

Proof. Let's consider the surface case ($d = 3$). The case of curves is similar and simpler. We also assume that δ_ε is supported in $[-\varepsilon, \varepsilon]$, like with (4.24) (the general case requiring only minor modifications). Consider a local chart $(u, v) \mapsto m(u, v)$ on $S = f^{-1}(0)$. Consider the equation

$$f(m(u, v) + tN(u, v)) = y$$

which we want to solve for t as a function of (u, v, y) in a neighborhood of some $u = u_0$, $v = v_0$ and $y = 0$. From the implicit function theorem, this is possible, because

$$\partial_t f(m + tN) = \langle \nabla f, N \rangle = -|\nabla f|,$$

which is not zero by assumption. Using the compactness of S, we can find a finite number of points $p_0 = m(u_0, v_0)$ and domains around $(u_0, v_0, 0) \in \mathbb{R}^3$ over which a function $t(m(u, v), y)$ such that $f(m+tN) = y$ is well-defined and such that the union of these domains forms an open set in $\mathbb{R}^3$ that contains S, and more generally contains the set $|f(p)| < y_0$ for y_0 small enough.

Taking $\varepsilon < y_0$, we can write

$$\int_{\mathbb{R}^d} \delta_\varepsilon \circ f(x) V(x) |\nabla f(x)| dx = \int_{|f|<y_0} \delta_\varepsilon \circ f(x) V(x) |\nabla f(x)| dx.$$

(Not assuming δ_ε to be compactly supported would add a small error to this identity, which is easily shown to be negligible when $\varepsilon \to 0$.)

We can decompose the integral over a partition of unity, which reduces the problem to the situation in which V is supported by one of the domains above. Working under this assumption, we make the change of variables $x(u, v, y) = m(u, v) + t(m(u, v), y) N(u, v)$ in this domain and let $J(u, v, t)$ be the associated Jacobian determinant, so that

$$\int_{|f|<y_0} \delta_\varepsilon \circ f(x) V(x) |\nabla f(x)| dx =$$

$$\int_{|y|<y_0} \delta_\varepsilon(y) V(x(u, v, y)) |\nabla f(x(u, v, y))| J(u, v, y) du dv dy.$$

By our assumptions, the integral

$$u(y) = \int V(x(u, v, y)) |\nabla f(x(u, v, y))| J(u, v, y) du dv$$

is continuous in y so that,

$$\lim_{\varepsilon \to 0} \int_{\mathbb{R}^d} \delta_\varepsilon \circ f(x) V(x) |\nabla f(x)| dx = u(0).$$

Now,

$$J(u, v, 0) = |\det(\partial_u m, \partial_v m, \partial_y t N)| = |\partial_u m \times \partial_v m| / |\nabla f(m)|$$

since $y = f(m + tN)$ implies $1 = t_y \langle \nabla f, N \rangle = -t_y |\nabla f|$. This implies that $|\nabla f|$ cancels in $u(0)$, which is equal to

$$u(0) = \int V(m(u, v)) |\partial_u m \times \partial_v m| du dv = \int_S V d\sigma,$$

which concludes the proof. □

The theorem is important in particular for numerical computations, because it replaces computations over a surface with computations over a grid that contains the surface.

The left-hand term of (4.27) is often written using the symbolic notation

$$\int_{\mathbb{R}^2} \delta_0 \circ f(x) V(x) |\nabla f(x)| dx.$$

The assumption that V is continuous is important (of course, we only need continuity near $f^{-1}(0)$). Take the following simple example with curves; let $f(u,v) = u^2 + v^2 - 1$, so that $f^{-1}(0) = S^1$, the unit circle and $V(u,v) = 1$ if $u^2 + v^2 \leq 1$ and 0 otherwise. Then

$$\int_{S^1} V\, dl = 2\pi$$

but

$$\lim_{\varepsilon \to 0} \int_{\mathbb{R}^2} \delta_\varepsilon \circ f(x) V(x) |\nabla f(x)|\, dx = \pi$$

(both integrals being easily computed in radial coordinates).

4.10 Gauss–Bonnet Theorem

The average of the Gauss curvature over a domain with piecewise geodesic boundary is provided by the Gauss–Bonnet formula [64]:

Theorem 4.26. *Let S be a regular surface and A be a domain on M such that ∂A is the union of N geodesics $\gamma^{(1)}, \ldots, \gamma^{(N)}$. Let $\varepsilon_i, i = 1, \ldots N$ be the sequence of consecutive angles between the curves at their intersection. Then*

$$\int_A K\, d\sigma = 2\pi - \sum_{i=1}^{N} \varepsilon_i. \tag{4.28}$$

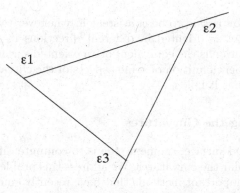

Fig. 4.3. The Gauss–Bonnet theorem in $\mathbb{R}^2$ reduces to the well-known property that the sum of consecutive angles in a polygon is 2π.

For example, when $N = 3$ (∂A is a "geodesic triangle"), we obtain the fact that the sum of the angles of a triangle is 2π minus the integral of the Gauss curvature over its interior. This is consistent with the sum being 2π in the plane which has zero Gauss curvature (Figure 4.3).

4.11 Triangulated Surfaces

4.11.1 Definition and Notation

Triangulated surfaces are the three-dimensional equivalent of polygons in two dimensions. This is most of the time how surfaces are stored on computers, and the kind of objects that must be handled in practical applications.

In full generality, a triangulated surface is a set of triangles, $F = \{f_1, \ldots, f_K\}$. Each f_k is a 3-tuple of vertices, $f_k = (v_{k1}, v_{k2}, v_{k3})$. One also wants to consider the set of all *distinct* vertices in the triangulation, $V = \{v_1, \ldots, v_N\}$, and the set of all *distinct* edges (pairs of vertices belonging to the same face), $E = \{e_1, \ldots, e_Q\}$.

The order of the vertices in each face is important and defines its orientation, which is invariant up to a cyclic permutation of the vertices. We will only consider *regular* triangulations, which are such that the intersection between two faces is either empty or an edge. This excludes situations in which the contact between two faces occurs at a vertex only, or in which some vertex belongs to the interior of an edge. The number

$$\chi = |V| - |E| + |F|$$

is a topological invariant of the surface called the Euler characteristic.

For a vertex v_i, we let F_i denote the faces that contain it, E_i the set of edges that contain it, and V_i the set of vertices (distinct from v_i) that belong to one of the faces in F_i. (V_i, E_i, F_i) represents the neighborhood of v_i in the triangulation.

The triangulation is said to be consistent if, whenever two faces intersect, their common vertex is oriented in different directions in the two faces. A consistent triangulation is the equivalent of an oriented surface. This property is needed for a proper definition of some aspects of the curvature and we shall therefore assume that it holds.

4.11.2 Estimating the Curvatures

Given a triangulated surface, the next step is to compute differential descriptors, and in particular the curvatures. We address this problem in this section, focusing on a few important methods that have recently emerged in the literature.

Taylor Expansions

The unit normal to an oriented triangle (v_1, v_2, v_3) is the vector

$$n = \frac{(v_2 - v_1) \times (v_3 - v_1)}{|(v_2 - v_1) \times (v_3 - v_1)|}.$$

Each face f_k in the triangulation therefore carries a uniquely defined normal, n_k. We can associate the normal to a specific point inside the face, for example its centroid $(v_1+v_2+v_3)/3$. (There are several possible definitions of the center of a triangle, however, including the circumcenter, which is the center of the circumscribed circle, the incenter, which is the center of the inscribed center, or the orthocenter, the intersection of the lines passing by the vertices and orthogonal to the opposite edge.)

In many cases, one wants to define normals also at the vertices. This can be done using a weighted average of the normals at the neighboring faces. If v_i is a vertex, define

$$n(v_i) = \frac{\sum_{f_k \in F_i} w_i(f_k) n_k}{|\sum_{f_k \in V_i} w_i(f_k) n_k|}$$

where $w_i(f_k)$ weights the importance of face f_k relative to vertex v_i. The simplest definition is the area, $|f_k|$, independent in this case of the chosen vertex. In [142], it is suggested to use the part of the face which is closest to v_i than to any of the other two vertices. It is the part of the face f_k delimited by the following four points: v_i, the two midpoints of the edges of f_k that contain v_i and the circumcenter of f_k if this point belongs to f_k, or the midpoint of the largest edge of f_k otherwise (the two cases are described in Figure 4.4). Such regions are called *Voronoï cells*. Let f_{ki} denote the part of face f_k which is associated to v_i in this way. One can use $w_i(f_k) = |f_{ki}|$.

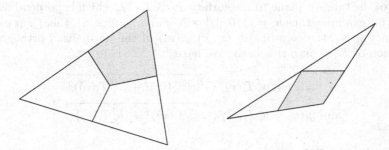

Fig. 4.4. Decomposition of triangles into Voronoï cells when the circumcenter is interior to the triangle (left) and when it is exterior (right).

Similarly, we can define a normal along an edge e to be a weighted average of the normals to the faces that intersect at e, using, for example, the areas of the faces as weights.

Having an estimation of the normal at each vertex allows for the approximation of the normal curvature of a curve on the surface passing through this vertex, which yields the second fundamental form. If $v \in V_i$, the path (v_i, v)

provides a discrete curve fragment passing by v_i. Define the tangent vector $T_i(v) = (v - v_i)/|v - v_i|$. Using Definition 4.23, one possible approximation of the second fundamental form at the midpoint between v_i and v in the direction $T_i(v)$ (which is also the normal curvature of the curve fragment at the midpoint) is

$$II_i(v) := -T_i(v)^T \left(\frac{n(v) - n(v_i)}{|v - v_i|} \right) = -\frac{(n(v) - n(v_i))^T (v - v_i)}{|v - v_i|^2}.$$

Also, using a Taylor expansion (assuming that $n(v)$ is the restriction to the vertices of a smooth function), one can prove that

$$\frac{(n(v) + n(v_i))^T (v - v_i)}{|v - v_i|^2} = O(|v - v_i|),$$

and adding this expression to the previous estimate for $II_i(v)$ yields the alternate formula [198]

$$II_i(v) := \frac{2n(v_i)^T (v - v_i)}{|v - v_i|^2}.$$

Since the matrix DN is symmetric in the tangent plane, it is described by three parameters in any orthonormal basis. Since each computation of the second fundamental form yields one linear equation, this requires at least three edges for the estimation, which is the minimum number provided by the triangulation. One possible way to proceed is to select an arbitrary basis (a_i, b_i) of the tangent plane to the surface at v_i, $T_{v_i}M$, which is perpendicular to n_{v_i} (for example, take $a_i = (1,0,0)^T \times N$ and $b_i = N \times a_i$), then, compute, for each $v \in N_i$, the coordinates $(x_i(v), y_i(v))$ of the normalized orthogonal projection of $T_i(v)$ onto this basis. We have

$$x_i(v) = a_i^T T_i(v)/\sqrt{(a_i^T T_i(v))^2 + (b_i^T T_i(v))^2}$$

and $y_i(v) = b_i^T T_i(v)/\sqrt{(a_i^T T_i(v))^2 + (b_i^T T_i(v))^2}.$

Then, letting $DN_i = \begin{pmatrix} \alpha_i & \gamma_i \\ \gamma_i & \beta_i \end{pmatrix}$ in this basis, we have the system of linear equations

$$\alpha_i x_i(v)^2 + 2\gamma_i x_i(v) y_i(v) + \beta_i y_i(v)^2 = -II_i(v), \; v \in N_i.$$

This is an over-constrained system, for which one can compute a least-squares solution. Once DN_i is computed, its trace, determinant and eigenvalues provide an estimation of the mean, Gaussian and principal curvatures.

A more direct approach for estimating the curvature from the second fundamental form has been proposed in [198]. Introduce, for continuous surfaces, the matrix (defined at a point p in the surface)

$$\Sigma_p = \frac{1}{2\pi} \int_0^{2\pi} \kappa_N(T_\theta) T_\theta T_\theta^T \, d\theta$$

where T_θ is the rotation (within the tangent plane) of an arbitrary reference vector $T \in T_p M$. A direct computation of this integral (using the basis $(T, T_{\pi/2})$) shows that

$$\Sigma_p = \frac{3}{8} DN(p) - \frac{1}{8} \det(DN(p)) DN(p)^{-1},$$

the last term being the co-matrix of $DN(p)$ (therefore also defined when $DN(p)$ is singular).

This implies that the eigenvalues of Σ_p are $\lambda_1 = -(3\kappa_1(p) - \kappa_2(p))/8$ and $\lambda_2 = -(3\kappa_2(p) - \kappa_1(p))/8$, which can be used to compute the curvatures, and that the eigenvectors of Σ_p coincide with those of $DN(p)$ and therefore provide the principal directions.

Returning to the discrete case, the curvatures at vertex v_i can therefore be estimated from an approximation Σ_i of Σ_{v_i}. Such an approximation is provided by the simple formula

$$\Sigma_i = - \sum_{v \in N_i} w_i(v) II_i(v) T_i(v) T_i(v)^T / \sum_{v \in V_i} w_i(v)$$

where $w_i(v) = (w_i(f^+(v)) + w_i(f^-(f)))/2$, $f^+(v)$ and $f^-(v)$ being the faces that intersect at v.

Gauss–Bonnet and Area Minimization

In the previous section, the curvature computations were based on Taylor expansions of the formulae that apply on smooth surfaces. More recently [63], an increased focus has been made on obtaining expressions that derive from intrinsic properties of surfaces that can be extended to polyhedral surfaces.

The right-hand side of equation (4.28) in the Gauss-Bonnet theorem can still be defined on polyhedral surfaces. This fact is used in [142] to provide an approximation of the Gauss curvature, using, for a vertex v_i in the triangulation, the region A_i formed by the union of the Voronoï cells around v_i (Figure 4.4). The expression is very simple because, in both cases in Figure 4.4, the exterior angles in A_i coincide with the angle of the corresponding face at v_i. Denoting, for $f \in F_i$, by $\theta_i(f)$ the angle of the face f at v_i, we see that the right-hand side of (4.28) is given by $2\pi - \sum_{f \in F_i} \theta_i(f)$. Approximating K by a constant over A_i, we get the formula

$$K_i = \frac{1}{|A_i|} \left(2\pi - \sum_{f \in F_i} \theta_i(f) \right).$$

The area, $|A_i|$ is also easy to compute. It is the sum of the areas of the shaded regions in Figure 4.4, over all faces that contain v_i. Let as above $\theta_i(f)$

be the angle at v_i for a face f. Let $v_i'(f)$ and $v_i''(f)$ be the other two vertices of f so that $v_i, v_i'(f)$ and $v_i''(f)$ are ordered consistently with the orientation of f. Let $e_i'(f)$ be the edge opposed to $v_i'(f)$ in f and $e_i''(f)$ the edge opposed to $v_i''(f)$ (we will later denote by $e_i(f)$ the edge opposed to v_i). Finally, let $\theta_i'(f)$ and $\theta_i''(f)$ respectively denote the angles at $v_i'(f)$ and $v_i''(f)$. Then, the area, $a_i(f)$, of the shaded region in Figure 4.4 is given by

$$
a_i(f) = \begin{cases}
\frac{1}{8}(|e_i'(f)|^2 \operatorname{ctn}(\theta_i'(f)) + |e_i''(f)|^2 \operatorname{ctn}(\theta_i''(f))) & \text{(non-obtuse case)}, \\
\operatorname{area}(f)/2 & \text{(obtuse case, } \theta_i(f) \geq \pi/2), \\
\operatorname{area}(f)/4 & \text{(obtuse case, } \theta_i(f) < \pi/2),
\end{cases}
$$

Given this, $|A_i|$ is the sum of these areas over $f \in F_i$. When there is no obtuse triangle around v_i, the area $|A_i|$ has another simple expression [142]. For $e \in E_i$ (edges stemming from v_i), let $\alpha(e)$ and $\beta(e)$ be the angles at vertices opposed to e in the triangles that intersect at e. Then

$$
|A_i| = \frac{1}{8} \sum_{e \in E_i} (\operatorname{ctn} \alpha(e) + \operatorname{ctn} \beta(e)) |e|^2.
$$

Consider now the mean curvature. First, let's consider an important interpretation of this curvature as a "gradient" of the surface area. Let S be a surface and $h : S \to \mathbb{R}^3$ be a (smooth) vector field on S. Assume that $h = 0$ on the boundary of S (if S has one). Define the surface S_ε as the one obtained by displacing each $p \in S$ along the vector $\varepsilon h(p)$. Then,

$$
\frac{d}{d\varepsilon} \operatorname{area}(S_\varepsilon)_{|\varepsilon=0} = 2 \int_S H(p) h(p)^T n(p) d\sigma(p).
$$

We can make the same construction with a discrete surface Σ by associating to each vertex v_i a small displacement $\varepsilon h_i \in \mathbb{R}^3$, and computing the derivative of the area of the obtained surface Σ_ε. Approximating the right-hand side in the above formula, we will then identify:

$$
\frac{d}{d\varepsilon} \operatorname{area}(\Sigma_\varepsilon)_{|\varepsilon=0} = 2 \sum_{i=1}^{N} h_i^T (H_i n_i) |A_i| \tag{4.29}
$$

where A_i is the neighborhood attributed to v_i and $H_i n_i$ can then be interpreted as the discretized product of the mean curvature with the normal at v_i.

Given that the area of a triangle with vertices v_1, v_2, v_3 is given by the half-norm of the cross product $(v_2 - v_1) \times (v_3 - v_1)$, the left-hand side in (4.29) is

$$\frac{1}{2} \sum_{f \in F} ((h_2(f) - h_1(f)) \times (v_3(f) - v_1(f)) +$$

$$(v_2(f) - v_1(f)) \times (h_3(f) - h_1(f)))^T n(f)$$

$$= \frac{1}{2} \sum_{f \in F} (h_1(f) \times (v_2(f) - v_3(f)) +$$

$$h_2(f) \times (v_3(f) - v_1(f)) + h_3(f) \times (v_2(f) - v_1(f)))^T n(f)$$

where $v_1(f), v_2(f)$ and $v_3(f)$ are ordered vertices of f, $h_1(f), h_2(f)$ and $h_3(f)$ the associated displacements and $n(f)$ is the normal to f,

$$n(f) = \frac{(v_2(f) - v_1(f)) \times (v_3(f) - v_1(f))}{|(v_2(f) - v_1(f)) \times (v_3(f) - v_1(f))|}.$$

For $f \in V_i$, let $e_i(f)$ be the oriented edge opposite to v_i. Using the relation $(x \times y)^T z = x^T (y \times z)$ and reordering the sums, we can write

$$\frac{d}{d\varepsilon} \text{area}(\Sigma_\varepsilon)|_{\varepsilon=0} = \frac{1}{2} \sum_{i=1}^{N} h_i^T \left(\sum_{f \in F_i} e_i(f) \times n(f) \right).$$

So this provides a definition of the discrete mean curvature at v_i:

$$H_i n_i = \frac{1}{4|A_i|} \sum_{f \in F_i} e_i(f) \times n(f).$$

Reordering this sum over edges and explicitly computing the cross product leads to the equivalent expression [142]

$$H_i n_i = \frac{1}{4|A_i|} \sum_{e \in E_i} (\text{ctn}\, \alpha(e) + \text{ctn}\, \beta(e)) e$$

where $\alpha(e)$ and $\beta(e)$ are as before the angles at the vertices opposite to e in each of the faces that contain e (e being oriented from v_i to the other vertex).

Note that this computation provides an estimate of the normal and the mean curvature together.

Curvature Measures

There is another way to interpret curvature on a surface that can be generalized to the non-smooth case, leading to another formula for curvature approximation on triangulated surfaces. On smooth surfaces, this is related to the volume of parallel sets.

Let M be a smooth oriented surface in $\mathbb{R}^3$ and N be its unit normal. For $B \subset M$ and $r > 0$, consider the set

$$\Omega_r(B) = \{p + tN(p), p \in B \cap M, t \in (0, r)\} .$$

If M is a compact manifold (without boundary), there exists $r_0 > 0$ such that the mapping $\varphi : (p, t) \to p + tN(p)$ is one-to-one on $M \times (0, r)$ for $r < r_0$. Let's prove this statement, assuming that it is false and reaching a contradiction. One would be able to find sequences p_n, p'_n, t_n and t'_n such that $\max(t_n, t'_n) \to 0$ and for all n

$$p_n + t_n N(p_n) = p'_n + t'_n N(p'_n).$$

This implies

$$p_n - p'_n = (t'_n - t_n)N(p'_n) + t_n(N(p'_n) - N(p_n))$$

so that (assuming the normals are C^1), for some constant C

$$|p_n - p'_n| \le |t'_n - t_n| + Ct_n|p_n - p'_n|,$$

which implies that $|p_n - p'_n| \to 0$. Because M is compact, one can (using a subsequence if needed) assume that p_n, and therefore p'_n, converges to some $p \in M$.

So take a local chart at p and let $(u, v) \mapsto m(u, v)$ be the parametrization on some open set $U \subset \mathbb{R}^2$. We have

$$\partial_u \varphi = \partial_u m + t\dot{N}_u, \ \partial_v \varphi = \partial_v m + t\dot{N}_v, \ \dot{\varphi}_t = N$$

so that

$$\begin{aligned}
\det(D\varphi) &= (\partial_u \varphi \times \partial_v \varphi)^T \partial_t \varphi \\
&= (\partial_u m \times \partial_v m)^T N + t(\partial_u m \times \dot{N}_v + \dot{N}_u \times \partial_v m)^T N \\
&\quad + t^2 (\dot{N}_u \times \dot{N}_v)^T N.
\end{aligned}$$

We have $\partial_u m \times \partial_v m = |\partial_u m \times \partial_v m| \, N$. Moreover, for any linear operator A on $\mathbb{R}^3$ and any basis (u_1, u_2, u_3) in $\mathbb{R}^3$, we have (the proof being left to the reader)

$$(u_1 \times u_2)^T A u_3 + (u_2 \times u_3)^T A u_1 + (u_3 \times u_1)^T A u_2 = \det(u_1, u_2, u_3)\text{trace}(A).$$

Applying this to

$$(\partial_u m \times \dot{N}_v + \dot{N}_u \times \partial_v m)^T N = (N \times \partial_u m)^T \dot{N}_v + (\partial_v m \times N)^T \dot{N}_u$$

with $\dot{N}_u = DN\partial m_u$, $\dot{N}_v = DN\partial m_v$, taking $A = DN$ on $T_p M$ and $AN = 0$, we get

$$(\partial_u m \times \dot{N}_v + \dot{N}_u \times \partial_v m)^T N = |\partial_u m \times \partial_v m|\text{trace}(DN(m)) = -2|\partial_u m \times \partial_v m|H(m).$$

Finally, since $\dot{N}_u$ and $\dot{N}_v$ are tangent to M at m, $(\dot{N}_u \times \dot{N}_v)^T N$ is the two-dimensional determinant of $[DN\partial_u m, DN\partial_v m]$, therefore equal to $K(m)|\partial_u m \times \partial_v m|$. We therefore have

$$\det(D\varphi) = (1 - 2tH(m) + t^2 K(m))|\partial_u m \times \partial_v m|. \tag{4.30}$$

Since the area measure $d\sigma$ is equal to $|\partial_u m \times \partial_v m| du dv$, we obtain the fundamental equation

$$\det(D\varphi) du dv = (1 - 2tH(m) + t^2 K(m)) d\sigma. \tag{4.31}$$

From equation (4.31) directly follows the fact that, for $r < r_0$,

$$|\Omega_r(B)| = \int_0^r \int_{B \cap M} (1 - 2tH + t^2 K) d\sigma dt$$

$$= r|B \cap M| - r^2 \int_{B \cap M} H d\sigma + \frac{r^3}{3} \int_{B \cap M} K d\sigma,$$

which proves that $\Omega_r(B)$ is a third-degree polynomial in r, with coefficients provided by the volume of $B \cap M$ and the integrals of the mean and Gauss curvatures over $B \cap M$. This provides a new interpretation of the curvature integrals. We now proceed to an extension of this identity to the discrete case, providing in this way a new definition of the curvatures in this case.

Our definition of Ω_r relies on the existence of the normal to M, which we have defined so far only for smooth manifolds. This definition can be extended to much more general sets. There are two equivalent points of view for doing this.

Assuming that M is smooth and closed (so that M is the boundary of a domain $\Omega \subset \mathbb{R}^3$), the first point of view is based on the fact that, for small r, $\Omega_r(B)$ is also the set of all $p \in \Omega^c$ that are at distance smaller than r to M, with closest point in B. This suggests an extension to a larger family of sets M as we now describe [80, 81]. A set $M \subset \mathbb{R}^2$ such that there exists $r_0 > 0$, such that every element $p \in \Omega^c$ with $\text{dist}(p, M) < r_0$ has a unique closest point in M, is called a *set of positive reach*. As we have seen, sets of positive reach include smooth closed manifolds. But they form a much larger class of sets, including, for example, arbitrary closed convex sets. For sets of positive reach, Federer [80] has proved that, if one defines $\Omega_r(B)$ as the set of points in Ω^c that are at distance less than r to M, with closest point in $B \cap M$, then the volume $|\Omega_r(B)|$ is still polynomial in r, taking the form

$$|\Omega_r(B)| = r\mu_0(M, B) - r^2 \mu_1(M, B) + \frac{r^3}{3}\mu_2(M, B). \tag{4.32}$$

In particular, $\mu_1(M, B)$ and $\mu_2(M, B)$ are generalizations of the integrals of the curvatures on B, and are called the mean and Gauss *curvature measures* on M. They have the important property of being additive, satisfying in particular

$$\mu_i(M \cup M', B) = \mu_i(M, B) + \mu_i(M', B) - \mu_i(M \cap M', B). \qquad (4.33)$$

Although it already is a rich class of sets, sets of positive reach do not include non-convex polyhedrons, so the construction cannot be immediately extended to triangulated surfaces. But formula (4.33) provides the key for this extension. Indeed, one can define a union of sets of positive reach as a set M that can be decomposed into

$$M = \cup_{j \in J} M_j$$

where each M_j has positive reach and any nonempty intersection of M_j's has positive reach [227]. Then, iterating (4.33) (using the exclusion–inclusion formula), we can set

$$\mu_k(M, B) = \sum_{I \subset J} (-1)^{|I|-1} \mu_k \left(\bigcap_{j \in I} M_j, B \right) \qquad (4.34)$$

the left-hand side being well-defined from the hypotheses. This is a valid definition of the right-hand side because it can be shown that the result does not depend on the chosen decomposition of M, which is not unique. This extension now includes all polyhedrons (and triangulated surfaces).

The second point of view gives an alternative interpretation of the curvature measures, based on the *normal bundle* to a set M. This requires a general definition of tangent and normal vectors to an arbitrary set $M \subset \mathbb{R}^2$ [81]. For tangents, the definition is quite natural: a vector v is tangent to M at a point p in M if one can find points x in M that are arbitrarily close to p and such that v is arbitrarily close to the line $\mathbb{R}^+(x - p)$. This is formalized in the following definition:

Definition 4.27. *If $M \subset \mathbb{R}^d$, and $p \in M$, the tangent set to M at p is the set $T_p M$ of vectors $v \in \mathbb{R}^d$ such that, for any $\varepsilon > 0$, there exist $x \in M$ and $r > 0$ such that $|x - p| < \varepsilon$ and $|v - r(x - p)| < \varepsilon$.*

It is easy to check that this coincides with our previous definition of tangent vectors on smooth curves or surfaces. The definition also applies to open subsets in $\mathbb{R}^d$; we shall more precisely be concerned with the situation in which M is included in the boundary of an open set Ω. In this case, for $m \in M$, $T_m \Omega$ contains vectors in $T_m M$ and all vectors v such that $m + \varepsilon v \in \Omega$ for small ε (vectors that point to the interior of Ω).

In view of studying triangulated surfaces, some simple particular cases are of interest for us. Start with the trivial case of M being a single point, $M = \{a\}$. It is clear from the definition that any tangent vector to M must satisfy $|v| < \varepsilon$ for any $\varepsilon > 0$ so that $T_M = \{0\}$.

Now consider a smooth regular curve $\gamma : [0,1] \to \mathbb{R}^3$, $M = \gamma([0,1])$, and $a = \gamma(0)$. Then, any $x \in M$ close to a is equal to $\gamma(t)$ for $t \simeq 0$, and a tangent vector v at a must be such that $v \simeq r(\gamma(t) - \gamma(0))$ with $r > 0$, so that $T_a M$ is the half-line $\mathbb{R}^+ \gamma'(0)$. By the same argument, if $b = \gamma(1)$, $T_b M = \mathbb{R}^- \gamma'(1)$.

As a last example, of obvious importance for our discussion, consider the case in which M a triangle and a is on its boundary; then, $T_a M$ simply is the set of vectors v such that $a + v$ points towards the interior of M.

From tangents, we now can define normals.

Definition 4.28. *Let $M \subset \mathbb{R}^3$. For $p \in M$, the normal vectors to M at p form the set $N_p M$, containing all vectors n such that $n^T v \leq 0$ for $v \in T_p M$. The normal bundle of M is the set $NM \subset \mathbb{R}^3 \times \mathbb{R}^3$ defined by*

$$NM = \{(p,n), p \in M, n \in N_p M, |n| = 1\}.$$

When $M \subset \partial\Omega$, we can also consider

$$NM^+ = \{(p,n), p \in M, n \in N_p \Omega, |n| = 1\}.$$

This corresponds to normals to M pointing outward from Ω.

The normal bundle is the structure on which the new curvature measures will be defined. Let's describe it for the previous examples. First, if M is a smooth closed oriented surface in $\mathbb{R}^3$, NM is simply the set $\{(p, N(p)), p \in M\} \cup \{(p, -N(p)), p \in M\}$. If M is a closed curve, with regular parametrization $s \mapsto \gamma(s)$, then $NM = \{(\gamma(s), n) : n^T \dot\gamma(s) = 0, |n| = 1\}$ (this can be thought of as a tube centered around γ).

If $M = \{a\}$, then (since $T_a M = \{0\}$), $NM = \{a\} \times S^2$, where S^2 is the unit two-dimensional sphere.

When M is an open curve, parametrized by γ, the set $N_a M$ for $a = \gamma(0)$ is the half-sphere $S^2 \cap \{n^T \dot\gamma(0) \leq 0\}$, while, for $b = \gamma(1)$, it is $N_b M = S^2 \cap \{n^T \dot\gamma(1) \leq 0\}$.

Finally, if M is a triangle, and a is on an edge, but not at a vertex, $N_a M$ is the half-circle, intersection of the unit circle orthogonal to the edge and of the half-space $\{n^T \nu \geq 0\}$ where ν is normal to the edge, in the triangle plane, and pointing towards the interior of the triangle. If a is a vertex, the set $N_a M$ is the "spherical corner", formed by the intersection of the unit sphere and the two half-planes $\{n^T e \leq 0\}$ and $\{n^T e' \leq 0\}$ where e and e' are the two edges stemming from a (oriented from a to the other vertex).

The interesting fact in the previous examples is that, in each case, NM was a 2-D structure, i.e., it could always be parametrized with two parameters. In fact, NM is a two-dimensional surface in a space of dimension 6.

For a smooth surface, we have proved the identity (valid within a chart)

$$|\Omega_r(B)| = \int_0^r \int_U |\det(\partial_u m + t\dot N_u, \partial_v m + t\dot N_v, N)| du\,dv\,dt \qquad (4.35)$$

$$= r|B \cap M| - r^2 \int_{B \cap M} H d\sigma + \frac{r^3}{3} \int_{B \cap M} K d\sigma.$$

We now make a slight shift in the interpretation of the function $\varphi(m,t) = m + tN(m)$, defined on $M \times (0, r_0)$ and consider it as a function $\varphi(m,n,t) = m + tn$ defined on $NM \times (0, r_0)$. For smooth oriented surfaces, this is equivalent if φ is restricted to the positive normal, i.e., NM^+, since the latter is uniquely determined by m. But we have seen cases for which the normal was just partially specified by p, and this new definition of φ also applies to this case. From now on, we assume that (u, v) are local coordinates on NM^+.

We now want to interpret the integral over U in (4.35) as an integral over NM^+. We need for this to compute the area form on NM^+, which is given by $d(u, v)dudv$ where

$$d(u,v)^2 = (|\partial_u m|^2 + |\dot{n}_u|^2)(|\partial_v m|^2 + |\dot{n}_v|^2) - (\partial_u m^T \partial_v m + \dot{n}_u^T \dot{n}_v)^2.$$

One can check (we skip the proof) that the ratio $Q = |\det(\partial_u m + t\dot{n}_u, \partial_v m + t\dot{n}_v, n)|/d(u,v)$ is invariant by a change of local chart on NM^+ and is therefore an intrinsic quantity. When M is a smooth surface, this can be easily computed, since we can assume that $\partial_u m$ and $\partial_v m$ correspond to the principal directions, yielding

$$Q = \frac{(1 + t\kappa_1)(1 + t\kappa_2)}{\sqrt{(1 + \kappa_1^2)(1 + \kappa_2^2)}}.$$

Returning to the general case, we have, by definition of Q:

$$\int_U |\det(\partial_u m + t\dot{n}_u, \partial_v m + t\dot{n}_v, n)|dudv = \int_A Q d\sigma$$

where A is the subset of NM^+ which corresponds to the local parametrization.

Assume that r_0 can be chosen so that Q does not vanish for $t \in (0, r_0)$. (That such an r_0 exists relates to the assumption of M having positive reach.) In this case, $\det(m_u + t n_u, m_v + t n_v, N)$ has constant sign, and one can expand Q in powers of t, yielding, for some functions S_0, S_1, S_2

$$\int_0^r \int_A Q d\sigma = r \int_A S_0 d\sigma - r^2 \int_A S_1 d\sigma + \frac{r^3}{3} \int_A S_2 d\sigma.$$

The functions S_k therefore provide densities for "generalized" curvature measures, defined on NM^+ (instead of on M).

Now assume that $M \subset \partial\Omega$ for some open set $\Omega \subset \mathbb{R}^2$. We can project the generalized curvature measures on M, for $B \subset \mathbb{R}^3$,

$$\mu_k(\Omega, B) = \int_{NM^+} \chi_B(m) S_k(m, n) d\sigma.$$

We integrate on NM^+ to reflect the fact that we have used oriented normals in our initial computation.

Now consider the case in which M is a triangulated surface. First let $B \cap M$ be included in the interior of a face: since M coincides there with a smooth surface with zero curvature, we have $\mu_0(M, B) = |B \cap M|$, $\mu_1(M, B) = \mu_2(M, B) = 0$.

Now let $B \cap M$ be included in the interior of a convex (salient) edge, e. At $p \in B$, normal vectors to Ω form the arc of the unit circle perpendicular to the edge delimited by the two normals to the neighboring faces. Let's denote the angle formed by these normals by $\beta_N(e)$. Now, on $B \cap M$, the normal bundle can be parametrized by $p = u(e/|e|) + n(v)$ where $n(v)$ is the normal to p that makes an angle v with one of the face normals. Using the fact that e and n are orthogonal, one finds $d(u, v) = 1$ and $|\det(\partial_u m, \dot{t}\dot{n}_v, n)| = t$. This implies that $\mu_0 = \mu_2 = 0$ and $\mu_1 = \beta_N(e)\text{length}(B \cap M)$.

To consider the case of concave edges (for which Ω does not have positive reach on B), what we can do is split Ω into two parts on each side of the bisecting angle between the faces at e and apply the formula (letting Ω_1 and Ω_2 be the two sections)

$$
\begin{aligned}
\mu_k(\Omega, B) &= \mu_k(\Omega_1, B) + \mu_k(\Omega_2, B) - \mu_k(\Omega_1 \cap \Omega_2, B) \\
&= ((\pi - \beta_N)/2 + (\pi - \beta_N(e))/2 - \pi)\text{length}(B \cap M) \\
&= -\beta_N\text{length}(B \cap M)
\end{aligned}
$$

where $\beta_N(e)$ is again the angle between the normal to the faces meeting at e, taken between 0 and π.

Finally, assume that $B \cap M$ is a vertex, say v. First note that, when M has positive reach, the volume of $\Omega_r(M, v)$ cannot be larger than that of the ball centered at v with radius r and is therefore an $O(r^3)$. From formula (4.34), this is also true when M is a triangulated surface. This implies that only the last term (the Gauss curvature measure) can be non-zero. The computation of this term is simplified if we also remark that the Gauss curvature measure does not change if we replace (locally at the vertex) Ω by $\Omega^c \cap \partial\Omega$, which corresponds to changing the orientation on M. This is because the function S_2 is an even function of the normal. Using this property, we get

$$
2\mu_2(\Omega, B) = \mu_2(\Omega, B) + \mu_2(M \cup \Omega^c, B) = \mu_2(\mathbb{R}^3, B) + \mu_2(M, B).
$$

Since $\mu_2(\mathbb{R}^3, B) = 0$, it remains to compute $\mu_2(M, B)$. Let $F_1, \ldots, F_q$ represent the faces containing v. We want to apply the inclusion/exclusion formula to

$$
\mu_2(M, B) = \mu_2\left(\bigcup_{i=1}^q F_i, B\right).
$$

For this, we need to compute the Gauss curvatures in three special cases. The simplest case is $\mu_2(\{v\}, B)$. In this case, we can parametrize the normal bundle by $(m(u, u'), n(u, u')) = (v, n(u, u'))$ where (u, u') is a parametrization of the unit sphere, for which we assume that $\dot{n}_u$ and $\dot{n}_{u'}$ are orthogonal with unit norm. In this case, $d(u, u') = 1$ and $|\det(t\dot{n}_u, t\dot{n}_{u'}, n)| = t^2$. This implies that $S_2 = 1$ and $\mu_2 = 4\pi$ (μ_2 is three times the volume of the sphere).

Now, let e be a segment having v as one of its extremities. Assume without loss of generality that e is supported by the first axis of coordinates and $v = 0$. We can parametrize Ne at v with $(u, u') \mapsto (v, n(u, u'))$ where $n(u, u')$ parametrizes the half-sphere that is normal to M at v. This provides $\mu_2(e, B) = 2\pi$.

The last case is a triangle F with vertex v. Let θ be the angle at v. Here, the normal bundle at v is the part of the unit sphere which is contained between the two hyperplanes normal to each edge of F incident at v, for which the volume is $2(\pi - \theta)/3$ so that $\mu_2(F, B) = 2(\pi - \theta)$.

Now, it remains to apply the inclusion/exclusion formula. This formula starts with $\sum_{i=1}^{q} \mu_2(F_i, B)$ which is $2q\pi - 2\sum_i \theta_i$. Then comes the sum of the measures associated to the intersection of two faces: this intersection is an edge for the q pairs of adjacent faces, and just $\{v\}$ for the $\binom{q}{2} - q$ remaining ones. This yields the contribution $2q\pi - 4\binom{q}{2}\pi$. We finally need to sum all terms for intersections of three or more sets, which is always equal to $\{v\}$. This is

$$4\pi \sum_{k \geq 3} (-1)^{k-1} \binom{q}{k} = 4\pi \left(1 - q + \binom{q}{2}\right),$$

where we used the fact that

$$\sum_{k \geq 0} (-1)^{k-1} \binom{q}{k} = (1 - 1)^q = 0.$$

Summing all the terms, we obtain $\mu_2(M, B) = 4\pi - 2\sum_{i=1}^{q} \theta_i$ so that

$$\mu_2(\Omega, B) = 2\pi - \sum_{i=1}^{q} \theta_i.$$

We have therefore obtained the curvature measures associated to an oriented triangulated surface [51]. For any set $B \in \mathbb{R}^3$, they are:

- The mean curvature measure:

$$\mu_1(M, B) = \sum_{e \in E} \text{length}(B \cap e)\varepsilon_e \beta_N(e)$$

where β_N is the angle between the normals to the faces at e and $\varepsilon_e = 1$ if the edge is convex and -1 otherwise.
- The Gauss curvature measure:

$$\mu_2(M, B) = \sum_{v \in V \cap B} \mu_2(M, v)$$

with

$$\mu_2(M, v) = 2\pi - \sum_{f \in F_v} \theta_v(f).$$

Using these expressions, we can make approximations of the curvature at a given vertex by letting

$$\mu_1(M, B) \simeq |B| H_i \text{ and } \mu_2(M, B) \simeq |B| K_i$$

for a vertex v_i in the triangulation. Taking $B = A_i$, as defined in the previous section, we obtain the same approximation of the Gauss curvature as the one obtained from the discretization of Gauss–Bonnet theorem. The formulae for the mean curvature differ, however.

Discrete Gradient and Laplace–Beltrami Operators

We conclude this section on triangulated surfaces by a computation of the discrete equivalent of the gradient and Laplacian on surfaces.

Let S be a triangulated surface, $V = \{v_1, \ldots, v_N\}$ and $F = \{f_1, \ldots, f_K\}$ denoting, respectively, the sets of vertices and faces of S. To simplify the discussion, we will assume that the surface has no boundary

A function ψ defined on S assigns a value $\psi(v_i)$ to each vertex, and the gradient of ψ will be defined as a vector indexed over faces. To compute it, we first focus on a face, $f = f_k$ for some $k \in \{1, \ldots, K\}$ that we will drop from the notation until further notice. Let (v_1, v_2, v_3) be the vertices of f (ordered consistently with the orientation), and let $e_1 = v_3 - v_2$, $e_2 = v_1 - v_3$ and $e_3 = v_2 - v_1$. Let $c = (v_1 + v_2 + v_3)/3$ be the center of the face.

We define the gradient $u = \nabla_S \psi(f)$ on the face by $u = \alpha_1 e_1 + \alpha_2 e_2$ such that $u^T(v_k - c) = \psi(v_k) - \psi(c)$ for $k = 1, 2, 3$, with $\psi(c) := (\psi(v_1) + \psi(v_2) + \psi(v_3))/3$. Since this implies that $u^T(v_k - v_l) = \psi(v_k) - \psi(v_l)$, this gives

$$\psi(v_3) - \psi(v_2) = (\alpha_1 e_1 + \alpha_2 e_2)^T e_1,$$
$$\psi(v_1) - \psi(v_3) = (\alpha_1 e_1 + \alpha_2 e_2)^T e_2.$$

Let ψ_f be the column vector $[\psi(v_1), \psi(v_2), \psi(v_3)]^T$, M the 2 by 3 matrix

$$M = \begin{pmatrix} 0 & -1 & 1 \\ 1 & 0 & -1 \end{pmatrix}$$

and G_f the matrix

$$G_f = \begin{pmatrix} |e_1|^2 & e_1^T e_2 \\ e_1^T e_2 & |e_2|^2 \end{pmatrix}.$$

With this notation, the previous system is $M\psi_f = G_f \alpha$. We therefore have

$$u = [e_1, e_2]\alpha = [e_1, e_2] G_f^{-1} M \psi_f.$$

We first notice that $\det G_f = |e_1|^2 |e_2|^2 - (e_1^T e_2)^2 = (|e_1||e_2| \sin \theta_3)^2$ where θ_3 is the angle at v_3. It is therefore equal to $4a(f)^2$ where $a(f)$ is the area of f. Given this, we can write:

$$G_f^{-1} M \psi_f = \det(G_f)^{-1} \begin{pmatrix} |e_2|^2 & -e_1^T e_2 \\ -e_1^T e_2 & |e_1|^2 \end{pmatrix} \begin{pmatrix} 0 & -1 & 1 \\ 1 & 0 & -1 \end{pmatrix} \psi_f$$

$$= \det(G_f)^{-1} \begin{pmatrix} -e_2^T e_1 & -e_2^T e_2 & -e_2^T e_3 \\ e_1^T e_1 & e_1^T e_2 & e_1^T e_3 \end{pmatrix} \psi_f, \qquad (4.36)$$

in which we have used the identity $e_3 = -e_1 - e_2$. Introducing the vector

$$h_\psi(f) = \psi(v_1) e_1 + \psi(v_2) e_2 + \psi(v_3) e_3$$

and the matrix

$$D_f = e_1 e_2^T - e_2^T e_1 = e_3^T e_1 - e_1^T e_3 = e_2^T e_3 - e_3^T e_1,$$

a little computation yields

$$\nabla_S \psi(f) = \frac{D_f}{4a(f)^2} h_\psi(f).$$

We now pass to the computation of the discrete Laplace–Beltrami operator, that we define via the discrete analog of the property

$$\int_S |\nabla_S \psi|^2 d\sigma_S = -\int_S \psi \Delta_S \psi d\sigma_S$$

that characterizes the operator on smooth surfaces without boundary. For triangulated surfaces, we will identify $\Delta_S \psi$ via

$$\sum_{k=1}^K |\nabla_S \psi(f_k)|^2 a(f_k) = -\sum_{i=1}^N \psi(v_i)(\Delta_S \psi)(v_i)|A_i|.$$

where $|A_i|$ is the area attributed to vertex v_i (using, for example, Voronoï cells).

For a given face f, we can write (using the previous notation): $|\nabla_S \psi(f)|^2 = \alpha^T G_f \alpha = \psi_f^T M^T G_f^{-1} M \psi_f$. Applying M^T to (4.36), we get

$$M^T G_f^{-1} M = \det(G_f)^{-1} \begin{pmatrix} |e_1|^2 & e_1^T e_2 & e_1^T e_3 \\ e_1^T e_2 & |e_2|^2 & e_2^T e_3 \\ e_1^T e_3 & e_2^T e_3 & |e_3|^2 \end{pmatrix}.$$

Let Σ_f denote this last matrix. We can write:

$$\sum_{k=1}^K \frac{\psi_{f_k}^T \Sigma_{f_k} \psi_{f_k}}{4a(f_k)} =$$

$$\frac{1}{4} \sum_{i=1}^N \psi(v_i) \sum_{f \in F_i} (|e_i(f)|^2 \psi(v_i) + e_i(f)^T e_i'(f) \psi(v_i'(f)) + e_i(f)^T e_i''(f) \psi(v_i''(f)))/a(f),$$

where $v_i'(f)$ and $v_i''(f)$ are the other two vertices of f (in addition to v_i) and $e_i(f), e_i'(f)$ and $e_i''(f)$ are, respectively, the edges opposed to $v_i, v_i'(f)$ and $v_i''(f)$ in face f. This implies that one should define

$$\Delta_S \psi(v_i) =$$
$$-\frac{1}{4|A_i|} \sum_{f \in F_i} (|e_i(f)|^2 \psi(v_i) + e_i(f)^T e_i'(f) \psi(v_i'(f)) + e_i(f)^T e_i''(f) \psi(v_i''(f)))/a(f).$$

One can rewrite this discrete Laplacian in terms of angles. Denoting as before by $\theta_i'(f)$ and $\theta_i''(f)$ the angles at $v_i'(f)$ and $v_i''(f)$, one has

$$e_i(f)^T e_i'(f) = -\cos\theta_i''(f)|e_i(f)|\,|e_i'(f)| = -2\operatorname{ctn}\theta_i''(f)a(f)$$

Similarly, $e_i(f)^T e_i''(f) = -2\operatorname{ctn}\theta_i'(f)a(f)$ and, since the sum of the edges is 0,

$$|e_i(f)|^2 = -e_i(f)^T (e_i'(f) + e_i''(f)) = 2(\operatorname{ctn}\theta_i'(f) + \operatorname{ctn}\theta_i''(f))a(f).$$

One can therefore write

$$\Delta_S \psi(v_i) = \frac{1}{2|A_i|} \sum_{f \in F_i} (\operatorname{ctn}\theta_i''(f)(\psi(v_i'(f)) - \psi(v_i)) + \operatorname{ctn}\theta_i'(f)(\psi(v_i''(f)) - \psi(v_i))),$$

which provides a discrete definition of the Laplace–Beltrami operator on S.

5

Isocontours and Isosurfaces

In this chapter, we start discussing methods for the extraction of shapes (curves or surfaces) from discrete image data. Most of the methods we discuss proceed by energy minimization and are implemented using curve evolution equations. Such equations will be discussed in Chapter 6. In this first, short, chapter, we discuss what is probably the simplest option that is available to extract a curve or a surface from an image, which is to define it implicitly based on the image values.

If $f : \mathbb{R}^d \to \mathbb{R}$ is the image (with $d = 2$ or 3), this corresponds to defining the shape as the level set

$$S_\lambda = \{m : f(m) = \lambda\}$$

for a properly chosen threshold, λ. As we know, if the gradient of f does not vanish on S_λ, this provides a smooth curve or surface (or a union of such).

When f is defined on a discrete grid, the problem is more difficult. The value λ may never be observable on the image, and a correct extraction of a discrete S_λ (a polygon or a triangulated surface) requires some interpolation. We start with a discussion of the two-dimensional case, which will help in addressing the computation of isosurfaces, which is more intricate.

5.1 Computing Isocontours

We consider here a two-dimensional grid, $\mathcal{G}$ which is formed with points $p(s,t) = (x_s, y_t)$, where $(x_s, s = 1, \ldots, M)$ is a discretization of the horizontal axis and $(y_t, t = 1, \ldots, N)$ a discretization of the vertical axis. We assume that a discretization of a smooth function f is observed, via the collection

$$(f_{st} = f(p(s,t)), s = 1, \ldots, M, t = 1, \ldots, N).$$

The problem is to compute the isocontour $(f = \lambda)$ for a given λ, in the form of a polygon or a union of polygons. Without loss of generality, we can and will assume $\lambda = 0$ in the following discussion.

Since the exact function f is not observed, some interpolation must be done, and we will use bilinear interpolation for this. This means that the true (unobserved) f will be replaced by the interpolation (that we still denote f, with some abuse of notation) which is defined as follows. Let $C(s,t)$ denote the cell (square) with vertices $p(s + \varepsilon_1, t + \varepsilon_2), \varepsilon_i \in \{0,1\}, i = 1, 2$. Then, for $p = x, y \in C(s,t)$, let

$$f(p) = \sum_{\varepsilon_1,\varepsilon_2=0}^{1} \prod_{i=1}^{2} (\varepsilon_i r_i(p) + (1 - \varepsilon_i)(1 - r_i(p))) f_{s+\varepsilon_1, t+\varepsilon_2} \tag{5.1}$$

with $r_1(p) = x - x_s, r_2(p) = y - y_t$.

Obviously, the set $(f = 0)$ is the union of its intersections with each cell in the grid, so that we can restrict to these intersections. Within a cell, f is given by (5.1), and the set $(f = 0)$ is either empty, or a line segment, or one or two branches of a hyperbola. This is because, introducing the coordinates $\xi = (x - x_s)/(x_{s+1} - x_s)$ and $\eta = (y - y_s)/(y_{s+1} - y_s)$, we can rewrite $f(p)$ in the cell as (up to a positive multiplicative constant):

$$f(p) = f_{--}(1 - \xi)(1 - \eta) + f_{+-}\xi(1 - \eta) + f_{-+}(1 - \xi)\eta + f_{++}\xi\eta$$

$$= \rho\left(\left(\xi + \frac{f_{-+} - f_{--}}{\rho}\right)\left(\eta + \frac{f_{+-} - f_{--}}{\rho}\right) - \frac{f_{++}f_{--} - f_{-+}f_{+-}}{\rho^2} \right)$$

if $\rho := f_{++} - f_{+-} - f_{-+} + f_{--} \neq 0$ and

$$f(p) = (f_{+-} - f_{--})\xi + (f_{-+} - f_{--})\eta + f_{--}$$

if $\rho = 0$. In this formula, f_{++}, f_{+-}, f_{-+} and f_{--} are the values of f at the vertices of the cell.

We will approximate the intersection by line segments intersecting the edges of the cell. There can be 0, 1 or 2 such line segments, and we now discuss when these situations occur. An important remark is that, because the bilinear interpolation is linear when restricted to the edges of the cell, there is at most one intersection of the set $(f = 0)$ and each edge, and this is only possible when f takes different signs at each of the edge end-points. When this occurs, the points on the edges at which $f = 0$ can be easily computed by solving a linear equation. They will form the vertices of the polygonal line.

(a) If all f_{++}, f_{+-}, f_{-+} and f_{--} have the same sign: there is no intersection with the edges, and therefore no intersection with the cell.

(b) Three of the values have the same sign, the last one having the opposite sign: there are two vertices in the cell, and one edge connecting them.

(c) Two values have the same sign on one edge and two have the opposite sign on the opposite edge. Here also, there are two vertices and one edge.

(d) The function changes sign on all the edges. There are four vertices and two edges. There are two subcases, letting $\delta = f_{++}f_{--} - f_{-+}f_{+-}$.

    (i) If $\delta > 0$, then one edge links the vertex on $(\xi = 0)$ to the one on $(\eta = 1)$, and the other the vertex on $(\eta = 0)$ to the one on $(\xi = 1)$.

(ii) If $\delta < 0$, then one edge links the vertex on ($\xi = 0$) to the one on ($\eta = 0$), and the other the vertex on ($\eta = 1$) to the one on ($\xi = 1$).

Cases (a), (b) and (c) can be decided based on the signs of f only. Case (d) is called ambiguous because it requires the exact numerical values of f. There are a few additional exceptional cases that are left aside in this discussion. When $f = 0$ at one of the vertices of the cell, this vertex also is in the polygonal line. It connects to other vertices at opposite edges of the cell, unless one of the cell edges that contain it is included in the polygon. There is no ambiguous situation in that case.

Case (d) with $\delta = 0$ is more of a problem, because it corresponds to a situation in which the interpolated surface is the intersection of of two lines and therefore has a singular point. One cannot lift this ambiguity, and one of the options (i) and (ii) should be selected. The selection cannot be completely arbitrary because this could create holes in the reconstructed polygons. One possible rule is to take one option (say (i)) when $\rho > 0$ and the other one when $\rho < 0$. The combination of case (d) and $\delta = \rho = 0$ implies that $f = 0$ at all vertices of the cell which therefore should be in the final polygon, but there is an unsolvable ambiguity on how they should be connected.

There is another way to handle case (d), disregarding δ, based, as we just discussed, on the sign of ρ, yielding

(d)' In case (d) above, take solution (i) if $\rho > 0$ and (ii) otherwise.

The resulting algorithm is simpler, because, in case (d), the sign of ρ can be computed directly based on the signs of f on the vertices of the cell. It does not correspond to the bilinear approximation anymore, but this approximation was somewhat arbitrary anyway. It does break the symmetry of the solution, in the sense that, if f is replaced by $-f$, the isocontours computed using (d)' will differ. This is illustrated in Figure 5.1.

In addition to allowing for the segmentation of specific shapes from images, when the interior of the shape is, say, darker than its exterior, isocontours have been used as basic components of image processing algorithms that are contrast invariant in the sense of mathematical morphology. A good introduction to this and to the related literature can be found in [38, 39].

Finally, let's note that isocontours can be easily oriented in accordance with our convention for implicit contours, by simply ensuring that grid points with negative values of f lie on the left of each oriented edge.

5.2 Computing Isosurfaces

We now pass to the case of level sets for functions defined over three dimensions, and describe the construction of triangulated isosurfaces. Although the problem is in principle similar to the two-dimensional case, the solution is notably more complex, mainly because of the large number of ambiguous

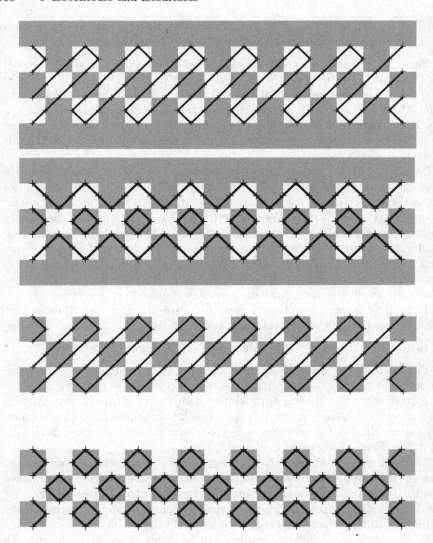

Fig. 5.1. Isocontouring a checkerboard strip using exact bilinear rule (d) (first row) and sign-based rule (d)' (second row). Note that the solutions are different, although both are possible isocontours for the image. Gray levels are switched in the last two rows, without changing the solution for rule (d) (third row) and significantly altering it for rule (d)' (fourth row), yielding a third plausible solution.

situations in the determination of the boundary. There is indeed a large literature on the subject, and the reader can refer (for example) to [26] for a recent bibliography.

The three-dimensional generalization of the algorithm that we have presented for isocontouring is called *marching cubes* [130], and progressively builds a triangulation by exploring every grid cell on which the function changes sign. We will use notation similar to the previous section, and let $\mathcal{G}$ be a regular three-dimensional grid, with grid coordinates $p(s, t, u) = (x_s, y_t, z_u)$ where $s = 1, \ldots, M, t = 1, \ldots, N, u = 1, \ldots, P$. Denote by $f_{stu} = f(p(s, t, u))$ the observed values of f on the grid. Like in two dimensions, we assume that f extends to the continuum with a tri-linear interpolation as follows: Let $C(s, t, u)$ denote the cube (cell) with vertices $p(s + \varepsilon_1, t + \varepsilon_2, q + \varepsilon_3), \varepsilon_i \in \{0, 1\}, i = 1, 2, 3$. Then, for $p = x, y, z \in C(s, t, u)$, let

$$f(p) = \sum_{\varepsilon_1, \varepsilon_2, \varepsilon_3 = 0}^{2} \prod_{i=1}^{3} (\varepsilon_i r_i(p) + (1 - \varepsilon_i)(1 - r_i(p))) f_{s+\varepsilon_1, t+\varepsilon_2, q+\varepsilon_3}$$

with $r_1(p) = x - x_s, r_2(p) = y - y_t, r_3(p) = z - z_u$.

The determination of the vertices of the triangulation is like in two dimensions: the intersections of the level set $f = 0$ and the edges of the cubes $C(s, t, u)$ can be computed by solving a simple linear equation; it exists only if f takes different signs at the end-points of the edges, and there can be at most one intersection. The difficulty is how to group these vertices into faces that provide a topologically consistent triangulation.

The main contribution of the marching cubes algorithm is to provide a method in which each cube is considered independently, yielding a reasonably simple implementation. The method works by inspection of the signs of f at the eight vertices of the cube. Like in two dimensions, there are some easy cases. The simplest is when all signs are the same, in which case the triangulation has no node on the cube. Other simple configurations are when the cube vertices of positive sign do not separate the other vertices in two or more regions and vice-versa. In this case, the triangulation has to separate the cube into two parts. There are, up to sign and space symmetry and up to rotation, six such cases, which are provided in Figure 5.2.

Such triangulations can be efficiently described by labeling the vertices and the edges of the cube, like in Figure 5.3. We can describe a sign configuration on the cube by listing the vertices which have a positive sign. We can also describe each triangulation by listing, for each triangle, the three edges it intersects. Figure 5.2 therefore describes the six triangulations

$\{1\} : [(1, 4, 9)]$
$\{1, 2\} : [(2, 4, 9), (2, 4, 10)]$
$\{2, 5, 6\} : [(1, 2, 9), (2, 8, 9), (2, 8, 6)]$
$\{1, 2, 5, 6\} : [(2, 6, 4), (4, 6, 8)]$
$\{2, 3, 4, 7\} : [(1, 10, 6), (1, 6, 7), (1, 7, 4), (4, 7, 12)]$
$\{1, 5, 6, 7\} : [(1, 10, 11), (1, 11, 8), (8, 11, 7), (4, 1, 8)]$

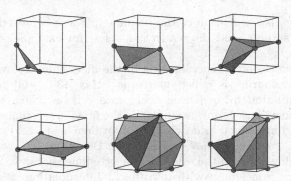

Fig. 5.2. Two-component (non-ambiguous) cases for the marching cubes algorithm.

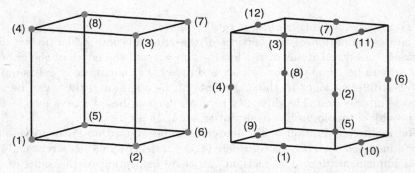

Fig. 5.3. Labels for the vertices (left) and edges (right) of the cube

Like in two dimensions, the cases when the signs form more than two connected components on the cube are problematic. They are ambiguous, because the way the surface crosses the tube cannot be decided from the sign pattern alone. One needs to rely on more information (i.e., the actual values of f at the nodes) to decide how to triangulate the surface within the cube, in order to avoid creating topological inconsistencies.

Take, for example, the case in which the cube vertices labeled (1) and (3) have signs distinct from the rest. Then, there are two possible ways (described in Figure 5.4) in which the surface can cross the cube.

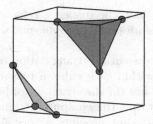
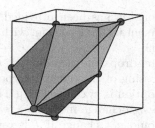

Fig. 5.4. Two triangulations associated to the {3, 8} sign configuration
[(2, 3, 11), (1, 4, 9)] and [(4, 9, 3), (3, 9, 11), (9, 11, 2), (1, 2, 9)].

Another kind of ambiguous configuration comes when two vertices in two opposite corners are isolated from the rest. Consider, for example, the situation when vertices 1 and 7 are positive while the rest are negative. Then the surface can do two things: either cut out the corners of the cube, or create a tunnel within the cube (see Figure 5.5).

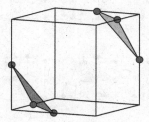
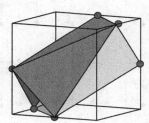

Fig. 5.5. Two triangulations associated to the {1, 7} sign configuration
[(1, 4, 9), (7, 6, 11)] and [(1, 4, 11),(1, 11, 6), (1, 9, 6), (9, 6, 7), (9, 4, 7), (4, 11, 7)].

There have been successive attempts to improve the marching cubes algorithm from its original version ([130], in which the discussion was incomplete) [157, 155, 209, 151, 45] and untying the ambiguous cases. In addition to the two cases described in Figures 5.4 and 5.5, five other ambiguous sign configurations can be listed, arising from combinations of these two basic cases. A complete description of all possible cases has been provided in [45], together with disambiguation rules. An extensive theoretical and numerical analysis of the algorithm has been provided in [156] to which the reader is referred for complementary information, with the listing of all possible topologies within the cube.

If one drops the requirement to provide an accurate triangulation of the zero-crossing of the linear interpolation of f within each cube, a reasonably simple option is available [151]. This approach has the disadvantage of breaking the sign-change invariance (which ensures that the computed triangulation should not change if f is replaced by $-f$), but provides a very simple algorithm, still based on the signs of f on the vertices (it can be seen as a generalization of (d)' in our discussion of the two-dimensional case). This results in 23 different cases (up to rotation invariance), listed in Figure 5.6. This had to be compared to the 15 cases initially proposed in [130], which was invariant by sign change, but created topological errors.

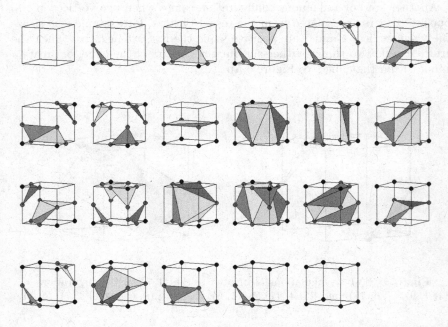

Fig. 5.6. Twenty-three configurations for consistent within-cube triangulation based on vertex signs. Dotted vertices correspond to positive values of the function.

An alternative to the marching cubes algorithm replaces cubic cells by tetrahedrons before computing the triangulation, which, when properly handled [42], provides a simpler and stabler procedure.

Extracting surfaces as level sets of functions is important even when the original data is not a three-dimensional image from which the region of interest is an isosurface. For example, when the original data is a set of unstructured points that roughly belong to the surface (i.e., they are subject to small errors) some of the commonly used algorithms that reconstruct the surface first reduce to the isosurface problem, trying to infer the signed distance function to the surface, at least in a neighborhood of the observed points. The approach used in [116] is to first approximate the tangent plane to the surface and then build the signed distance function. A similar goal is pursued in [6], using an approach based on computational topology.

Marching cubes (or tetrahedrons) have the drawback of providing a very large number of triangles, sometimes with very acute angles. Simplifying meshes is also the subject of a large literature, but this will not be addressed here (see, for example [74]).

6

Evolving Curves and Surfaces

In this chapter, we discuss how to represent curve or surface evolution using partial differential equations. This connects to fundamental mathematical results, some of them beyond the scope of this book, but also has important practical implications, especially when implementing optimization algorithms over curves and surfaces. One important example will be active contours, which will be described in the next chapter.

6.1 Curve Evolution

Evolution for curves is somewhat simpler to address than for surfaces, because global parametrizations of a closed or open contour are always available. We consider, in this section, curves that depend on time, which will be denoted by t, the curve parameter being denoted by u (or s for arc length). A time-dependent curve is a function of two variables

$$m : [0, \Delta] \times [0, 1] \to \mathbb{R}^2$$
$$(t, u) \quad \mapsto m(t, u).$$

We therefore assume that the domain over which the curve is parametrized (the interval $[0, \Delta]$) is fixed. The curve at time t will be denoted $m^t : u \mapsto m(t, u)$. Its length will be denoted L^t, and the arc length $s^t : [0, 1] \to [0, L^t]$. The curvature at a point p will be $\kappa^t(p)$; $T^t(p)$ and $N^t(p)$ will be the unit tangent and normals at p. Exponents t must not be confused with subscripts t which refer to differentiation in time.

We consider curves evolving according to differential equations of the kind:

$$\partial_t m(t, u) = A(t, u) T^t(m(t, u)) + B(t, u) N^t(m(t, u)). \tag{6.1}$$

In this equation, $A(t, u)$ and $B(t, u)$ depend on the curve at time t and are scalar functions that depend on the parameter u. Most of the time, they will involve local properties of the curve at u (like the curvature).

The decomposition of the evolution into tangent and normal terms is useful, because each of them is associated to different properties of the evolution. The normal term is directly related to the geometric evolution of the curve, as implied by the following lemma:

Lemma 6.1 ([75]). *Assume that m is C^2 in space, C^1 in time, regular at all times $t \in [0, \Delta]$ and satisfies the equation*

$$\partial_t m = AT + BN.$$

Then, there exists a time-dependent change of parameter on m, denoted ψ, such $\tilde{m}(t, u) := m(t, \psi(t, u))$ is solution of

$$\partial_t \tilde{m} = \tilde{B} N$$

with $\tilde{B}(t, u) = B(t, \psi(t, u))$.

Proof. Let $u \mapsto \psi(t, u)$ be as in the lemma. The evolution of $\tilde{m}(t, u) = m(t, \psi(t, u))$ is

$$\partial_t \tilde{m}(t, u) = \partial_t m(t, \psi(t, u)) + (\partial_t \psi) \dot{m}_u(t, \psi(t, u))$$
$$= (A(t, \psi(t, u)) + \partial_t \psi |\dot{m}_u|)T + B(t, \psi(t, u))N.$$

We therefore need to show that there exists ψ such that

$$A(t, \psi) + \partial_t \psi |\dot{m}_u|(t, \psi) = 0.$$

This results from the general theory of ordinary differential equations (cf. Appendix C). For fixed u, let $\xi(t) = \psi(t, u)$. It must satisfy the ordinary differential equation

$$\partial_t \xi = -A(t, \xi)/|\dot{m}_u|(t, \xi).$$

Existence and uniqueness of the solution is ensured by the fact that A and $\dot{m}_u$ are C^1. For $\dot{m}_u$, this is true because m is assumed to be C^2 in space, and for A, it suffices to remark that $A = \dot{m}_t^T T$ and therefore is C^1 with respect to u.   □

This lemma implies that the evolution of the curve is essentially captured by the function B, where A is only involved in changes of parameters.

Very often, the curve variation at time t only depends on the curve at the same time, so that there exist transformations $m \mapsto (\alpha_m, \beta_m)$ with $\alpha_m, \beta_m : [0, \Delta] \to \mathbb{R}$ such that

$$A(t, u) = \alpha_{m^t}(u) \text{ and } B(t, u) = \beta_{m^t}(u).$$

(α and β could also be made to depend on time.)

As usual, we say that these functions are geometric if they are invariant under changes of parameter, so that

$$\beta_{m\circ\psi} \circ \psi = \beta_m$$

for any change of parameter ψ (and similarly for α, although this is of less interest). In view of Lemma 6.1, we see that if β_m is geometric, the evolution

$$\partial_t m = \alpha_m T + \beta_m N$$

can be transformed, after a change of parameter, to

$$\partial_t \tilde{m} = \beta_{\tilde{m}} N.$$

We now discuss the evolution of contours and domain integrals associated to curves evolving according to (6.1). The results are summarized in the following proposition.

Proposition 6.2. *Let* $V : \mathbb{R}^2 \to \mathbb{R}$ *be a smooth function, and assume that* $m : (t.u) \mapsto m(t, u)$ *satisfies* (6.1), *has continuous partial derivatives* $\ddot{m}_{tu}$, *and is such that* $u \mapsto m(t, u)$ *is a* C^2 *regular curve for all* $t \in [0, t_0]$. *Define*

$$F(t) = \int_{m^t} V dl = \int_0^{\Delta} V(m(t, u)) |\dot{m}_u| du.$$

Then

$$\partial_t F = [V(m) \langle \partial_t m, T^t \rangle]_0^{\Delta} + \int_{m^t} (\langle \nabla V, N^t \rangle - \kappa^t) \langle N^t, \partial_t m \rangle dl \qquad (6.2)$$

(the first term vanishing if the curve is closed). Here κ^t *is the curvature of* $m(t, \cdot)$.

Assuming that $m(t, .)$ *is simple (and closed) for* $t \in [0, t_0]$, *and letting* Ω_t *denote its interior, define*

$$G(t) = \int_{\Omega_t} V(x) dx.$$

Then

$$\partial_t G = - \int_{m^t} V \langle N^t, \partial_t m \rangle dl. \qquad (6.3)$$

Finally, let $W : \mathbb{R}^2 \to \mathbb{R}^2$ *be a smooth vector field, and*

$$H(t) = \int_{m^t} \langle W, N^t \rangle dl$$

where m^t *can be open or closed. Then*

$$\partial_t H = -[\det(W(m^t), \partial_t m)]_0^{\Delta} + \int_{m^t} \mathrm{div}(W) \langle N^t, \partial_t m \rangle dl. \qquad (6.4)$$

Proof. First consider F. We have $\partial_t |\dot{\mu}_u|^2 = 2\langle \ddot{m}_{tu}, \dot{m}_u \rangle$ and since this is also equal to $2 |\dot{m}_u| \partial_t |\dot{m}_u|$, we can write

$$\partial_t F = \int_0^\Delta \langle \nabla V(m), \partial_t m \rangle |\dot{m}_u| du + \int_0^\Delta V(m) \langle \ddot{m}_{tu}, T \rangle du$$

$$= \int_0^\Delta \langle \nabla V(m), \dot{m}_t \rangle |\dot{m}_u| du + [V(m) \langle \dot{m}_t, T \rangle]_0^\Delta$$

$$\quad - \int_0^\Delta (\langle \nabla V(m), \dot{m}_u \rangle \langle T, \dot{m}_t \rangle + V(m) \langle \dot{m}_t, \dot{T}_u \rangle) du$$

$$= [V(m) \langle \dot{m}_t, T \rangle]_0^\Delta + \int_0^1 \langle \nabla V(m) - \langle \nabla V(m), T \rangle T, \dot{m}_t \rangle |\dot{m}_u| du$$

$$\quad - \int_0^1 \kappa V(m) \langle \dot{m}_t, N \rangle |\dot{m}_u| du$$

$$= [\langle \dot{m}_t, T \rangle]_0^\Delta + \int_m \langle \langle \nabla V(m), N \rangle N - \kappa N, \dot{m}_t \rangle dl.$$

This proves (6.2). We now prove (6.3) and (6.4), first remarking that the former is a consequence of the latter. Indeed, introduce φ such that $V = \text{div}\varphi$, taking, for example,

$$2\varphi(x_1, x_2) = \left(\int^{x_1} V(x_1', x_2) dx_1', \int^{x_2} V(x_1, x_2') dx_2' \right).$$

Then, from the divergence theorem

$$G(t) = - \int_{m^t} \langle \varphi(m^t), N^t \rangle dl$$

so that (6.3) is deduced from (6.4) and the fact that m is assumed to be closed.

For (6.4), we can write

$$H(t) = - \int_0^\Delta \det(W(m), \dot{m}_u) du$$

so that

$$\partial_t H = - \int_0^\Delta \det(DW(m) \partial_t m, \dot{m}_u) du - \int_0^\Delta \det(W(m), \partial_t \dot{m}_u) du$$

$$= - \int_0^\Delta \det(DW(m) \partial_t m, \dot{m}_u) du - [\det(W(m), \partial_t m)]_0^\Delta$$

$$\quad + \int_0^\Delta \det(DW(m) \dot{m}_u, \partial_t m) du$$

and the conclusion comes from the identity, true for any 2 by 2 matrix A and vectors e_1, e_2,

$$\det(A e_1, e_2) + \det(e_1, A e_2) = \text{trace}(A) \det(e_1, e_2)$$

applied to $A = DW(m)$, $e_1 = \partial_t m$ and $e_2 = \dot{m}_u$. □

6.1.1 Grassfires

As a first example of (6.1), let's consider the simplest case for which $A = 0$ and $B = 1$. This corresponds to

$$\partial_t m = N \tag{6.5}$$

which (taking as usual the inward normal) means that m shrinks towards its interior at unit speed. Such evolutions are often called *grassfires* because the evolving curve would look like the boundary of a lawn around which fire is set at time 0, with the fire propagating inward. They are closely related to medial axes (defined in chapter 2), because the skeleton is the location at which two separate fronts meet.

It is quite easy to study (6.5) and prove that solutions exist in small time when starting with a smooth curve, but that singularities are developed in finite time.

First, let's assume that a solution is given, in the form $m(t, u)$ for $t \in [0, t_0]$, $m(0, .)$ being a simple closed curve, and $m(t, \cdot)$ being regular and C^2 for all t in this interval. We first prove that the normals remain unchanged: $N(t, u) = N(0, u)$ for all $u \in [0, \Delta]$. Since $|N| = 1$ at all times, we have $\dot{N}_t^T N = 0$. Moreover

$$\ddot{m}_{tu} = \dot{N}_u = -|\dot{m}_u| \kappa T = -\kappa \dot{m}_u.$$

This implies

$$0 = \partial_t(\dot{m}_u^T N) = \ddot{m}_{tu}^T N + \dot{m}_u^T \dot{N}_t = \dot{m}_u^T \dot{N}_t$$

Therefore, $\dot{N}_t = 0$, since it is perpendicular to both the normal and the tangent; so the normal is constant over time. Given this, the integration of the evolution equation is straightforward and yields

$$m(t, u) = m(0, u) + tN(0, u) \tag{6.6}$$

and it is easy to show that this provides a solution of (6.5).

In this computation, we have used the assumption that m is smooth (in u) for all $t \in [0, t_0]$. We now check that this fails to be true in finite time, whatever the initial curve is. Indeed, as long as the computation is valid, we have, computing the derivative of (6.6) (assuming that the curve is initially parametrized with arc length)

$$\dot{m}_u(t, u) = (1 - t\kappa(0, u))T(0, u).$$

In particular, if $\kappa(0, u) > 0$, then, for $t = 1/\kappa(0, u)$, we have $\dot{m}_u = 0$, and the curve is not regular anymore (the previous discussion becomes invalid at this point). Note that there must exist points of positive curvature on the curve, since, for simple positively oriented curves, the integral of the curvature (the rotation index) is 2π. The curvature for small t can be computed using

$$\dot{T}_u(t, u) = |\dot{m}_u| \kappa(t, u)N(t, u).$$

But, since N is constant in time, so is T, which implies

$$\dot{T}_u(t,u) = \dot{T}_u(0,u) = \kappa(0,u)N(0,u).$$

Therefore

$$\kappa(t,u) = \frac{\kappa(0,u)}{1 - t\kappa(0,u)}.$$

This implies that the curvature tends to infinity at the point u_0 of highest curvature in the initial curve, when t tends to $t_0 = 1/\kappa(0,u_0)$.

Even after t_0, we can still define a curve m using (6.6). A very detailed description of what happens immediately after t_0 in the neighborhood of u_0 can be done: this is part of the theory of singularities, and it can be shown that the curve crosses itself, the singularity at u_0 forking into two new singularities, providing a shape called "swallow tail" (see Figure 6.1). There are other types of singularity which can be created over time; for example, non-contiguous arcs of the original curve may meet and the region split into two parts (see second example in Figure 6.1), creating two new singularities that will evolve.

Fig. 6.1. Curve evolution with the grassfire equation. Left: evolution according to (6.6). Right: same evolution after removing the burned-out parts.

Returning to the grassfire analogy, however, it makes sense to require that grass which is already burned does not burn again. So, in the grassfire model, the swallow tail part, and other curve portions after self-intersection should not be included in the evolving curve (Figure 6.1). An important remark is that both evolutions can be seen as solutions of the original equation (6.5). Its solutions are therefore not uniquely defined once singularities appear.

The location of the points of self-intersection (the first one being at u_0) is interesting since these points belong to the medial axis. (Note that there will be several such points at larger times, each one starting at a local maximum of the curvature.) Tracking them over time provides a plausible computation of the skeleton [192, 191, 190, 200].

6.1.2 Curvature Motion

We now take, in (6.1), $A = 0$ and $B = \kappa^t$, the curvature of m^t. This gives the equation:

$$\partial_t m = \kappa^t(m(t, u))N^t(m(t, u)). \tag{6.7}$$

Note that, because κ and N change signs together when the orientation is changed, this evolution does not depend on the orientation (the previous one did). The following theorem provides a detailed description of how a curve evolves under this equation.

Theorem 6.3 ([88], [100]). *Assume that $u \mapsto m_0(u)$ is a regular, C^2 closed simple curve in $\mathbb{R}^2$. Then the previous equation has a solution over an interval $[0, t_0]$. The curve remains simple during the evolution. It first becomes convex, then shrinks to a point while its shape becomes circular.*

We will not prove this theorem (the proof is quite involved). The interested reader can refer to [47] for a proof and more results on related evolution. The following simple remarks can help, however, in understanding why such a result holds.

The first of these remarks is that (6.7) can be interpreted as gradient descent for the function $m \mapsto L(m) = \text{length}(m)$. Indeed, applying Proposition 6.2 with $V = 1$ yields (for any evolving curve $\varepsilon \mapsto \mu(\varepsilon, .)$ with $\mu(0, .) = m$)

$$\partial_\varepsilon L(\mu(\varepsilon, .)) = - \int_{\mu^\varepsilon} \kappa^{\mu^\varepsilon} \langle \dot\mu_\varepsilon, N^{\mu^\varepsilon} \rangle dl.$$

Taking this at $\varepsilon = 0$, we see that $(-\kappa N)$ is the gradient of the length functional for the metric (cf. Appendix D)

$$\langle h, \tilde h \rangle_m = \int_m \langle h, \tilde h \rangle dl.$$

This implies that the length decreases over time (curvature motion is also called the curve-shortening flow) and we have in fact (if m satisfies (6.7), and dropping the time superscript)

$$\partial_t L(m) = -\int_0^1 \kappa^2 |\dot{m}_u| du = -\int_0^L \kappa^2 ds.$$

The variation of the enclosed area is also interesting to compute. Letting $A(t)$ denote the enclosed area of $m(t, .)$, we can again apply Proposition 6.2 with $V = 1$ to obtain

$$\partial_t A = -\int_{m^t} \langle N^t, \kappa^t N^t \rangle dl = -\int_{m^t} \kappa^t dl = -2\pi.$$

The area of the curve therefore decreases at constant speed, 2π. This also shows that the curve disappears in finite time (initial area divided by 2π).

The last interesting quantity is the isoperimetric ratio, which is given by $r = A/L^2$. The isoperimetric inequality (1.7) states that this ratio is always smaller than $1/(4\pi)$, and equal to this value only if the curve is a circle. We have

$$\dot{r}_t = \frac{1}{L^3}\left(L\dot{A}_t - 2A\dot{L}_t\right)$$

$$= \frac{2A}{L^3}\left(-\pi\frac{L}{A} + \int_0^L \kappa^2 ds\right).$$

Such a quantity has been analyzed in [87], in which it is shown that

$$\pi\frac{L}{A} \leq \int_0^L \kappa^2 ds$$

as soon as the curve is convex. When this is true, this implies that the isoperimetric ratio increases during the evolution. Since the ratio is at a maximum for circles, this explains why the curve becomes circular (this does not explain while the curve becomes convex, which is the most difficult part of theorem 6.3).

The interesting aspect of such a flow is that it can be seen as a tool for smoothing curves, since it asymptotically provides a circle. However, this smoothing is combined with an asymptotic reduction to a dot which is a somewhat unwanted behavior. One way to deal with this is to simply let the curve evolve over a time t and rescale it to its original area. The evolution can also be compensated in real time so that the area remains constant: it suffices to use the equation

$$\dot{m}_t = (\kappa - 2\pi/L)N. \tag{6.8}$$

For such curves, the previous computations show that the area satisfies

$$\dot{A}_t = -\int_0^1 (\kappa - 2\pi/L)|\dot{m}_u| du = 0$$

and for the length

$$\dot{L}_t = -\int_0^1 \kappa(\kappa - 2\pi/L)|\dot{m}_u|du$$

$$= -\int_0^1 (\kappa - 2\pi/L)^2 |\dot{m}_u|du.$$

So the length of the evolving curve decreases, unless κ is constant (equal to $2\pi/L$), in which case the considered curve is a circle. For the isoperimetric ratio, we have

$$\dot{r}_t = \frac{1}{L^3}(L\dot{A}_t - 2A\dot{L}_t)$$

$$= \frac{2A}{L^3}\int_0^L (\kappa - 2\pi/L)^2 ds.$$

This therefore also increases unless the curve is a circle.

6.1.3 Implicit Representation of the Curvature Motion

Equation (6.7) may look simple when expressed in terms of geometric quantities, but it is a rather complicated partial differential equation when seen in a fixed parametrization, since it can be rewritten

$$\partial_t m = \frac{\ddot{m}_{uu}}{|\dot{m}_u|^2} - \frac{\ddot{m}_{uu}^T \dot{m}_u}{|\dot{m}_u|^4}\dot{m}_u. \qquad (6.9)$$

The direct numerical implementation (by finite differences) of this equation is somewhat unstable (it clearly involves divisions by very small numbers). One obtains a much stabler algorithm if an implicit parametrization is used [184, 183, 163].

For this purpose, assume that the initial curve is the zero level set of a smooth function f^0, so that

$$\mathcal{R}_{m^0} = \left\{ p \in \mathbb{R}^2, f^0(p) = 0 \right\}.$$

The principle of an implicit implementation is to make f^0 evolve over time so that its zero level set evolves according to (6.7).

Introduce a time-dependent function $(t, p) \mapsto f(t, p)$, with $f(0, .) = f^0$. In order that the zero level set of $f(t, .)$ coincides with the curve $m(t, .)$ defined above, we need $f(t, m(t, u)) = 0$ for all t and all $u \in [0, 1]$, which implies, after differentiation:

$$\partial_t f + \nabla f(m)^T \dot{m}_t = 0.$$

Using $\nabla f(m) = -|\nabla f(m)|N$, $\dot{m}_t = \kappa N$ and formula (1.24) for the curvature, this yields the equation

$$\partial_t f = |\nabla f| \mathrm{div} \frac{\nabla f}{|\nabla f|}. \qquad (6.10)$$

This equation (which is an anisotropic diffusion) is very stable and easy to implement (see next section). Figure 6.2 provides some examples of evolving curves. This equation has also proved itself important in image processing as a way to smooth images while preserving edges (see [76, 4, 5]).

Fig. 6.2. Curves evolving according to the curve-shortening flow. *First row:* three examples of superimposed evolving shapes. *Second row:* details of the evolution of the third shape (spiral) in first row.

6.1.4 More on the Implementation

We now describe a numerical implementation of equation (6.10). It is the simplest one, although not the most efficient (see [183]).

Initializing the Process

We must initialize the function f^0 so that its zero level set coincides with the initial curve $m(0,.)$. A very simple solution to this is provided by the signed distance function $f^0(m) = \varepsilon d(m, m(0,.))$ with $\varepsilon = 1$ outside $m(0,.)$ and $\varepsilon = -1$ inside $m(0,.)$. The distance function can be computed efficiently using the algorithm described in Section F.4 of Appendix F.

The determination of the points belonging to the interior of the initial curve can be done using standard algorithms in computational geometry ([166]). The following algorithm is applicable when the curve is discretized finely enough so that no hole exists in its outline and it does not meet the boundary of the image. Pixels belonging to the curve are assumed to be labeled with 0. Let the initial number of labels be Nlab = 1.

- Label the first column and the first row of the image as 1

- *First scan:* Scan each row, labeling each pixel like its predecessor in the row, unless this predecessor is a zero. In this case, look at the predecessor in the column: if it has a non-zero label, use this label, otherwise, create a new label, (Nlab + 1), to label the pixel, and add 1 to Nlab.
- *Second scan:* The previous scan results in over-segmented regions. The last step consists in merging labels by running a backward scan (starting from the last pixel of the last row). Two labels are merged when they are attributed to two neighbor pixels (not separated by a zero).

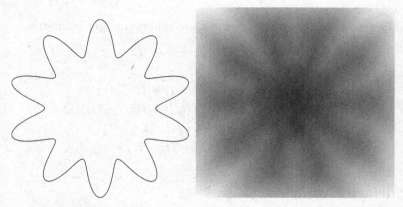

Fig. 6.3. Signed distance map. *Left:* original curve. *Right:* associated signed distance map.

Finite-Difference Scheme

We now describe how (6.10) can be discretized. Let δt be the time step. The time-discretized evolution algorithm is

$$f^{(n+1)} = f^{(n)} + \delta t \left| \nabla f^{(n)} \right| \operatorname{div} \left(\frac{\nabla f^{(n)}}{\left| \nabla f^{(n)} \right|} \right).$$

δt is the "time" discretization step.

We need to compute finite-difference approximations of the derivatives, involving a "space" discretization step δx. We first compute an alternative expression for $\operatorname{div} \left(\frac{\nabla f}{|\nabla f|} \right)$. We must compute:

$$\partial_x \left(\frac{\dot{f}_x}{\sqrt{\dot{f}_x^2 + \dot{f}_y^2}} \right) + \partial_y \left(\frac{\dot{f}_y}{\sqrt{\dot{f}_x^2 + \dot{f}_y^2}} \right).$$

This is:

$$\frac{\ddot{f}_{xx}}{\sqrt{\dot{f}_x^2 + \dot{f}_y^2}} - \frac{\dot{f}_x^2 \ddot{f}_{xx} + \dot{f}_x \dot{f}_y \ddot{f}_{xy}}{(\dot{f}_x^2 + \dot{f}_y^2)^{3/2}}$$

$$+ \frac{\ddot{f}_{yy}}{\sqrt{\dot{f}_x^2 + \dot{f}_y^2}} - \frac{\dot{f}_y^2 \ddot{f}_{yy} + \dot{f}_x \dot{f}_y \ddot{f}_{xy}}{(\dot{f}_x^2 + \dot{f}_y^2)^{3/2}}$$

$$= \frac{\ddot{f}_{xx} \dot{f}_y^2 - 2\dot{f}_x \dot{f}_y \ddot{f}_{xy} + \ddot{f}_{yy} \dot{f}_x^2}{(\dot{f}_x^2 + \dot{f}_y^2)^{3/2}}.$$

To discretize the derivatives, one can use symmetric approximations:

$$f_x(i,j) = (f(i+1,j) - f(i-1,j))/2$$
$$f_y(i,j) = (f(i,j+1) - f(i,j-1))/2$$
$$f_{xx}(i,j) = (f(i+1,j) - 2f(i,j) + f(i-1,j))$$
$$f_{yy}(i,j) = (f(i,j+1) - 2f(i,j) + f(i,j-1))$$
$$f_{xy}(i,j) = (f(i+1,j+1) - f(i-1,j+1)$$
$$- f(i+1,j-1) + f(i-1,j-1))/4.$$

6.2 Surface Evolution

Describing surfaces that evolve over time via a differential equation is not as simple as with curves, because surfaces are, to start with, harder to describe than curves. We can certainly describe the evolution of parametrized surfaces as we did with curves, writing

$$\partial_t m(t,u,v) = \tau(t,u,v) + B(t,u,v)N^t(u,v) \tag{6.11}$$

where $\tau(t,u,v)$ is tangent to m^t at $m(t,u,v)$ (i.e. in the plane generated by $\partial_u m$ and $\partial_v m$ and B is a scalar function).

For a general surface, the simplest way to proceed is to parametrize the evolving surfaces using another regular surface, typically the surface at time $t = 0$. Letting S_t be the surface a time t, we assume that, for each time t, there exists a function $q \mapsto \mu(t,q)$, defined over the initial surface S_0 (or any other reference regular surface) such that $S_t = \mu(t, S_0)$ and, for all $q \in S_0$, $D\mu(q)$ has rank 2 as a map from $T_q S_0$ to $\mathbb{R}^2$.

The set S_t might not be a regular surface (it can have multiple points, for example) so that the tangent planes and normal to S_t at $\mu(t,q)$ may not be defined. However, for any $q \in S_0$, one can find a parametrized region in S_0 around q (with parametrization $(u,v) = m(u,v)$) which maps into a regular parametrized patch in S_t which has a well-defined tangent plane and normal, that will be respectively denoted L_q^t and N_q^t. We can therefore generalize (6.11) by considering surface evolutions of the kind

$$\partial_t \mu(t, q) = \tau(t, q) + B(t, q) N_q^t \tag{6.12}$$

with $\dot{\tau}(t, q) \in L_q^t$.

If $m_0 : (u, v) \to m_0(u, v)$ is a local chart at q, we get a local parametrization

$$m(t, u, v) = \mu(t, m_0(u, v)) \tag{6.13}$$

on S_t, with

$$\partial_t m(t, u, v) = \tau(t, m_0(u, v)) + B(t, m_0(u, v)) N^t(m_0(u, v)).$$

If S_t happens to be a regular surface at all times, then $L_q^t = T_{\mu(t,q)} S_t$, and we can think of (6.12) as an equation of the form

$$\partial_t p = \tilde{\tau}(t, p) + \tilde{B}(t, p) N^t(p) \tag{6.14}$$

with $\tilde{\tau}(t, p) \in T_p S^t$.

Like with curves, the geometric evolution is only captured by B. Let's sketch the justification of this fact. If μ evolves according to (6.12), and $q \mapsto \varphi(t, q)$ is a global change of variable of S_0 (a C^1 invertible map from S_0 to S_0 with a C^1 inverse), then we can consider $\tilde{\mu}(t, q) = \mu(t, \varphi(t, q))$. Then

$$\partial_t \tilde{\mu} = D\mu \circ \varphi \, \partial_t \varphi + \tau \circ \varphi + B \circ \varphi N^t \circ \varphi.$$

So, if

$$\partial_t \varphi = -(D\mu^{-1}\tau) \circ \varphi \tag{6.15}$$

then $\tilde{\mu}$ follows (6.12) with $\tau = 0$. But (6.15) defines the flow associated to an ordinary differential equation (cf. Appendix C) on S_0 and therefore provides a change of parameter.

We have the equivalent of Proposition 6.2:

Proposition 6.4. *Let $V : \mathbb{R}^3 \to \mathbb{R}$ be a smooth function, and assume that a time-dependent surface S^t is defined by (6.12), and admits C^2 local parametrizations which are such that $\dot{m}_u$ and $\dot{m}_v$ are C^1 in time. Define*

$$F(t) = \int_{S^t} V \, d\sigma^t$$

where σ^t is the area form on S^t. Then

$$\partial_t F = -\int_{\partial S^t} V \langle n^t, \partial_t \mu \rangle dl$$

$$+ \int_{S^t} \left(-2VH^t + \langle \nabla V, N^t \rangle \right) \langle N^t, \partial_t \mu \rangle d\sigma^t, \tag{6.16}$$

where H^t is the mean curvature on S^t and n^t is the inward normal to $\partial_t S^t$.

Assuming that S^t coincides with the boundary of an open subset $\Omega_t \subset \mathbb{R}^3$, define

$$G(t) = \int_{\Omega_t} V(x)dx.$$

Then

$$\partial_t G = -\int_{S^t} V\langle N^t, \partial_t \mu\rangle d\sigma^t. \tag{6.17}$$

Finally, let $W : \mathbb{R}^3 \to \mathbb{R}^3$ be a smooth vector field, and

$$L(t) = \int_{S^t} \langle W, N^t\rangle dl$$

(where S^t can be open or closed). Then

$$\partial_t L = -\int_{\partial S^t} \langle h^t, \tau^t \times W\rangle d\sigma^t + \int_{S^t} \mathrm{div}(W)\langle N^t, \partial_t \mu\rangle d\sigma^t \tag{6.18}$$

where τ^t is the unit tangent to ∂S^t, oriented so that (τ^t, n^t, N^t) forms a direct orthonormal basis.

Proof. To analyze the variations of F, take a family $((U_i, m_i^t), i = 1, \ldots, n)$ of positively oriented local parametrizations of S^t (the U_i's being independent of time), as given by (6.13), and an adapted partition of unity $(\omega_i^0, i = 1, \ldots, n)$ on S^0. Define, for $t \geq 0$ $\omega_i^t(m_i^t(u, v)) = \omega_i^0(m_i^0(u, v))$, to obtain a partition of unity on S^t and write

$$F(t) = \sum_{i=1}^N \int_{U_i} \omega_i^0(m_i^0(u, v))V(m_i^t(u, v))|\partial_u m_i^t \times \partial_v m_i^t|dudv. \tag{6.19}$$

We now focus on the variation of one of the terms of the sum; if (U, m^t) is a local patch on S^t, $\omega^0 \circ m^0$ a scalar function on U, define

$$f(t) = \int_U \omega^0(m^0(u, v))V(m^t(u, v))|\partial_u m^t \times \partial_v m^t|dudv.$$

Let $h^t = \partial_t m^t$; we have, $\partial_t V(m^t(u, v)) = \langle \nabla V(m^t(u, v)), h^t\rangle$. Also,

$$\partial_t(|\partial_u m^t \times \partial_v m^t|) = \langle \partial_u h^t \times \partial_v m^t + \partial_u m^t \times \partial_v h^t, N^t\rangle$$

where N^t, the normal in the chart, is given by

$$N^t(u, v) = \frac{\partial_u m^t \times \partial_v m^t}{|\partial_u m^t \times \partial_v m^t|}.$$

Rearranging the cross products, we get

$$\partial_t(|\partial_u m^t \times \partial_v m^t|) = \langle \partial_u h^t, \partial_v m^t \times N^t\rangle + \langle \partial_v h^t, N^t \times \partial_u m^t\rangle.$$

Let's apply Lemma 4.16, using, as in the lemma,

$$\rho(A) = \text{trace}((\text{Id} - NN^T)A)$$

for an operator $A : T_pS \to \mathbb{R}^3$. We therefore have

$$\partial_t(|\partial_u m^t \times \partial_v m^t|) = \rho(Dh^t(m^t(u,v)))|\partial_u m^t \times \partial_v m^t|$$

and

$$\partial_t f = \int_U \omega^0 \circ m^0(V \circ m^t \rho(Dh^t \circ m^t) + \langle \nabla V \circ m, h^t \rangle|\partial_u m^t \times \partial_v m^t| du dv$$

$$= \int_{m^t(U) \cap S^t} \omega^t(V\rho(Dh^t) + \langle \nabla V, h^t \rangle)d\sigma^t$$

with $\omega^t \circ m^t = \omega^0 \circ m^0$.

We now can apply this to all the terms in (6.19) to obtain

$$\partial_t F = \int_S (\rho(Dh^t)V + \langle \nabla V, h^t \rangle)d\sigma^t. \tag{6.20}$$

(With some abuse of notation, we use the same letter, h^t, to denote the point velocity in the chart and on the manifold.) We now eliminate Dh in (6.20), which will require an integration by parts via the divergence theorem. We decompose h^t into a tangent and a normal part, namely

$$h^t = h^t_T + \eta^t N^t,$$

with $\eta^t = \langle h^t, N^t \rangle$ (and therefore $h^t_T(p) \in T_p S^t$), so that, by definition of ρ and equation (4.20),

$$\rho(Dh^t) = \text{div}'_{S^t}(h^t) = \text{div}_{S^t} h^t_T - 2\eta^t H^t. \tag{6.21}$$

Applying Green's formula, letting n^t be the inward normal to ∂S^t, we get

$$\int_{S^t} V\rho(h^t_T)d\sigma^t = -\int_{\partial S^t} V\langle h^t_T, n^t \rangle dl - \int_{S^t} \langle \nabla_{S^t} V, h^t_T \rangle d\sigma^t,$$

since $\text{div}_{S^t}(Vh^t_T) = V\text{div}_{S^t}(h^t_T) + \langle \nabla_{S^t} V, h^t_T \rangle$. This in turn implies that

$$\partial_t F = -\int_{\partial S^t} V\langle h^t_T, n^t \rangle dl + \int_{S^t} (-2V\eta^t H^t - \langle \nabla V, h^t_T \rangle + \langle \nabla V, h^t \rangle)d\sigma^t,$$

which yields, since $\langle \nabla_{S^t} V, h^t \rangle = \langle \nabla_{S^t} V, h^t_T \rangle + \eta^t \langle \nabla_{S^t} V, N^t \rangle$, and using $\eta^t = \langle h^t, N^t \rangle$, $\langle h^t_T, n^t \rangle = \langle h^t, n^t \rangle$:

$$\partial_t F = -\int_{\partial S^t} V\langle h^t, n^t \rangle dl + \int_{S^t} (-2VH^t + \langle \nabla V, N^t \rangle)\langle N^t, h^t \rangle d\sigma^t \tag{6.22}$$

as needed.

Like with curves, (6.17) is a consequence of (6.18) and of the divergence theorem. We therefore directly prove (6.18). As above, we can decompose the integral over charts using a partition of unity. So we introduce the variation on a single chart and consider, for a scalar function $\omega^0 \circ m^0$,

$$\ell(t) = \int_U \omega^0 \circ m^0 \langle W(m^t), \partial_u m^t \times \partial_v m^t \rangle du dv$$

$$= \int_U \omega^0 \circ m^0 \det(W(m^t), \partial_u m^t, \partial_v m^t) du dv.$$

Computing time derivatives, we get, using again $h^t = \partial_t m^t$,

$$\partial_t \ell = \int_U \omega^0 \circ m^0 \langle DW(m^t) h^t, \partial_u m^t \times \partial_v m^t \rangle du dv$$

$$+ \int_U \omega^0 \circ m^0 \langle W(m^t), \partial_u h^t \times \partial_v m^t + \partial_u m^t \times \partial_v h^t \rangle du dv.$$

From Lemma 4.16, we know that the normal component of $\zeta := \partial_u h^t \times \partial_v m^t + \partial_u m^t \times \partial_v h^t$ is

$$\langle \zeta, N^t \rangle = \rho(Dh^t) |\partial_u m^t \times \partial_v m^t|.$$

To compute the tangential component, note that

$$\langle \zeta, \partial_u m^t \rangle = \langle \partial_u h^t \times \partial_v m^t, \partial_u m^t \rangle = -\langle \partial_u h^t, N^t \rangle |\partial_u m^t \times \partial_v m^t|$$

and similarly

$$\langle \zeta, \partial_v m^t \rangle = -\langle \partial_v h^t, N^t \rangle |\partial_u m^t \times \partial_v m^t|.$$

This implies that, for any $\xi \in T_p S$, we have

$$\langle \zeta, \xi \rangle = -\langle Dh^t(p)\xi, N^t \rangle |\partial_u m^t \times \partial_v m^t|.$$

So, if we decompose $W = W_T + W_N N$ in a tangent and a normal components to S, we get

$$\langle W, \partial_u h^t \times \partial_v m^t + \partial_u m^t \times \partial_v h^t \rangle =$$
$$(W_N \rho(Dh^t) - \langle Dh^t W_T, N^t \rangle) |\partial_u m^t \times \partial_v m^t|.$$

This gives (taking as above $\omega^t \circ m^t = \omega^0 \circ m^0$)

$$\partial_t \ell = \int_U \omega^t \circ m^t \langle DW(m^t) h^t, N^t \rangle |\partial_u m^t \times \partial_v m^t| du dv$$

$$+ \int_U \omega^t \circ m^t (W_N(m^t) \rho(Dh^t) - \langle Dh^t W_T(m^t), N^t \rangle)$$

$$|\partial_u m^t \times \partial_v m^t| du dv.$$

This yields (after summing over a partition of the unity)

$$\partial_t L(t) = \int_{S^t} \langle DWh^t + \rho(Dh^t)W - Dh^t W_T, N^t \rangle d\sigma^t.$$

Using (6.21), we get

$$\partial_t L(t) = - \int_{\partial S^t} W_N \langle h^t, n^t \rangle dl$$
$$+ \int_{S^t} \left(\langle DWh^t, N^t \rangle - 2\eta^t H^t W_N - \langle \nabla_S W_N, h_T^t \rangle - \langle Dh^t W_T, N^t \rangle \right) d\sigma^t. \tag{6.23}$$

We have (by definition of the gradient)

$$\langle \nabla_S W_N, h_T^t \rangle = D(W_N) h_T^t = \langle DWh_T^t, N^t \rangle + \langle DN^t h_T^t, W \rangle. \tag{6.24}$$

We will also use

$$\begin{aligned} \text{div}_S(\eta^t W_T) &= \langle \nabla \eta^t, W_T \rangle + \eta^t \text{div}_S W_T \\ &= \langle Dh^t W_T, N^t \rangle + \langle DN^t W_T, h^t \rangle + \eta^t \text{div}_S W_T \end{aligned} \tag{6.25}$$

and

$$\langle DN^t W_T, h^t \rangle = \langle DN^t W_T, h_T^t \rangle = \langle W_T, DN^t h_T^t \rangle = \langle W, DN^t h_T^t \rangle$$

to write

$$\begin{aligned} \langle DWh^t, N^t \rangle &- 2\eta^t H^t W_N - \langle \nabla_S W_N, h_T^t \rangle - \langle Dh^t W_T, N^t \rangle \\ &= \eta^t(\langle DWN^t, N^t \rangle + \text{div}_S W_T - 2H^t W_N) - \text{div}_S(\eta^t W_T) \\ &= \eta^t(\langle DWN^t, N^t \rangle + \rho(DW)) - \text{div}_S(\eta^t W_T) \\ &= \eta^t \text{div}(W) - \text{div}_S(\eta^t W_T). \end{aligned}$$

Using this in (6.23) and applying the divergence theorem yields

$$\partial_t L(t) = - \int_{\partial S^t} (\langle W, N^t \rangle \langle h^t, n^t \rangle - \langle W, n^t \rangle \langle h^t, N^t \rangle) dl$$
$$+ \int_{S^t} \text{div}(W) \langle h^t, N^t \rangle. \tag{6.26}$$

It now suffices to remark that

$$\langle h^t, \tau^t \times W \rangle = \det(\tau^t, W, h^t) = \langle W, N^t \rangle \langle h^t, n^t \rangle - \langle W, n^t \rangle \langle h^t, N^t \rangle$$

to retrieve (6.18).

$$\square$$

6.2.1 Mean Curvature Surface Flow

The equivalent of the curve-shortening flow for surfaces is the gradient descent for the area functional, the variation of which being given by (6.16) with $V = 1$. The gradient of the area, for the dot product

$$\langle h, \tilde{h} \rangle_S = \int_S \langle h, \tilde{h} \rangle d\sigma_S$$

therefore is $-2HN$, yielding the gradient descent evolution, called mean curvature flow

$$\partial_t p = 2H(p)N(p). \tag{6.27}$$

This is a regularizing flow for surfaces, and solutions exist in short time. It may happen, however, that singularities form in the evolving surface, even when starting with relatively simple ones, which did not happen with the curve-shortening flow.

The mean curvature flow in implicit form has exactly the same expression as the curve-shortening flow. Indeed, if the evolving surface is defined (at time t) by $f(t,p) = 0$ (with $\nabla f \neq 0$ on the surface and f differentiable in time), then, using equation (4.22) and the same argument as the one we made with curves, f satisfies the evolution equation

$$\partial_t f = |\nabla f| \mathrm{div}\Big(\frac{\nabla f}{|\nabla f|}\Big). \tag{6.28}$$

This gives an alternate formulation of the mean curvature flow, that has the advantage to carry over the solution beyond singularities and to allow for changes of topology. This is a diffusion that smoothes f tangentially to the surface, since

$$|\nabla f| \mathrm{div}\Big(\frac{\nabla f}{|\nabla f|}\Big) = \Delta f - \langle N, D^2 f N \rangle.$$

6.3 Gradient Flows

The curvature and mean curvature flows for curves and surfaces are special cases of gradient flows that evolve curves along steepest descent directions relative to a given objective function (the length or the area), for a given metric (the L^2 metric relative to the arc length or area form).

This approach can be applied to any objective function that has a gradient relative to a chosen dot product; considering as we did curves over some interval, or surfaces parametrized over some reference surface, we can consider energies $E(m)$ (or $E(\mu)$) that have a gradient ∇E with respect to some metric, and implement

$$\partial_t m = -\nabla E(m).$$

which can obviously put into the form (6.1) or (6.12).

As an example beyond mean curvature flows, let's consider plane curves and the energy given by the integral of the squared curvature

$$E(m) = \frac{1}{2} \int_m \kappa^2 dl \qquad (6.29)$$

which is often called the bending energy of the curve. To compute the gradient, let's introduce a small perturbation $m(\varepsilon, .)$ of a curve m_0 depending of a parameter ε, such that $m(0, .) = m_0$. We want to compute $\partial_\varepsilon E(m(\varepsilon, .))$.

A first remark that will simplify the computation is that, because E is parametrization invariant, we can assume that $\partial_\varepsilon m$ is oriented along the normal to $m(\varepsilon, .)$. We can indeed always restrict ourselves to this case using an ε-dependent change of variable that does not affect the value of the energy.

A second simplification will come by introducing the variation with respect to arc length along a curve m, denoted ∂_s^m defined by

$$\partial_s^m = \frac{1}{|\partial_u m|} \partial_u \qquad (6.30)$$

when m is parametrized by u (the definition being parametrization independent). We will drop the m superscript from the notation and only write ∂_s, although it is important to remember that this operator depends on the curve that supports it.

Assuming that m depends on ε, computing the derivative in ε yields

$$\partial_\varepsilon \partial_s = \partial_s \partial_\varepsilon - \left\langle \frac{\partial_u \partial_\varepsilon m}{|\partial_u m|}, T \right\rangle \partial_s.$$

If $\partial_\varepsilon m = BN$, this gives

$$\partial_\varepsilon \partial_s = \partial_s \partial_\varepsilon + \kappa B \partial_s. \qquad (6.31)$$

Using this identity, we get

$$\partial_\varepsilon \partial_s m = \partial_s(BN) + \kappa BT = (\partial_s B)N.$$

and

$$\partial_\varepsilon \partial_s^2 m = \partial_s(\partial_\varepsilon \partial_s m) + \kappa B \partial_s^2 m$$
$$= -(\partial_s B)\kappa T + (\partial_s^2 B + \kappa^2 B)N.$$

Since

$$\partial_\varepsilon \partial_s^2 m = \kappa \partial_\varepsilon N + (\partial_\varepsilon \kappa)N$$

and $\partial_\varepsilon N$ is perpendicular to N, we find

$$\partial_\varepsilon \kappa = \partial_s^2 B + \kappa^2 B \text{ when } \partial_\varepsilon m = BN. \qquad (6.32)$$

(We find in passing that, the evolution of the curvature with (6.7) is $\partial_t \kappa = \partial_s^2 \kappa + \kappa^3$.)

Combining these results and Proposition 6.2, we can easily compute the variation of E in ε (assuming the m is closed to simplify), yielding

$$\partial_\varepsilon E = \int_m (\partial_s^2 B + \kappa^2 B)\kappa dl - \int_m \kappa^3 B dl$$

$$= \int_m (\kappa \partial_s^2 B + \kappa^3 B/2) dl$$

$$= \int_m (\partial_s^2 \kappa + \kappa^3/2) B dl$$

which provides the gradient of E. Minimizing E by itself leads to the uninteresting solution of a curve blowing up to infinity; indeed, since scaling a curve m by a factor a divides its curvature by a, we have $E(am) = E(m)/a$. However, minimizing E over curves with constant length, or the related problem of minimizing

$$E_\lambda(m) = E(m) + \lambda \text{length}(m) \tag{6.33}$$

has been widely studied, in dimension 2 or larger, and curves for which the gradient of E_λ vanishes for some positive λ are called elasticae. We have, using the same notation

$$\partial_\varepsilon E_\lambda = \int_m (\partial_s^2 \kappa + \kappa^3/2 - \lambda \kappa) B dl \tag{6.34}$$

and the related gradient descent algorithm

$$\partial_t m = -(\partial_s^2 \kappa + \kappa^3/2 - \lambda \kappa) N$$

provides a well-defined and converging evolution [128, 72].

The counterpart of (6.29) for surfaces is the Willmore energy

$$E(S) = \int_S H^2 d\sigma, \tag{6.35}$$

and the associated minimizing flow is called the Willmore flow. It is given by

$$\partial_t p = (\Delta_S H + 2H(H^2 - K))N \tag{6.36}$$

(see [218]).

All the flows above were defined as gradient descent for the metric given by the L^2 norm relative to the arc length or the area form. As remarked in Appendix D, changing the metric can induce fundamental changes in the resulting algorithms. For example, as proposed in [197], gradient flows of curves associated to Sobolev metrics

$$\langle h, \tilde{h} \rangle_m = \int \langle Ah, \tilde{h} \rangle ds$$

where A is a differential operator (assumed to be symmetric and positive) can create interesting variants of the original gradient flows. More precisely, they transform an L^2 flow that would originally take the form

$$\partial_t m = -\nabla_{L^2}(m)$$

into the flow

$$\partial_t m = -A^{-1}(\nabla_{L^2}(m))$$

where A^{-1} is the inverse of A and therefore a smoothing operator. This results in an evolution that favors smooth changes.

In fact, for closed plane curves, many interesting operators can be represented in terms of Fourier series. For a curve of length L, we can consider the normalized arc length $\tilde{s} = s/L$ defined over $[0,1]$. Now, a periodic function f defined over $[0,1]$ can be decomposed (if smooth enough) in terms of its Fourier series

$$f(\tilde{s}) = \sum_{k \in \mathbb{Z}} c_k(f) e^{2i\pi k \tilde{s}}$$

with

$$c_k(f) = \int_0^1 f(u) e^{-2i\pi ku} du.$$

To an even sequence, $(a_k, k \in \mathbb{Z})$, of positive numbers, one associates an operator

$$Af(\tilde{s}) = \sum_{k \in \mathbb{Z}} a_k c_k(f) e^{2i\pi k \tilde{s}}.$$

This operator is defined over functions f such that

$$\sum_{k \in \mathbb{Z}} a_k^2 |c_k(f)|^2 < \infty.$$

The inverse operator is then directly defined by

$$A^{-1}f(\tilde{s}) = \sum_{k \in \mathbb{Z}} \frac{c_k(g)}{a_k} e^{2i\pi k \tilde{s}}.$$

For example, the differential operator $Af = -2\partial_{\tilde{s}}^2$ is such that

$$c_k(Af) = 4\pi^2 k^2 c_k(f)$$

as can be easily computed by integration by parts, and is therefore associated to $a_k = 4\pi^2 k^2$. The operator $Af = -\partial_{\tilde{s}}^2 + \lambda id$ (with $\lambda > 0$) corresponds to $a_k = 4\pi^2 k^2 + \lambda$, which is positive (implying that A is invertible). Computations like this, and the fact that Fourier coefficients and Fourier series are discretized by the fast Fourier transform and its inverse, lead to very simple variations of a gradient flow associated to a given objective function (see Section 6.4.11).

6.4 Active Contours

6.4.1 Introduction

Active contour methods let curves or surfaces evolve in order to minimize a segmentation energy (i.e., an energy that measures the alignment of the contour or surface with the boundary of a shape present in the image). They provide an important class of curve (and surface) evolution based on specific gradient flows.

We start with a parametrization dependent formulation (in two dimensions), which is the original method introduced in [122]. We will assume that a function $p \mapsto V(p)$ is defined for $p \in \mathbb{R}^2$. It is assumed that V is small where a contour is likely to be present in the image (based on some measure of discontinuity within an image, for example) and the goal of the method is to ensure that the evolving contour settles along regions where V is small while remaining a smooth closed curve.

To a smooth parametrized curve $m : [0, 1] \to \mathbb{R}^2$, we associate the energy

$$E(m) = \alpha \int_0^1 |\ddot{m}_{uu}|^2 du + \beta \int_0^1 |\dot{m}_u|^2 du + \gamma \int_0^1 V(m(u)) du. \qquad (6.37)$$

Minimizing this energy results in a compromise between smoothness constraints (provided by the first two integrals) and the fact that m aligns with discontinuities, which comes from the last integral. The minimization is made subject to constraints at the extremities. Typical constraints are either:

(i) $m(0)$ and $m(1)$ are fixed, together with $\dot{m}_u(0)$ and $\dot{m}_u(1)$.
(ii) $m(0) = m(1)$, $\dot{m}_u(0) = \dot{m}_u(1)$ (closed curves).

We will assume that one of these two conditions is enforced in the following computation.

6.4.2 First Variation and Gradient Descent

Given a curve $m_0(u)$, $u \in [0, 1]$, we evaluate the impact on the cost function E of a small variation $\varepsilon \mapsto m(\varepsilon, .)$. If the extremities ($m(0)$ and $m(1)$) are fixed, we will have $\partial_\varepsilon m(\varepsilon, 0) = \partial_\varepsilon m(\varepsilon, 1) = 0$. The closedness condition requires $\partial_\varepsilon m(\varepsilon, 0) = \partial_\varepsilon m(1, 0)$. Letting $h = \partial_\varepsilon m$, we have

$$\partial_\varepsilon E = 2\alpha \int_0^1 \langle \ddot{m}_{uu}, \ddot{h}_{uu} \rangle du + 2\beta \int_0^1 \langle \dot{m}_u, \dot{h}_u \rangle du$$
$$+ \gamma \int_0^1 \langle \nabla V(m(u)), h(u) \rangle du.$$

Assume that m is C^4. We perform two integrations by parts of the first integral and one of the second to obtain:

$$\partial_\varepsilon E = -2\alpha[\langle m_{uuu}^{(3)}, h\rangle]_0^1 + 2\alpha[\langle \dot{m}_{uu}, \dot{h}_u\rangle]_0^1 + 2\beta[\langle \dot{m}_u, h\rangle]_0^1$$
$$+ 2\alpha \int_0^1 \langle m_{uuuu}^{(4)}, h\rangle du - 2\beta \int_0^1 \langle \ddot{m}_{uu}, h\rangle du$$
$$+ \gamma \int_0^1 \langle \nabla V(m(u)), h(u)\rangle du. \tag{6.38}$$

The boundary terms (first line) disappear for both types of boundary conditions. We can therefore write

$$\partial_\varepsilon E = \varepsilon \int_0^1 \langle \nabla E(m), h\rangle du \tag{6.39}$$

with

$$\nabla E(m) = 2\alpha m_{uuuu}^{(4)} - 2\beta \ddot{m}_{uu} + \gamma \nabla V(m).$$

Note that, in this formula, $\nabla E(m)$ is the variational gradient, therefore a function $u \mapsto \nabla E(m)(u)$, whereas $\nabla V(m)$ is the ordinary gradient (a vector). Using the L^2 metric on $[0, 1]$, we get the following gradient descent evolution:

$$\dot{m}_t = -2\alpha m_{uuuu}^{(4)}(t, .) + 2\beta \ddot{m}_{uu}(t, .) - \gamma \nabla V(m(t, .)). \tag{6.40}$$

6.4.3 Numerical Implementation

The discretization of (6.40) is relatively straightforward. Assume that m is discretized into a sequence of points $(x_1, y_1), \dots (x_n, y_n)$, stacked into a matrix M with n rows and two columns.

The finite-difference derivatives with respect to the parameter u are linear operations on M. For example, $M' = D_1 M$ with (in the case of closed curves)

$$D_1 = \frac{1}{2\delta u} \begin{pmatrix} 0 & 1 & 0 & 0 & \cdots & 0 & 0 & 0 & -1 \\ -1 & 0 & 1 & 0 & \cdots & 0 & 0 & 0 & 0 \\ \cdots \\ 0 & 0 & 0 & 0 & \cdots & 0 & -1 & 0 & 1 \\ 1 & 0 & 0 & 0 & \cdots & 0 & 0 & -1 & 0 \end{pmatrix}.$$

where δu is the discretization step. The second derivative is obtained with a tridiagonal matrix, with $1, -2, 1$ on the diagonal (and periodic assignment of values at the extremities), divided by $(\delta u)^2$. The fourth derivative is pentadiagonal, with values $1, -4, 6, -4, 1$ divided by $(\delta u)^4$. Therefore, the expression

$$-2\alpha m_{uuuu}^{(4)} + 2\beta \ddot{m}_{uu}$$

is discretized in the form $A.M_t$ where A is a pentadiagonal matrix which depends only on α and β.

The function V is rarely known analytically, and most often discretized on a grid ($V = V_{ij}, i, j = 1, \dots, N$). To compute $\nabla V(m)$, the partial derivatives of V must be estimated at points (x, y) which are not necessarily on the grid. The bilinear interpolation of V at a point $(x, y) \in \mathbb{R}^2$ is

$$\hat{V}(x, y) = (1 - \alpha_x)(1 - \alpha_y)V_{ij} + (1 - \alpha_x)\alpha_y V_{ij+1}$$
$$+ \alpha_x(1 - \alpha_y)V_{i+1j} + \alpha_x\alpha_y V_{i+1j+1},$$

where i and j are such that $i\delta x \leq x < (i+1)\delta x$ and $j\delta x \leq y < (j+1)\delta x$, and $\alpha_x = (x - i\delta x)/\delta x$, $\alpha_y = (y - j\delta x)/\delta x$ (with a spatial discretization step δx). This implies that

$$\partial_x \hat{V} = (1 - \alpha_y)(V_{i+1j} - V_{ij}) + \alpha_y(V_{i+1j} - V_{ij+1}),$$

except along lines $x = i\delta x$ at which $\hat{V}$ is not differentiable. One can use central finite difference at such points, i.e.,

$$\partial_1^0 V_{ij} = (V_{i+1j} - V_{i-1j})/(2\delta x).$$

Similar formulae can be used for derivatives with respect to the second variable. This yields a discrete version of the gradient of V, that we will denote $\nabla V(M)$, which is nonlinear in M.

The direct discretization of (6.40), using a time step δt, yields

$$M_{t+1} = (I + \delta t\, A)M_t + \delta t\, \nabla V(M_t).$$

This algorithm will typically converge to a local minimum of E when δt is small enough.

Its numerical behavior can, however, be significantly improved by using a so-called semi-implicit scheme (the previous scheme being called explicit), given by

$$(I - \delta t\, A)M_{t+1} = M_t + \delta t\, \nabla V(M_t). \tag{6.41}$$

This requires the computation of the inverse of $(I - \delta t A)$, which can be costly, but this operation has to be done only once, since A does not depend on M.

The discretization needs to be adapted when the distance between two consecutive points in M becomes too large. In this case, the estimation of derivatives becomes unreliable. When this occurs, one should rediscretize the evolving curve. This creates a new matrix M (and also requires one to recompute $(I - \delta t\, A)^{-1}$ if the number of points has changed).

6.4.4 Initialization and Other Driving Techniques

It is important to understand that the global minimum of E is generally not interesting. For closed contours, it is a degenerate curve reduced to a point (the result is zero in this case). So, the method must start with a reasonable initial curve and let it evolve to the closest local minimum. This is why active contours are often run in an interactive way, the user initializing the process with a coarse curve around the region of interest (using a graphic interface), and the evolution providing the final contour, hopefully a fine outline of the shape. Since, as we will see, the snake energy generally shrinks curves, the initial contour has to be drawn outside the targeted shape.

A "balloon" technique [50] allows the user to only specify an initial region *within* the shape, which is often easier and can be sometimes done automatically (because the interior of a shape is often more specific than its exterior). The idea is to complete the evolution (6.40) with an outward normal force, yielding the equation

$$\dot{m}_t = -2\alpha m^{(4)}_{uuuu} + 2\beta \dot{m}_{uu} - \gamma \nabla_m V - pN, \qquad (6.42)$$

p being a small positive constant and N the inward normal to the contour. Once the curve stabilizes (generally slightly outside the shape because of the effect of the pressure), the algorithm must be continued with $p = 0$ to provide a correct alignment with the shape boundary. Note that this normal force can be obtained also from a variational approach, adding to the energy an area term

$$-p \int_{\Omega_m} dx$$

where Ω_m is the interior of the curve m.

Because the "gradient force" $\nabla V(m)$ only carries information near points of high gradient, some improvement can be obtained by first extending it to a vector field v defined on the whole image, and use v instead of $\nabla V(m)$ in (6.42). This results in the gradient vector flow (GVF) method developed in [219], in which v is built by solving an auxiliary variational problem: minimize

$$\int_{\Omega} (\text{trace}(Dv^T Dv) + |\nabla V| \, |\nabla V - v|^2) dm.$$

Thus, $v \simeq \nabla V$ when $|\nabla V|$ is large, and is a smooth interpolation otherwise. Extrapolating the driving force to the whole image in this way allows the algorithm to be more robust to initial data. It also provides improved convergence results, allowing in particular the snake to enter into concave regions in the shape, which is much harder with the original formulation.

6.4.5 Evolution of the Parametrization

The snake energy is not geometric: it depends on the selected parametrization, which is therefore also optimized in the process. We will discuss purely geometric methods in the next section, but it is interesting to study how the snake energy optimizes the parametrization, which can be associated, in numerical implementation, to the density of points in the discretized curve.

Let s_m be the (Euclidean) arc length of a curve m, $ds_m = |\dot{m}_u| du$ and $L = L_m$ be its length. We let $\psi : [0, L] \to [0, 1]$ be the inverse change of parameter $\psi(s_m(u)) = u$. We also let $\bar{m}(s) = m(\psi(s))$ be the arc-length parametrization of m. We have $\dot{m}_u \circ \psi = \dot{\bar{m}}_s / \dot{\psi}_s$ which implies in turn

$$\dot{\psi}_s \ddot{m}_{uu} \circ \psi = \frac{\ddot{\bar{m}}_{ss}}{\dot{\psi}_s} - \frac{\ddot{\psi}_{ss}}{\dot{\psi}_s^2} \dot{\bar{m}}_s .$$

Since $\bar{m}$ is parametrized with arc length, we have

$$|\dot{m}_u \circ \psi|^2 = \frac{1}{\dot{\psi}_s^2} , \quad |\ddot{m}_{uu} \circ \psi|^2 = \frac{\kappa^2}{\dot{\psi}_s^4} + \frac{\ddot{\psi}_{ss}^2}{\dot{\psi}_s^6}$$

where κ is the curvature.

We can make the change of variable $u \to \psi(u)$ in $E(m)$, which yields

$$E(m) = \int_0^L \left(\alpha \frac{\ddot{\psi}_{ss}^2}{\dot{\psi}_s^5} + \alpha \frac{\kappa^2}{\dot{\psi}_s^3} + \frac{\beta}{\dot{\psi}_s} + \gamma V(\bar{m}) \dot{\psi}_s \right) ds. \qquad (6.43)$$

To separate the length-shrinking effect from the rest, we renormalize the geometric and parametric quantities. We let $\tilde{\kappa}(s) = L\kappa(Ls)$ (which is the curvature of m rescaled by a factor $1/L$) and $\chi(s) = \psi(Ls)$. To lighten the expression, we also let $\tilde{V}(s) = V(\bar{m}(Ls))$. These functions are all defined on $[0, 1]$, and a linear change of variables in the previous energy yields (with $\dot{\chi}_s(s) = L\dot{\psi}_s(Ls)$ and $\ddot{\chi}_{ss}(s) = L^2 \ddot{\psi}_{ss}(Ls)$),

$$E(m) = L^2 \int_0^1 \left(\alpha \frac{\ddot{\chi}_{ss}^2}{\dot{\chi}_s^5} + \alpha \frac{\tilde{\kappa}^2}{\dot{\chi}_s^3} + \frac{\beta}{\dot{\chi}_s} \right) ds + \gamma \int_0^1 \dot{\chi}_s \tilde{V} ds.$$

The first integral is scale-independent (it only depends on the rescaled curve) and is multiplied by L^2. This therefore shows the length reduction effect of the smoothing part of the snake energy. When L tends to 0, $\tilde{V}$ becomes equal to $V(m(0))$ and the limiting integral has the same value since $\int_0^1 \dot{\chi}_s = 1$. This is therefore minimal when $m(0)$ is the point of lowest value for V, and we see that the global minimum of the energy is a dot at the minimum of V (not a very interesting solution).

Beside the shrinking effect, it is interesting to analyze how the parametrization is optimized for a fixed geometry (i.e., fixed L and $\tilde{\kappa}$, which also implies that $\tilde{V}$ is fixed). Recall that $1/\dot{\chi}_s$ is proportional to ds/du, the speed of the parametrized curve. The term $\tilde{\kappa}^2/\dot{\chi}_s^3$ shows that this speed is more penalized when the curvature is high. For a discrete curve, this implies that points have a tendency to accumulate at corners. On the other hand, the term $\dot{\chi}_s \tilde{V}$ creates more sparsity in the regions where V is large.

We now specialize to the case $\alpha = 0$, for which the computation can be pushed further. In this case, we have

$$E(m) = \int_0^L \left(\frac{\beta}{\dot{\psi}_s} + \gamma V(\bar{m}) \dot{\psi}_s \right) ds .$$

We can optimize the parametrization with fixed geometry. We therefore need to optimize for $\dot\psi_s$ with the constraint $\int_0^L \dot\psi_s = 1$ and $\dot\psi_s > 0$.

First we can remark that E is a convex function of $\dot\psi_s$, minimized over a convex set. This implies, in particular, uniqueness of the minimum if it exists. We disregard the positivity constraint, which, as we will see, will be automatically satisfied by the solution. Using Lagrange multipliers, we obtain the Lagrangian

$$\int_0^L \left(\frac{\beta}{\dot\psi_s} + \gamma V(\bar m)\dot\psi_s \right) ds - \lambda \int_0^L \dot\psi_s ds \,.$$

A variation of this with respect to $\dot\psi_s$ yields the equation

$$\frac{\beta}{\dot\psi_s^2} = \gamma V(\bar m) - \lambda \,.$$

The solution must therefore take the form $\dot\psi_s(s) = \dfrac{\sqrt{\beta}}{\sqrt{\gamma V(\bar m(s)) - \lambda}}$, for a suitable value of λ, which must be smaller than $\lambda_* = \gamma \min_s V(\bar m(s))$ and such that $\int_0^L \dot\psi_s(s)ds = 1$.

Let's prove that such a λ exists, in the case when m is closed and V and m are smooth. Consider the function

$$f : \lambda \mapsto \int_0^L \frac{\sqrt{\beta}}{\sqrt{\gamma V(\bar m(s)) - \lambda}} d\lambda$$

defined on $(-\infty, \lambda_*)$. It is obviously continuous, and tends to 0 when λ tends to $-\infty$. Now, let $\lambda \to \lambda_*$ and take s_0 such that $\lambda_* = \gamma V(\bar m(s_0))$. Since λ_* is a minimum and we assume that V and $\bar m$ are smooth, the difference $\gamma V(\bar m(s)) - \gamma V(\bar m(s_0))$ must be an $O((s - s_0)^2)$ or smaller when s is close to s_0. But this implies that the integral diverges to $+\infty$ when $\lambda \to \lambda_*$. Therefore, a value of λ exists with $f(\lambda) = 1$. (If m is an open curve, the above argument is valid with the additional assumption that the minimum of V is not attained at one of the extremities of m.)

From the computation above, we see that the optimal parametrization must satisfy

$$\dot\psi_s^{-2} - (\gamma/\beta)V(\bar m) = \text{cst} \,.$$

Consider now the following class of problems: minimize E under the constraint $\dot\psi_s^{-2} - (\gamma/\beta)V(\bar m) = \mu$ for some constant μ. The previous discussion shows that the solution of the original problem is also a solution of this constrained problem for some value of μ (one says that the two classes of problems are equivalent). However, considering the latter problem, we see that it boils down to the purely geometric problem: minimize

$$\int_0^L W(\bar{m}(s))ds$$

with $W(p) = \sqrt{\beta}(\beta\mu + 2\gamma V(p))/\sqrt{\beta\mu + \gamma V(p)}$. This new variational problem fits in the category of geodesic active contours, which are addressed in the next section.

6.4.6 Geometric Methods

To obtain a geometric formulation of (6.37), it suffices to assume that it applies to arc-length parametrized curves, leading to

$$E(m) = \int_0^L (\alpha\kappa^2(s) + \beta)ds + \gamma \int_0^L V(m(s))ds. \tag{6.44}$$

The first term is the elastica energy that we defined in (6.33). Combining (6.34) and Proposition 6.2, we see that the gradient flow associated to E is

$$\partial_t m = \left(-\alpha(\partial_s^2\kappa + \kappa^3/2) + (\beta + \gamma V)\kappa - \langle \nabla V, N \rangle \right)N. \tag{6.45}$$

Geodesic Active Contours

The case $\alpha = 0$ corresponds to what has been called *geodesic active contours* [41, 178], which correspond to minimizing

$$E(m) = \int_0^L W(m(s))ds \tag{6.46}$$

(letting $W = \beta + \gamma V$) with the associated evolution

$$\partial_t m = (W(m)\kappa - \langle \nabla W(m), N \rangle)N. \tag{6.47}$$

This equation can be conveniently implemented using level sets. So, consider a function $(t, p) \mapsto F(t, p)$, such that its zero level set at fixed time t is a curve denoted $u \mapsto m(t, u)$. The equation $F(t, m(t, u)) = 0$ is by definition valid for all t and all u. Computing its derivative with respect to t yields

$$\partial_t F(t, m(t, u)) + \langle \nabla F(t, m(t, u)), \partial_t m(t, u) \rangle = 0$$

where ∇F is computed at fixed time (with respect to the p variable).

Assume, as is the case here, that the evolution of m is prescribed, in the form

$$\partial_t m(t, u) = \beta(m(t, u))N(t, u).$$

This yields, using $N = -\nabla F/|\nabla F|$,

$$\partial_t F(t, m(t, u)) = \beta(m(t, u))|\nabla F(m(t, u))|.$$

Assume that β, which describes the speed of the normal evolution of γ, can be extended to the whole domain Ω (it is *a priori* only defined on the evolving curve). Then, we can introduce the global evolution

$$\partial_t F(t, p) = \beta(p) \, |\nabla F(p)| \, .$$

This equation being valid for all p, it is *a fortiori* valid on the zero level set of f. Therefore, if this level set did coincide with the initial curve $m(0, .)$, it will coincide at all times with the curve $m(t, .)$ which satisfies the evolution equation $\partial_t m = \beta N$ (implicitly assuming that the evolution is well-defined over the considered time interval).

Returning to geodesic active contours, there is a natural extension for the function β, namely

$$\beta(p) = W(p)\mathrm{div}\left(\frac{\nabla F(p)}{|\nabla F(p)|}\right) + \left\langle \nabla W(p), \frac{\nabla F(p)}{|\nabla f(p)|} \right\rangle.$$

This therefore directly yields the partial differential equation:

$$\partial_t F = W \mathrm{div}\left(\frac{\nabla F}{|\nabla F|}\right)|\nabla F| + \langle \nabla W, \nabla F \rangle.$$

Implementation details are similar to those provided in Section 6.1.4.

6.4.7 Controlled Curve Evolution

For completeness, we mention another curve evolution method for contour estimation [61], which consists in using a smoothing evolution equation like in section 6.1.3, with an additional factor which controls the speed of the evolution, and essentially stops it in the presence of contours.

The curve evolution equation, in this case, takes the form

$$\partial_t m = V(m)\kappa N \tag{6.48}$$

where V is now the stopping function. This does not derive anymore from a variational approach, but we can see that when V is constant, the evolution is like (6.10), but that points stop moving when V is close to 0. So V retains the same interpretation as before, as a function which vanishes near regions of high gradient in the image.

The level-set formulation of this equation is

$$\partial_t F = V(t, \cdot)|\nabla F|\mathrm{div}\frac{\nabla F}{|\nabla F|} \, .$$

6.4.8 Geometric Active Surfaces

The transcription of (6.44) to surfaces is

$$E(S) = \int_S (\alpha H^2(p) + \beta)dp + \gamma \int_S V(p)dp, \qquad (6.49)$$

with associated gradient flow (using (6.36) and proposition 6.4)

$$\partial_t p = \big(\alpha(\Delta_S H + 2H(H^2 - K)) + 2(\beta + \gamma V)H - \langle \nabla V, N \rangle\big)N$$

if the surface has no boundary.

The case $\alpha = 0$ simplifies to (letting $W = \beta + \gamma V$)

$$\partial_t p = \big(2WH - \langle \nabla W, N \rangle\big)N$$

for a surface without a boundary and to

$$\partial_t p = \big(2WH - \langle \nabla W, N \rangle\big)N + \delta_{\partial S}(p)Wn_S$$

if the surface has a boundary.

For closed surfaces, a level-set evolution can be derived similarly to curves, yielding [178]

$$\partial_t F = |\nabla F|\Big(W \operatorname{div} \frac{\nabla F}{|\nabla F|} + \Big\langle \nabla W, \frac{\nabla F}{|\nabla F|} \Big\rangle\Big).$$

6.4.9 Designing the V Function

For most applications, boundaries of shapes in images correspond to rapid variation in image intensity, which results in large values of the image gradient. It is therefore natural to use functions V related to the gradient, like

$$V = -|\nabla I|$$

(recall that active contours align with small V). Designing bounded V's is, however, numerically preferable, a possible choice being given by

$$V = \frac{1}{1 + |\nabla I|/C}$$

for some constant C.

Another option is to rely on an edge detector [134, 37], and start from a binary image indicating edge points. Letting E denote the set of detected edges, we can define V to be the distance map to E

$$V(m) = \inf \{d(m,e), e \in E\}$$

(see Appendix F for quick implementations). Another commonly used approach is to convolve the binary edge image with a Gaussian kernel and let

$$V(m) = \sum_{e \in E}(1 - e^{-|g-m|^2/2\sigma^2}).$$

6.4.10 Inside/Outside Optimization

When information is available on the image values inside and outside a closed shape, it is possible to add area or volume integrals to the geometric formulations. Letting Ω denote the interior of the shape, the additional term typically takes the form ($\tilde{V}_{in}$ and $\tilde{V}_{out}$ being defined on some fixed bounded set $D \subset \mathbb{R}^d$, the image domain, and measuring the plausibility for a given point to belong in the interior or the exterior of the shape)

$$\int_{\Omega} \tilde{V}_{in}(x)dx + \int_{D\backslash\Omega^c} \tilde{V}_{out}(x)dx = \int_{\Omega} (\tilde{V}_{in}(x) - \tilde{V}_{out})dx + \int_{D} \tilde{V}_{out}(x)dx,$$

which reduces to (disregarding the last integral which is constant)

$$\int_{\Omega} \tilde{V}(x)dx$$

with $\tilde{V} = \tilde{V}_{in}(x) - \tilde{V}_{out}$. From Propositions 6.2 and 6.4, this simply adds the term $\tilde{V}N$ to the gradient descent algorithms.

A simple example is when the image is expected to average around a constant, say c_{in} over the interior, and around c_{out} over the exterior of the shape, leading to choosing

$$\tilde{V}_{in}(x) = \lambda_{in}(I(x) - c_{in})^2 \text{ and } \tilde{V}_{out}(x) = \lambda_{out}(I(x) - c_{out})^2.$$

This assumption is made in the Chan–Vese segmentation model [43], which defines (in two dimensions)

$$E(m) = \mu \, \text{length}(m) + \nu \, \text{area}(\Omega_m)$$
$$+ \int_{\Omega_m} \left(\lambda_{in}(I(x) - c_{in})^2 - \lambda_{out}(I(x) - c_{out})^2 \right)dx, \quad (6.50)$$

where Ω_m is the interior of the shape (with an obvious generalization to surfaces in three dimensions). This energy is a simplified version of the Mumford–Shah functional [154], which is designed for the approximation of an observed image by a piecewise smooth function (here the approximation is by a piecewise constant function, and the contour separates the image into exactly two regions). The associated shape evolution is

$$\partial_t m = (\mu\kappa + \nu + \lambda_{in}(I(x) - c_{in})^2 - \lambda_{out}(I(x) - c_{out})^2)N \quad (6.51)$$

with a level-set formulation

$$\partial_t F = |\nabla F| \left(\mu \text{div}\left(\frac{\nabla F}{|\nabla F|}\right) + \nu + \lambda_{in}(I(x) - c_{in})^2 - \lambda_{out}(I(x) - c_{out})^2 \right). \quad (6.52)$$

Instead of being seen as a function of a curve, energies which, like (6.50), can be put in the form

$$E(m) = \int_m V(p)dl + \int_{\Omega_m} \tilde{V}(x)dx, \tag{6.53}$$

can also be considered as functions of the domain Ω, simply setting

$$E(\Omega) = \int_{\partial\Omega} V(p)dl + \int_\Omega \tilde{V}(x)dx. \tag{6.54}$$

The interesting feature of this point of view is that this energy applies to domains that are more general than the interior of a curve, allowing for multiple connected components, for example, which can be useful when the topology of the targeted set is not known in advance. The associated level-set evolution

$$\partial_t F = |\nabla F|\left(V\mathrm{div}\left(\frac{\nabla F}{|\nabla F|}\right) + \langle \nabla V, \nabla F\rangle + \tilde{V}\right) \tag{6.55}$$

allows for such changes of topology, but it is also interesting to express (6.54) directly in terms of level sets of a function. To be able to handle line integrals, we will need to use the δ function approximation, like in Section 4.9.1, and using the notation of this section,

$$E(\Omega) \simeq \int_{\mathbb{R}^d} \delta_\varepsilon(F(x))V(x)|\nabla F(x)|dl + \int_{\mathbb{R}^d} (1 - H_\varepsilon(F(x)))\tilde{V}(x)dx \tag{6.56}$$

when $\Omega = \{F \leq 0\}$ (V must be continuous for this approximation to be valid). Letting $E_\varepsilon(F)$ denote the right-hand side in (6.56), we can directly compute a gradient descent evolution for F by studying the variations of E_ε.

So let $F(\alpha, .)$ be a variation of a a given function $F_0 = F(0, .)$. The computation of $\partial_\alpha E_\varepsilon(F(\alpha, .))$ gives, letting $h = \partial_\alpha F$, and assuming it to be compactly supported

$$\partial_\alpha E_\varepsilon = \int_{\mathbb{R}^d} \delta'_\varepsilon(F)V|\nabla F|hdx + \int_{\mathbb{R}^d} \delta_\varepsilon(F)V\left\langle \nabla h, \frac{\nabla F}{|\nabla F|}\right\rangle dx$$

$$- \int \delta_\varepsilon(F)\tilde{V}hdx$$

$$= -\int_{\mathbb{R}^d} \delta_\varepsilon(F)\left\langle \nabla V, \frac{\nabla F}{|\nabla F|}\right\rangle hdx - \int_{\mathbb{R}^d} \delta_\varepsilon(F)Vhdiv\left(\frac{\nabla F}{|\nabla F|}\right)$$

$$- \int \delta_\varepsilon(F)\tilde{V}hdx$$

where we have used the divergence theorem and the fact that

$$\mathrm{div}(\delta_\varepsilon(F)Vh\frac{\nabla F}{|\nabla F|}) = \delta'_\varepsilon(F)Vh|\nabla F| + \delta_\varepsilon(f)h\left\langle \nabla V, \frac{\nabla F}{|\nabla F|}\right\rangle$$

$$+ \delta_\varepsilon(F)V\left\langle \nabla h, \frac{\nabla F}{|\nabla F|}\right\rangle + \delta_\varepsilon(F)Vhdiv\left(\frac{\nabla F}{|\nabla F|}\right).$$

This provides the gradient descent evolution

$$\partial_t F = \delta_\varepsilon(F)\Big(V\operatorname{div}\big(\frac{\nabla F}{|\nabla F|}\big) + \langle \nabla V, \nabla F\rangle + \tilde{V}\Big) \tag{6.57}$$

which is similar to (6.55) with the first $|\nabla F|$ replaced to $\delta_\varepsilon(F)$.

Returning to the Chan–Vese energy (6.50), (6.57) becomes

$$\partial_t F = \delta_\varepsilon(F)\Big(\operatorname{div}\big(\frac{\nabla F}{|\nabla F|}\big) + \langle \nabla V, \nabla F\rangle$$
$$+ \lambda_{in}(I(x) - c_{in})^2 - \lambda_{out}(I(x) - c_{out})^2\Big). \tag{6.58}$$

The numbers c_{in} and c_{out} that characterize the image values inside and outside the shape are not always known in advance. In such a situation, it is natural to also minimize (6.50) with respect to c_{in} and c_{out}. For this to be possible, one needs to limit the size of the integration domain for the outside integral. When Ω is fixed, their optimal values are easily computed and are given by

$$c_{in} = \frac{\int_\Omega I(x)dx}{\operatorname{area}(\Omega)} \text{ and } c_{out} = \frac{\int_{D\setminus\Omega} I(x)dx}{\operatorname{area}(D\setminus\Omega)}.$$

If the ε-approximation using the delta function is used, the appropriate values are

$$c_{in} = \frac{\int_D(1 - H_\varepsilon(F(x)))I(x)dx}{\int_D(1 - H_\varepsilon(F(x)))dx} \text{ and } c_{out} = \frac{\int_D H_\varepsilon(F(x))I(x)dx}{\int_D H_\varepsilon(F(x))dx}.$$

Minimization in F, c_{in} and c_{out} can be implemented by running (6.55) or (6.57) while periodically updating c_{in} and c_{out} according to these expressions. The same approach extends to more general situations where the minimized energy takes the form

$$E(\Omega, \theta) = \int_{\partial\Omega} V(p, \theta)dl + \int_\Omega \tilde{V}(x, \theta)dx, \tag{6.59}$$

where the functions V and $\tilde{V}$ depend on an additional parameter θ, and one can run, say (6.55), by periodically updating θ in order to minimize E with fixed Ω.

6.4.11 Sobolev Active Contours

We now follow-up on our discussion at the end of Section 6.3, and describe the interesting variants of the active contour algorithms introduced in [197]. Using the notation of Section 6.3, the so-called Sobolev active contours modify an evolution like

$$\partial_t m = (\kappa V - \langle V, N\rangle + \tilde{V})N$$

(which corresponds to the minimization of (6.53)) and transform it into

$$\partial_t m = A^{-1}\big((\kappa V - \langle V, N\rangle + \tilde{V})N\big) \tag{6.60}$$

where A is a differential operator, or more generally an operator $f \mapsto Af$ such that the kth Fourier coefficient of Af is equal to $a_{|k|}c_k(f)$, where a_k is a positive sequence of numbers that tend to infinity (e.g., $a_k = 4\pi^2 k^2 + \lambda$) and $c_k(f)$ is the Fourier coefficient of f. When implementing (6.60), high-order Fourier coefficients of $(\kappa V - \langle V, N\rangle + \tilde{V})N$ are divided by large numbers, which results in attenuating high-frequency motion and focusing on low frequency variations, yielding global changes in the shape. The resulting algorithm makes the evolution less sensitive to noise (which typically induces high-frequency motion), and also accelerates convergence because large moves are easier to make. The counterpart is that small details in the shape boundary are harder to acquire, and may require running the algorithm over longer periods. Examples illustrating the robustness of Sobolev active contours are provided in Figure 6.4.

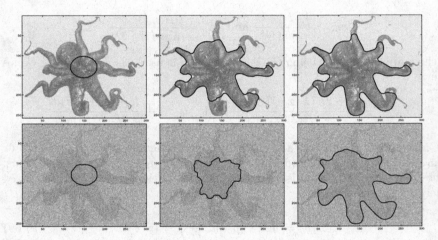

Fig. 6.4. Comparison between geometric and Sobolev active contours. In each row, the left image is the initial contour for both algorithms, the center one is the result of geometric active contours once the evolution stabilizes, and the right one is for Sobolev active contours. The first row presents a clean image and the second a noisy one. In both cases, geometric active contours get stuck at an unsatisfactory local minimum (especially in the noisy case), while Sobolev active contours reach a much more accurate solution.

7

Deformable templates

Deformable templates represent shapes as deformations of a given prototype, or template. Describing a shape therefore requires providing the following information:

(1) A description of the template.
(2) A description of the relation between the shape and the template.

This has multiple interesting aspects. The first one is that the template needs to be specified only once, for a whole family of curves. Describing the variation usually results in a simpler representation, typically involving a small number of parameters. The conciseness of the description is important for detection or tracking algorithms in which the shape is a variable, since it reduces the number of degrees of freedom. Another aspect is that small-dimensional representations are more easily amenable to probabilistic modeling, leading, as we will see, to interesting statistical shape models.

The methods that we describe provide a parametrized family of shapes, $m(\theta), \theta \in \Theta$ where Θ is a parameter set. Most of the time, Θ will be some subset of $\mathbb{R}^d$ but it can also be infinite-dimensional. We will always assume that $0 \in \Theta$ and that $m(0)$ represents the template.

To fix the ideas, we will assume that $m(\theta)$ is a parametrized curve $t \mapsto m(t, \theta)$ defined over a fixed interval $[a, b]$. Other situations can easily be transposed from this one. For example, one commonly uses configurations of *labeled* points, or landmarks, with $m(\theta) = (m_1(\theta), \ldots, m_N(\theta))$ as a finite-dimensional descriptor of a shape. Transposition from two dimensions to three dimensions is also easy.

7.1 Linear Representations

We start with a description of linear methods, in which

$$m(\theta) = m(0) + \sum_{k=1}^{n} \theta_k u_k$$

where u_k is a variation of $m(0)$: for example, if $m(0)$ is a closed curve, u_k is defined on $[a, b]$, taking values in $\mathbb{R}^d$ with $u_k(a) = u_k(b)$. If $m(0)$ is a configuration of points, u_k is a list of two-dimensional vectors.

The issue in this context is obviously how to choose the u_k's. We will here provide two examples, the first one based on a deterministic approach, and the second relying on statistical learning.

7.1.1 Energetic Representation

The framework developed in this section characterizes an object using a "small-deformation" model partially inspired by elasticity or mechanics. The object will therefore be described, not by what it looks like, but by how it deforms. This representation is inspired by the one developed in [165] for face recognition. It also includes the principal warps described in [28] as a particular case, and provides an interesting way of decomposing shape variations in a basis that depends on the geometry of the considered shape.

For such a shape m, we will consider small variations, represented by transformations $h \mapsto F(m, h)$. For example, one can take $F(m, h) = m + h$ when this makes sense. We assume that the small variations, h, belong to a Hilbert space H, with dot product $\langle \cdot, \cdot \rangle_m$, possibly depending on m.

Associate to h some deformation energy, denoted $E(h)$. Consider a time-dependent variation, $t \mapsto h(t)$ to which we attribute the total energy:

$$J(h) = \frac{1}{2} \int \|\dot{h}_t(t)\|_m^2 dt + \int E(h(t)) dt.$$

In the absence of external interaction, the trajectories prescribed are the extremals of the Lagrangian $\|\dot{h}_t\|_m^2/2 - E(h)$, characterized by

$$\ddot{h}_{tt} + \nabla E(h(t)) = 0$$

where ∇E is the Hilbert gradient, defined by

$$\partial_\varepsilon E(h + \varepsilon w)_{|_{\varepsilon=0}} = \langle \nabla E(h), w \rangle_m.$$

We make the assumption that this gradient exists. In fact, because we only analyze small variations, we will assume that a second derivative exists at $h = 0$, i.e., we assume that, for some symmetric operator Σ_m,

$$\nabla E(h) = \Sigma_m h + o(\|h\|).$$

Typically, we will have $E \geq 0$ with $E(0) = 0$, which ensures that Σ_m is a non-negative operator. The linearized equation for h now becomes

$$\ddot{h}_{tt} + \Sigma_m h = 0. \tag{7.1}$$

This equation has a simple solution when Σ_m is diagonalizable. Making this assumption (which is always true in finite dimension), letting $f_1, f_2, \ldots$ be the eigenvectors and $\lambda_1, \lambda_2, \ldots$ the corresponding eigenvalues (in decreasing order), solutions of (7.1) take the form

$$h(t) = \sum_{k \geq 1} \alpha^{(k)}(t) f_k$$

with $\ddot{\alpha}_{tt}^{(k)} + \lambda_k \alpha^{(k)}$ so that $\alpha^{(k)}$ oscillates with frequency $\omega_k = 1/\sqrt{\lambda_k}$.

This *modal representation* [181] (the ω_k are called the vibration modes of m) can be used to describe and compare shapes. One can use them as features and decide that similar shapes should have similar vibration modes. It is also possible to use this model for a template-based representation: let m be a template, with a modal decomposition as before, and represent small variations as

$$(\alpha_1, \ldots, \alpha_N) \to \tilde{m} = F(m, \sum_{k=1}^{N} \alpha_k f_k)$$

which has a linearized deformation energy given by $\sum_k \lambda_k \alpha_k^2$.

Consider a first example using plane curves. Let $m(.) = m(0, .)$ be the prototype and Ω_m its interior. We assume that m is parametrized with arc length. A deformation of m can be represented as a vector field $s \mapsto h(s)N(s)$, where h is a scalar function and N the unit normal to m. The deformed template is $s \mapsto m(s) + h(s)N(s)$. A simple choice for E is

$$E(h) = \frac{1}{2} \int_0^L h_s^2 ds$$

for which $\Sigma_m h = -\ddot{h}_{ss}$ and equation (7.1) is

$$\ddot{h}_{tt} = \ddot{h}_{ss}$$

which is the classical wave equation in one dimension. Since this equation does not depend on the prototype, m, it is not really interesting for our purposes, and we need to consider energies that depend more on geometric properties of m. Probably, the next simplest choice can be

$$E(h) = \frac{1}{2} \int_0^L \rho_m(s) \dot{h}_s^2 ds$$

where ρ_m is some function defined along m, for example $\rho_m = 1 + \kappa_m^2$ (where κ_m is the curvature along m). In this case, we get $\Sigma_m h = -2\partial_s(\rho h_s)$. The vibration modes are the eigenvectors of this inhomogeneous diffusion operator along the curve.

This approach is obviously subject to many variants. Another example is to consider discrete shapes, represented by a finite collection of landmarks. So a shape is now a finite collection $m = (x_1, \ldots, x_N)$ with each $x_i \in \mathbb{R}^2$. Given displacements $h = (h_1, \ldots, h_N)$, define $h^{(m)}(x)$, for $x \in \mathbb{R}^2$ by

$$h^{(m)}(x) = \sum_{i=1}^{N} g(|x - x_i|^2)\alpha_i$$

with $g(t) = e^{-t/2\sigma^2}$, where the $\alpha_i's$ are chosen so that $h(x_i) = h_i$. Then, we can define

$$E_m(h) = \int_{\mathbb{R}^2} |h^{(m)}(x)|^2 dx$$

$$= \sum_{i,j=1}^{N} \langle \alpha_i, \alpha_j \rangle \int_{\mathbb{R}^2} g(|x_i - x|^2)g(|x - x_j|^2)dx$$

$$= \sum_{i,j=1}^{N} c_{ij}(m)\langle \alpha_i, \alpha_j \rangle$$

with

$$c_{ij} = \int_{\mathbb{R}^2} e^{-\frac{|x_i - x|^2}{2\sigma^2} - \frac{|x_i - x|^2}{2\sigma^2}} dx = \pi\sigma^2 e^{-\frac{|x_i - x_j|^2}{4\sigma^2}}.$$

Finally, noting that, from the constraints, $\alpha = S(m)^{-1}h$ with $s_{ij}(m) = g(|x_i - x_j|^2)$, we have

$$E_m(h) = \mathbf{1}_d^T h^T S(m)^{-1} C(m) S(m)^{-1} h \mathbf{1}_d$$

where, in this expression, h is organized in a N by d matrix and $\mathbf{1}_d$ is the d-dimensional vector with all coordinates equal to 1. The modal decomposition will, in this case, be provided by eigenvalues and eigenvectors of $S(m)^{-1}C(m)S(m)^{-1}$.

The principal warp representation [28] is very similar to this one, and corresponds to

$$E_m(h) = \mathbf{1}_d^T h^T S(m)^{-1} h \mathbf{1}_d. \tag{7.2}$$

It is also associated to some energy computed as a function of $h^{(m)}$, as will be clear to the reader after the description of reproducing kernel Hilbert spaces in chapter 9. One can also define

$$E_m(h) = \int_{\mathbb{R}^2} \text{trace}((Dh^{(m)})^T Dh^{(m)})dx$$

or some other function of $(Dh^{(m)})^T Dh^{(m)}$, which corresponds to elastic energies. Closed-form computation can still be done as a function of $h_1, \ldots, h_N$, and provides a representation similar to the one introduced in [181, 165].

7.2 Probabilistic Decompositions

7.2.1 Deformation Axes

One can build another kind of modal decompositions using a database of shapes and principal component analysis (PCA).

Choose a shape representation like in Chapter 1, or, like in [54], based on a finite collection of points (landmarks) placed along (or within) the shape. The methods that follow provide a quite general framework, including infinite-dimensional representations, when they can be embedded in a Hilbert space.

Let's assume that we work with parametric curves. A database is assumed to contain N shapes, that we will consider as versions of the same object, or class of objects. We shall denote them $m^{(k)}(.)$, $k = 1, \ldots, N$, and they are assumed to all be defined on the same interval, I. The average is given by

$$\overline{m}(u) = \frac{1}{N} \sum_{k=1}^{N} m^{(k)}(u)$$

A PCA (cf. Appendix E) applied to $m^{(k)}$, $k = 1, \ldots, N$, with the L^2 dot product provides a finite-dimensional approximation called the *active shape representation*

$$m^{(k)}(u) = \overline{m}(u) + \sum_{i=1}^{p} \alpha_{ki} e^{(i)}(u) \tag{7.3}$$

where the principal directions $e^{(1)}, \ldots, e^{(p)}$ provide deformation modes along which the shape has the most variations.

This provides a new, small-dimensional, curve representation, in terms of variations of the template $\overline{m}$. One can use it, for example, to detect shapes in an image, since this requires the estimation of only p parameters.

One must be aware, when using this method, of the validity of the PCA approach, which is a linear method. It is important to discuss whether it is meaningful to compute linear combinations of deformation vectors. Obviously, once the data is represented by an array of numbers, one can always do this kind of computation. But the question to ask is whether one can safely go back, that is, whether one can associate a valid shape (which can be interpreted as an object of the same category as the initial dataset) to any such linear combination. The answer, in general, is yes, provided the coefficients in the decomposition are not too large; because of this, PCA-based decompositions can often be considered as first-order linear approximations for more complex, nonlinear, variations.

Returning to shapes, one can always declare that a curve is a linear object, belonging to some specified functional space. Although linear combinations are valid there, the results in doing so can often be unsatisfactory. As a simple example, assume that the objects that are considered all have triangular shapes. PCA has no mechanism to ensure that the shapes remain triangular after decomposition on a few principal components. Most often, the representation will be very poor, as far as shape interpretation is concerned.

In fact, shape decomposition must always be, in one way or another, coupled with some feature alignment on the dataset. In [54], this is implicit, since the approach is based on landmarks that have been properly selected by hand.

To deal with general curves, it is important to preprocess the parametrizations to ensure that they are consistent, in the sense that points with the same parameter have similar geometric properties. The curves cannot, in particular, all be assumed to be arc-length parametrized. One way to proceed is to assume that the parametrization is arc length for only one curve, say $m^{(0)}$. For the other curves, say $m^{(k)}, k = 1, \ldots, N$, we want to make a change of parametrization, $\varphi^{(k)}$, such that $m^{(k)}(\varphi^{(k)}(s)) = m_0(s) + \delta^{(k)}(s)$ with $\delta^{(k)}$ as small as possible. Methods to achieve such simultaneous parametrizations implement curve-matching algorithms. They will be presented later in Chapter 10.

In addition to aligning the parametrization, it is important to also ensure that the geometries are aligned, with respect to linear transformations (like rotations, translations, scaling). All these operations have the effect of representing all the shapes in the same "coordinate system", within which linear methods will be more likely to perform well.

This framework can be used to generate stochastic models of shapes. We can use the expression

$$m(u) = \overline{m}(u) + \sum_{i=1}^{p} \alpha_i e^{(i)}(u)$$

and generate random curves m by using randomly generated α_i's. Based on the statistical interpretation of PCA, the α_i's are uncorrelated, and their respective variances are the eigenvalues λ_i^2 that correspond to the eigenvector $e^{(i)}$. Simple models generate the α_i's as independent Gaussian variables with variance λ_i^2, or uniformly distributed on $[-\sqrt{3}\lambda_i, \sqrt{3}\lambda_i]$.

7.3 Stochastic Deformations Models

7.3.1 Generalities

The previous approaches analyzed variations directly in the shape representation. We now discuss a point of view which first models deformations as a generic process, before applying it to the template.

We consider here the (numerically important) situation in which the deformed curves are polygons. Restricting ourselves to this finitely generated family will simplify the mathematical formulation of the theory.

The template will therefore be represented as a list of contiguous line segments, and we will model a deformation as a process that can act on each line segment separately. The whole approach is a special case of Grenander's theory of deformable templates, and we refer to [101, 103, 102, 104] for more references and information. The general principles of deformable templates assume that an "object" can be built by assembling elementary components (called generators), with specified composition rules. In the case we consider here, generators are line segments and composition rules imply that exactly two segments are joined at their extremities. The second step in the theory is to specify a group action that can affect the generators, under the constraints of maintaining the composition rules. In our example, the group will consist of collections of planar similitudes.

7.3.2 Representation and Deformations of Planar Polygonal Shapes

The formulae being much simpler when expressed with complex notation, we identify a point $p = (x, y)$ in the plane to the complex number $x + iy$, that we also denote p. A polygonal line can either be defined by the ordered list of its vertices, say $s_0, \ldots, s_N \in \mathbb{C}$ or, equivalently, by one vertex s_0 and the sequence of vectors $v_k = s_{k+1} - s_k$, $k = 0, \ldots, N - 1$. The latter representation has the advantage that the sequence $(v_0, \ldots, v_{N-1})$ is a translation-invariant representation of the polygon. A polygonal line modulo translations will therefore be denoted $\pi = (v_0, \ldots, v_{N-1})$. The polygonal line is a polygon if it is closed, i.e., if and only if $v_0 + \cdots + v_{N-1} = 0$. A polygonal line with origin s_0 will be denoted (s_0, π).

A polygonal line can be deformed by a sequence of rotations and scalings applied separately to each edge v_k. In $\mathbb{C}$, such a transformation is just a complex multiplication. Therefore, a deformation is associated to an N-tuple of non-vanishing complex numbers $\mathbf{z} = (z_0, \ldots, z_{N-1})$, the action of $\mathbf{z}$ on π being

$$\mathbf{z}.\pi = (z_0 v_0, \ldots, z_{N-1} v_{N-1}). \tag{7.4}$$

This defines a group action (cf. Appendix B.5) of $G = (\mathbb{C} \setminus \{0\})^N$ on the set of polygonal lines with N vertices.

In this group, some transformations play a particular role. Introduce the set

$$\Delta = \{\mathbf{z} \in G, \mathbf{z} = z(1, \ldots, 1), z \in \mathbb{C}\}$$

(the diagonal in G). An element in Δ provides a single similitude applied simultaneously to all the edges, i.e., Δ represents the actions of similitudes on polygons. Similarly, the set

$$\Delta_0 = \{\mathbf{z} \in G, \mathbf{z} = z(1, \ldots, 1), z \in \mathbb{C}.|z| = 1\}$$

represents the action of rotations.

A polygonal line modulo similitudes (resp. rotations) can be represented as an orbit $\Delta.\pi$ (resp. $\Delta_0.\pi$). We can define the quotient groups G/Δ and G/Δ_0, namely the sets of orbits $\Delta.\mathbf{z}$ (resp. $\Delta_0.\mathbf{z}$) for $\mathbf{z} \in G$ (they have a group structure because G is commutative). One obtains a well-posed action of, say, G/Δ on polygonal lines modulo similitudes, by defining

$$(\mathbf{z}.\Delta).(\Delta.\pi) = \Delta.(\mathbf{z}.\pi).$$

Given a polygon, π, we define $F(\pi)$ to be the set of group elements $\mathbf{z}$ in G that transform π into another polygon, namely

$$F(\pi) = \{\mathbf{z} \in G, z_0 v_0 + \cdots + z_{N-1} v_{N-1} = 0\}. \tag{7.5}$$

Note that $F(\pi)$ is not a subgroup of G.

We can use this representation to provide a model for random polygonal lines. It suffices for this to choose a template π and a random variable ζ on G and to take $\zeta.\pi$ to obtain a random polygonal line. So what we need is a probability model on G.

In fact, since we are interested in shapes, we will restrict ourselves to closed lines. Therefore, given $\pi = (v_0, \ldots, v_{N-1})$, the model will have to be supported on the set $F(\pi)$ defined in (7.5).

Consider the function:

$$E(\mathbf{z}) = (\alpha/2) \sum_{k=0}^{N-1} |z_k - 1|^2 + (\beta/2) \sum_{k=0}^{N-1} |z_k - z_{k-1}|^2.$$

The first term is large when $\mathbf{z}$ is far from the identity, and the second one penalizes strong variations between consecutive z_i's. Here and in the following, we let $z_{-1} = z_{N-1}$.

We want to choose a probability on G which is small when E is large. A natural choice would be to take the measure with density proportional to $\exp(-E(\mathbf{z}))/\prod_{k=1}^{N-1} |z_k|$ with respect to Lebesgue's measure on $\mathbb{C}^{N-1}$. This is the "Gibbs measure", with energy E, relative to the Haar measure, $\prod_{k=1}^{N-1} dz_k/|z_k|$ which is the uniform measure on G. Such a choice is of interest in that it gives a very small probability to small values of $|z_k|$, which is consistent with the fact that the z_k's are non-vanishing on G.

Unfortunately, this model leads to intractable computations, and we will rely on the simpler, but less accurate, model with density proportional to $\exp(-E(\mathbf{z}))$. This choice will greatly simplify the simulation algorithms, and in particular, the handling of the closedness constraint, which has not been taken into account yet.

With $\pi = (v_0, \ldots, v_{N-1})$, this constraint is expressed by

$$\sum_k v_k z_k = 0.$$

The model we are interested in therefore is the previous one, conditional to this constraint. This can be analyzed by using a discrete Fourier transform. Denote

$$u_l = \hat{z}_l = \frac{1}{\sqrt{N}} \sum_{k=0}^{N-1} z_k e^{-2i\pi \frac{kl}{N}}.$$

One can easily prove that E can be written

$$E(\mathbf{z}) = \alpha |u_0 - \sqrt{N}|^2 + \sum_{l=1}^{N-1} \left(\alpha + 2\beta(1 - \cos \frac{2\pi l}{N}) \right) |u_l|^2,$$

and that the constraint becomes

$$\sum_{l=0}^{N-1} \hat{v}_l u_l = 0$$

with $\hat{v}_l = \frac{1}{\sqrt{N}} \sum_{k=0}^{N-1} v_k e^{-2i\pi \frac{kl}{N}}$. Note that, since π is closed, we have $\hat{v}_0 = 0$.

Denote $w_0 = \sqrt{\alpha}(u_0 - \sqrt{N})$, and, for $l \geq 1$, $w_l = \sqrt{\alpha + 2\beta(1 - \cos\frac{2\pi l}{N})}u_l$, so that

$$E(\mathbf{z}) = \sum_{l=0}^{N-1} |w_l|^2.$$

Without the constraint, the previous computation implies that the real and imaginary parts of $w_0, \ldots, w_{N-1}$ are mutually independent standard Gaussian variables: they therefore can be easily simulated, and the value of $z_0, \ldots z_{N-1}$ directly computed after an inverse Fourier transform. Conditioning on closedness only slightly complicates the procedure. Replacing u_l by its expression as a function of $w_1, \ldots, w_{N-1}$, and using $\hat{v}_0 = 0$, the constraint can be written in the form

$$\sum_{l=0}^{N-1} c_l w_l = 0.$$

with $c_0 = 0$ ($c_p = \sqrt{\alpha + 2\beta(1 - \cos\frac{2\pi l}{N})}$). The following standard lemma from the theory of Gaussian variables solves our problem.

Lemma 7.1. *Let $\mathbf{w}$ be a standard Gaussian vector in $\mathbb{R}^{2N}$, and let V be a vector subspace of $\mathbb{R}^{2N}$. Let Π_V be the orthogonal projection on V. Then, the random variable $\Pi_V(\mathbf{w})$ follows the conditional distribution given that $\mathbf{w} \in V$.*

Assume that $\mathbf{c} = (c_0, \ldots, c_{N-1})$ has been normalized so that $\sum |c_i|^2 = 1$. To sample closed random polygonal lines, it suffices to sample a standard Gaussian $\mathbf{w}^*$ in $\mathbb{C}^N$, and set

$$\mathbf{w} = \mathbf{w}^* - \left(\sum_{l=0}^{N-1} c_l w_l^*\right).\mathbf{c}.$$

Some examples of random shapes simulated with this process are provided in Figure 7.1.

7.4 Segmentation with Deformable Templates

Using deformable templates in shape segmentation algorithms incorporates much stronger constraints than with active contours, which only implement the fact that shapes can be assumed to be smooth. If one knows the kind of

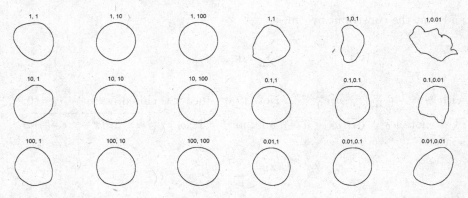

Fig. 7.1. Deformations of a circle (with different values for α and β).

shapes that are to be detected, one obviously gains in robustness and accuracy by using a segmentation method that looks for small variation of an average shape in this category.

Detection algorithms can be associated to the models provided in Sections 7.2 and 7.3. Let's start with Section 7.2 with a representation that takes the form, denoting $\alpha = (\alpha_1, \ldots \alpha_{p_0})$:

$$m^\alpha = \overline{m} + \sum_{i=1}^{p_0} \alpha_i K^{(i)} \,,$$

for some template $\overline{m}$ and vector fields $K^{(i)}$. This information also comes with the variance of α_i, denoted λ_i^2.

The pose of the shape within the image is also unknown. It will be associated to a rigid or affine deformation g applied to m^α. The problem is then to find g and α such that $g m^\alpha$ is close to regions of low deformation energy within the image.

One can use a variational approach for this purpose. Like in Section 6.4.9, one starts with the definition of a potential V which is small when close to contours. One can then define

$$E(g, \alpha) = \sum_{i=1}^{n} \frac{\alpha_i^2}{\lambda_i^2} + \beta \int_I V(g.m^\alpha(u))du.$$

The derivatives of E are

$$\frac{\partial E}{\partial \alpha_i} = 2\frac{\alpha_i}{\lambda_i^2} + \beta \int_I \langle \nabla V(g.m^\alpha(u)), g.K^i(u) \rangle du$$

and

$$\frac{\partial E}{\partial g} = \int_I m^\alpha(u) \nabla V(g.m^\alpha(u))^T du$$

(it is a matrix). Similar computation can be made for variants of the definition of the cost function. One can, for example add a penalty (like $|\log\det(g)|$) to penalize shapes that are too small or too large. One can also replace the quadratic term in α_i by boundedness constraints, like $|\alpha_i| < \sqrt{3}\lambda_i$.

If scaling to very small curves is penalized, it is plausible that, contrary to the case of active contours, the global minimum of E provides an acceptable solution. However, from a practical point of view, minimizing E is a difficult problem, with many local minima. It is therefore still necessary to start the algorithm with a good guess of the initial curve.

Consider now the representation of Section 7.3. We will use the same notation, a shape being modeled by a polygon π with N edges denoted $(v_0, \ldots, v_{N-1})$. A deformation is represented by N complex numbers $\mathbf{z} = (z_0, \ldots, z_{N-1})$, with the action

$$\mathbf{z}.\pi = (z_0 v_0, \ldots, z_{N-1} v_{N-1}).$$

We have denoted by Δ (resp. Δ_0) the set of $\mathbf{z}$'s for which all z_i's coincide (resp. coincide and have modulus 1); these subgroups of $G = (\mathbb{C} \setminus \{0\})^N$ correspond to plane similitudes (resp. plane rotations).

We denote by $[\mathbf{z}]$ and $[\mathbf{z}]_0$ the classes of $\mathbf{z}$ modulo Δ and Δ_0. Similarly, when π is a polygon, we denote by $[\pi]$ and $[\pi]_0$ the classes of π modulo Δ and Δ_0. For example,

$$[\mathbf{z}] = \{c.\mathbf{z}, c \in \Delta\}$$

and

$$[\pi] = \{\mathbf{z}.\pi, \mathbf{z} \in \Delta\}.$$

A template should be considered as a polygon modulo Δ or Δ_0 (depending on whether scale invariance is required), whereas a shape embedded in an image should be a polygon with an origin. Let $\bar{\pi}$ denote the template, although we should use the notation $[\bar{\pi}]$ or $[\bar{\pi}]_0$. Introduce a function $V(.)$ defined over the image, that is large for points that are far from image contours. The quantity that can be minimized is

$$Q(\mathbf{z}, s_0) = E([\mathbf{z}]) + \int_{(s_0 + \mathbf{z}.\bar{\pi})} V(m)dm.$$

with $s_0 \in \mathbb{C}$ and $\mathbf{z} \in G$. The deformation energy E is a function defined on G/Δ (or equivalently a function defined on G, invariant to similitude transformations), that measure the difference between $\mathbf{z}$ and a similitude. For example, with $\mathbf{z} = (z_0, \ldots, z_{N-1})$, and $z_k = r_k e^{i\theta_k}$, one can take

$$E([\mathbf{z}]) = \sum_{k=1}^{N} (\log r_k - \log r_{k-1})^2 + \sum_{k=1}^{N} \arg(e^{i\theta_k - i\theta_{k-1}})^2.$$

Here, we have defined $\arg z$, for $z \neq 0$ as the unique $\theta \in]-\pi, \pi]$ such that $z = re^{i\theta}$ with $r > 0$. We also use the convention $r_N = r_0, \theta_N = \theta_0$ for the last term of the sum (assuming we are dealing with closed curves).

If scale invariance is relaxed, a simpler choice is

$$E([\mathbf{z}]_0) = \sum_{k=1}^{N} |z_k - z_{k-1}|^2.$$

Note that for closed curves, it must be ensured that $\mathbf{z}\bar{\pi}$ remains closed, which induces the additional constraint, taking $\pi = (v_0, \ldots, v_{N-1})$:

$$\sum_{k=0}^{N-1} z_k v_k = 0.$$

It is interesting to compute the continuum limit of this energy. Still using complex numbers, we consider a C^1 template curve $m : I \to \mathbb{C}$, where $I = [0, L]$ is an interval, with arc-length parametrization. For a given N, we consider the polygon $\pi = (v_0, \ldots, v_{N-1})$, with

$$v_k = m(\frac{kL}{N}) - m(\frac{(k-1)L}{N}) \simeq \frac{L}{N} \dot{m}_s(\frac{(k-1)L}{N}).$$

A deformation, represented by $\mathbf{z} = (z_0, \ldots, z_{N-1})$ will also be assumed to come from a continuous curve ζ defined on $[0, 1]$ with $z_k = \zeta(k/N)$. The continuum equivalent of $\pi \mapsto \pi.\mathbf{z}$ can then be written

$$\dot{m}_s(s) \mapsto \zeta(s/L)\dot{m}_s(s)$$

which leads us to define an action of non-vanishing complex-valued curves ζ on closed curves by

$$(\zeta.m)(s) = \int_0^s \zeta(u/L)\dot{m}_s(u)du.$$

In the rotation-invariant case, the energy of the action should be given by the limit of

$$\sum_{k=1}^{N-1} |z_k - z_{k-1}|^2.$$

Using the fact that $z_k - z_{k-1} \simeq \dot{\zeta}_u((k-1)/N)/N$, we have the continuum equivalent

$$N \sum_{k=1}^{N-1} |z_k - z_{k-1}|^2 \to \int_0^1 |\dot{\zeta}_s(s)|^2 du.$$

This is the H^1 norm of the deformation generator along the curve.

8

Ordinary Differential Equations and Groups of Diffeomorphisms

8.1 Introduction

This chapter introduces spaces of diffeomorphisms, and describe how ordinary differential equations provide a convenient way of generating deformations.

We fix an open subset Ω in $\mathbb{R}^d$. A deformation is a function φ which assigns to each point $x \in \Omega$ a displaced position $y = \varphi(x) \in \Omega$. There are two undesired behaviors that we would like to forbid:

- The deformation should not create holes: every point $y \in \Omega$ should be the image of some point $x \in \Omega$, i.e., φ should be onto.
- Folds are also prohibited: two distinct points x and x' in Ω should not target the same point $y \in \Omega$, i.e., φ must be one-to-one.

Thus deformations must be bijections of Ω. In addition, we require some smoothness for φ. The next definition recalls some previously discussed vocabulary:

Definition 8.1. *A homeomorphism of Ω is a continuous bijection $\varphi : \Omega \to \Omega$ such that its inverse, φ^{-1} is continuous.*

A diffeomorphism of Ω is a continuously differentiable homeomorphism $\varphi : \Omega \to \Omega$ such that φ^{-1} is continuously differentiable.

In fact, using the inverse mapping theorem, one shows that a continuously differentiable homeomorphism is a diffeomorphism. From now on, most of the deformations we shall consider will be diffeomorphisms of some open set $\Omega \subset \mathbb{R}^d$. If φ and φ' are diffeomorphisms, then $\varphi \circ \varphi'$ is a diffeomorphism, and so is φ^{-1} by definition. Diffeomorphisms of Ω form a group for the composition of functions, often denoted $\mathrm{Diff}(\Omega)$.

We will be concerned with specific subgroups of $\mathrm{Diff}(\Omega)$ associated to additional smoothness requirements. Using the notation of Appendix A, we define, for a multivariate, vector-valued function f, $\|f\|_{p,\infty}$ as the sum of the supremum norms of the partial derivatives of order less than or equal to p of the components of f.

Definition 8.2. *Let $p \geq 1$. We define $G_{p,\infty}(\Omega)$ as the set of diffeomorphisms of Ω, φ, such that*

$$\max(\|\varphi - id\|_{p,\infty}, \|\varphi^{-1} - id\|_{p,\infty}) < \infty.$$

We will write $G_{p,\infty}$ instead of $G_{p,\infty}(\Omega)$ whenever Ω is clear from the context. We show that $G_{p,\infty}$ is a subgroup of $\mathrm{Diff}(\Omega)$. The facts that $id \in G_{p,\infty}$ and that $\varphi \in G_{p,\infty} \Rightarrow \varphi^{-1} \in G_{p,\infty}$ is obvious. Stability for composition will be clear from the following lemma. For $k \in \{1, \ldots, d\}$, denote by $\partial_k f$ the partial derivative of f with respect to the kth coordinate. For any p-tuple $J = (k_1, \ldots, k_p)$, we will let

$$\partial_J f = \partial_{k_1} \ldots \partial_{k_p} f. \tag{8.1}$$

(Note that indices can be repeated in J and that the operator does not depend on how the elements of J have been ordered.) We say that a q-tuple I is included in J ($I \subset J$) if $I = (k_{i_1}, \ldots, k_{i_q})$ with $1 \leq i_1 < \cdots < i_q \leq p$, and define the set

$$\partial_{(J)} f = \{\partial_I f, I \subset J\}. \tag{8.2}$$

We have the following lemma.

Lemma 8.3. *Let $g : \Omega \to \Omega$ and $f : \Omega \to \mathbb{R}$ be C^p functions on Ω, and $J = (k_1, \ldots, k_p) \subset \{1, \ldots, d\}^p$. Then $f \circ g$ is C^p and*

$$\partial_J(f \circ g) = \sum_{I, |I| \leq |J|} ((\partial_I f) \circ g) B_{I,J}(\partial_{(J)} g) \tag{8.3}$$

where $B_{I,J}(\partial_{(J)} g)$ is a linear combination (with coefficients that do not depend on f or g) of terms of the form $\partial_{I_1} g_{i_1} \ldots \partial_{I_q} g_{i_q}$ where each I_j is included in J, each element of J belongs in exactly one of the I_j, and g_k denotes the kth component of g.

Proof. We can show (8.3) by induction (assuming the result true for J and showing that it remains valid with one more derivative). We have

$$\partial_J \partial_k(f \circ g) = \partial_J(Df \circ g\, \partial_k g)$$

$$= \sum_{j=1}^{d} \partial_J(\partial_j f \circ g\, \partial_k g_j)$$

$$= \sum_{j=1}^{d} \sum_{I \subset J} \partial_I(\partial_j f \circ g)\, \partial_{J \setminus I} \partial_k g_j$$

$$= \sum_{j=1}^{d} \sum_{I \subset J} \sum_{\tilde{I}, |\tilde{I}| \leq |I|} (\partial_{\tilde{I}} \partial_j f) \circ g\, B_{\tilde{I}, I}(\partial_{(I)} g)\, \partial_{J \setminus I} \partial_k g_j$$

$$= \sum_{j=1}^{d} \sum_{\tilde{I}, |\tilde{I}| \leq |J|} (\partial_{\tilde{I}} \partial_j f) \circ g \sum_{I \subset J, |I| \geq |\tilde{I}|} B_{\tilde{I}, I}(\partial_{(I)} g)\, \partial_{J \setminus I} \partial_k g_j.$$

This new expression can now be seen to take the form described in the lemma, with J replaced by $J \cup \{k\}$. In this computation, we have used the product rule

$$\partial_J(uv) = \sum_{I \subset J} \partial_I u \, \partial_{J \setminus I} v \tag{8.4}$$

which can be shown by induction, the proof being left to the reader.   □

From this lemma directly follows:

Proposition 8.4. *Let $g : \Omega \to \Omega$ and $f : \Omega \to \mathbb{R}$ be C^p functions on Ω. Then*

$$\|f \circ g\|_{p,\infty} \leq C(p,d)\|f\|_{p,\infty} \sum_{q_1,\dots,q_j : q_1 + \cdots + q_j = p} \|g\|_{q_1,\infty} \cdots \|g\|_{q_j,\infty} \tag{8.5}$$

for some fixed constant $C(p,d)$ depending on p and on the dimension.

We also have:

Corollary 8.5. $G_{p,\infty}(\Omega)$ *is a subgroup of* $\mathrm{Diff}(\Omega)$.

Diffeomorphisms act on various structures supported by Ω. Consider, for example, a function $I : \Omega \to \mathbb{R}$ (an image) and a diffeomorphism φ of Ω. The associated deformation creates a new image I' on Ω by letting $I'(y)$ be the value of I at the position x which has been moved to y, i.e., $I'(y) = I(\varphi^{-1}(y))$ or $I' = I \circ \varphi^{-1}$. We will be specifically interested in the inverse problem of estimating the best diffeomorphism from the output of its action. For example, the image-matching problem consists in finding an algorithm which, given two functions I and I' on Ω, is able to recover a plausible diffeomorphism φ such that $I' = I \circ \varphi^{-1}$.

To be able to develop these algorithms, we will need a computational construction of diffeomorphisms (this is not provided by Definition 8.2). In order to motivate our general construction, we start with a direct, but limited, way of building diffeomorphisms, by small perturbations of the identity.

Proposition 8.6. *Let $u \in C^1(\Omega, \mathbb{R}^d)$, and assume that,*

(i) $u(x)$ and $Du(x)$ tend to 0 when x tend to infinity.
(ii) There exists $\delta_0 > 0$ such that $x + \delta u(x) \in \Omega$ for all $x \in \Omega$ and $\delta < \delta_0$.

Then, for small enough ε, $\varphi : x \mapsto x + \varepsilon u(x)$ is a diffeomorphism of Ω.

Condition (ii) is true, for example, if $\mathrm{dist}(x, \Omega^c) > \mathrm{cst}|u(x)|$ for $x \in \Omega$. Another case is when Ω can be defined as the domain $F < 0$ for some function F with $\nabla F \neq 0$ on $F = 0$, and such that $\langle \nabla F, u \rangle < 0$ near the domain $F = 0$.

Proof. The function φ is obviously continuously differentiable, and takes values in Ω as soon as as $\varepsilon < \delta \|u\|_\infty$.

Since Du is continuous and tends to 0 at infinity, it is bounded and there exists a constant C such that $|u(x) - u(x')| \leq C\,|x - x'|$. If $\varphi(x) = \varphi(x')$ we have

$$|x - x'| = \varepsilon\,|u(x) - u(x')| \leq C\varepsilon\,|x - x'|$$

which implies $x = x'$ as soon as one takes $\varepsilon < 1/C$, and φ is one-to-one in this case.

Let's show that φ is onto. Take $y \in \Omega$ and $\delta < \delta_0$ such that $B(y, \delta) \subset \Omega$. Consider the function ψ_y, defined on $B(0, \delta)$ by $\psi_y(\eta) = -\varepsilon u(y + \eta)$. If $\varepsilon < \delta\,\|u\|_\infty$, we have $\psi_y(\eta) \in B(0, \delta)$ and the inequality

$$|\psi_y(\eta) - \psi_y(\eta')| \leq \varepsilon C\,|\eta - \eta'|.$$

If $\varepsilon C < 1$, ψ_y is contractive, and the fixed-point theorem (Theorem C.2) implies that there exists $\eta \in B(0, \delta)$ such that $\psi_y(\eta) = \eta$. But in this case,

$$\varphi(y + \eta) = y + \eta + \varepsilon u(y + \eta) = y + \eta - \psi_y(\eta) = y$$

so that $y \in \varphi(\Omega)$ and φ is onto.

It remains to prove that φ^{-1} is continuous. Take $y, y' \in \Omega$ and let η_y and $\eta_{y'}$ be the fixed points of ψ_y and $\psi_{y'}$ in $B(y, \delta)$ and $B(y', \delta)$ respectively (taking a small enough δ). We have

$$|\varphi^{-1}(y) - \varphi^{-1}(y')| = |\eta_y + y - \eta_{y'} - y'| \leq |\eta_y - \eta_{y'}| + |y - y'|$$

so that it suffices to prove that η_y is close to $\eta_{y'}$ when y is close to y'. We have

$$\begin{aligned}
|\eta_y - \eta_{y'}| &= |\psi_y(\eta_y) - \psi_{y'}(\eta_{y'})| \\
&= \varepsilon|u(y + \eta_y) - u(y' + \eta_{y'})| \\
&\leq C\varepsilon(|y - y'| + |\eta_y - \eta_{y'}|)
\end{aligned}$$

so that, if $C\varepsilon < 1$,

$$|\eta_y - \eta_{y'}| \leq \frac{C\varepsilon}{1 - C\varepsilon}|y - y'|.$$

which proves the continuity of φ^{-1}. $\square$

We therefore know how to build small deformations. Of course, we cannot be satisfied with this, since they correspond to a rather limited class of diffeomorphisms. However, we can use them to generate large deformations, because diffeomorphisms can be combined using composition.

Thus, let $\varepsilon_0 > 0$ and $u_1, \dots, u_n, \dots$ be vector fields on Ω which are such that, for $\varepsilon < \varepsilon_0$, $id + \varepsilon u_i$ is a diffeomorphism of Ω. Consider

$$\varphi_n = (id + \varepsilon u_n) \circ \cdots \circ (id + \varepsilon u_1).$$

We have

$$\varphi_{n+1} = (id + \varepsilon u_n) \circ \varphi_n = \varphi_n + \varepsilon u_n \circ \varphi_n$$

which can also be written $(\varphi_{n+1} - \varphi_n)/\varepsilon = u_n \circ \varphi_n$. Fixing $x \in \Omega$ and letting $x_0 = x$, $x_n = \varphi_n(x)$, we have the relation $(x_{n+1} - x_n)/\varepsilon = u_n(x_n)$. This can be viewed as a discretization of a differential equation, of the form (introducing a continuous time variable t):

$$\partial_t x(t) = u(t, x(t)).$$

This motivates the rest of this chapter, which will be devoted to building diffeomorphisms as flows associated to ordinary differential equations (ODEs).

8.2 Flows and Groups of Diffeomorphisms

8.2.1 Definitions

We let $\Omega \subset \mathbb{R}^d$ be open and bounded. We denote by $C_0^1(\Omega, \mathbb{R}^d)$ the Banach space of continuously differentiable vector fields v on Ω such that v and Dv vanish on $\partial\Omega$ and at infinity. Elements $v \in C_0^1(\Omega, \mathbb{R}^d)$ can be considered as defined on $\mathbb{R}^d$ by setting $v(x) = 0$ if $x \notin \Omega$.

We define the set $\mathcal{X}_1^1(T, \Omega)$ of absolutely integrable functions from $[0, T]$ to $C_0^1(\Omega, \mathbb{R}^d)$. An element of $\mathcal{X}_1^1(T, \Omega)$ is a time-dependent vector field, $(v(t, .), t \in [0, 1])$ such that, for each t, $v(t) := v(t, .) \in C_0^1(\Omega, \mathbb{R}^d)$ and

$$\|v\|_{\mathcal{X}_1^1, T} := \int_0^T \|v(t)\|_{1,\infty} \, dt < \infty \tag{8.6}$$

where $\|v\|_{1,\infty} = \|v\|_\infty + \|Dv\|_\infty$. More generally we define $\mathcal{X}_p^1(T, \Omega)$ like $\mathcal{X}_1^1$, simply replacing the $(1, \infty)$ norm by the (p, ∞) norm.

For $v \in \mathcal{X}_1^1(T, \Omega)$, we consider the ordinary differential equation $dy/dt = v(t, y)$. Since $\|v\|_{1,\infty}$ is an upper bound for the Lipschitz constant of v, the general results proved in Appendix C imply that this equation has a unique solution over $[0, T]$ given any initial condition $y(s) = x$, and that the associated flow, φ_{st}^v, defined by

$$\partial_t \varphi_{st}^v = v(t) \circ \varphi_{st}^v, \tag{8.7}$$

and $\varphi_{ss}^v = id$, is a homeomorphism of Ω, which is Lipschitz with

$$|\varphi_{sr}^v(x) - \varphi_{sr}^v(y)| \leq |x - y| \exp\left(\int_s^t \|v(r)\|_{1,\infty} \, dr\right) \tag{8.8}$$

(cf. equation (C.6)).

In fact, φ_{st}^v is also a diffeomorphism, as we now prove.

Theorem 8.7. Let $v \in \mathcal{X}_1^1(T, \Omega)$. The associated flow, φ_{st}^v, is at all times a diffeomorphism of Ω.

Proof. To guess what we aim at, we assume first that the result is true and formally differentiate the equation

$$\partial_t \varphi_{st}^v(x) = v(t, \varphi_{st}^v(x))$$

to obtain, applying the chain rule,

$$\partial_t D\varphi_{st}^v(x).h = Dv(t, \varphi_{st}^v(x)) D\varphi_{st}^v(x).h.$$

Here, the notation $Dv(t, .)$ refers to a differentiation in the space variables (not t). This computation suggests introducing the linear differential equation

$$\partial_t W(t) = Dv(t, \varphi_{st}^v(x))W(t)$$

with initial conditions $W(s) = h$. By corollary C.4, this equation also admits a unique solution on $[0, T]$. The previous computation indicates that this solution should be such that $W(t) = D\varphi_{st}^v(x).h$, and we now prove this result. Define

$$a(t, h) = \varphi_{st}^v(x + h) - \varphi_{st}^v(x) - W(t).$$

We need to show that $a(t, h) = o(|h|)$ when $h \to 0$. For $\alpha > 0$, define

$$\mu(t, \alpha) = \max \{|Dv(t, x) - Dv(t, y)| : x, y \in \Omega, |x - y| \leq \alpha\}.$$

The function $x \mapsto Dv(t, x)$ is continuous and tends to 0 at infinity (or on $\partial\Omega$) and therefore is uniformly continuous, which is equivalent to the fact that $\mu(t, \alpha) \to 0$ when $\alpha \to 0$. We can write

$$a(t, h) = \int_s^t (v(r, \varphi_{sr}^v(x + h)) - v(r, \varphi_{sr}^v(x))) \, dr - \int_s^t Dv(r, \varphi_{sr}^v(x))W(r)dr$$

$$= \int_s^t Dv(r, \varphi_{sr}^v(x))a(r, h)dr$$

$$+ \int_s^t (v(r, \varphi_{sr}^v(x + h)) - v(r, \varphi_{sr}^v(x))$$

$$- \varepsilon Dv(r, \varphi_{sr}^v(x)) (\varphi_{sr}^v(x + h) - \varphi_{sr}^v(x))) \, dr.$$

We have, for all $x, y \in \Omega$:

$$|v(t, y) - v(t, x) - Dv(t, x)(y - x)| \leq \mu(t, |x - y|) |x - y|.$$

This inequality, combined with equation (8.8), yields

$$|a(t, h)| \leq \int_s^t \|v(r)\|_{1,\infty} |a(r, h)| \, dr + C(v) |h| \int_0^T \mu(r, C(v) |h|)dr$$

for a constant $C(v)$ which only depends on v. To conclude the proof using Gronwall's lemma, we need the fact that

$$\lim_{\alpha \to 0} \int_0^T \mu(r, \alpha) dr = 0.$$

This is a consequence of the fact that $\mu(r, \alpha) \to 0$ for fixed r when $\alpha \to 0$ and of the upper bound $\mu(r, \alpha) \le 2 \|v\|_{1,\infty}$ which allows us to apply the dominated convergence theorem. $\qquad\square$

We have incidentally proved the following important fact:

Proposition 8.8. Let $v \in \mathcal{X}_1^1(T, \Omega)$. Then for fixed $x \in \Omega$, $D\varphi_{st}^v(x)$ is the solution of the linear differential equation

$$\partial_t W(t) = Dv(t, \varphi_{st}^v(x)) W(t) \tag{8.9}$$

with initial condition $W(s) = \mathrm{Id}_d$.

In fact, this result can be extended to p derivatives.

Theorem 8.9. If $p \ge 1$ and $\int_0^t \|v(t)\|_{p,\infty} dt < \infty$, then φ_{st}^v is p times differentiable and for all $q \le p$

$$\partial_t D^q \varphi_{st}^v = D^q (v(t) \circ \varphi_{st}^v). \tag{8.10}$$

Moreover, there exist constants C, C' (independent of v) such that,

$$\sup_{s \in [0,1]} \|\varphi_{st}^v - id\|_{p,\infty} \le C e^{C' \int_0^1 \|v(t)\|_{p,\infty} dt}. \tag{8.11}$$

Equation (8.10) says that, for any iterated partial differential operator (as defined in (8.1)), we have

$$\partial_t \partial_J \varphi_{st}^v = \partial_J (v(t) \circ \varphi_{st}^v). \tag{8.12}$$

The right-hand side takes the form provided in Lemma 8.3. It depends on all partial derivatives $\partial_I \varphi_{st}^v$ for $I \subset J$, but the condition given in Lemma 8.3 implies that it depends linearly on $\partial_J \varphi_{st}^v$. This implies that, if derivatives of φ_{st}^v of order smaller that p have been computed, derivatives of order p can be computed by solving a non-homogeneous linear differential equation, with vanishing initial conditions for $p \ge 2$. To prove Theorem 8.9, proceed by induction, assuming equation (8.12) when $|J| < p$, and using Lemma 8.3 to see that Proposition 8.8 can be applied to get one more derivative (details are left to the reader).

Variations with Respect to the Vector Field

It will be quite important, in the following, to be also able to characterize the effects that a variation of the time-dependent vector v may have on the induced flow. For this purpose, we fix $v \in \mathcal{X}_1^1(T, \Omega)$, and $h \in \mathcal{X}_1^1(T, \Omega)$, and proceed to the computation of $\partial_\varepsilon \varphi_{st}^{v+\varepsilon h}$ at $\varepsilon = 0$. The argument is similar to the one in the previous section: first, make a guess by formal differentiation of the original ODE, then proceed to a rigorous argument to show that the guessed expression is correct. So, consider the equation

$$\partial_t \varphi_{st}^{v+\varepsilon h} = v(t) \circ \varphi_{st}^{v+\varepsilon h} + \varepsilon h(t) \circ \varphi_{st}^{v+\varepsilon h}$$

and formally compute its derivative with respect to ε. This yields

$$\partial_t \partial_\varepsilon \varphi_{st}^{v+\varepsilon h} = h(t) \circ \varphi_{st}^{v+\varepsilon h} + D(v(t) + \varepsilon h(t)) \circ \varphi_{st}^{v+\varepsilon h}(\partial_\varepsilon \varphi_{st}^{v+\varepsilon h})$$

and, letting $\varepsilon = 0$,

$$\partial_t \Big(\frac{\partial}{\partial \varepsilon} \varphi_{st}^{v+\varepsilon h} \big|_{\varepsilon=0} \Big) = h(t) \circ \varphi_{st}^v + (Dv(t) \circ \varphi_{st}^v) \Big(\frac{d}{d\varepsilon} \varphi_{st}^{v+\varepsilon h} \big|_{\varepsilon=0} \Big)$$

which naturally leads to introducing the solution of the differential equation

$$\partial_t W(t) = h(t) \circ \varphi_{st}^v + (Dv(t) \circ \varphi_{st}^v) W(t) \tag{8.13}$$

with initial condition $W(s) = 0$. We now set

$$a^\varepsilon(t) = \left(\varphi_{st}^{v+\varepsilon h}(x) - \varphi_{st}^v(x) \right) / \varepsilon - W(t)$$

and express it in the form

$$a^\varepsilon(t) = \int_s^t Dv(u, \varphi_{su}^v) a^\varepsilon(u) du + \int_s^t (h(u, \varphi_{su}^{v+\varepsilon h}(x)) - h(u, \varphi_{su}^v(x))) du$$

$$+ \frac{1}{\varepsilon} \int_s^t \big(v(u, \varphi_{su}^{v+\varepsilon h}(x)) - v(u, \varphi_{su}^v(x))$$

$$- \varepsilon\, Dv(u, \varphi_{su}^v(x)) \big(\varphi_{su}^{v+\varepsilon h}(x) - \varphi_{su}^v(x) \big) \big)\, du.$$

The proof can proceed exactly as in Theorem 8.7, provided it has been shown that $\left| \varphi_{su}^{v+\varepsilon h}(x) - \varphi_{su}^v(x) \right| = O(\varepsilon)$ which is again a direct consequence of Gronwall's lemma and of the inequality

$$\frac{1}{\varepsilon} \left| \varphi_{st}^{v+\varepsilon h}(x) - \varphi_{st}^v(x) \right|$$

$$\le \int_s^t \|v(u)\|_{1,\infty} \left(\frac{1}{\varepsilon} \left| \varphi_{su}^{v+\varepsilon h}(x) - \varphi_{su}^v(x) \right| \right) du + \int_s^t \|h(u)\|_\infty\, du.$$

Equation (8.13) is the same as equation (8.9), with the additional term $h(t) \circ \varphi_{st}^v$. This implies that the solution of (8.13) may be expressed as a function of the solution of (8.9) by variation of the constant, i.e., it takes the form

$$W(t) = D\varphi_{st}^v(x) A(t)$$

with $A(s) = 0$; $A(t)$ may be identified by writing

$$h(t) \circ \varphi_{st}^v + Dv(t) \circ \varphi_{st}^v W(t) = \partial_t W(t) = D\varphi_{st}^v(x)\partial_t A(t) + Dv(t) \circ \varphi_{st}^v W(t)$$

so that

$$\partial_t A(t) = (D\varphi_{st}^v(x))^{-1} h(t) \circ \varphi_{st}^v(x) = (D\varphi_{ts}^v h(t)) \circ \varphi_{st}^v(x).$$

This implies that

$$A(t) = \int_s^t (D\varphi_{us}^v h(u)) \circ \varphi_{su}^v(x)du$$

and

$$W(t) = \int_s^t D\varphi_{st}^v(x)(D\varphi_{us}^v h(u)) \circ \varphi_{su}^v(x)du$$

which, using the chain rule, can be written

$$W(t) = \int_s^t (D\varphi_{ut}^v h(u)) \circ \varphi_{su}^v(x)du$$

We summarize this discussion with the following theorem:

Theorem 8.10. *Let $v, h \in \mathcal{X}_1^1(T, \Omega)$. Then, for $x \in \Omega$*

$$\partial_\varepsilon \varphi_{st}^{v+\varepsilon h}(x)_{|\varepsilon=0} = \int_s^t (D\varphi_{ut}^v h(u)) \circ \varphi_{su}^v(x)du. \tag{8.14}$$

A direct consequence of this theorem is the fact that φ_{st}^v depends continuously on v. But one can be more specific: we have the inequalities

$$\left|\varphi_{st}^v(x) - \varphi_{st}^{v'}(x)\right| \leq \left|\int_s^t \left(v(u, \varphi_{su}^v(x)) - v'(u, \varphi_{su}^{v'}(x))\right) du\right|$$

$$\leq \left|\int_s^t v(u, \varphi_{su}^v(x)) - v'(u, \varphi_{su}^v(x))du\right|$$

$$+ \int_s^t \left|v'(u, \varphi_{su}^v(x)) - v'(u, \varphi_{su}^{v'}(x))\right| du$$

$$\leq \left|\int_s^t v(u, \varphi_{su}^v(x)) - v'(u, \varphi_{su}^v(x))du\right|$$

$$+ \int_s^t \|v'(u)\|_{1,\infty} \left|\varphi_{su}^v(x) - \varphi_{su}^{v'}(x)\right| du.$$

We now apply inequality (C.4) in Theorem C.8. We take

$$u(\tau, x) = \left|\varphi_{ss+\tau}^v(x) - \varphi_{ss+\tau}^{v'}(x)\right|,$$

$\alpha(\tau) = \|v'(s+\tau)\|_{1,\infty}$ and

$$c(\tau, x) = \left| \int_s^{s+\tau} (v(\theta, \varphi_{s\theta}^v(x)) - v'(\theta, \varphi_{s\theta}^v(x))) \, d\theta \right|$$

yielding the inequality

$$\left| \varphi_{st}^v(x) - \varphi_{st}^{v'}(x) \right| \le$$

$$c(t - s, x) + \int_s^t c(\theta - s, x) \| v'(\theta) \|_{1,\infty} \exp \left(\int_\theta^t \| v'(\theta') \|_{1,\infty} \, d\theta' \right) d\theta. \quad (8.15)$$

The upper bound

$$c(\tau, x) \le \int_s^{s+\tau} \| v(\theta, \cdot) - v'(\theta, \cdot) \|_\infty d\theta$$

directly yields the inequality

$$\| \varphi_{st}^v - \varphi_{st}^{v'} \|_\infty \le$$

$$\left(1 + \int_s^t \| v'(\theta) \|_{1,\infty} \exp \left(\int_\theta^t \| v'(\theta') \|_{1,\infty} \, d\theta' \right) d\theta \right) \int_s^t \| v(\theta) - v'(\theta) \|_\infty d\theta$$

$$(8.16)$$

which proves that $v \mapsto \varphi_{st}^v$ is Lipschitz in v for $\| \cdot \|_{\mathcal{X}^1,T}$. Even more interestingly, (8.15) implies the convergence of $\varphi_{st}^{v^n}$ for weakly converging sequences.

Indeed, consider the function $v' \mapsto \int_s^t v'(\theta) \circ \varphi_{s\theta}^v d\theta$. It is a linear form on $\mathcal{X}^1(T, \Omega)$ which is obviously continuous, since

$$\left| \int_s^t v'(\theta) \circ \varphi_{s\theta}^v d\theta \right| \le \int_s^t \| v'(\theta) \|_\infty \, d\theta \le \| v' \|_{\mathcal{X}^1,T}.$$

Thus, consider a sequence $v^n \in \mathcal{X}_1^1(T, \Omega)$ which is bounded in $\mathcal{X}_1^1(T, \Omega)$ and weakly converges to v. If we let $c^n(\tau, x)$ be equal to $c(\tau, x)$ with v^n instead of v', weak continuity implies that $c^n(\tau, x) \to 0$ when n tends to infinity. We have $c^n(\tau, x) \le \| v^n \|_{\mathcal{X}_1,T} + \| v \|_{\mathcal{X}_1,T}$, so that it is uniformly bounded and the dominated convergence theorem can be applied to equation (8.15) to obtain the fact that, for all $s, t \in [0, T]$, for all $x \in \Omega$ $\varphi_{st}^{v^n}(x) \to \varphi_{st}^v(x)$.

When Ω is bounded, the convergence is in fact uniform in x. Indeed, we have shown that the Lipschitz constant of $\varphi_{st}^{v^n}$ could be expressed as a function of $\| v^n \|_{\mathcal{X}_1,T}$. This ensures the existence of a constant C such that, for all $n > 0$, and $x, y \in \Omega$

$$\left| \varphi_{st}^{v^n}(x) - \varphi_{st}^{v^n}(y) \right| \le C \, |x - y| .$$

This implies that the family $\varphi_{st}^{v^n}$ is equicontinuous and a similar argument shows that it is bounded, so that Ascoli's theorem [221] implies that $(\varphi_{st}^{v^n}, n \ge 0)$ is relatively compact for the uniform convergence. But the limit of any subsequence that converges uniformly must be φ_{st}^v since it is already the pointwise limit of the whole sequence. This implies that the uniform limit exists and is equal to φ_{st}^v. If Ω is not bounded, the same argument implies that the limit is uniform on compact sets [95]. Thus, we have just proved the following theorem:

Theorem 8.11. *If $v \in \mathcal{X}_1^1(T, \Omega)$ and v^n is a bounded sequence in $\mathcal{X}_1^1(T, \Omega)$ which weakly converges to v in $\mathcal{X}^1(T, \Omega)$, then, for all $s, t \in [0, T]$, for every compact subset $Q \subset \overline{\Omega}$*

$$\lim_{n \to \infty} \max_{x \in Q} |\varphi_{st}^{v^n}(x) - \varphi_{st}^v(x)| = 0.$$

Equation (8.16) and Theorem 8.11 can be generalized to higher derivatives in φ. It suffices for this to apply the analysis leading to Theorem 8.10 to the equations provided in (8.10) or (8.12). This would lead to a generalization of (8.16) in the form

$$\|\varphi_{st}^v - \varphi_{st}^{v'}\|_{p,\infty} \leq C \left(\int_s^t \|v'(u)\|_{p+1,\infty} \right) \int_s^t \|v - v'\|_{p,\infty}, \tag{8.17}$$

and uniform convergence on compact sets for derivatives of $\varphi_{st}^{v^n}$ up to order p, as soon as v^n weakly converges in $\mathcal{X}_p^1(T, \Omega)$. Note that the control of the variation in (p, ∞) norm between the flow, requires the finiteness of the integrated $(p + 1, \infty)$ norm of at least one of the time-dependent vector fields.

8.2.2 Admissible Banach Spaces

The previous results are true *a fortiori* for vector fields v belonging to Banach or Hilbert spaces that are smaller than $C_0^1(\Omega, \mathbb{R}^d)$ (Hilbert spaces are particularly interesting because of their special structure). We formalize this with the following definitions.

Definition 8.12. *A Banach space $V \subset C_0^1(\Omega, \mathbb{R}^d)$ is admissible if it is (canonically) embedded in $C_0^1(\Omega, \mathbb{R}^d)$, i.e., there exists a constant C such that, for all $v \in V$,*

$$\|v\|_V \geq C \|v\|_{1,\infty}. \tag{8.18}$$

If V is admissible, we denote by $\mathcal{X}_V^1(\Omega)$ the set of time-dependent vector fields, $(v(t), t \in [0, 1])$ such that, for each t, $v(t) \in V$ and

$$\|v\|_{\mathcal{X}_V^1} := \int_0^1 \|v(t)\|_V \, dt < \infty.$$

If the interval $[0, 1]$ is replaced by $[0, T]$, we will use the notation $\mathcal{X}_V^1(T, \Omega)$ and $\|v\|_{\mathcal{X}_V^1, T}$.

8.2.3 Induced Group of Diffeomorphisms

Definition 8.13. *If $V \subset C_0^1(\Omega, \mathbb{R}^d)$ is admissible, we denote by*

$$G_V = \left\{ \varphi_{01}^v, v \in \mathcal{X}_V^1(\Omega) \right\}$$

the set of diffeomorphisms provided by flows associated to elements $v \in \mathcal{X}_V^1(\Omega)$ at time 1.

Theorem 8.14. *G_V is included in $G_{1,\infty}$ and is a group for the composition of functions.*

Proof. The identity function belongs to G: it corresponds, for example, to φ_{01}^v when $v = 0$. If $\psi = \varphi_{01}^v$ and $\psi' = \varphi_{01}^{v'}$, with $v, v' \in \mathcal{X}_V^1$, then $\psi' \circ \psi = \varphi_{01}^w$ with $w(t) = v(2t)$ for $t \in [0, 1/2]$ and $w(t) = v'(2t - 1)$ for $t \in]1/2, 1]$ (details are left to the reader) and w belongs to $\mathcal{X}_V^1(\Omega)$. Similarly, if $\psi = \varphi_{01}^v$, then $\psi^{-1} = \psi_{01}^w$ with $w(t) = -v(1-t)$. Indeed, we have

$$\varphi_{0,1-t}^w(y) = y - \int_0^{1-t} v(1-s) \circ \varphi_{0s}^w(y) ds = y + \int_1^t v(s) \circ \varphi_{0,1-s}^w ds$$

which implies (by the uniqueness theorem) that $\varphi_{0,1-t}^w(y) = \varphi_{1t}^v(y)$ and in particular $\varphi_{01}^w = \varphi_{10}^v$. This proves that G_V is a group. The inclusion in $G_{1,\infty}$ is an immediate consequences of Gronwall's lemma, the definition of φ_{st}^v and Proposition 8.8.   □

Thus, by selecting a certain Banach space V, we can in turn specify a group of diffeomorphisms. In particular, elements in G_V inherit the smoothness properties of elements of V. Theorem 8.9 indeed implies that G_V is a subgroup of $G_{p,\infty}$ as soon as V is embedded in $C_0^p(\Omega, \mathbb{R}^d)$.

8.2.4 A Distance on G_V

Let V be an admissible Banach space. For ψ and ψ' in G_V, we let

$$d_V(\psi, \psi') = \inf_{v \in \mathcal{X}_V^1(\Omega)} \left\{ \|v\|_{\mathcal{X}_V^1}, \psi' = \psi \circ \varphi_{01}^v \right\}. \tag{8.19}$$

We have the following theorem:

Theorem 8.15 (Trouvé). *The function d_V is a distance on G_V, and (G_V, d_V) is a complete metric space.*

Recall that d_V is a distance if it is symmetrical, satisfies the triangle inequality $d_V(\psi, \psi') \leq d_V(\psi, \psi'') + d_V(\psi'', \psi')$ and is such that $d_V(\psi, \psi') = 0$ if and only if $\psi = \psi'$.

Proof. Note that the set over which the infimum is computed is not empty: if $\psi, \psi' \in G_V$, then $\psi^{-1} \circ \psi' \in G_V$ (since G_V is a group) and therefore can be written under the form φ_{01}^v for some $v \in \mathcal{X}_V^1(\Omega)$.

Let us start with the symmetry: fix $\varepsilon > 0$ and v such that $\|v\|_{\mathcal{X}_V^1} \leq d(\psi, \psi') + \varepsilon$ and $\psi' = \psi \circ \varphi_{01}^v$. This implies that $\psi = \psi' \circ \varphi_{10}^v$, but we know (from the proof of Theorem 8.14) that $\varphi_{10}^v = \varphi_{01}^w$ with $w(t) = -v(1-t)$. Since $\|w\|_{\mathcal{X}_V^1} = \|v\|_{\mathcal{X}_V^1}$, we have, from the definition of d_V:

$$d_V(\psi', \psi) \leq \|w\|_{\mathcal{X}_V^1} \leq d_V(\psi, \psi') + \varepsilon$$

and since this is true for every ε, we have $d_V(\psi', \psi) \leq d_V(\psi, \psi')$. Inverting the roles of ψ and ψ' yields $d_V(\psi', \psi) = d_V(\psi, \psi')$. For the triangular inequality, let v and v' be such that $\|v\|_{\mathcal{X}_V^1} \leq d(\psi, \psi'') + \varepsilon$, $\|v'\|_{\mathcal{X}_V^1} \leq d(\psi'', \psi') + \varepsilon$, $\psi'' = \psi \circ \varphi_{01}^v$ and $\psi' = \psi'' \circ \varphi_{01}^{v'}$. We thus have $\psi' = \psi \circ \varphi_{01}^v \circ \varphi_{01}^{v'}$ and we know, still from the proof of Theorem 8.14, that $\varphi_{01}^v \circ \varphi_{01}^{v'} = \varphi_{01}^w$ with $w(t) = v'(2t)$ for $t \in [0, 1/2]$ and $w(t) = v(2t - 1)$ for $t \in (1/2, 1]$. But, in this case, $\|w\|_{\mathcal{X}_V^1} = \|v\|_{\mathcal{X}_V^1} + \|v'\|_{\mathcal{X}_V^1}$ so that

$$d(\psi, \psi') \leq \|w\|_{\mathcal{X}_V^1} \leq d(\psi, \psi'') + d(\psi'', \psi') + 2\varepsilon$$

which implies the triangular inequality, since this is true for every $\varepsilon > 0$.

We obviously have $d(\psi, \psi) = 0$ since $\varphi_{01}^0 = id$. Assume that $d(\psi, \psi') = 0$. This implies that there exists a sequence v_n such that $\|v_n\|_{\mathcal{X}_V^1} \to 0$ and $\psi' \circ \psi^{-1} = \varphi_{01}^{v_n}$. The continuity of $v \mapsto \varphi_{st}^v$ implies that $\varphi_{01}^{v_n} \to \varphi_{01}^0 = id$ so that $\psi = \psi'$.

Let us now check that we indeed have a complete metric space. Let ψ^n be a Cauchy sequence for d_V, so that, for any $\varepsilon > 0$, there exists n_0 such that, for any $n \geq n_0$, $d_V(\varphi^n, \psi^{n_0}) \leq \varepsilon$. Taking recursively $\varepsilon = 2^{-n}$, it is possible to extract a subsequence ψ^{n_k} of ψ^n such that

$$\sum_{k=0}^{\infty} d_V(\psi^{n_k}, \psi^{n_{k+1}}) < \infty.$$

Since a Cauchy sequence converges whenever one of its subsequences does, it is sufficient to show that ψ^{n_k} has a limit.

From the definition of d_V, there exists, for every $k \geq 0$, an element v^k in $\mathcal{X}_V^1(\Omega)$ such that $\psi^{n_{k+1}} = \varphi_{01}^{v^k} \circ \psi^{n_k}$ and

$$\|v^k\|_{\mathcal{X}_V^1} \leq d_V(\psi^{n_k}, \psi^{n_{k+1}}) + 2^{-k-1}.$$

Let us define a time-dependent vector field v by $v(t) = 2v^0(2t)$ for $t \in [0, 1/2[$, $v(t) = 4v^1(4t - 2)$ for $t \in [1/2, 3/4[$, and so on: to define the general term, introduce the dyadic sequence of times $t_0 = 0$ and $t_{k+1} = t_k + 2^{-k-1}$ and let

$$v(t) = 2^{k+1} v^k(2^{k+1}(t - t_k))$$

for $t \in [t_k, t_{k+1}[$. Since t_k tends to 1 when $t \to \infty$, this defines $v(t)$ on $[0, 1)$, and we fix $v(1) = 0$. We have

$$
\begin{aligned}
\|v(t)\|_{\mathcal{X}_V^1} &= \sum_{k=0}^{\infty} 2^{k+1} \int_{t_k}^{t_{k+1}} \|v^k(2^{k+1}(t - t_k))\|_V \, dt \\
&= \sum_{k=0}^{\infty} \int_0^1 \|v^k(t)\|_V \, dt \\
&\leq 1 + \sum_{k=0}^{\infty} d_V(\psi^{n_k}, \psi^{n_{k+1}})
\end{aligned}
$$

so that $v \in \mathcal{X}_V^1(\Omega)$. Now, consider the associated flow φ_{0t}^v: it is obtained by first integrating $2v^0(2t)$ between $[0, 1/2[$, which yields $\varphi_{0,1/2}^v = \varphi_{01}^{v^0}$. Iterating this, we have

$$
\varphi_{0t_{k+1}}^v = \varphi_{01}^{v^k} \circ \cdots \circ \varphi_{01}^{v^0},
$$

so that

$$
\psi^{n_{k+1}} = \varphi_{0t_{k+1}}^v \circ \psi^{n_0}.
$$

Let $\psi^{\infty} = \varphi_{01}^v \circ \psi^{n_0}$. We also have $\psi^{\infty} = \varphi_{t_k 1}^v \circ \psi^{n_k}$. Since $\varphi_{t_k 1}^v = \varphi_{01}^{w^k}$ with $w^k(t) = v((t - t_k)/(1 - t_k))/(1 - t_k)$, and

$$
\|w\|_{\mathcal{X}_V^1} = \int_{t_k}^1 \|v(t)\|_V \, dt
$$

we obtain the fact that $d_V(\psi^{n_k}, \psi^{\infty}) \to 0$ which completes the proof of Theorem 8.15. $\qquad \square$

8.2.5 Properties of the Distance

We first introduce the set of square integrable (in time) time-dependent vector fields:

Definition 8.16. *Let V be an admissible Banach space. We define $\mathcal{X}_V^2(\Omega)$ as the set of time-dependent vector fields $v = (v(t), t \in [0, 1])$ such that, for each t, $v_t \in V$ and*

$$
\int_0^1 \|v(t)\|_V^2 \, dt < \infty.
$$

We state without proof the important result (in which one identifies time-dependent vector fields that coincide for almost all t):

Proposition 8.17. *$\mathcal{X}_V^2(\Omega)$ is a Banach space with norm*

$$
\|v\|_{\mathcal{X}_V^2} = \left(\int_0^1 \|v_t\|_V^2 \, dt \right)^{1/2}.
$$

Moreover, if V is a Hilbert space, then $\mathcal{X}_V^2(\Omega)$ is Hilbert too with

$$\langle v, w \rangle_{\mathcal{X}_V^2} = \int_0^1 \langle v_t, w_t \rangle_V dt.$$

Because $\left(\int_0^1 \|v(t)\|_V \, dt \right)^2 \leq \int_0^1 \|v(t)\|_V^2 \, dt$, we have $\mathcal{X}_V^2(\Omega) \subset \mathcal{X}_V^1(\Omega)$ and if $v \in \mathcal{X}_V^2(\Omega)$, $\|v\|_{\mathcal{X}_V^1} \leq \|v\|_{\mathcal{X}_V^2}$. The computation of d_V can be reduced to a minimization over $\mathcal{X}_V^2$ by the following theorem.

Theorem 8.18. *If V is admissible and $\psi, \psi' \in G_V$, we have*

$$d_V(\psi, \psi') = \inf_{v \in \mathcal{X}_V^2(\Omega)} \left\{ \|v\|_{\mathcal{X}_V^2}, \psi' = \psi \circ \varphi_{01}^v \right\}. \tag{8.20}$$

Proof. Let $\delta_V(\psi, \psi')$ be given by (8.20) and d_V be given by (8.19). Since d_V is the infimum over a larger set than δ_V, and minimizes a quantity which is always smaller, we have $d_V(\psi, \psi') \leq \delta_V(\psi, \psi')$ and we now proceed to proving the reverse inequality. For this, consider $v \in \mathcal{X}_V^1(\Omega)$ such that $\psi' = \psi \circ \varphi_{01}^v$. It suffices to prove that, for any $\varepsilon > 0$ there exists a $w \in \mathcal{X}^2(\Omega)$ such that $\psi' = \psi \circ \varphi_{01}^w$ and $\|w\|_{\mathcal{X}_V^2} \leq \|v\|_{\mathcal{X}_V^1} + \varepsilon$. The important remark for this purpose is that, if α is a differentiable increasing function from $[0,1]$ onto $[0,1]$ (which implies $\alpha(0) = 0$ and $\alpha(1) = 1$), then,

$$\varphi_{0\alpha(t)}^v(x) = x + \int_0^{\alpha(t)} v(u, \varphi_{0u}(x)) du$$

$$= x + \int_0^t \dot{\alpha}_s(s) v(\alpha(s), \varphi_{0\alpha(s)}(x)) ds$$

so that the flow generated by $w = \dot{\alpha}_s(s) v(\alpha(s))$ is $\varphi_{0\alpha(t)}^v$ and therefore coincides with φ_{01}^v at $t = 1$. We have

$$\|w\|_{\mathcal{X}_V^1} = \int_0^1 \dot{\alpha}_s(s) \|v(\alpha(s))\|_V \, ds = \int_0^1 \|v(t)\|_V \, dt = \|v\|_{\mathcal{X}_V^1}$$

so that this time change does not affect the minimization in (8.19). However, we have, denoting by $\beta(t)$ the inverse of $\alpha(t)$,

$$\|w\|_{\mathcal{X}_V^2}^2 = \int_0^1 \dot{\alpha}_s(s)^2 \|v(\alpha(s))\|_V^2 \, ds = \int_0^1 \dot{\alpha}_t(\beta(t)) \|v(t)\|_V^2 \, dt$$

so that this transformation can be used to reduce $\|v\|_{\mathcal{X}_V^2}$. If $\|v(t)\|_V$ never vanishes, we can choose $\dot{\alpha}_s(\beta(t)) = c/\|v(t)\|_V$ or equivalently $\dot{\beta}_t(t) = \|v(t)\|_V /c$ with c chosen so that

$$\int_0^1 \dot{\beta}_t dt = 1$$

which yields $c = \|v\|_{\mathcal{X}_V^1}$. This gives

$$\|w\|_{\mathcal{X}_V^2}^2 = c \int_0^1 \|v(t)\|_V \, dt = \|v\|_{\mathcal{X}_V^1}^2 \, .$$

which is exactly the kind of result we are after. In the general case, we let, for some $\eta > 0$, $\dot{\alpha}_s(\beta(t)) = c/(\eta + \|v(t)\|_V)$ which yields $\dot{\beta}_t(t) = (\eta + \|v(t)\|_V)/c$ and $c = \eta + \|v\|_{\mathcal{X}_V^1}$. This gives

$$\|w\|_{\mathcal{X}_V^2}^2 = c \int_0^1 \frac{\|v(t)\|_V^2}{\eta + \|v(t)\|_V} dt \leq c \|v\|_{\mathcal{X}_V^1} = (\|v\|_{\mathcal{X}_V^1} + \eta) \|v\|_{\mathcal{X}_V^1}$$

By choosing η small enough, we can always arrange that $\|w\|_{\mathcal{X}_V^2} \leq \|v\|_{\mathcal{X}_V^1} + \varepsilon$ which is what we wanted to prove.   □

A consequence of this result is the following fact.

Corollary 8.19. *If the infimum in (8.20) is attained at some* $v \in \mathcal{X}_V^2(\Omega)$, *then* $\|v(t)\|_V$ *is constant.*

Indeed, let v achieve the minimum in (8.20): we have

$$d_V(\psi, \psi') = \|v\|_{\mathcal{X}_V^2} \geq \|v\|_{\mathcal{X}_V^1}$$

but $\|v\|_{\mathcal{X}_V^1} \geq d_V(\psi, \psi')$ by definition. Thus, we must have $\|v\|_{\mathcal{X}_V^2} = \|v\|_{\mathcal{X}_V^1}$, which corresponds to the equality case in the Schwartz inequality, which can only be achieved by (almost everywhere) constant functions.

Corollary 8.19 is usefully completed by the following theorem:

Theorem 8.20. *If* V *is Hilbert and admissible, and* $\psi, \psi' \in G_V$, *there exists* $v \in \mathcal{X}_V^2(\Omega)$ *such that*

$$d_V(\psi, \psi') = \|v\|_{\mathcal{X}_V^2}$$

and $\psi' = \varphi_{01}^v \circ \psi$.

Proof. By Proposition 8.17, $\mathcal{X}^2(\Omega)$ is a Hilbert space. Let us fix a minimizing sequence for $d_V(\psi, \psi')$, i.e., a sequence $v^n \in \mathcal{X}_V^2(\Omega)$ such that $\|v^n\|_{\mathcal{X}_V^2} \to d_V(\psi, \psi')$ and $\psi' = \varphi_{01}^{v^n} \circ \psi$. This implies that $(\|v^n\|_{\mathcal{X}_V^2})$ is bounded and by Theorem A.16, one can extract a subsequence of v^n (which we still denote by v^n) which weakly converges to some $v \in \mathcal{X}_V^2(\Omega)$, such that

$$\|v\|_{\mathcal{X}_V^2} \leq \liminf \|v^n\|_{\mathcal{X}_V^2} = d_V(\psi, \psi').$$

We now apply Theorem 8.11 (using the fact that weak convergence in $\mathcal{X}_V^2(\Omega)$ implies weak convergence in the bigger space $\mathcal{X}(1, \Omega)$) and obtain the fact that φ^{v^n} converges to φ^v so that $\psi' = \varphi_{01}^v \circ \psi$ remains true: this proves Theorem 8.20.   □

9

Building Admissible Spaces

We have defined in the previous chapter a category of admissible spaces V that drive the construction of groups of diffeomorphisms as flows associated to ordinary differential equations with velocities in V. We now show how such spaces can be explicitly constructed, focusing on Hilbert spaces. This construction is fundamental, because it is intimately related to computational methods involving flows of diffeomorphisms. We will in particular introduce the notion of *reproducing kernels* associated to an admissible space, which will provide our main computational tool. We introduce this in the next section.

9.1 Reproducing Kernel Hilbert Spaces

9.1.1 The Scalar Case

Although we build diffeomorphisms from Hilbert spaces of vector fields, it will be easier to introduce reproducing kernel Hilbert spaces for scalar-valued functions, which has its own interest anyway [13, 215, 71, 9, 10].

Let $\Omega \subset \mathbb{R}^d$. Consider a Hilbert space V included in $L^2(\Omega, \mathbb{R})$. We assume that elements of V are smooth enough, and require the inclusion and the canonical embedding of V in $C^0(\Omega, \mathbb{R})$. For example, it suffices (from Morrey's theorem, see Theorem A.12) that $V \subset H^m(\Omega, \mathbb{R})$ for $m > k/2$. This implies that there exists a constant C such that, for all $v \in V$,

$$\|v\|_\infty \leq C \|v\|_V .$$

We make another assumption on V. *We assume that a relation of the kind* $\sum_{i=1}^N \alpha_i v(x_i) = 0$ *cannot be true for every* $v \in V$ *unless* $\alpha_1 = \cdots = \alpha_N = 0$, $(x_1, \ldots, x_N)$ being an arbitrary family of distinct points in Ω. This is true, for example, if V contains functions supported on arbitrary compact sets.

Each x in Ω specifies a linear form δ_x defined by $(\delta_x \,|\, v) = v(x)$ for $x \in V$. We have

$$|(\delta_x \,|\, v)| \leq \|v\|_\infty \leq C\,\|v\|_V$$

so that δ_x is continuous on V. This implies, by Riesz's theorem (see Theorem A.10) that there exists an element K_x in V such that, for every $v \in V$,

$$v(x) = \langle K_x,\, v \rangle_V. \tag{9.1}$$

Since it belongs to V, K_x is a continuous function $y \mapsto K_x(y)$. This defines a function of two variables, denoted $K : \Omega \times \Omega \to \mathbb{R}$ by $K(y, x) = K_x(y)$. We will preferably use the notation $K(., x)$ instead of K_x.

This function K has several interesting properties. First, applying equation (9.1) to $v = K(., y)$ yields

$$K(x, y) = \langle K(., x),\, K(., y) \rangle_V.$$

Since the last term is symmetric, we have $K(x, y) = K(y, x)$, and because of the obtained identity, K is called the *reproducing kernel* of V.

A second property is the fact that K is positive definite, in the sense that, for any family $x_1, \ldots, x_N \in V$ and any sequence $\alpha_1, \ldots, \alpha_N$ in $\mathbb{R}$, the double sum

$$\sum_{i,j=1}^{N} \alpha_i \alpha_j K(x_i, x_j)$$

is non-negative, and vanishes if and only if all α_i equal 0. Indeed, by the reproducing property, this sum may be written $\left\| \sum_{i=1}^{N} \alpha_i K(., x_i) \right\|_V^2$ and this is positive. If it vanishes, then $\sum_{i=1}^{N} \alpha_i K(., x_i) = 0$, which implies, by equation (9.1), that, for every $v \in V$, one has $\sum_{i=1}^{N} \alpha_i v(x_i) = 0$, and our assumption on V implies that $\alpha_1 = \cdots = \alpha_N = 0$.

Scalar Spline Interpolation

As a first (and important) example of application of kernels, we discuss the following interpolation problem [215].

($\mathcal{S}$) Fix a family of distinct points $x_1, \ldots, x_N$ in Ω. Find a function $v \in V$ of minimal norm satisfying the constraints $v(x_i) = \lambda_i$, where $\lambda_1, \ldots, \lambda_N \in \mathbb{R}$ are prescribed values.

To solve this problem, define V_0 to be the set of v's for which the constraints vanish:

$$V_0 = \{v \in V : v(x_i) = 0, i = 1, \ldots, N\}.$$

Using the kernel K, we may write

$$V_0 = \left\{v \in V : \langle K(., x_i),\, v \rangle = 0, i = 1, \ldots, N\right\}$$

so that

$$V_0 = \mathrm{vect}\,\{K(., x_1), \ldots, K(., x_N)\}^\perp$$

the orthogonal being taken for the V inner product. We therefore have the following first result.

Lemma 9.1. *If there exists a solution $\hat{v}$ of problem $\mathcal{S}$, then $\hat{v} \in V_0^\perp = vect\{K(.,x_1),\ldots,K(.,x_N)\}$. Moreover, if $\hat{v} \in V_0^\perp$ is a solution of $\mathcal{S}$ restricted to this set, then it is a solution of $\mathcal{S}$ on V.*

Proof. Let $\hat{v}$ be this solution, and let v^* be its orthogonal projection on $vect\{K(.,x_1),\ldots,K(.,x_N)\}$. From the properties of orthogonal projections, we have $\hat{v} - v^* \in V_0$, which implies, by the definition of V_0 that $\hat{v}(x_i) = v^*(x_i)$ for $i = 1,\ldots,N$. But, since $\|v^*\|_V \leq \|\hat{v}\|_V$ (by the variational characterization of the projection), and $\|\hat{v}\|_V \leq \|v^*\|_V$ by assumption, both norms are equal, which is only possible when $\hat{v} = v^*$. Therefore, $\hat{v} \in V_0^\perp$ and the proof of the first assertion is complete.

Now, if $\hat{v}$ is a solution of $\mathcal{S}$ in which V is replaced by $V_0^\perp$, and if v is any function in V which satisfies the constraints, then $v - \hat{v} \in V_0$ and $\|v\|_V^2 = \|\hat{v}\|_V^2 + \|v - \hat{v}\|_V^2 \geq \|\hat{v}\|_V^2$ which shows that $\hat{v}$ is a solution of the initial problem. $\square$

This lemma allows us to restrict the search for a solution of $\mathcal{S}$ to the set of linear combinations of $K(.,x_1),\ldots,K(.,x_N)$ which places us in a convenient finite-dimensional situation. We look for $\hat{v}$ in the form

$$\hat{v}(x) = \sum_{i=1}^N \alpha_i K(x,x_i)$$

and we introduce the $N \times N$ matrix S with coefficients $s_{ij} = K(x_i,x_j)$. The whole problem may now be reformulated as a function of the vector $\alpha = (\alpha_1,\ldots,\alpha_N)^T$ (a column vector) and of the matrix S. Indeed, by the reproducing property of K, we have

$$\|\hat{v}\|_V^2 = \sum_{i=1}^N \alpha_i \alpha_j K(x_i,x_j) = \alpha^T S \alpha \qquad (9.2)$$

and each constraint may be written $\lambda_i = v(x_i) = \sum_{j=1}^N \alpha_j K(x_i,x_j)$, so that, letting $\lambda = (\lambda_1,\ldots,\lambda_N)^T$, the whole system of constraints may be expressed as $S\alpha = \lambda$.

Our hypotheses imply that S is invertible; indeed, if $S\alpha = 0$, then $\alpha^T S \alpha = 0$ which, by equation (9.2) and the positive definiteness of K, is only possible when $\alpha = 0$ (we assume that the x_i are distinct). Therefore, there is only one $\hat{v}$ in $V_0^\perp$ which satisfies the constraints, and it corresponds to $\alpha = S^{-1}\lambda$. These results are summarized in the next theorem.

Theorem 9.2. *Problem $\mathcal{S}$ has a unique solution in V, given by*

$$\hat{v}(x) = \sum_{i=1}^N K(x,x_i)\alpha_i$$

with

$$\begin{pmatrix} \alpha_1 \\ \vdots \\ \alpha_N \end{pmatrix} = \begin{pmatrix} K(x_1,x_1) & \dots & K(x_1,x_N) \\ \vdots & \vdots & \vdots \\ K(x_N,x_1) & \dots & K(x_N,x_N) \end{pmatrix}^{-1} \begin{pmatrix} \lambda_1 \\ \vdots \\ \lambda_N \end{pmatrix}.$$

Another important variant of the same problem comes when the hard constraints $v(x_i) = \lambda_i$ are replaced by soft constraints, in the form of a penalty function added to the minimized norm. This may be expressed as the minimization of a function of the form

$$E(v) = \|v\|_V^2 + C \sum_{i=1}^N \varphi(|v(x_i) - y_i|)$$

for some increasing, convex function on $[0, +\infty[$ and $C > 0$. Since the second term of E does not depend on the projection of v on V_0, Lemma 9.1 remains valid, reducing again the problem to finding v of the kind

$$v(x) = \sum_{i=1}^N K(x, x_i)\alpha_i$$

for which

$$E(v) = \sum_{i,j=1}^N \alpha_i \alpha_j K(x_i, x_j) + C \sum_{i=1}^N \varphi \left(\left| \sum_{j=1}^N K(x_i, x_j)\alpha_j - \lambda_i \right| \right).$$

Assume, to simplify, that φ is differentiable and $\varphi'(0) = 0$. We have, letting $\psi(x) = \text{sign}(x)\varphi'(x)$,

$$\frac{\partial E}{\partial \alpha_j} = 2 \sum_{i=1}^N \alpha_i K(x_i, x_j) + C \sum_{i=1}^N K(x_i, x_j)\psi \left(\left| \sum_{l=1}^N K(x_i, x_l)\alpha_l - \lambda_i \right| \right).$$

Assuming, still, that the x_i are distinct, we can apply S^{-1} to the system $\frac{\partial E}{\partial \alpha_j} = 0, j = 1, \dots, N$, which characterizes the minimum, yielding

$$2\alpha_i + C\psi \left(\left| \sum_{j=1}^N K(x_i, x_j)\alpha_j - \lambda_i \right| \right) = 0. \tag{9.3}$$

This suggests an algorithm (which is a form of gradient descent) to minimize E, that we can describe as follows:

Algorithm 1 (General Spline smoothing)

Step 0: Start with an initial guess α^0 for α.

Step t.1 Compute, for $i = 1, \ldots, N$,

$$\alpha_i^{t+1} = (1 - \gamma)\alpha_i^t - \frac{\gamma C}{2} \psi \left(\left| \sum_{j=1}^N K(x_i, x_j)\alpha_j - \lambda_i \right| \right)$$

where γ is a fixed, small enough, real number.
Step t.2 Use a convergence test: if positive, stop, otherwise increment t and restart step t.

This algorithm will converge if γ is small enough. However, the particular case of $\varphi(x) = x^2$ is much simpler to solve, since in this case $\psi(x) = 2x$ and equation (9.3) becomes

$$2\alpha_i + 2C \left(\sum_{j=1}^N K(x_i, x_j)\alpha_j - \lambda_i \right) = 0.$$

The solution of this equation is $\alpha = (S + \mathrm{Id}/C)^{-1}\lambda$, yielding a result very similar to theorem 9.2:

Theorem 9.3. *The minimum over V of*

$$\|v\|_V^2 + C \sum_{i=1}^N |v(x_i) - \lambda_i|^2$$

is attained at

$$\hat{v}(x) = \sum_{i=1}^N K(x, x_i)\alpha_i$$

with

$$\begin{pmatrix} \alpha_1 \\ \vdots \\ \alpha_N \end{pmatrix} = (S + \mathrm{Id}/C)^{-1} \begin{pmatrix} \lambda_1 \\ \vdots \\ \lambda_N \end{pmatrix}$$

and

$$S = \begin{pmatrix} K(x_1, x_1) & \ldots & K(x_1, x_N) \\ \vdots & \vdots & \vdots \\ K(x_N, x_1) & \ldots & K(x_N, x_N) \end{pmatrix}.$$

9.1.2 The Vector Case

In the previous section, elements of V were functions from Ω to $\mathbb{R}$. When working with deformations, which is our goal here, functions of interest describe displacements of points in Ω and therefore must be vector valued. This leads us to address the problem of spline approximation for vector fields in Ω, which, as will be seen, is handled quite similarly to the scalar case.

So, in this section, V is a Hilbert space, canonically embedded in $L^2(\Omega, \mathbb{R}^d)$ and in $C^0(\Omega, \mathbb{R}^d)$ (which is *a fortiori* true if V is admissible). Fixing $x \in \Omega$, the evaluation function $v \mapsto v(x)$ is a continuous linear map from V to $\mathbb{R}^d$. This implies that, for any $a \in \mathbb{R}^d$, the function $v \mapsto a^T v(x)$ is a continuous linear functional on V. We will denote this linear form by $a \otimes \delta_x$, so that

$$(a \otimes \delta_x \mid v) = a^T v(x). \tag{9.4}$$

From Riesz's theorem, there exists a unique element, denoted K_x^a in V such that, for any $v \in V$

$$\langle K_x^a, v \rangle_V = a^T v(x). \tag{9.5}$$

The map $a \mapsto K_x^a$ is linear from $\mathbb{R}^d$ to V (this is because $a \mapsto a^T v(x)$ is linear and because of the uniqueness of the Riesz representation). Therefore, for $y \in \Omega$, the map $a \mapsto K_x^a(y)$ is linear from $\mathbb{R}^d$ to $\mathbb{R}^d$. This implies that there exists a matrix, that we will denote $K(y, x)$, such that, for $a \in \mathbb{R}^d$, $x, y \in \Omega$, $K_x^a(y) = K(y, x)a$. Like in the scalar case, we will preferably denote $K_x^a = K(., x)a$.

The kernel K here being matrix-valued, the reproducing property is

$$\langle K(., x)a, K(., y)b \rangle_V = a^T K_y^b(x) = a^T K(x, y)b.$$

From the symmetry of the first term, we obtain the fact that, for all $a, b \in \mathbb{R}^d$, $a^T K(x, y)b = b^T K(y, x)a$ which implies that $K(y, x) = K(x, y)^T$.

To ensure the positivity of K, we make an assumption similar to the scalar case:

Assumption 1 *If $x_1, \ldots, x_N \in \Omega$ and $\alpha_1, \ldots, \alpha_N \in \mathbb{R}^d$ are such that, for all $v \in V$, $\alpha_1^T v(x_1) + \cdots + \alpha_N^T v(x_N) = 0$, then $\alpha_1 = \cdots = \alpha_N = 0$.*

Under this assumption, it is easy to prove that, for all $\alpha_1, \ldots, \alpha_N \in \mathbb{R}^d$,

$$\sum_{i,j=1}^{N} \alpha_i^T K(x_i, x_j)\alpha_j \geq 0$$

with equality if and only if all α_i vanish.

Vector Spline Interpolation

The interpolation problem in the vector case is

(S_d) Given $x_1, \ldots, x_N$ in Ω, $\lambda_1, \ldots, \lambda_N$ in $\mathbb{R}^d$, find v in V, with minimum norm such that $v(x_i) = \lambda_i$.

As before, we let

$$V_0 = \{v \in V : v(x_i) = 0, i = 1, \ldots, N\}.$$

Then, Lemma 9.1 remains valid (we omit the proof which duplicates the scalar case):

Lemma 9.4. *If there exists a solution $\hat{v}$ of problem $\mathcal{S}_d$, then $\hat{v} \in V_0^\perp$. Moreover, if $\hat{v} \in V_0^\perp$ is a solution of $\mathcal{S}$ restricted to this set, then it is a solution of $\mathcal{S}$ on V.*

The characterization of $V_0^\perp$ is similar to the scalar case:

Lemma 9.5.

$$V_0^\perp = \left\{ v = \sum_{i=1}^{N} K(.,x_i)\alpha_i, \alpha_1, \ldots, \alpha_N \in \mathbb{R}^d \right\}.$$

Thus, a vector field v belongs to $V_0^\perp$ if and only if there exists $\alpha_1, \ldots, \alpha_N$ in $\mathbb{R}^d$ such that, for all $x \in \Omega$,

$$v(x) = \sum_{i=1}^{N} K(x,x_i)\alpha_i.$$

This expression is formally similar to the scalar case, the only differences being that $K(x,x_i)$ is a matrix and α_i is a vector.

Proof. It is clear that $w \in V_0$ if and only if, for any $\alpha_1, \ldots, \alpha_N$, one has

$$\sum_{i=1}^{N} \alpha_i^T v(x_i) = 0.$$

Thus $w \in V_0$ if and only if $\langle v, w \rangle_V = 0$ for all v of the kind $v = \sum_{i=1}^{N} K(.,x_i)\alpha_i$. Thus

$$V_0 = \left\{ v = \sum_{i=1}^{N} K(.,x_i)\alpha_i, \alpha_1, \ldots, \alpha_N \in \mathbb{R}^d \right\}^\perp$$

and since $\left\{ v = \sum_{i=1}^{N} K(.,x_i)\alpha_i, \alpha_1, \ldots, \alpha_N \in \mathbb{R}^d \right\}$ is finite-dimensional, hence closed, one has

$$V_0^\perp = \left\{ v = \sum_{i=1}^{N} K(.,x_i)\alpha_i, \alpha_1, \ldots, \alpha_N \in \mathbb{R}^d \right\}.$$

$\square$

When $v = \sum_{j=1}^{N} K(.,x_j)\alpha_j \in V_0^\perp$, the constraint $v(x_i) = \lambda_i$ yields

$$\sum_{j=1}^{N} K(x_i,x_j)\alpha_j = \lambda_i.$$

Since we also have, in this case,

$$\|v\|_V^2 = \sum_{i,j=1}^{N} \alpha_i{}^T K(x_i, x_j) \alpha_j$$

the whole problem can be rewritten quite concisely with matrices, introducing the notation

$$S = S(x_1, \ldots, x_N) = \begin{pmatrix} K(x_1, x_1) & \ldots & K(x_1, x_N) \\ \vdots & \vdots & \vdots \\ K(x_N, x_1) & \ldots & K(x_N, x_N) \end{pmatrix} \tag{9.6}$$

which is now a block matrix of size $Nd \times Nd$,

$$\alpha = \begin{pmatrix} \alpha_1 \\ \vdots \\ \alpha_N \end{pmatrix}, \quad \lambda = \begin{pmatrix} \lambda_1 \\ \vdots \\ \lambda_N \end{pmatrix}$$

each α_i, λ_i being d-dimensional column vectors.. The whole set of constraints now becomes $S\alpha = \lambda$ and $\|v\|_V^2 = \alpha^T S \alpha$. Thus, replacing numbers by blocks, the problem has exactly the same structure as in the scalar case, and we can repeat the results we have obtained.

Theorem 9.6 (Interpolating splines). *Problem (S_d) has a unique solution in V, given by*

$$\hat{v}(x) = \sum_{i=1}^{N} K(x, x_i) \alpha_i$$

with

$$\begin{pmatrix} \alpha_1 \\ \vdots \\ \alpha_N \end{pmatrix} = \begin{pmatrix} K(x_1, x_1) & \ldots & K(x_1, x_N) \\ \vdots & \vdots & \vdots \\ K(x_N, x_1) & \vdots & K(x_N, x_N) \end{pmatrix}^{-1} \begin{pmatrix} \lambda_1 \\ \ldots \\ \lambda_N \end{pmatrix}.$$

Theorem 9.7 (Smoothing splines). *The minimum over V of*

$$\|v\|_V^2 + C \sum_{i=1}^{N} |v(x_i) - \lambda_i|^2$$

is attained at

$$\hat{v}(x) = \sum_{i=1}^{N} K(x, x_i) \alpha_i$$

with

$$\begin{pmatrix} \alpha_1 \\ \vdots \\ \alpha_N \end{pmatrix} = (S + \mathrm{Id}/C)^{-1} \begin{pmatrix} \lambda_1 \\ \vdots \\ \lambda_N \end{pmatrix}$$

and $S = S(x_1, \ldots, x_N)$ given at equation (9.6).

Finally, Algorithm 1 remains the same, replacing scalars by vectors.

9.2 Building V from Operators

One way to define a Hilbert space V of smooth functions or vector fields is to use inner products associated to operators. In this framework, an inner product, defined from the action of an operator, is defined on a subspace of $L^2(\Omega, \mathbb{R})$ or $L^2(\Omega, \mathbb{R}^d)$. This subspace typically contains smooth functions (allowing for classical definitions of differential operators), but is not complete for the induced norm. The subspace is then extended to a larger one, which provides our Hilbert space V. This is called the Friedrichs extension of an operator. Since it is not restricted to subspaces of $L^2(\Omega, \mathbb{R})$, we make the following presentation with an arbitrary Hilbert space H.

To start, we need a subspace D, included in H and dense in this space, and an operator (i.e., a linear functional), $L : D \to H$. Our typical application will be with $D = C_K^\infty(\Omega, \mathbb{R}^d)$ (the set of C^∞ functions with compact support in Ω) and $H = L^2(\Omega, \mathbb{R}^d)$. In such a case, L may be chosen as a differential operator of any degree, since derivatives of C^∞ functions with compact support obviously belong to L^2. However, L will be assumed to satisfy an additional monotonicity constraint:

Assumption 2 *The operator L is assumed to be symmetric and strongly monotonic on D, which means that there exists a constant $c > 0$ such that, for all $u \in D$,*

$$\langle u, Lu \rangle_H \geq c \langle u, u \rangle_H \tag{9.7}$$

and for all $u, v \in D$

$$\langle u, Lv \rangle_H = \langle Lu, v \rangle_H. \tag{9.8}$$

An example of strongly monotonic operator on $C_K^\infty(\Omega, \mathbb{R})$ is given by $Lu = -\Delta u + \lambda u$ where Δ is the Laplacian: $\Delta u = \sum_{i=1}^k \frac{\partial^2 u}{\partial x_i^2}$. Indeed, in this case, and when u has compact support, an integration by parts yields

$$-\int_\Omega \Delta u(x) u(x) dx = \sum_{i=1}^k \int_\Omega \left(\frac{\partial u}{\partial x_i} \right)^2 dx \geq 0$$

so that $\langle u, Lu \rangle_H \geq \lambda \langle u, u \rangle_H$.

Returning to the general case, the operator L induces an inner product on D, defined by

$$\langle u, v \rangle_L = \langle u, Lv \rangle_H.$$

Assumption 2 ensures the symmetry of this product and its positive definiteness. But D is not complete for $\|.\|_L$, and we need to enlarge it (and simultaneously extend L) to obtain a Hilbert space. Note that there always exists an extension of a pre-Hilbertian structure (such as the one we have on D) to an abstract Hilbert space. Our concern here is whether this extension is a subspace embedded in H.

Theorem 9.8 (Freidrich's extension). *The dot product $\langle . , . \rangle_L$ can be extended from D to a dense subspace $V \subset H$, such that D is dense subspace in V for $\|.\|_L$. The operator L can also be extended to an operator $\hat{L} : V \to V^*$. The extensions have the properties that:*

- *V is continuously embedded in H.*
- *If $u, v \in D$, $\langle u, v \rangle_V = \langle Lu, v \rangle_H = (\hat{L}u \,|\, v)$.*
- *V is a Hilbert space for $\langle \cdot , \cdot \rangle_V$.*

So $\hat{L}$ is just the duality operator on V that identifies vectors to linear forms according to the Riesz representation theorem (Appendix A). We have used the notation (cf. Appendix A) $(m \,|\, v) = m(v)$ for a linear form $m \in V^*$ and $v \in V$. The fact that $\hat{L}$ is an extension of L comes modulo the identification $H = H^*$. Indeed, we have $V \subset H = H^* \subset V^*$ (by the "duality paradox"), so that, L, defined on $D \subset V$ can be seen as an operator with values in H^*.

Definition 9.9. *The operator $\hat{L}$ defined in Theorem 9.8 is called the energetic extension of L. Its restriction to the space*

$$D_L = \left\{ u \in V : \hat{L}u \in H^* = H \right\}$$

is called the Freidrich's extension of L.

The energetic extension of L will still be denoted L in the following.

We will not prove Theorem 9.8, but the interested reader may refer to [228]. The Freidrich extension has other interesting properties:

Theorem 9.10. *$L : V_L \to H$ is bijective and self-adjoint ($\langle Lu, v \rangle_H = \langle u, Lv \rangle_H$ for all $u, v \in V_L$).*
Its inverse, $L^{-1} : H \to H$ is continuous and self-adjoint.
If the embedding $V \subset H$ is compact, then $L^{-1} : H \to H$ is compact.

In the following, we will mainly be interested in embeddings stronger than the L^2 embedding implied by the monotony assumption. It is important that such embeddings are conserved by the extension whenever they are true in the initial space D. This is stated in the next proposition.

Proposition 9.11. *Let D, V and H be as in Theorem 9.8, and B be a Banach space such that $D \subset B \subset H$ and B is canonically embedded in H (there exists $c_1 > 0$ such that $\|u\|_B \geq c_1 \|u\|_H$). Assume that there exists a constant c_2 such that, for all $u \in D$, $\sqrt{\langle Lu, u \rangle_H} \geq c_2 \|u\|_B$. Then $V \subset B$ and $\|u\|_V \geq c_2 \|u\|_B$ for all $u \in V$.*
In particular, if B is compactly embedded in H, then $L^{-1} : H \to H$ is compact.

Proof. Let $u \in V$. Since D is dense in V, there exists a sequence $u_n \in D$ such that $\|u_n - u\|_V \to 0$. Thus u_n is a Cauchy sequence in V and, by our assumption, it is also a Cauchy sequence in B, so that there exists $u' \in B$ such that $\|u_n - u'\|_B$ tends to 0. But since V and B are both embedded in H, we have $\|u_n - u\|_H \to 0$ and $\|u_n - u'\|_H \to 0$ which implies that $u = u'$. Thus u belongs to B, and since $\|u_n\|_V$ and $\|u_n\|_B$ respectively converge to $\|u\|_V$ and $\|u\|_B$, passing to the limit in the inequality $\|u_n\|_V \geq c_2 \|u_n\|_B$ completes the proof of Proposition 9.11.   □

We now describe the relation between L and the reproducing kernel of V discussed in the previous section. We first consider the scalar case and assume that $D = C_K^\infty(\Omega)$ and $H = L^2(\Omega, \mathbb{R})$. We need the continuity of the evaluation of functions in V, which here leads to assuming that there exists a constant such that, for all $u \in D$,

$$\langle Lu, u \rangle_{L^2} \geq c \|u\|_\infty^2 \tag{9.9}$$

since, as we have seen, this implies that V will be continuously embedded in $C^0(\Omega, \mathbb{R})$. Thus, the kernel K is well-defined on V, with the property that $\langle K(.,x), v \rangle_V = v(x) = (\delta_x \,|\, v)$. The linear form δ_x belongs to V^* and, by Theorem 9.8, which implies that $\langle K(.,x), v \rangle_V = (\hat{L}K(.,x) \,|\, v)$, we have $\hat{L}K(.,x) = \delta_x$, or

$$K(.,x) = \hat{L}^{-1}\delta_x.$$

This exhibits a direct relationship between the self-reproducing kernel on V and the energetic extension of L. For this reason, we will denote (with some abuse of notation) $K = \hat{L}^{-1} : V^* \to V$ and write $K_x = K\delta_x$. The same analysis is valid in the vector case, in particular when V is admissible, with the identification

$$K(.,x)a = L^{-1}(a \otimes \delta_x). \tag{9.10}$$

There is another consequence, which stems from the fact that $C^0(\Omega, \mathbb{R}^d)$ (or $\mathbb{R}$) is compactly embedded in $L^2(\Omega, \mathbb{R}^d)$ (or $\mathbb{R}$). Theorem 9.10 indeed implies that L^{-1} is a compact, self-adjoint operator. Such operators have the important property to admit an orthonormal sequence of eigenvectors: more precisely, there exists a decreasing sequence, (ρ_n) of positive numbers, which is either finite or tends to 0, and an orthonormal sequence φ_n in $L^2(\Omega, \mathbb{R}^d)$, such that, for $u \in L^2(\Omega, \mathbb{R}^d)$

$$L^{-1}u = \sum_{n=1}^\infty \rho_n \langle u, \varphi_n \rangle_{L^2} \varphi_n.$$

This directly characterizes V_L as the set

$$V_L = \left\{ u \in L^2(\Omega, \mathbb{R}^d) : \sum_{n=1}^\infty \frac{\langle u, \varphi_n \rangle_{L^2}^2}{\rho_n^2} < \infty \right\}$$

and for $u \in V_L$, we have

$$Lu = \sum_{n=1}^{\infty} \rho_n^{-1} \langle u, \varphi_n \rangle_{L^2} \varphi_n$$

so that, for $u, v \in V_L$

$$\langle u, v \rangle_V = \langle Lu, v \rangle_{L^2} = \sum_{n=1}^{\infty} \rho_n^{-1} \langle u, \varphi_n \rangle_{L^2} \langle u, \varphi_n \rangle_{L^2}.$$

This indicates that V should be given by

$$V = \left\{ u \in L^2(\Omega, \mathbb{R}^d) : \sum_{n=1}^{\infty} \rho_n^{-1} \langle u, \varphi_n \rangle_{L^2}^2 < \infty \right\}.$$

This is indeed the case; V_L is dense in this set: if $u \in V$, then $u_N = \sum_{n=1}^{N} \langle u, \varphi_n \rangle \varphi_n$ belongs to V_L and $\|u_N - u\|_V \to 0$. We summarize what we have just obtained in the following theorem.

Theorem 9.12. *Assume that $D = C_K^{\infty}(\Omega, \mathbb{R}^d)$, $H = L^2(\Omega, \mathbb{R}^d)$ and $L : D \to H$ is symmetric and satisfies*

$$\langle Lu, u \rangle_{L^2} \geq c \|u\|_{\infty}^2$$

for some constant $c > 0$. Then the energetic space of L, V, is continuously embedded in $C^0(\Omega, \mathbb{R}^d)$ and its reproducing kernel is $K(., x)a = \hat{L}^{-1}(a \otimes \delta_x)$. Moreover, there exists an orthonormal basis, (φ_n) in $L^2(\Omega, \mathbb{R}^d)$ and a decreasing sequence of positive numbers, (ρ_n), which tends to 0 such that

$$V = \left\{ u \in L^2(\Omega, \mathbb{R}^d) : \sum_{n=1}^{\infty} \rho_n^{-1} \langle u, \varphi_n \rangle_{L^2}^2 < \infty \right\}.$$

Moreover,

$$Lu = \sum_{n=1}^{\infty} \rho_n^{-1} \langle u, \varphi_n \rangle_{L^2} \varphi_n$$

whenever

$$\sum_{n=1}^{\infty} \left(\frac{\langle u, \varphi_n \rangle_{L^2}}{\rho_n} \right)^2 < \infty.$$

9.3 Invariance of the Inner Product

We haven't discussed so far under which criteria the inner product (or the operator L) should be selected. One important criterion, in particular for shape recognition, is equivariance or invariance with respect to rigid transformations. We consider here the situation $\Omega = \mathbb{R}^d$. This analysis will also lead to explicit constructions of operators and kernels.

9.3.1 Invariance: the Operator Side

To formulate the invariance requirements, let's return to the interpolation problem for vector fields. Let $x_1, \ldots, x_N \in \Omega$ and vectors $v_1, \ldots, v_N$ be given. We want the optimal interpolation h to be invariant under an orthogonal change of coordinates. This means that, if A is a rotation and $b \in \mathbb{R}^d$, and if we consider the problem associated to the points $Ax_1 + b, \ldots, Ax_N + b$ and the constraints $Av_1, \ldots, Av_N$, the obtained function $h^{A,b}$ must satisfy $h^{A,b}(Ax+b) = Ah(x)$. This can also be written $h^{A,b}(x) = Ah(A^{-1}x - A^{-1}b) := ((A, b) \star h)(x)$.

The transformation $h \to (A, b) \star h$ is an action of the group of translations-rotations on functions defined on Ω. To ensure that $h^{A,b} = (A, b) \star h$, it is clearly sufficient to enforce that the norm in V is invariant by this transformation: $\|(A, b) \star h\|_V = \|h\|_V$.

So, we want the action of translations and rotations to be isometric on V. Assume that $\|h\|_V^2 = (Lh \,|\, h)$ for some operator L, and define the operator $(A, b) \star L$ by the fact that, for all $h, \tilde{h} \in V$,

$$\left(L((A, b) \star h) \,|\, (A, b) \star \tilde{h} \right) =: \left(((A, b) \star L)h \,|\, \tilde{h} \right).$$

With this notation, the invariance condition can be written in terms of L, namely: $(A, b) \star L = L$.

Let h_k denote the kth coordinate of h, and $(Lh)_k$ the kth coordinate of Lh. We can write the matrix operator L in the form

$$(Lh)_k = \sum_{l=1}^{d} L_{kl} h_l$$

where each L_{kl} is a scalar operator. We consider the case when L_{kl} is a pseudo-differential operator in the sense that, for any C^∞ scalar function u, the Fourier transform of $L_{kl}u$ takes the form, for $\xi \in \mathbb{R}^d$:

$$\mathcal{F}(L_{kl}u)(\xi) = f_{kl}(\xi)\hat{u}(\xi)$$

for some scalar function f_{kl}. Here and in the following, the Fourier transform of a function u is denoted either $\hat{u}$ or $\mathcal{F}(u)$ (see Section A.8).

An interesting case is when f_{kl} is a polynomial, which corresponds, from standard properties of the Fourier transform, to L_{kl} being a differential operator. Also, by the isometric property of Fourier transforms,

$$\left(Lh \,|\, h \right) = \sum_{k=1}^{d} \left((Lh)_k \,|\, h_k \right)$$

$$= \sum_{k,l=1}^{d} \left(L_{kl} h_l \,|\, h_k \right)$$

$$= \sum_{k,l=1}^{d} \left(\mathcal{F}\left(L_{kl} h_l \right) \,|\, \mathcal{F}\left(h_k \right) \right)$$

$$= \sum_{k,l=1}^{d} \left(f_{kl} \hat{h}_l \,|\, \hat{h}_k \right).$$

Note that, since we want L to be a positive operator, we need the matrix $f(\xi) = (f_{kl}(\xi))$ to be a complex, positive, Hermitian matrix. Moreover, since we want $\left(Lh \,|\, h \right) \geq c\left(h \,|\, h \right)$ for some $c > 0$, it is natural to enforce that the smallest eigenvalue of $f(\xi)$ is uniformly bounded from below (by c).

We now compute $\mathcal{F}\left((A, b) \star h \right)$ as a function of $\hat{h}$. We have

$$\mathcal{F}\left((A, b) \star h \right)(\xi) = \int_{\mathbb{R}^d} e^{-2\iota\pi\xi^T x} (A, b) \star h(x) dx$$

$$= \int_{\mathbb{R}^d} e^{-2\iota\pi\xi^T x} Ah(A^{-1}x - A^{-1}b) dx$$

$$= \int_{\mathbb{R}^d} e^{-2\iota\pi\xi^T (Ay+b)} Ah(y) dy$$

$$= e^{-2i\pi\xi^T b} Ah(A^T \xi)$$

where we have used the fact that $\det A = 1$ (since A is a rotation matrix) in the change of variables. We can therefore write, denoting by φ_b the function $\xi \mapsto \exp(-2\iota\pi\xi^T b)$

$$\left(L((A,b) \star h) \,|\, (A,b) \star h \right) = \sum_{k,l=1}^{d} \sum_{k',l'=1}^{d} \left(\varphi_b f_{kl} a_{ll'} \hat{h}_{l'} \circ A^T \,|\, \varphi_b a_{kk'} \hat{h}_{k'} \circ A^T \right)$$

$$= \sum_{k,l=1}^{d} \sum_{k',l'=1}^{d} \left((f_{kl} \circ A) a_{ll'} \hat{h}_{l'} \,|\, a_{kk'} \hat{h}_{k'} \right).$$

We have used the facts that $\det A = 1$ and $A^T = A^{-1}$. The function φ_b cancels in the complex product:

$$(u \,|\, v) = \int_{\mathbb{R}^d} u\bar{v} d\xi$$

where $\bar{v}$ is the complex conjugate of v.

Reordering the last equation, we obtain

$$\left(L((A,b) \star h) \,|\, (A,b) \star h \right) = \sum_{k',l'=1}^{d} \left(\sum_{k,l=1}^{d} a_{kk'} (f_{kl} \circ A) a_{ll'} \hat{h}_{l'} \,|\, \hat{h}_{k'} \right)$$

which yields our invariance condition in terms of the f_{kl}'s: for all k, l

$$f_{kl} = \sum_{k',l'=1}^{d} a_{k'k}(f_{k'l'} \circ A)a_{l'l}$$

or, representing f as a complex matrix, $A^T f(A\xi)A = f(\xi)$ for all ξ and all rotation matrices A.

We now investigate the consequences of this constraint. First consider the case $\xi = 0$. The identity $A^T f(0)A = f(0)$ for all rotations A implies, by a standard linear algebra theorem, that $f(0) = \alpha \mathrm{Id}$ for some $\alpha \in \mathbb{R}$. Let's now fix $\xi \neq 0$, and let $\eta_1 = \xi/|\xi|$. Extend η_1 into an orthonormal basis, $(\eta_1, \ldots, \eta_d)$ and let $(e_1, \ldots, e_d)$ be the canonical basis of $\mathbb{R}^d$. Define A_0 by $A_0 \eta_i = e_i$; this yields $A_0 \xi = |\xi| e_1$ and $f(\xi) = A_0^T f(|\xi| e_1)A_0 = M_0$. Now, for any rotation A that leaves η_1 invariant, we also have $f(\xi) = A^T A_0^T f(|\xi| e_1)A_0 A$ since $A_0 A \eta_1 = e_1$. This yields the fact that, for any A such that $A\eta_1 = \eta_1$, we have $A^T M_0 A = M_0$.

This property implies, in particular, that $A^T M_0 \eta_1 = M_0 \eta_1$ for all A such that $A\eta_1 = \eta_1$. Taking a decomposition of $M_0 \eta_1$ on η_1 and $\eta_1^\perp$, it is easy to see that this implies that $M_0 \eta_1 = \alpha \eta_1$ for some α. So η_1 is an eigenvector of M_0, which implies that the space $\eta_1^\perp$ is closed under the action of M_0. But the fact that $A^T M_0 A = M_0$ for any rotation A of $\eta_1^\perp$ implies that M_0 restricted to $\eta_1^\perp$ is a homothety, i.e., that there exists λ such that $M_0 \eta_i = \lambda \eta_i$ for $i \geq 2$.

Using the decomposition $u = \langle u, \eta_1 \rangle \eta_1 + u - \langle u, \eta_1 \rangle \eta_1$, we can write

$$M_0 u = \alpha \langle u, \eta_1 \rangle \eta_1 + \lambda(u - \langle u, \eta_1 \rangle \eta_1) = ((\alpha - \lambda)\eta_1 \eta_1^T + \lambda \mathrm{Id})u,$$

which finally yields the fact that (letting $\mu = \alpha - \lambda$ and using the fact that the eigenvalues of $M_0 = A^T f_0(|\xi| e_1)A$ are the same as those of $f_0(|\xi| e_1)$ and therefore only depend on $|\xi|$)

$$f(\xi) = \mu(|\xi|)\frac{\xi \xi^T}{|\xi|^2} + \lambda(|\xi|)\mathrm{Id}.$$

Note that, since we know that $f(0) = \lambda \mathrm{Id}$, we must have $\mu(0) = 0$. Also, we must ensure that $f(\xi)$ is a positive matrix for all ξ. Since its eigenvalues are $\lambda(|\xi|)$ and $\lambda(|\xi|) + \mu(|\xi|)$, we obtain the condition that $\lambda > \max(-\mu, 0)$. We summarize all this in the following theorem.

Theorem 9.13. *Let L be a matricial differential operator such that there exists a matrix-valued function f such that, for any $u \in C_K^\infty(\mathbb{R}^d, \mathbb{R}^d)$, we have*

$$\mathcal{F}(Lu)_k(\xi) = \sum_{l=1}^{d} f_{kl}(\xi)\hat{u}(\xi).$$

Then L is rotation invariant if and only if f takes the form

$$f(\xi) = \mu(|\xi|)\frac{\xi\xi^T}{|\xi|^2} + \lambda(|\xi|)\mathrm{Id}$$

for some functions λ and μ such that $\mu(0) = 0$ and $\lambda > \max(-\mu, 0)$.

Consider the case $\mu = 0$: this gives $f_{kl} = 0$ if $k \neq l$ and $f_{kk}(\xi) = \lambda(|\xi|)$ for all k, which implies that $L_{kl} = 0$ for $k \neq l$ and the L_{kk}'s coincide with a single scalar operator L_0, i.e.,

$$Lh = (L_0 h_1, \ldots, L_0 h_d).$$

This is the type of operator which is mostly used in practice. If one wants L_0 to be a differential operator, $\lambda(|\xi|)$ must be a polynomial in the coefficients of ξ, which is only possible if it takes the form

$$\lambda(|\xi|) = \sum_{q=0}^{p} \lambda_q |\xi|^{2q}.$$

In this case, the corresponding operator is $L_0 u = \sum_{q=0}^{p} \lambda_q (-1)^q \Delta^q u$. For example, if $\lambda(|\xi|) = (\alpha + |\xi|^2)^2$, we get $L_0 = (\alpha \mathrm{Id}_d - \Delta)^2$.

Let's consider an example of valid differential operator with $\mu \neq 0$, taking $\mu(|\xi|) = \alpha |\xi|^2$ and λ as above. This yields the operator

$$L_{kl} u = -\alpha \frac{\partial^2 u}{\partial x_k \partial x_l} + \delta_{kl} \sum_{q=0}^{p} \lambda_q (-1)^q \Delta^q u$$

so that

$$Lh = -\alpha \nabla \mathrm{div} h + \sum_{q=0}^{p} \lambda_q (-1)^q \Delta^q h.$$

Similarly, taking $\mu(\xi) = \alpha |\xi|^2 \lambda(\xi)$ yields the operator

$$Lh = \sum_{q=0}^{p} \lambda_q (-1)^q \Delta^q (h - \nabla \mathrm{div} h).$$

9.3.2 Invariance: the Kernel Side

The kernel (also called the Green function) that corresponds to an operator associated to a given f has a simple expression in the Fourier domain. It is given by

$$\mathcal{F}(K_{kl} u)(\xi) = g_{kl}(\xi)\hat{u}(\xi)$$

with $g = f^{-1}$ (the inverse matrix), as can be checked by a direct computation. Note that we need $K(a \otimes \delta_x)$ to be well-defined. The kth component of $a \otimes \delta_x$ being $a_k \delta_x$, we can write $\mathcal{F}((a \otimes \delta_x))_k(\xi) = a_k \exp(-i\xi^T x)$ so that

$$\mathcal{F}\left(K(a \otimes \delta_x)\right)_k = \sum_{l=1}^{d} g_{kl}(\xi) a_l \exp(-i\xi^T x).$$

A sufficient condition for this to provide a well-defined function after Fourier inversion is that all g_{kl} are integrable. Assuming this, we have

$$(K(a \otimes \delta_x)(y))_k = \sum_{l=1}^{d} \tilde{g}_{kl}(y - x) a_l$$

where $\tilde{g}_{kl}$ is the inverse Fourier transform of g_{kl}.

When $f(\xi) = \lambda(\xi)\text{Id} + \mu(\xi)\xi\xi^T/|\xi|^2$, we can check that

$$f^{-1}(\xi) = \lambda^{-1}(\xi)\left(\text{Id} - \frac{\mu(\xi)}{\lambda(\xi) + \mu(\xi)}\xi\xi^T/|\xi|^2\right).$$

Let $\rho(\xi) = \lambda^{-1}(|\xi|)$ and $K_\rho(x, y) = \tilde{\rho}(x - y)$ ($\tilde{\rho}$ denoting the inverse Fourier transform of ρ). Similarly, let

$$\psi(\xi) = -\frac{\xi\xi^T \mu(|\xi|)}{|\xi^2|(\lambda(|\xi|) + \mu(|\xi|))}$$

and let A_ψ denote the operator associated to ψ in the Fourier domain. This operator does not need to be a kernel operator and is not necessarily positive. Using the fact that convolution reduces to multiplication with Fourier transforms, we can therefore decompose the application of K to a smooth vector field f as:

$$Kf = K_\rho(f + A_\psi f).$$

Of course, A_ψ cannot be arbitrarily chosen, given the conditions obtained in Theorem 9.13. In particular, we need $\lambda + \mu > 0$ and $\mu(0) = 0$; these conditions require that $\psi(\xi) + \xi\xi^T/|\xi|^2$ is a positive matrix, and $\psi(0) = 0$. These conditions are automatically satisfied if ψ itself is a positive matrix, and a very simple example for this can be obtained by taking (with $\alpha > 0$) $\mu = -\alpha\lambda|\xi|^2/(1 + \alpha|\xi|^2)$ which implies that $\psi(\xi) = \alpha\xi\xi^T$. In this case, $A_\psi(f) = -\alpha\nabla(div f)$, and

$$Kf = K_\rho(f - \alpha\nabla(\text{div} f)).$$

Using integration by parts, it is easy to prove that

$$\int_{\mathbb{R}^d} f^T K f dx = \int_{\mathbb{R}^d} \int_{\mathbb{R}^d} K_\rho(x, y)(f^T(x)f(y) + \alpha\text{div} f(x)\text{div} f(y))dxdy.$$

If we take $\mu(\xi) = \alpha|\xi|^2\lambda(\xi)$ as in the previous section, the corresponding ψ is

$$\psi(\xi) = -\frac{\alpha\xi\xi^T}{1 + \alpha|\xi|^2}$$

and $A_\psi f = \alpha(\mathrm{Id} - \alpha\Delta)^{-1}(\nabla\mathrm{div} f)$ resulting in

$$Kf = K_\rho(f - \alpha(\mathrm{Id} - \alpha\Delta)^{-1}(\nabla\mathrm{div} f)).$$

Note that the inverse Fourier transform of a radial function (i.e., a function $\rho(\xi)$ that only depends on $|\xi|$) is also a radial function. To see this, we can write

$$\begin{aligned}
\tilde{\rho}(x) &= \int_{\mathbb{R}^d} \rho(|\xi|)e^{i\xi^T x}d\xi \\
&= \int_0^\infty \rho(t)t^{d-1}\int_{S^{d-1}} e^{i\eta^T tx}ds(\eta) \\
&= \int_0^\infty \rho(t)t^{d-1}\int_{S^{d-1}} e^{i\eta_1 t|x|}ds(\eta)
\end{aligned}$$

where S^{d-1} is the unit sphere in $\mathbb{R}^d$; the first change of variable was $\xi = t\eta$ and the last identity comes from the fact that the integral is invariant by rotation of x, so that we could take x parallel to the first axis of coordinates. The last integral can in turn be expressed in terms of the Bessel function $J_{d/2}$ [160], yielding an expression which will not be detailed (or used) here. This implies that K_ρ above is also a radial kernel, with $K_\rho(x, y) = \gamma(|x - y|)$ for some function γ.

There is a nice dimension-independent characterization of scalar radial kernels [179].

Proposition 9.14. *The scalar kernel $K(x, y) = \gamma(|x - y|)$ is positive definite for all dimensions d if and only there exists a positive measure μ on $]0, +\infty[$ such that*

$$\gamma(t) = \int_0^{+\infty} e^{-t^2 u}d\mu(u).$$

This includes, in particular, the case of all functions of the form

$$\gamma(t) = \int_0^{+\infty} e^{-t^2 u} f(u)du \tag{9.11}$$

for a positive function f.

Translation-invariant kernels (not necessarily radial), of the kind $K(x, y) = \Gamma(x - y)$ can be characterized in a similar way, by Bochner theorem.

Proposition 9.15. *The kernel $\chi(x, y) = \Gamma(x - y)$ is positive definite if and only if there exists a positive, symmetric measure μ on $\mathbb{R}^d$ such that*

$$\Gamma(x) = \int_{\mathbb{R}^d} e^{-2\iota\pi x^T u}d\mu(u).$$

9.3.3 Examples of Radial Kernels

Letting μ be a Dirac measure ($\mu = \delta_{\sigma^{-2}}$) in equation (9.11) yields $\gamma(t) = e^{-\frac{t^2}{\sigma^2}}$.
The associated kernel

$$K(x,y) = e^{-\frac{|x-y|^2}{\sigma^2}} \mathrm{Id}_d$$

is the *Gaussian kernel* on $\mathbb{R}^d$ and is one of the most commonly used for spline
smoothing.

We can also use Proposition 9.14 with $f(u) = e^{-u}$. This provides the
Cauchy kernel

$$K(x,y) = \frac{1}{1 + |x-y|^2} \mathrm{Id}.$$

Other choices can be made: if f is the indicator function of the interval
$[0,1]$, then

$$\gamma(t) = \int_0^1 e^{-t^2 u} du = \frac{1 - e^{-t^2}}{t^2}$$

with the corresponding kernel. If one takes $f(u) = \exp(-u^2/2)$, then

$$\gamma(t) = \int_0^\infty e^{-u^2/2 - ut^2} du$$

$$= e^{t^4/2} \int_0^\infty e^{-(u+t^2)^2/2} du$$

$$= \frac{\sqrt{\pi}}{2} e^{t^4/2} \mathrm{erfc}(t^2)$$

where $\mathrm{erfc}(q)$ is the probability that a standard Gaussian distribution is larger
than q in absolute value,

$$\mathrm{erfc}(q) = \frac{2}{\sqrt{\pi}} \int_q^\infty e^{-u^2/2} du.$$

(This function is widely tabulated.)

We now consider scalar kernels that correspond to differential operators
that are polynomial in the Laplacian. Using the inverse Fourier form, they
correspond to kernels given by $K(x,y) = \Gamma(x-y)$ with

$$\Gamma(z) = \int_{\mathbb{R}^d} \frac{e^{iz^T \xi}}{P(|\xi|^2)} d\xi$$

for some polynomial P such that $P(t) > 0$ for $t \geq 0$. Of particular interest is
the case $P(t) = (1+t)^k$ for some positive integer k, which corresponds to the
operator $(\mathrm{Id} - \Delta)^k$, because the associated kernel can be explicitly computed,
at least in odd dimensions. Note that $1/P(|\xi|^2)$ must be integrable, which in
this particular case means $k \geq (d+1)/2$.

To compute Γ, we can assume (by rotation invariance) that z is on the positive side of first axis of coordinates, i.e., $z = (|z|, 0, \ldots, 0)$. Write $\xi = (t, \eta)$, with $t \in \mathbb{R}$ and $\eta \in \mathbb{R}^{d-1}$ so that

$$\Gamma(z) = \int_{-\infty}^{+\infty} e^{it|z|} \int_{\mathbb{R}^{d-1}} (1 + t^2 + |\eta|^2)^{-k} d\eta dt.$$

Making the change of variable $\eta = \sqrt{1 + t^2}\zeta$ (so that $d\eta = (1 + t^2)^{(d-1)/2} d\zeta$) expresses $\Gamma(z)$ as the product of two integrals,

$$\Gamma(z) = \int_{-\infty}^{+\infty} \frac{e^{it|z|}}{(1 + t^2)^{k-(d-1)/2}} dt \int_{\mathbb{R}^{d-1}} (1 + |\zeta|^2)^{-k} d\zeta.$$

The second integral is a constant $c_d(k)$ (it can be explicitly computed, but we will not use its exact value here; it suffices to define the kernel up to a multiplicative constant which can be arbitrarily chosen). The first integral can also be computed exactly, using the method of residues [176], as provided by the following lemma, which we give without proof.

Lemma 9.16. *Given an integer $c \geq 1$, we have*

$$\int_{-\infty}^{+\infty} \frac{e^{it|z|}}{(1 + t^2)^{c+1}} dt = \frac{\pi e^{-|z|}}{4^c c!} \sum_{l=0}^{c} a(c, l)|z|^l \tag{9.12}$$

with $a(c, l) = 2^l(2c - l) \ldots (c + 1 - l)/l!$.

Ignoring the constants, this yields the kernel (letting $c = k - (d+1)/2$)

$$K_c(x, y) = e^{-|x-y|} \sum_{l=1}^{c} a(c, l)|x - y|^l.$$

For $c = 0$, $K_0(x, y) = \exp(-|x - y|)$ is called the Laplacian or Abel's kernel. From lemma 9.16, we get

$$K_1(x, y) = 2(1 + |x - y|) \exp(-|x - y|)$$
$$K_2(x, y) = 4(3 + 3|x - y| + |x - y|^2) \exp(-|x - y|)$$
$$K_3(x, y) = 8(15 + 15|x - y| + 6|x - y|^2 + |x - y|^3) \exp(-|x - y|)$$
$$K_4(x, y) = 16(105 + 105|x - y| + 45|x - y|^2 + 10|x - y|^3 + |x - y|^4)$$
$$\exp(-|x - y|).$$

Note that K_1 is differentiable (with respect to each variable), and K_2 is twice differentiable. More generally K_c has c derivatives, still with respect to each variable. As usual, the kernel can be scaled, replacing x and y by x/σ and y/σ. For a given c, this corresponds (up to a multiplicative constant) to the operator

$$L = (\mathrm{Id} - \sigma\Delta)^{c+(d+1)/2}.$$

If the dimension d is even, the computation above is not valid since $(d-1)/2$ is not an integer. There is no explicit expression for the kernel associated to $(\mathrm{Id} - \Delta)^c$ when c is an integer. However, since a valid radial kernel in dimension d is clearly also valid for smaller dimensions, Laplacian and higher-order kernels, K_c, can also be used when d is even (but they do not correspond to differential operators anymore).

The previous collection of kernels can be extended by the following series of combination rules. It is indeed obvious that the addition of two kernels is a kernel, as well as the multiplication of a kernel by a positive number. A kernel can also be scaled by a positive factor: $K(x,y) \rightarrow K(x/a, y/a)$. The composition of two kernels is also a kernel, i.e.,

$$(K_1 * K_2)(x,y) = \int K_1(x,z)K_2(z,y)dz.$$

Also, the direct multiplication of two kernels is a kernel (i.e., $(K_1.K_2)(x,y) :=$ $K_1(x,y)K_2(x,y)$). So, for example, in dimensions 1 or 3, the kernel defined by $K(x,y) = \gamma(|x-y|^2)$ with

$$\gamma(t) = (1 + \sqrt{t})e^{-\sqrt{t}-t/2}$$

is a valid kernel, since it is the direct multiplication of the Gaussian and power of Laplacian kernels.

9.4 Mercer's Theorem

The interest of the discussion above lies in the fact that it makes it possible to define the Hilbert space V from a positive definite kernel. We gave a description of kernels using Fourier transforms, but another way to achieve this (in particular when Ω is bounded) is by using Mercer's theorem [174].

Theorem 9.17. *We take $\Omega = \mathbb{R}^d$. Let $K : \Omega \times \Omega \rightarrow \mathbb{R}$ be a continuous, positive definite kernel, such that*

$$\int_{\Omega \times \Omega} K(x,y)^2 dx dy < \infty.$$

Then, there exists an orthonormal sequence of functions in $L^2(\Omega, \mathbb{R})$, φ_1, $\varphi_2, \ldots$ and a decreasing sequence (ρ_n) which tends to 0 when n tends to ∞ such that

$$K(x,y) = \sum_{n=1}^{\infty} \rho_n \varphi_n(x) \varphi_n(y).$$

Let the conditions of Mercer's Theorem be true, and define the Hilbert space V by

$$V = \left\{ v \in L^2(\Omega, \mathbb{R}) : \sum_{n=1}^{\infty} \langle v, \varphi_n \rangle_{L^2}^2 < \infty \right\}.$$

Define, for $v, w \in V$:

$$\langle v, w \rangle_V = \sum_{n=1}^{\infty} \rho_n^{-1} \langle v, \varphi_n \rangle_{L_2} \langle w, \varphi_n \rangle_{L_2}.$$

Note that, for $v \in V$, there is a pointwise convergence of the series

$$v(x) = \sum_{n=1}^{\infty} \langle v, \varphi_n \rangle_{L^2} \varphi_n(x)$$

since

$$\left(\sum_{n=p}^{m} \langle v, \varphi_n \rangle_{L^2} \varphi_n(x) \right)^2 \leq \sum_{n=p}^{m} \rho_n^{-1} \langle v, \varphi_n \rangle_{L^2}^2 \sum_{n=p}^{m} \rho_n (\varphi_n(x))^2$$

and both terms in the upper bound can be made arbitrarily small (recall that $\sum_n \rho_n \varphi_n(x)^2 = K(x, x) < \infty$). Similarly,

$$v(x) - v(y) = \sum_{n=1}^{\infty} \langle v, \varphi_n \rangle_{L^2} (\varphi_n(x) - \varphi_n(y))$$

so that

$$(v(x) - v(y))^2 \leq \sum_{n=1}^{\infty} \rho_n^{-1} \langle v, \varphi_n \rangle_{L^2}^2 \sum_{n=1}^{\infty} \rho_n (\varphi_n(x) - \varphi_n(y))^2$$

$$= \|v\|_V^2 \left(K(x, x) - 2K(x, y) + K(y, y) \right)$$

and v is continuous. Then,

$$\langle \varphi_m, K(., x) \rangle_{L^2} = \sum_{n=1}^{\infty} \rho_n \varphi_n(x) \langle \varphi_m, \varphi_n \rangle_{L^2} = \rho_m \varphi_m(x)$$

so that

$$\sum_{n=1}^{\infty} \rho_n^{-1} \langle \varphi_n, K_x \rangle_{L^2}^2 = \sum_{n=1}^{\infty} \rho_n \varphi_n(x)^2 = K(x, x) < \infty$$

which implies $K_x \in V$ and a similar computation shows that $\langle K_x, K_y \rangle_V = K(x, y)$ and that K is reproducing.

Finally, if $v \in V$,

$$\langle v, K_x \rangle_V = \sum_{n=1}^{\infty} \rho_n^{-1} \langle v, \varphi_n \rangle_{L^2} \langle \varphi_n, K_x \rangle_{L^2} = \sum_{n=1}^{\infty} \langle v, \varphi_n \rangle_{L^2} \varphi_n(x) = v(x)$$

so that K_x corresponds to the Riesz representation of the evaluation functional on V.

9.5 Thin-Plate Interpolation

Thin-plate theory corresponds to the situation in which the operator L is some power of the Laplacian. As a tool for shape analysis, it was originally introduced by Bookstein [29]. Consider the following bilinear form:

$$\langle f, g \rangle_L = \int_{\mathbb{R}^2} \Delta f \Delta g \, dx = \int_{\mathbb{R}^2} f \Delta^2 g \, dx, \tag{9.13}$$

which corresponds to the operator $L = \Delta^2$.

We need to define it on a somewhat unusual Hilbert space. We consider the Beppo–Levi space $\mathcal{H}_1$ of all functions in $\mathbb{R}^d$ with square integrable second derivatives, which have a bounded gradient at infinity. In this space, $\|f\|_L = 0$ is equivalent to the fact that f is affine, i.e., $f(x) = a^T x + b$, for some $a \in \mathbb{R}^d$ and $b \in \mathbb{R}$. The Hilbert space we consider is the space of $f \in \mathcal{H}_1$ modulo affine functions, i.e., the set of equivalent classes

$$[f] = \{g : g(x) = f(x) + a^T x + b, a \in \mathbb{R}^d, b \in \mathbb{R}\}$$

for $f \in \mathcal{H}_1$. Obviously the norm associated to (9.13) is constant over the set $[f]$, and $\|f\| = 0$ if and only if $[f] = [0]$.

This space also has a kernel, although the analysis has to be different from what we have done previously, since the evaluation functional is not defined on $\mathcal{H}$ (functions are only known up to the addition of an affine term). However, the following is true [140]. Let $U(r) = (1/8\pi)r^2 \log r$ if the dimension, d, is 2, and $U(r) = (1/16\pi)r^3$ for $d = 3$. Then, for all $u \in \mathcal{H}_1$, there exist $a_u \in \mathbb{R}^d$ and $b_u \in \mathbb{R}$ such that

$$u(x) = \int_{\mathbb{R}^d} U(|x - y|)\Delta^2 u(y) dy + a_u^T x + b_u.$$

We will denote by $U(., x)$ the function $y \mapsto U(|x - y|)$.

The spline interpolation problem must be addressed in a different way in this context. Fix, as in Section 9.1.1, landmarks $x_1, \ldots, x_N$ and scalar constraints $(c_1, \ldots, c_N)$. Again, the constraint $h(x_i) = c_i$ has no meaning in $\mathcal{H}$, but the constraint $\langle U(\cdot - x_i), h \rangle_{\mathcal{H}} = c_i$ does, and means that there exist a_h and b_h such that

$$h(x_i) + a_h^T x_i + b_h = c_i,$$

i.e., h satisfies the constraints up to an affine term.

Denote, as before, $S_{ij} = U(|x_i - x_j|)$. The function h, which is optimal under the constraints $\langle U(\cdot - x_i), h \rangle_{\mathcal{H}} = c_i, i = 1, \ldots N$, must therefore take the form

$$h(x) = \sum_{i=1}^{N} \alpha_i U(|x - x_i|) + a^T x + b.$$

The corresponding interpolation problem is: minimize $\|h\|$ under the constraint that there exist a, b such that $h(x_i) + a_h^T x_i + b_h = c_i, i = 1, \ldots, N$. The inexact-matching problem simply consists in minimizing

$$\|h\|_{\mathcal{H}}^2 + \lambda \sum_{i=1}^{N} (h(x_i) + a^T x_i + b - c_i)^2$$

with respect to h and a, b.

Replacing h by its expression in terms of the α's, a and b yields the finite-dimensional problem: minimize

$$\alpha^T S \alpha$$

under the constraint $S\alpha + Q\gamma = c$ with $\gamma = (a_1, \ldots, a_d, b)^T$ (with size $(d + 1) \times 1$) and Q, with size $N \times (d + 1)$ given by (letting $x_i = (x_i^1, \ldots, x_i^d)$):

$$Q = \begin{pmatrix} x_1^1 & \cdots & x_1^d & 1 \\ \vdots & \vdots & \vdots \\ x_N^1 & \cdots & x_N^d & 1 \end{pmatrix}.$$

The optimal (α, γ) can be computed by identifying the gradient to 0. One obtains

$$\hat{\gamma} = \left(Q^T S^{-1} Q \right)^{-1} Q^T S^{-1} c$$

and $\hat{\alpha} = S^{-1}(c - Q\hat{\gamma})$.

For the inexact-matching problem, one minimizes

$$\alpha^T S \alpha + \lambda (S\alpha + a^T x_i + b - c)^T (S\alpha + a^T x_i + b - c),$$

and the solution is provided by the same formulae, simply replacing S by $S_\lambda = S + (1/\lambda)\mathrm{Id}$.

When the function h (and the c_i) take d-dimensional values (e.g., correspond to displacements), the above computation has to be applied to each coordinate, which simply corresponds to using a diagonal operator, each component equal to Δ^2, in the definition of the dot product. This is equivalent to using the diagonal scalar kernel associated to U.

9.6 Asymptotically Affine Kernels

We return to vector fields, and discuss how affine components can be combined with any kernel. We assume here that $\Omega = \mathbb{R}^d$. We would like to consider spaces V that contain vector fields with an affine behavior at infinity. Note that the spaces V that we have considered so far, either by completion of C^∞ compactly supported functions or using kernels defined by Fourier transforms only contain functions that vanish at infinity. We recall the definition of a function that vanishes at infinity:

Definition 9.18. *A function $f : \Omega \to \mathbb{R}^d$ is said to vanish at infinity if and only if, for all $\varepsilon > 0$, there exists $A > 0$ such that $|f(x)| < \varepsilon$ whenever $x \in \Omega$ and $|x| > A$.*

Here, we let V be a Hilbert space of vector fields that vanish at infinity and define

$$V_{\text{aff}} = \left\{ w : \exists w_0 \in V, A \in \mathcal{M}_d(\mathbb{R}) \text{ and } b \in \mathbb{R}^d \text{ with } w(x) = w_0(x) + Ax + b \right\}.$$

We have the following important fact:

Proposition 9.19. *If V is a Hilbert space of continuous vector fields that vanish at infinity, then the decomposition $w(x) = w_0(x) + Ax + b$ for $w \in V_{\text{aff}}$ is unique.*

Proof. Using differences, it suffices to prove this for $w = 0$, and so, if w_0, A and b are such that, for all x, $w_0(x) + Ax + b = 0$, then, for any fixed x and $t > 0$, $w_0(tx) + tAx + b = 0$ so that $Ax = -(w_0(tx) + b)/t$. Since the term in right-hand side tends to 0 when t tends to infinity, we get $Ax = 0$ for all x so that $A = 0$. Now, for all t, x, we get $b = -w_0(tx)$, which implies $b = 0$ (since w_0 vanishes at infinity), and therefore also $w_0 = 0$. $\square$

So we can speak of the affine part of an element of V_{aff}. Given the inner product in V, we can define

$$\langle w, \tilde{w} \rangle_{V_{\text{aff}}} = \langle w_0, \tilde{w}_0 \rangle_V + \langle A, \tilde{A} \rangle + \langle b, \tilde{b} \rangle$$

where $\langle A, \tilde{A} \rangle$ is some inner product between matrices (e.g., trace($A^T \tilde{A}$)) and $\langle b, \tilde{b} \rangle$ some inner product between vectors (e.g., $b^T \tilde{b}$). With this product V_{aff} is obviously a Hilbert space. We want to compute its kernel, K_{aff}, as a function of the kernel K of V (assuming that V is reproducing). Given $x \in \mathbb{R}^d$ and $a \in \mathbb{R}^d$, we need to express $a^T w(x)$ in the form $\langle K_{\text{aff}}(., x)a, w \rangle_{V_{\text{aff}}}$. Using the decomposition, we have

$$a^T w(x) = a^T w_0(x) + a^T Ax + a^T b$$
$$= \langle K(., x)a, w_0 \rangle_V + \langle (ax^T)^\sharp, A \rangle + \langle a^\sharp, b \rangle$$

where, for a matrix M, we define $M^\sharp$ by the identity $\langle M^\sharp, A \rangle = \text{trace}(M^T A)$ for all A and for a vector z, $z^\sharp$ is such that $\langle z^*, b \rangle = z^T b$ for all b. From the definition of the extended inner product, we can define and compute

$$K_{\text{aff}}(y, x)a = K(y, x)a + (ax^T)^\sharp y + a^\sharp.$$

In particular, when $\langle A, \tilde{A} \rangle = \lambda \,\text{trace}(A^T \tilde{A})$ and $\langle b, \tilde{b} \rangle = \mu b^T \tilde{b}$, we get

$$K_{\text{aff}}(y, x)a = K(y, x)a + ax^T y + a = (K(x, y) + (x^T y/\lambda + 1/\mu)\text{Id}_d)a.$$

This provide an immediate extension of spline interpolation of vector fields which includes affine transformations, by just replacing K by K_{aff}. For example, exact interpolation with constraints $v(x_i) = c_i$ is obtained by letting

$$v(x) = \sum_{i=1}^{N} K_{aff}(x_i, x_j)\alpha_j,$$

the vectors α_j being obtained by solving the system

$$\sum_{j=1}^{N} (K(x_i, x_j)\alpha_j + (x_i^T x_j/\lambda + 1/\mu)\alpha_j) = c_i, \text{ for } i = 1, \ldots, N. \tag{9.14}$$

The case $\lambda = \mu \to 0$ is particularly interesting, since this corresponds to relaxing the penalty on the affine displacement, and we obtain in this way an affine invariance similar to thin plates. In fact, the solution of (9.14) has a limit v^* which is a solution of the affine invariant interpolation problem: minimize $\|v\|_V^2$ under the constraint that there exist A and b with $v(x_i) = c_i - Ax_i - b$ for $i = 1, \ldots, N$. Indeed, (9.14) provides the unique solutions, v^λ, A^λ and b^λ, of the problem: minimize, with respect to v, A, b

$$\|v\|_V^2 + \lambda(\|A\|^2 + \|b\|^2)$$

under the constraints $v(x_i) = c_i - Ax_i - b$ for $i = 1, \ldots, N$. Because $\|v^\lambda\|_V$ is always smaller that the norm of the optimal v for the problem without affine component, we know that it is bounded. Thus, to prove that v^λ converges to v^*, it suffices to prove that any weakly converging subsequence of v^λ has v^* as a limit. So take such a sequence, v^{λ_n}, and let v^0 be its weak limit. Because $v \mapsto v(x)$ is a continuous linear form, we have $v^{\lambda_n}(x_i) \to v(x_i)$ for all i. Because of this, and because there exists a solution A^λ, b^λ to the overconstrained system $A^\lambda x_i + b^\lambda = c_i - v^\lambda(x_i)$, we can conclude that there exist A^0 and b^0 such that $A^0 x_i + b^0 = c_i - v^0(x_i)$ (because the condition for the existence of solutions passes to the limit). So v^0 satisfies the constraints of the limit problem. Moreover, for any λ, we have

$$\|v^*\|_V^2 + \lambda(\|A^*\|^2 + \|b^*\|^2) \geq \|v^\lambda\|_V^2 + \lambda(\|A^\lambda\|^2 + \|b^\lambda\|^2) \geq \|v^\lambda\|_V^2.$$

Since the liminf of the last term is larger than $\|v^0\|_V^2$, and the limit of the left-hand side is $\|v^*\|_V^2$, we can conclude that $v_0 = v^*$ since v^* is the unique solution of the limit problem.

10

Deformable Objects and Matching Functionals

10.1 General Principles

In the previous two chapters, we introduced and studied basic tools related to deformations and their mathematical representation using diffeomorphisms. In this chapter, we start investigating relations between deformations and the objects they affect, which we will call deformable objects, and discuss the variations of matching functionals, which are cost functions that measure the quality of the registration between two deformable objects.

Let Ω be an open subset of $\mathbb{R}^d$ and G a group of diffeomorphisms on Ω. Consider a set $\mathcal{I}$ of structures of interest, on which G has an action: for every I in $\mathcal{I}$ and every $\varphi \in G$, the result of the action of φ on I is denoted $\varphi \cdot I$ and is a new element of $\mathcal{I}$. This requires (see Section B.5) that $id \cdot I = I$ and $\varphi \cdot (\psi \cdot I) = (\varphi \circ \psi) \cdot I$. Elements of $\mathcal{I}$ will be referred to as *deformable objects*.

A matching functional is based on a function $D : \mathcal{I} \times \mathcal{I} \to [0, +\infty)$ such that $D(I, I')$ measures the discrepancy between the two objects I and I', and is defined over G by

$$E_{I,I'}(\varphi) = D(\varphi \cdot I, I'). \tag{10.1}$$

So $E_{I,I'}(\varphi)$ measures the difference between the *target* object I' and the deformed one $\varphi \cdot I$. Because it is mapped on the target by the deformation, the object I will be often referred to as the *template* (and $\varphi \cdot I$ as the deformed template).

Even if our discussion of matching principles and algorithms is rather extensive, and occupies a large portion of this book, the size of the literature, and our choice of privileging methods that implement diffeomorphic matching will prevent us from providing an exhaustive account of the registration methods that have been proposed over the last decades. The interest reader can refer to a few starting points in order to complement the presentation that is made here, including [84, 17, 18, 28, 29, 8, 7, 171, 201, 94, 173, 15].

10.2 Variation with Respect to Diffeomorphisms

We will review, starting with the next section, a series of matching functionals that are adapted to different types of deformable objects (landmarks, images, curves, etc.). We will also compute the derivative of each of them with respect to the diffeomorphism φ.

For this purpose, we introduce a special form of differential which is adapted to variational problems over diffeomorphisms. This shape, or Eulerian differential, as we will call it, is a standard tool in shape optimization [60], and we will interpret it later on as a gradient for a specific Riemannian metric over diffeomorphisms.

Recall that we have defined $G_{p,\infty}$ to be the set of diffeomorphisms ψ such that

$$\max(\|\psi - id\|_{p,\infty}, \|\psi^{-1} - id\|_{p,\infty}) < \infty.$$

Definition 10.1. *A function $\varphi \mapsto U(\varphi)$ is (p, ∞)-compliant if it is defined for all φ in $G_{p,\infty}$.*

A (p, ∞)-compliant U is locally (p, ∞)-Lipschitz if, for all $\varphi \in G_{p,\infty}$, there exist positive numbers $\varepsilon(\varphi)$ and $C(\varphi)$ such that

$$|U(\psi) - U(\tilde{\psi})| \leq C(\varphi)\|\psi - \tilde{\psi}\|_{p.\infty}$$

whenever ψ and $\tilde{\psi}$ are diffeomorphisms such that

$$\max(\|\psi - \varphi\|_{p,\infty}, \|\tilde{\psi} - \varphi\|_{p,\infty}) < \varepsilon(\varphi).$$

Note that a (p, ∞)-compliant (resp. locally Lipschitz) U is (q, ∞)-compliant (resp. locally Lipschitz) for any q larger than p.

If V is an admissible vector space and $v \in V$, we will denote by φ_{0t}^v the flow associated to the equation

$$\partial_t y = v(y).$$

Note that this is the same notation as the flow associated to a differential equation $\partial_t y = v(t, y)$ where v is now a time-dependent vector field. This is not a conflict of notation if one agrees to identify vector fields, v, in V and the associated constant time-dependent vector field defined by $\tilde{v}(t, \cdot) = v$ for all t.

Definition 10.2. *Let V be an admissible Hilbert space continuously embedded in $C_0^p(\Omega, \mathbb{R}^d)$ (so that $G_V \subset G_{p,\infty}$). We say that a (p, ∞)-compliant function U over diffeomorphisms has an Eulerian differential in V at ψ if there exists a linear form $\bar{\partial}U(\psi) \in V^*$ such that, for all $v \in V$,*

$$\left(\bar{\partial}U(\psi) \mid v\right) = \partial_\varepsilon U(\varphi_{0\varepsilon}^v \circ \psi)_{|\varepsilon=0}. \tag{10.2}$$

If the Eulerian differential exists, the V-Eulerian gradient of U at ψ, denoted $\overline{\nabla}^V U(\varphi) \in V$, is defined by

$$\langle \overline{\nabla}^V U(\varphi), v \rangle_V = (\bar{\partial} U(\varphi) \mid v). \tag{10.3}$$

In this case, $\overline{\nabla}^V U(\varphi) = K \bar{\partial} U(\varphi)$ where K is the kernel of V.

The following proposition indicates when equation (10.2) remains valid with time-dependent vector fields v.

Proposition 10.3. *Let V be an admissible Hilbert space continuously embedded in $C_0^{p+1}(\Omega, \mathbb{R}^d)$. Let V and U be like in Definition 10.2. If U is (p, ∞)-Lipschitz and has a V-Eulerian differential at ψ and if $v(t, .)$ is a time-dependent vector field such that*

$$\lim_{\varepsilon \to \infty} \frac{1}{\varepsilon} \int_0^\varepsilon \|v(t, \cdot) - v(0, \cdot)\|_V dt = 0,$$

then

$$(\bar{\partial} U(\psi) \mid v(0, \cdot)) = \partial_\varepsilon U(\varphi_{0\varepsilon}^v \circ \psi)_{|\varepsilon=0}. \tag{10.4}$$

Proof. Letting $v_0 = v(0, \cdot)$, we need to prove that

$$\frac{1}{\varepsilon}(U(\varphi_{0\varepsilon}^v \circ \psi) - U(\varphi_{0\varepsilon}^{v_0} \circ \psi)) \to 0$$

when $\varepsilon \to 0$. From proposition 8.4, we know that if $\psi, \varphi, \tilde{\varphi}$ are in $G_{p,\infty}$, there exists a constant $C_p(\psi)$ such that

$$\|\varphi \circ \psi - \tilde{\varphi} \circ \psi\|_{p,\infty} \le C_p(\psi)\|\varphi - \tilde{\varphi}\|_{p,\infty}.$$

Now, since U is Lipschitz, we have, for small enough ε,

$$\begin{aligned}
|U(\varphi_{0\varepsilon}^v \circ \psi) - U(\varphi_{0\varepsilon}^{v_0} \circ \psi)| &\le C(\psi)\|\varphi_{0\varepsilon}^v \circ \psi - \tilde{\varphi}_{0\varepsilon}^v \circ \psi\|_{p,\infty} \\
&\le C(\psi)C_p(\psi)\|\varphi_{0\varepsilon}^v - \varphi_{0\varepsilon}^{v_0}\|_{p,\infty} \\
&\le C(\psi)C_p(\psi)\tilde{C}(v) \int_0^\varepsilon \|v(t, \cdot) - v_0\|_V dt
\end{aligned}$$

where $\tilde{C}(v_0)$ depends on $\|v_0\|_{p+1,\infty}$ and is provided in equation (8.17). This proves the proposition. $\qquad\square$

To the Eulerian gradient of U, we associate a "gradient descent" process (that we will formally interpret later as a Riemannian gradient descent for a suitable metric, see Section 12.4.3) which generates a time-dependent element of G by setting

$$\partial_t \varphi(t, x) = -\overline{\nabla}^V U(\varphi(t))(\varphi(t, x)) \tag{10.5}$$

As long as $\int_0^t \left\|\overline{\nabla}^V U(\varphi(s))\right\|_V ds$ is finite, this generates a time-dependent element of G_V. This is therefore an evolution within the group of diffeomorphisms, an important property. Assuming that Proposition 10.3 applies (which essentially requires $\overline{\nabla}U(\varphi)$ to be continuous in φ), we can write

$$\partial_t U(\varphi(t)) = \left\langle \overline{\nabla}^V U(\varphi(t)), \partial_t \varphi \right\rangle = -\left\| \overline{\nabla}^V U(\varphi(t)) \right\|_V^2$$

so that $U(\varphi(t))$ decreases with time.

We now make a formal comparison with more conventional derivatives of U. Directional derivatives in infinite dimensions are usually represented as Gâteaux derivatives, in the form

$$\left(\frac{\delta U}{\delta \varphi}(\psi) \,\middle|\, h \right) = \partial_\varepsilon U(\psi + \varepsilon h)_{|_{\varepsilon=0}}, \tag{10.6}$$

assuming that the right-hand side exists and is a *linear function* of h. (For this to make sense, we must also make sure that h is chosen so that $\psi + \varepsilon h \in G$ for small enough ε.) A slight variation of this is to let

$$\left(\frac{\delta U}{\delta \varphi}(\psi) \,\middle|\, h \right) = \partial_\varepsilon U(\varphi(\varepsilon, \cdot))_{|_{\varepsilon=0}}, \tag{10.7}$$

with $h = \partial_\varepsilon \varphi(0, \cdot)$, $\psi = \varphi(0, \cdot)$ and $\varphi(\varepsilon, \cdot)$ belongs to a certain class of time-dependent diffeomorphisms. Taking $\varphi(\varepsilon, \cdot) = \varphi_{0\varepsilon}^v \circ \psi$ with $v \in V$ yields $h = v \circ \psi$ and the identity (assuming that both sides are well-defined)

$$\left(\bar{\partial} U(\psi) \,\middle|\, v \right) = \left(\frac{\delta U}{\delta \varphi}(\psi) \,\middle|\, v \circ \psi \right). \tag{10.8}$$

There is another interesting relation between $\delta U/\delta \varphi$ and the Eulerian differential in the following particular case, that includes matching functionals. Assume that U takes the form

$$U(\varphi) = U_I(\varphi) = Z(\varphi \cdot I)$$

where I is a fixed deformable object. It is often the case that derivatives of U_I at $\psi = id$ are easier to compute than at non-trivial ψ. We have

$$\begin{aligned}
\left(\bar{\partial} U_I(\psi) \,\middle|\, v \right) &= \partial_\varepsilon U_I(\varphi_{0\varepsilon}^v \circ \psi)_{|_{\varepsilon=0}} \\
&= \partial_\varepsilon Z((\varphi_{0\varepsilon}^v \circ \psi) \cdot I)_{|_{\varepsilon=0}} \\
&= \partial_\varepsilon Z(\varphi_{0\varepsilon}^v \cdot (\psi \cdot I))_{|_{\varepsilon=0}} \\
&= \left(\bar{\partial} U_{\psi \cdot I}(id) \,\middle|\, v \right).
\end{aligned}$$

This says that Eulerian differentials of U_I (for arbitrary objects I) only have to be computed at $\psi = id$. Differentials at $\psi \neq id$ can be deduced immediately by replacing I by $\psi \cdot I$. Note also that $\bar{\partial} U(id)$ is formally identical to the Gâteaux derivative $(\delta U/\delta \varphi)(id)$.

We now present a series of matching functionals, adapted to different types of deformable objects, and compute their differentials.

10.3 Labeled Point Matching

The simplest way to represent a visual structure is with configurations of labeled points, or *landmarks* attached to the structure. Anatomical shapes or images are typical examples of structures on which landmarks can be easily defined; this includes specific locations in faces (corners of the eyes, tip of the nose, etc.), fingertips for hands, apex of the heart, etc. Many man-made objects, like cars or other vehicles, can be landmarked too. Finally, landmarks can represent the centers of simple objects, like cells in biological images.

In the labeled point-matching problem, objects are ordered collections of N points $x_1, \ldots, x_N \in \Omega$, where N is fixed. Diffeomorphisms act on such objects by:

$$\varphi \cdot (x_1, \ldots, x_N) = (\varphi(x_1), \ldots, \varphi(x_N)). \tag{10.9}$$

The problem is not to find correspondences between two objects, say $I = (x_1, \ldots, x_N)$ and $I' = (x'_1, \ldots, x'_N)$, since we know that x_i and x'_i are homologous, but to extrapolate these correspondences to the rest of the space. The simplest matching functional that we can consider for this purpose is

$$E_{x,x'}(\varphi) = \sum_{i=1}^{N} |x'_i - \varphi(x_i)|^2. \tag{10.10}$$

It is clear that E is (p, ∞)-Lipschitz with $p = 0$. The computation of $\delta E / \delta \varphi$ is straightforward and provides

$$\left(\frac{\delta E}{\delta \varphi} \mid h \right) = 2 \sum_{i=1}^{N} (\varphi(x_i) - x'_i)^T h(x_i). \tag{10.11}$$

This can be written

$$\frac{\delta E}{\delta \varphi} = 2 \sum_{i=1}^{N} (\varphi(x_i) - x'_i) \otimes \delta_{x_i}.$$

From (10.8), we have

$$\left(\bar{\partial} E(\varphi) \mid h \right) = 2 \sum_{i=1}^{N} (\varphi(x_i) - x'_i)^T h \circ \varphi(x_i). \tag{10.12}$$

or

$$\bar{\partial} E(\varphi) = 2 \sum_{i=1}^{N} (\varphi(x_i) - x'_i) \otimes \delta_{\varphi(x_i)}$$

and (10.3) gives

$$\bar{\nabla}^V E(\varphi) = 2 \sum_{i=1}^{N} K(., \varphi(x_i))(\varphi(x_i) - x'_i). \tag{10.13}$$

The gradient descent algorithm (10.5) takes a very simple form:

$$\partial_t \varphi(t, x) = -2 \sum_{i=1}^{N} K(\varphi(t, x), \varphi(t, x_i))(\varphi(t, x_i) - x_i'). \qquad (10.14)$$

This system can be solved in two steps: denote $y_i(t) = \varphi(t, x_i)$. Applying (10.14) at $x = x_j$ yields

$$\partial_t y_j = -2 \sum_{i=1}^{N} K(y_j, y_i)(y_i - x_i').$$

This is a differential system in $y_1, \ldots, y_N$. The first step is to solve it with initial conditions $y_j(0) = x_j$. Once this is done, the extrapolated value of $\varphi(t, x)$ for a general x is the solution of the differential equation

$$\partial_t y = -2 \sum_{i=1}^{N} K(y, y_i)(y_i - x_i')$$

initialized at $y(0) = x$. Figure 10.1 gives an example obtained with running this procedure, providing an illustration of the impact of the choice of the kernel for the solution. The last panel in Figure 10.1 also shows the limitations of this algorithm, in the sense that it is trying to move the points as much as possible in the direction of the target, while a more indirect path can be more efficient (these results should be compared to Figure 11.1 in Chapter 11).

10.4 Image Matching

Images, or more generally multivariate functions, are also important and widely used instances of deformable objects. They correspond to functions I defined on Ω with values in $\mathbb{R}$. Diffeomorphisms act on them by:

$$(\varphi \cdot I)(x) = I(\varphi^{-1}(x))$$

for $x \in \Omega$. Fixing two such functions I and I', the simplest matching functional which can be considered is the least-squares error,

$$E_{I,I'}(\varphi) = \int_{\Omega} \left| I \circ \varphi^{-1}(x) - I'(x) \right|^2 dx. \qquad (10.15)$$

In the following computation, we will assume that I is differentiable. The analysis without this assumption (e.g., for piecewise smooth images) is much more difficult, and the reader can refer to recent results in [210, 212, 211]. To compute $\bar{\partial} E(\varphi)$, we can first notice that, if $\varphi(\varepsilon, \cdot)$ is such that $\varphi(0, \cdot) = \psi$ and $\partial_\varepsilon \varphi(0) = h$, then

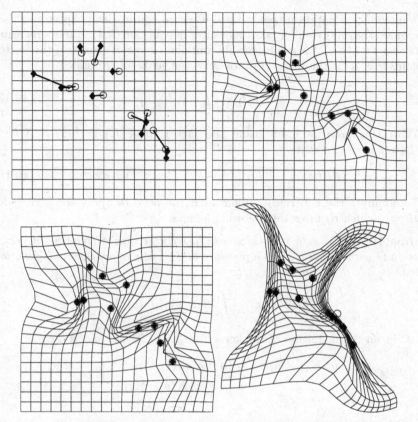

Fig. 10.1. Greedy landmark matching. Implementation of the gradient descent algorithm in (10.14), starting with $\varphi = id$, for the correspondences depicted in the upper-left image (diamonds moving to circles). The following three images provide the result after numerical convergence for Gaussian kernels $K(x, y) = \exp(-|x - y|^2/2\sigma^2)$ with $\sigma = 1, 2, 4$ in grid units. Larger σ induces increasing smoothness in the final solution, and deformations affecting a larger part of the space. As seen in the figure for $\sigma = 4$, the evolution can result in huge deformations.

$$\partial_\varepsilon(\varphi_\varepsilon^{-1})(0) = -D(\psi^{-1})h \circ \psi^{-1}. \tag{10.16}$$

This can be proved by computing the derivative of the equation $(\varphi(\varepsilon, \cdot))^{-1} \circ \varphi(\varepsilon, \cdot) = id$ at $\varepsilon = 0$. Using this, we can compute

$$\left(\bar{\partial}E(\varphi) \,|\, h\right) = -2 \int_\Omega (I \circ \varphi^{-1}(x) - I'(x)) \nabla I^T \circ \varphi^{-1} D(\varphi^{-1}) h \circ \varphi^{-1} dx.$$

and

$$\left(\bar{\partial}E(\varphi) \,|\, h\right) = -2 \int_\Omega (I \circ \varphi^{-1}(x) - I'(x)) \nabla I^T \circ \varphi^{-1} D(\varphi^{-1}) h dx.$$

Let's introduce a notation that we will use throughout this chapter and that generalizes the one given for point measures in equation (9.4). If μ is a measure on Ω and $z : \Omega \to \mathbb{R}^d$ a μ-measurable function (a vector density), the vector measure $(z \otimes \mu)$ is the linear form over vector fields on Ω defined by

$$(z \otimes \mu \mid h) = \int_\Omega \langle z, h \rangle d\mu. \tag{10.17}$$

With this notation, and using the fact that $\nabla(I \circ \varphi^{-1})^T = \nabla I^T \circ \varphi^{-1} D(\varphi^{-1})$, we can write

$$\bar{\partial} E(\varphi) = -2(I \circ \varphi^{-1} - I')\nabla(I \circ \varphi^{-1}) \otimes dx. \tag{10.18}$$

To compute the Eulerian gradient of E, we need to apply the kernel, K, to $\bar{\partial} E(\varphi)$, which requires the following lemma.

Lemma 10.4. *If V is a reproducing kernel Hilbert space (RKHS) of vector fields on Ω with kernel K, μ is a measure on Ω and z a μ-measurable function from Ω to $\mathbb{R}^d$, then, for all $x \in \Omega$,*

$$K(z \otimes \mu)(x) = \int_\Omega K(x, y) z(y) d\mu(y).$$

Proof. From the definition of the kernel, we have, for any $a \in \mathbb{R}^d$:

$$\begin{aligned}
a^T K(z \otimes \mu)(x) &= \big(a \otimes \delta_x \mid K(z \otimes \mu)\big) \\
&= \big(z \otimes \mu \mid K(a \otimes \delta_x)\big) \\
&= \big(z \otimes \mu \mid K(., x)a\big) \\
&= \int_\Omega z^T(y) K(y, x) a \, d\mu(y) \\
&= a^T \int_\Omega K(x, y) z(y) d\mu(y),
\end{aligned}$$

which proves Lemma 10.4. □

The expression of the Eulerian gradient of U is now given by Lemma 10.4:

$$\overline{\nabla}^V E(\varphi) = -2 \int_\Omega (I \circ \varphi^{-1}(y) - I'(y)) K(., y) \nabla(I \circ \varphi^{-1})(y) dy. \tag{10.19}$$

This provides the following greedy image-matching algorithm [48, 203].

Algorithm 2 (Greedy image matching) *Start with $\varphi(0) = id$ and solve the evolution equation*

$$\partial_t \varphi(t, y) = 2 \int_\Omega (J(t, x) - I'(x)) K(\varphi(t, y), x) \nabla J(t, x) dx \tag{10.20}$$

with $I(t, .) = I \circ (\varphi(t))^{-1}$.

This algorithm can also be written uniquely in terms of the evolving image, J, using $\partial_t J \circ \varphi + \langle J \circ \varphi, \partial_t \varphi \rangle = 0$. This yields

$$\partial_t J(t, y) = -2 \int_\Omega K(y, x)(J(t, x) - I'(x))\langle \nabla J(t, x), \nabla I(t, y)\rangle dx.$$

In contrast to what we did in the landmark case, this algorithm should not be run indefinitely (or until numerical convergence). The fundamental difference is that, in the landmark case, there is an infinity of solutions to the diffeomorphic interpolation problem, and the greedy algorithm simply progresses until it finds one of them and then stabilizes. In the case of images, it is perfectly possible (and even typical) that there is no solution to the matching problem, i.e., no diffeomorphism φ such that $I \circ \varphi^{-1} = I'$. In that case, Algorithm 2 will run indefinitely, creating huge deformations while trying to solve an impossible problem.

To decide when the evolution should be stopped, an interesting suggestion has been made in [203]. Define

$$v(t, x) = 2 \int_\Omega (J(t, x) - I'(x))K(y, x)\nabla J(t, x)dx$$

so that (10.20) reduces to $\partial_t \varphi = v(t) \circ \varphi$. As we know from Chapter 8, the smoothness of φ at time t can be controlled by

$$\int_0^t \|v(s)\|_V^2 ds,$$

the norm being explicitly given by

$$\|v(s)\|_V^2$$
$$= 2 \int_{\Omega \times \Omega} K(y, x)(J(s, x) - I'(x))(J(s, y) - I'(y))\langle \nabla J(s, x), \nabla J(s, y)\rangle dx dy.$$

Define, for some parameter λ

$$E(t) = \frac{1}{t} \int_0^t \|v(s)\|_V^2 ds + \lambda \int_\Omega (J(t, y) - I'(y))^2 dy.$$

Then, the stopping time proposed in [203] for Algorithm 2 is the first t at which $E(t)$ stops decreasing. Some experimental results using this algorithm and stopping rule are provided in Figure 10.2.

There are many other possible choices for a matching criterion, least squares being, as we wrote, the simplest one. Among other possibilities, comparison criteria involving histograms provide an interesting option, because they allow for contrast-invariant comparisons.

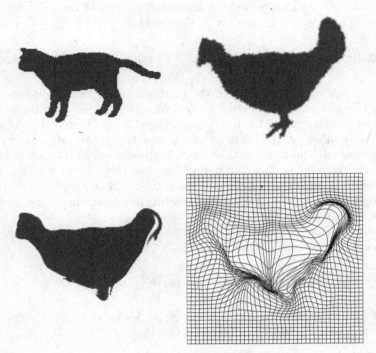

Fig. 10.2. Greedy image matching. Output of the algorithm when estimating a deformation of the first image to match the second one. The third image is the obtained deformation of the first one and the last provides the deformation applied to a grid.

Given a pair of images, I, I', associate to each $x \in \Omega$ and image values λ and λ' the local histogram $H_x(\lambda, \lambda')$, which counts the frequency of simultaneous occurrence of values λ in I and λ' in I' at the same location in a small window around x. One computationally feasible way to define it is to use the kernel estimator

$$H_{I,I'}(x, \lambda, \lambda') = \int_\Omega f(|I(y) - \lambda|) f(|I'(y) - \lambda'|) g(x, y) dy$$

in which f is a positive function such that $\int_{\mathbb{R}} f(t)dt = 1$, and f vanishes when t is far from 0, and $g \geq 0$ is such that for all x, $\int_\Omega g(x, y)dy = 1$ and $g(x, y)$ vanishes when y is far from x.

For each x, H_x is a bi-dimensional probability function, and there exist several ways for measuring the degree of dependence between its components. The simplest one, which is probably sufficient for most applications, is the correlation ratio, given by

$$C_{I,I'}(x) = 1 - \frac{\int_{\mathbb{R}^2} \lambda\lambda' H_{I,I'}(x,\lambda,\lambda')d\lambda d\lambda'}{\sqrt{\int_{\mathbb{R}^2} \lambda^2 H_{I,I'}(x,\lambda,\lambda')d\lambda d\lambda' \int_{\mathbb{R}^2} (\lambda')^2 H_{I,I'}(x,\lambda,\lambda')d\lambda d\lambda'}}.$$

It is then possible to define the matching function by

$$E_{I,I'}(\varphi) = \int_\Omega C_{I\circ\varphi^{-1},I'}(x)dx.$$

Variations of E with respect to φ require a lengthy (but elementary) computation. Some details can be found in [110]. A slightly simpler option is to use criteria based on the global histogram, which is defined by

$$H_{I,I'}(\lambda,\lambda') = \int_\Omega f(|I(y)-\lambda|)f(|I'(y)-\lambda'|)dy,$$

and the matching criterion simply is $E(\varphi) = C_{I\circ\varphi^{-1},I'}$ or, as introduced in [214, 133], the mutual information computed from the joint histogram.

10.5 Measure Matching

The running assumption in section 10.3 was that the point sets $(x_1, \ldots, x_N)$ were labeled, so that, when comparing two of them, correspondences were known and the problem was to extrapolate them to the whole space.

In some applications, correspondences are not given and need to be inferred as part of the matching problem. One way to handle this is to include them as new unknowns (in addition to the unknown diffeomorphisms), add extra terms to the energy that measure the quality of correspondences, and minimize the whole thing. Such an approach is taken, for example, in [172, 171].

Another point of view is to start with a representation of the point set that does not depend on how the points are ordered. A natural mathematical representation of a subset of $\mathbb{R}^d$ is by the uniform measure on this set, at least when it is well-defined. For a very general class of sets, this corresponds to the Hausdorff measure for the appropriate dimension [81], which, for finite sets, simply provides the sum of Dirac measures at each point, i.e., $x = (x_1, \ldots, x_N)$ is represented by

$$\mu_x = \sum_{i=1}^N \delta_{x_i}.$$

For us, this induces the issue of comparing measures using diffeomorphisms, which will be referred to as the measure-matching problem.

Like with all matching problems we are considering in this chapter, specifying measure matching requires, first, defining the action of diffeomorphisms on the considered objects, and second, using a good comparison criterion between two objects.

Let's start with the action of diffeomorphisms. The only fact we need here concerning measures is that they are linear forms acting on functions on $\mathbb{R}^d$ via

$$(\mu \,|\, f) = \int_{\mathbb{R}^d} f d\mu.$$

In particular, if μ_x is as above, then

$$(\mu_x \,|\, f) = \sum_{i=1}^{N} f(x_i).$$

If φ is a diffeomorphism of Ω and μ a measure, we define a new measure $\varphi \cdot \mu$ by

$$(\varphi \cdot \mu \,|\, f) = (\mu \,|\, f \circ \varphi).$$

It is straightforward to check that this provides a group action. If $\mu = \mu_x$, we have

$$(\varphi \cdot \mu_x \,|\, f) = \sum_{i=1}^{N} f \circ \varphi(x_i) = (\mu_{\varphi(x)} \,|\, f)$$

so that the transformation of the measure associated to a point set x is the measure associated to the transformed point set, which is reasonable.

When μ has a density with respect to Lebesgue's measure, say $\mu = zdx$, this action can be translated to a resulting transformation over densities as follows.

Proposition 10.5. *If $\mu = zdx$, where z is a positive, Lebesgue integrable function on $\Omega \subset \mathbb{R}^d$, and φ is a diffeomorphism of Ω, then $\varphi \cdot \mu = (\varphi \cdot z)dx$ with*

$$(\varphi \cdot z) = \det(D\varphi^{-1})z \circ \varphi^{-1}. \tag{10.21}$$

This is an immediate consequence of the definition of $\varphi \cdot \mu$ and of the change of variable formula (details are left to the reader). Note that the action of diffeomorphisms does not change the total mass of a positive measure, that is $(\varphi \cdot \mu)(\Omega) = \mu(\Omega)$ if φ is a diffeomorphism of Ω.

Now that we have defined the action, we need to choose a function $D(\mu, \mu')$ to compare two measures μ and μ'. Many such functions exist, especially when measures are normalized to have a unit mass, since this allows for the use of many comparison criteria defined in probability or information theory (like the Kullback–Leibler divergence [56]). A very general example is the Wasserstein distance [217, 170], which is associated to a positive, symmetric, cost function $\rho : \Omega \times \Omega \to [0, +\infty)$ and defined by,

$$d_\rho(\mu, \mu') = \inf_\nu \int_{\Omega^2} \rho(x, y)\nu(dx, dy) \tag{10.22}$$

where the minimization is over all ν with first marginal given by μ, and the second one by μ'. If μ and μ' are uniform measures on discrete point sets, i.e.,

$$\mu = \frac{1}{N} \sum_{k=1}^{N} \delta_{x_k}, \; \mu' = \frac{1}{M} \sum_{k=1}^{M} \delta_{x'_k}$$

then computing the Wasserstein distance reduces to minimizing

$$\sum_{k=1}^{N} \sum_{l=1}^{M} \rho(x_k, x'_l) \nu(x_k, x'_l)$$

subject to the constraints

$$\sum_{l=1}^{M} \nu(x_k, x'_l) = 1/N \text{ and } \sum_{k=1}^{N} \nu(x_k, x'_l) = 1/M.$$

This *linear assignment* problem is solved by finite-dimensional linear programming. If this is combined with diffeomorphic interpolation, i.e., if one tries to compute a diffeomorphism φ minimizing $d_\rho(\varphi \cdot x, x')$, this results in a formulation that mixes discrete and continuous optimization problems, similar to the methods introduced in [171]. The Wasserstein distance is also closely related to the mass transport problem, which can also be used to estimate diffeomorphisms, and will be addressed in the next chapter. For the moment, we focus on matching functionals associated to measures, and start with the case in which the compared measures are dense with respect to Lebesgue's measure, i.e., with the problem of matching densities.

10.5.1 Matching Densities

Since densities are scalar-valued functions, we can use standard norms to design matching functionals for them. As an example, let's take the simplest case of the L^2 norm, as we did with images. The difference with the image case is that the action is different, and has the interesting feature of involving the derivative of the diffeomorphism, via the Jacobian determinant.

So, let's consider the action $\varphi \cdot \zeta$ given by (10.21) and use the matching functional

$$E_{\zeta,\zeta'}(\varphi) = \int_\Omega (\varphi \cdot \zeta - \zeta')^2 dx.$$

Since we will need it for the differentiation of the Jacobian, we recall the following standard result on the derivative of the determinant.

Proposition 10.6. *Let $F(A) = \det(A)$ be defined over $\mathcal{M}_n(\mathbb{R})$, the space of all n by n matrices. Then, for any $A, H \in \mathcal{M}_n(\mathbb{R})$,*

$$DF(A)H = \text{trace}(\text{Com}(A)H) \tag{10.23}$$

where $\text{Com}(A)$ is the co-matrix of A, i.e., the matrix with (i,j) entry given by the determinant of A with the jth row and ith column removed.

In particular, when $A = \mathrm{Id}$, we have

$$DF(\mathrm{Id})H = \mathrm{trace}(H). \tag{10.24}$$

Proof. To prove this proposition, start with $A = \mathrm{Id}$ and use the facts that, if δ_{ij} is the matrix with 1 as the (i, j) entry and 0 everywhere else, then $\det(\mathrm{Id} + \varepsilon\delta_{ij}) = 1 + \varepsilon$ if $i = j$ and 1 otherwise, which directly gives (10.24). Then, prove the result for an invertible A using

$$\det(A + \varepsilon H) = \det(A)\det(\mathrm{Id} + \varepsilon A^{-1}H)$$

and the fact that, when A is invertible, $\det(A)A^{-1} = \mathrm{Com}(A)$. This also implies the result for a general (not necessarily invertible) A because the determinant is a polynomial in the entries of a matrix, and so are its partial derivatives, and the coefficients of these polynomials are fully determined by the values taken on the dense set of invertible matrices.   □

To compute the Eulerian differential of $E_{\zeta,\tilde{\zeta}}$ at any given φ, it will be convenient to use the trick described at the end of Section 10.2, starting with the computation of the differential at the identity and deducing from it the differential at any φ by replacing ζ by $\varphi \cdot \zeta$. We have

$$E_{\zeta,\zeta'}(\varphi) = \int_\Omega (\zeta \circ \varphi^{-1}\det(D\varphi^{-1}) - \zeta')^2 dx,$$

which can be easily checked to be $(1, \infty)$ locally Lipschitz.

If $\varphi(\varepsilon, \cdot)$ is a diffeomorphism that depends on a parameter ε, such that $\varphi(0, \cdot) = id$ and $\partial_\varepsilon\varphi(0, \cdot) = h$, then, at $\varepsilon = 0$, $\partial_\varepsilon\zeta \circ \varphi(\varepsilon, \cdot)^{-1} = -\langle\nabla\zeta, h\rangle$ and $\partial_\varepsilon\det(\varphi(\varepsilon, \cdot)^{-1}) = -\mathrm{trace}(Dh) = -\mathrm{div}h$. This implies

$$\partial_\varepsilon E_{\zeta,\zeta'}(\varphi_\varepsilon) = -2\int_\Omega (\zeta - \zeta')\mathrm{div}(\zeta h)dx$$

(since $\langle\nabla\zeta, h\rangle + \zeta\mathrm{div}h = \mathrm{div}(\zeta h)$). So this gives

$$(\partial_\varepsilon E_{\zeta,\zeta'}(id)\,|\,h) = -2\int_\Omega (\zeta - \zeta')\mathrm{div}(\zeta h)dx$$

and

$$(\partial_\varepsilon E_{\zeta,\zeta'}(\varphi)\,|\,h) = -2\int_\Omega (\varphi \cdot \zeta - \zeta')(\mathrm{div}((\varphi \cdot \zeta)h)dx. \tag{10.25}$$

We can use the divergence theorem to obtain an alternate expression (assuming that h vanishes on $\partial\Omega$ – or is tangent to it – or at infinity), yielding

$$(\partial_\varepsilon E_{\zeta,\zeta'}(\varphi)\,|\,h) = 2\int_\Omega \nabla(\varphi \cdot \zeta - \zeta')(\varphi \cdot \zeta)hdx \tag{10.26}$$

or

$$\partial_\varepsilon E_{\zeta,\zeta'}(\varphi) = 2(\varphi \cdot \zeta)\nabla(\varphi \cdot \zeta - \zeta') \otimes dx. \tag{10.27}$$

One can appreciate the symmetry of this expression with the one obtained with images in (10.18).

10.5.2 Dual RKHS Norms on Measures

One of the limitations of functional norms, like the L^2 one, is that they do not apply to singular objects, like the Dirac measures that motivated our study of the measure-matching problem. It is certainly possible to smooth out singular objects and transform them into densities that can be compared using the previous matching functional. For example, given a density function ρ (a Gaussian, for example) and a point set $(x_1, \ldots, x_N)$, we can compute a density

$$\zeta_x(y) = \sum_{k=1}^{N} \rho\Big(\frac{y - x_k}{\sigma}\Big), \tag{10.28}$$

where σ is a positive scale parameter (this is a standard kernel density estimator). One can then compare two point sets, say x and x' by comparing the associated ζ_x and $\zeta_{x'}$ using the previous method.

The representation in (10.28) is somewhat imperfect, in the sense that, for the natural actions we have defined, we have in general $\varphi \cdot \zeta_x \neq \zeta_{\varphi \cdot x}$: the density associated to a deformed point set is not the deformed density. If the goal is to compare two point sets, it makes more sense to use $\zeta_{\varphi \cdot x}$ instead of $\varphi \cdot \zeta_x$ as a density resulting from the deformation, and to rather use the cost function

$$E_{x,x'}(\varphi) = \int_{\mathbb{R}^d} (\zeta_{\varphi \cdot x} - \zeta_{x'})^2 dy, \tag{10.29}$$

which can be written, if $x = (x_1, \ldots, x_N)$ and $x' = (x'_1, \ldots, x'_M)$, and introducing the function

$$\xi(z, z') = \int_{\mathbb{R}^d} \rho\Big(\frac{y - z}{\sigma}\Big) \rho\Big(\frac{y - z'}{\sigma}\Big) dy, \tag{10.30}$$

as

$$E_{x,x'}(\varphi) = \sum_{k,l=1}^{N} \xi(\varphi(x_k), \varphi(x_l))$$

$$- 2 \sum_{k=1}^{N} \sum_{l=1}^{M} \xi(\varphi(x_k), x'_l) + \sum_{k,l=1}^{M} \xi(x'_k, x'_l). \tag{10.31}$$

Before computing the variations of this energy, we make the preliminary remark that the obtained expression is a particular case of what comes from a representation of measures as linear forms over RKHS's of scalar functions. Indeed, since measures are linear forms on functions, we can compute with them the dual of a function norm, given by

$$\|\mu\| = \sup \left\{ (\mu \mid f) : \|f\| = 1 \right\}. \tag{10.32}$$

Following [97], assume that the function norm in (10.32) is that of an RKHS. More precisely, let W be an RKHS of real-valued functions, so that we have an operator $\xi : W^* \to W$ with $\xi \delta_x := \xi(.,x)$ and with the identity $(\mu \mid f) = \langle \xi \mu, f \rangle_W$ for $\mu \in W^*$, $f \in W$. With this choice, (10.32) becomes

$$\|\mu\|_{W^*} = \sup \left\{ (\mu \mid f) : \|f\|_W = 1 \right\}$$
$$= \sup \left\{ \langle \xi \mu, f \rangle_W : \|f\|_W = 1 \right\}$$
$$= \|\xi \mu\|_W.$$

This implies that

$$\|\mu\|_{W^*}^2 = \langle \xi \mu, \xi \mu \rangle_W = (\mu \mid \xi \mu).$$

If μ is a measure, this expression is very simple and is given by

$$\|\mu\|_{W^*}^2 = \int \xi(x,y)d\mu(x)d\mu(y).$$

This is because $\xi \mu(x) = (\delta_x \mid \xi \mu) = (\mu \mid \xi \delta_x) = \int \xi(y,x)d\mu(y)$. So we can take

$$E_{\mu,\mu'}(\varphi) = \|\varphi \cdot \mu - \mu'\|_{W^*}^2. \qquad (10.33)$$

Expanding the norm, we get

$$E_{\mu,\mu'}(\varphi) = \langle \varphi \cdot \mu, \varphi \cdot \mu \rangle_{W^*} - 2\langle \varphi \cdot \mu, \mu' \rangle_{W^*} + \langle \mu', \mu' \rangle_{W^*}$$
$$= (\varphi \cdot \mu \mid \xi(\varphi \cdot \mu)) - 2(\varphi \cdot \mu \mid \xi \mu') + (\mu' \mid \xi \mu')$$
$$= \int \xi(\varphi(x), \varphi(y))d\mu(x)d\mu(y) - 2 \int \xi(\varphi(x), y)d\mu(x)d\mu'(y)$$
$$+ \int \xi(x,y)d\mu'(x)d\mu'(u).$$

We retrieve (10.31) when μ and μ' are sums of Dirac measures and ξ is chosen as in (10.30), but the RKHS formulation is more general.

It is now easy to compute $\bar{\partial} E_{\mu,\mu'}$ at $\varphi = id$. It is given by (letting $\nabla_1 \xi$ be the gradient of ξ with respect to its first variable and using the fact that ξ is symmetric)

$$\left(\bar{\partial} E_{\mu,\mu'}(id) \mid h \right) = 2 \int \nabla_1 \xi(x,y)^T h(x)d\mu(x)d\mu(y)$$
$$- 2 \int \nabla_1 \xi(x,z)^T h(x)d\mu(x)d\mu'(z)$$

which can be written as

$$\bar{\partial} E_{\mu,\mu'}(id) = 2 \left(\int \nabla_1 \xi(.,y)d\mu(y) - \int \nabla_1 \xi(.,z)d\mu'(z) \right) \otimes \mu. \qquad (10.34)$$

To obtain the Eulerian differential at a generic φ, it suffices to replace μ by $\varphi \cdot \mu$, which yields:

Proposition 10.7. *The Eulerian derivative and gradient of* (10.33) *are*

$$\bar{\partial} E_{\mu,\mu'}(\varphi) = 2 \left(\int \nabla_1 \xi(., \varphi(y)) d\mu(y) - \int \nabla_1 \xi(., z) d\mu'(z) \right) \otimes (\varphi.\mu) \quad (10.35)$$

and

$$\bar{\nabla}^V E_{\mu,\mu'}(\varphi)(.) = 2 \int K(., \varphi(x))$$

$$\left[\int \nabla_1 \xi(\varphi(x), \varphi(y)) d\mu(y) - \nabla_1 \xi(\varphi(x), z) d\mu'(z) \right] d\mu(x). \quad (10.36)$$

The derivative of the expression in (10.31) can be directly deduced from this expression. This leads to the following unlabeled point-matching evolution for point sets $x = (x_1, \ldots, x_N)$ and $x' = (x'_1, \ldots, x'_M)$:

$$\partial_t \varphi(z) = -2 \sum_{i=1}^{N} K(\varphi(z), \varphi(x_i))$$

$$\left[\sum_{j=1}^{N} \nabla_1 \xi(\varphi(x_i), \varphi(x_j)) - \sum_{h=1}^{M} \nabla_1 \xi(\varphi(x_i), x'_h) \right]. \quad (10.37)$$

Like in the case of labeled point sets, this equation may be solved in two stages: letting $z_i(t) = \varphi(t, x_i)$, first solve the system

$$\partial_t z_q = -2 \sum_{i=1}^{N} K(z_q, z_i) \left[\sum_{j=1}^{N} \nabla_1 \xi(z_i, z_j) - \sum_{h=1}^{M} \nabla_1 \xi(z_i, x'_h) \right].$$

Once this is done, the trajectory of an arbitrary point $z(t) = \varphi_t(z_0)$ is

$$\partial_t z = -2 \sum_{i=1}^{N} K(z, z_i) \left[\sum_{j=1}^{N} \nabla_1 \xi(z_i, z_j) - \sum_{h=1}^{M} \nabla_1 \xi(z_i, x'_h) \right].$$

10.6 Matching Curves and Surfaces

Curves in two dimensions and surfaces in three dimensions are probably the most natural representations of shapes, and their comparison using matching functionals is a fundamental issue. In this section, we discuss a series of representations that can be seen as extensions of the measure-matching methods that were described in the previous one. (This is not the unique way to compare such objects, and we will see a few more methods in the following chapters, especially for curves.)

Note that we are looking here for correspondences between points in curves and surfaces that derive from global diffeomorphisms of the ambient space. The curve- (or surface-) matching problems are often studied in the literature as trying to find diffeomorphic correspondences between points along the curve (or surface) only. Even if such restricted diffeomorphisms can generally be extended to diffeomorphisms of the whole space, the two approaches generally lead to very different algorithms. The search for correspondences within the structures is often implemented as a search for correspondences between parametrizations. This is easier for curves (looking, for example, for correspondences of the arc-length parametrizations), than for surfaces, which may not be topologically equivalent in the first place (a sphere cannot be matched to a torus); when matching topologically equivalent surfaces, special parametrizations, like conformal maps [52, 194] can be used. In this framework, once parametrizations are fixed, one can look for diffeomorphisms in parameter space that optimally align some well-chosen, preferably intrinsic, representation. In the case of curves, one can choose the representation $s \mapsto \kappa_\gamma(s)$, where κ_γ is the curvature of a curve γ, with the curve rescaled to have length 1 to fix the interval over which this representation is defined. One can then use image-matching functionals to compare them, i.e., find φ (a diffeomorphism of the unit interval) such that $\varphi \cdot \kappa_\gamma \simeq \kappa_{\gamma'}$.

But, as we wrote, the main focus in this chapter is the definition of matching functionals for deformable objects in $\mathbb{R}^d$, and we now address this problem with curves and surfaces.

10.6.1 Curve Matching with Measures

We can arguably make a parallel between point sets and curves in that labeled point sets correspond to parametrized curves and unlabeled point sets to curves modulo parametrization. In this regard we have a direct generalization of the labeled point-matching functional to parametrized curves (assumed to be defined over the same interval, say $[0, 1]$), simply given by

$$E_{\gamma,\gamma'}(\varphi) = \int_0^1 |\varphi(\gamma(u)) - \gamma'(u)|^2 du.$$

But being given two consistent parametrizations of the curves (to allow for direct comparisons as done above) almost never happens in practice. Interesting formulations of the curve matching problem should therefore consider curves modulo parametrization, so that the natural analogy is with unlabeled point sets. The counterpart of a uniform measure over a finite set of points is the uniform measure on the curve, defined by, if γ is parametrized over an interval $[a, b]$,

$$(\mu_\gamma \,|\, f) = \int_\gamma f \, dl = \int_a^b f(\gamma(u))|\dot{\gamma}_u(u)| \, du.$$

This is clearly a parametrization-independent representation. Now, if φ is a diffeomorphism, we have, by definition of the action of diffeomorphisms on measures

$$(\varphi \cdot \mu_\gamma \mid f) = \int_\gamma f \circ \varphi dl = \int_a^b f(\varphi(\gamma(u))) |\dot{\gamma}_u(u)| du.$$

However, we have

$$(\mu_{\varphi \cdot \gamma} \mid f) = \int_{\varphi(\gamma)} f dl = \int_a^b f(\varphi(\gamma(u))) |D\varphi(\gamma(u)) \dot{\gamma}_u(u)| du.$$

So, in contrast to point sets, for which we had $\varphi \cdot \mu_x = \mu_{\varphi(x)}$, the image of the measure associated to a curve is not the measure associated to the image of a curve. When the initial goal is to compare curves, and not measures, it is more natural to use the second definition, $\mu_{\varphi \cdot \gamma}$, rather than the first one. We can set, using the notation of the previous section, and introducing a target curve γ' defined on $[a', b']$,

$$
\begin{aligned}
E_{\gamma, \gamma'}(\varphi) &= \langle \mu_{\varphi \cdot \gamma}, \mu_{\varphi \cdot \gamma} \rangle_{W^*} - 2 \langle \mu_{\varphi \cdot \gamma}, \mu_{\gamma'} \rangle_{W^*} + \langle \mu_{\gamma'}, \mu_{\gamma'} \rangle_{W^*} \\
&= (\mu_{\varphi \cdot \gamma} \mid \xi(\mu_{\varphi \cdot \gamma})) - 2(\mu_{\varphi \cdot \gamma} \mid \xi \mu_{\gamma'}) + (\mu_{\gamma'} \mid \xi \mu_{\gamma'}) \\
&= \int_a^b \int_a^b \xi(\varphi(\gamma(u)), \varphi(\gamma(v))) |D\varphi(\gamma(u)) \dot{\gamma}_u(u)| |D\varphi(\gamma(v)) \dot{\gamma}_u(v)| du dv \\
&\quad - 2 \int_a^b \int_{a'}^{b'} \xi(\varphi(\gamma(u)), \gamma'(v)) |D\varphi(\gamma(u)) \dot{\gamma}_u(u)| |\dot{\gamma}_u'(v)| du dv \\
&\quad + \int_{a'}^{b'} \int_{a'}^{b'} \xi(\gamma'(u), \gamma'(v)) |\dot{\gamma}_u'(u)| |\dot{\gamma}_u'(v)| du dv.
\end{aligned}
$$

Since $E_{\gamma, \gamma'}$ is based on a dot product, we have, taking as usual $\varphi(\varepsilon, \cdot)$ such that $\varphi(0, \cdot) = id$ and $\partial_\varepsilon \varphi(0, \cdot) = h$,

$$\partial_\varepsilon E(\varphi(\varepsilon, \cdot)) = 2 \partial_\varepsilon \langle \mu_{\varphi(\varepsilon, \cdot) \cdot \gamma} - \mu_{\gamma'}, \mu_\gamma - \mu_{\gamma'} \rangle_{W^*} = 2 \partial_\varepsilon \langle \mu_{\varphi(\varepsilon, \cdot) \cdot \gamma}, \mu_\gamma - \mu_{\gamma'} \rangle_{W^*}.$$

So, introduce

$$\tilde{E}(\varphi) = \langle \mu_{\varphi \cdot \gamma}, \mu_\gamma - \mu_{\gamma'} \rangle_{W^*}$$

and let, for a given curve $\tilde{\gamma}$,

$$Z^{\tilde{\gamma}}(\cdot) = \int_{\tilde{\gamma}} \xi(\cdot, p) dl(p). \tag{10.38}$$

Let also $\zeta = Z^\gamma - Z^{\gamma'}$ and, for further use, $\zeta^\varphi = Z^{\varphi \cdot \gamma} - Z^{\gamma'}$. With this notation, we have

$$\tilde{E}(\varphi) = \int_{\varphi(\gamma)} \zeta(p) dl(p)$$

and we can use Proposition 6.2 to derive, letting p_0 and p_1 be the extremities of γ

$$\partial_\varepsilon \tilde{E}(\varphi(\varepsilon, \cdot))|_{\varepsilon=0} = \zeta(p_1)\langle h(p_1), T^\gamma(p_1)\rangle - \zeta(p_0)\langle h(p_0), T^\gamma(p_0)\rangle$$
$$+ \int_\gamma (\langle \nabla\zeta, N^\gamma\rangle - \zeta\kappa^\gamma)\langle h, N^\gamma\rangle dl.$$

Replacing γ by $\varphi \cdot \gamma$, this provides the expression of the Eulerian derivative of E at φ, namely

$$\frac{1}{2}\bar{\partial}_\varepsilon E_{\gamma,\gamma'}(\varphi) = \zeta^\varphi T^\gamma \otimes (\delta_{p_1} - \delta_{p_0})$$
$$+ (\langle \nabla\zeta^\varphi, N^{\varphi\cdot\gamma}\rangle - \zeta^\varphi\kappa^{\varphi\cdot\gamma})N^{\varphi\cdot\gamma} \otimes \mu_{\varphi\cdot\gamma}. \quad (10.39)$$

The Eulerian gradient on V therefore is

$$\frac{1}{2}\bar{\partial}_\varepsilon E_{\gamma,\gamma'}(\varphi) = K(\cdot, p_1)\zeta^\varphi(p_1)T^\gamma(p_1) - K(\cdot, p_0)\zeta^\varphi(p_0)T^\gamma(p_0)$$
$$+ \int_{\varphi\cdot\gamma} \Big(\langle \nabla\zeta^\varphi(p), N^{\varphi\cdot\gamma}(p)\rangle - \zeta^\varphi(p)\kappa^{\varphi\cdot\gamma}(p)\Big)K(\cdot, p)N^{\varphi\cdot\gamma}(p)dl(p). \quad (10.40)$$

To write this expression, we have implicitly assumed that γ is C^2. In fact, we can give an alternate expression for the Eulerian gradient that does not require this assumption, by directly computing the variation of $\tilde{E}(\varphi(\varepsilon, \cdot))$ without applying Proposition 6.2. This yields, using the fact that, if z is a function of a parameter ε, then $\partial_\varepsilon |z| = \langle \dot{z}_\varepsilon, z/|z|\rangle$,

$$\partial_\varepsilon |D\varphi(\varepsilon, \cdot)(\gamma)\dot{\gamma}_u|_{\varepsilon=0} = \langle T^\gamma, Dh(\gamma)\dot{\gamma}_u\rangle = \langle T^\gamma, Dh(\gamma)T^\gamma\rangle|\dot{\gamma}_u|$$

and $\partial_\varepsilon \tilde{E}(\varphi(\varepsilon, \cdot)) = \int_\gamma (\langle \nabla\zeta, h\rangle + \zeta\langle T^\gamma, DhT^\gamma\rangle)dl.$

The term involving Dh can be written in terms of V-dot products of h with derivatives of the kernel, K. We can indeed formally write, for $h \in V$ and $a, b \in \mathbb{R}^d$,

$$a^T Dh(x)b = \partial_\varepsilon a^T h(x + \varepsilon b)|_{\varepsilon=0}$$
$$= \partial_\varepsilon (\langle h, K(\cdot, x + \varepsilon b)a\rangle_V)|_{\varepsilon=0}$$
$$= \langle h, \partial_\varepsilon (K(\cdot, x + \varepsilon b)a)|_{\varepsilon=0}\rangle_V.$$

(Computing derivatives under the V dot product requires some justifications (and assumptions); see Lemma 11.10.) If M is a matrix-valued function defined on $\mathbb{R}^d$, with entries M^{ij}, and $b \in \mathbb{R}^d$, $DM(x)b$ denotes the matrix with entries

$$(DM(x)b)^{ij} = \langle \nabla M^{ij}(x), b\rangle,$$

i.e., $DM(x)b = \partial_\varepsilon M(x + \varepsilon b)|_{\varepsilon=0}$. We therefore have

$$\partial_\varepsilon K(.,x+\varepsilon b)a_{|_{\varepsilon=0}} = (D_2K(\cdot,x)b)a$$

where D_2K is the differential of K with respect to its second coordinate, and

$$a^T Dh(x)b = \langle h, (D_2K(.,x)b)a\rangle_V. \tag{10.41}$$

This gives

$$\partial_\varepsilon \tilde{E}(\varphi(\varepsilon,\cdot)) = \int_\gamma \Big(\langle K(\cdot,p)(\nabla\zeta(p), h\rangle_V$$
$$+ \langle \zeta(p)(D_2K(.,p)T^\gamma(p))T^\gamma(p), h\rangle_V\Big)dl(p)$$

and a new expression of the Eulerian gradient

$$\frac{1}{2}\bar{\nabla}^V E_{\gamma,\gamma'}(\varphi) = \int_{\varphi\cdot\gamma} \Big(K(\cdot,p)\nabla\zeta^\varphi(p)$$
$$+ \zeta^\varphi(p)(D_2K(.,p)T^{\varphi\cdot\gamma}(p))T^{\varphi\cdot\gamma}(p)\Big)dl(p). \tag{10.42}$$

To be complete, let's consider the variation of a discrete form of $E_{\gamma,\gamma'}(\varphi)$. If a curve γ is discretized with points $x_0,\ldots x_N$ (with $x_N = x_0$ if the curve is closed), one can define the discrete measure, still denoted μ_γ

$$(\mu_\gamma \mid f) = \sum_{i=1}^N f(c_i)|\tau_i|$$

with $c_i = (x_i + x_{i-1})/2$ and $\tau_i = x_i - x_{i-1}$. Use a similar expression for the measure associated to a discretization of $\varphi\cdot\gamma$, with $c_i^\varphi = (\varphi(x_i)+\varphi(x_{i-1}))/2$ and $\tau_i^\varphi = \varphi(x_i) - \varphi(x_{i-1})$. Finally, let γ' be discretized in $x_1',\ldots,x_M'$, and define

$$E_{\gamma,\gamma'}(\varphi) = \sum_{i,j=1}^N \xi(c_i^\varphi,c_j^\varphi)|\tau_i^\varphi||\tau_j^\varphi|$$
$$- 2\sum_{i=1}^N\sum_{j=1}^N \xi(c_i^\varphi,c_j')|\tau_i^\varphi||\tau_j'| + \sum_{i,j=1}^M \xi(c_i',c_j')|\tau_i'||\tau_j'|$$

in which we identify indices 1 and $N+1$ or $M+1$ (assuming closed curves). Note that this functional depends on $\varphi\cdot x$ and x'. The computation of the differential proceeds as above. Define, for a point set $\tilde{x} = (\tilde{x}_1,\ldots,\tilde{x}_Q)$

$$Z^{\tilde{x}}(\cdot) = \sum_{j=1}^Q \xi(\cdot,\tilde{c}_j)|\tilde{\tau}_j|,$$

and $\zeta = Z^x - Z^{x'}$, $\zeta^\varphi = Z^{\varphi \cdot x} - Z^{x'}$. We then obtain

$$\frac{1}{2}\bar{\partial}E_{\gamma,\gamma'}(id) = \sum_{i=1}^{N}(\nabla\zeta(c_i)|\tau_i| + \nabla\zeta(c_{i+1})\tau_{i+1}) \otimes \delta_{x_i}$$

$$- 2\sum_{i=1}^{N}\left(\zeta(c_{i+1})\frac{\tau_{i+1}}{|\tau_{i+1}|} - \zeta(c_i)\frac{\tau_i}{|\tau_i|}\right) \otimes \delta_{x_i}.$$

The Eulerian differential at $\varphi \neq id$ is obtained by replacing ζ, c_i, τ_i by $\zeta^\varphi, c_i^\varphi, \tau_i^\varphi$ and the Eulerian gradient by applying the V-kernel to it.

10.6.2 Curve Matching with Vector Measures

Instead of describing a curve with a measure, which is a linear form on functions, it is possible to represent it by a vector measure, which is a linear form on vector fields. Given a parametrized curve $\gamma : [a,b] \to \mathbb{R}^d$, we define a vector measure ν_γ, which associates to each vector field f on $\mathbb{R}^d$ a number $(\nu_\gamma \,|\, f)$ given by

$$(\nu_\gamma \,|\, f) = \int_a^b \dot{\gamma}_u(u)^T f \circ \gamma(u)du.$$

We have in fact $\nu_\gamma = T^\gamma \otimes \mu_\gamma$ where T^γ is the unit tangent to γ and μ_γ is the line measure along γ, as defined in the previous section. This definition is invariant by a change of parametrization, but depends on the orientation of γ. If φ is a diffeomorphism, we then have

$$\nu_{\varphi \cdot \gamma}(f) = \int_a^b (D\varphi(\gamma(u))\dot{\gamma}_u(u))^T f \circ \varphi(\gamma(u))du.$$

Like with scalar measures, we can use a dual norm for the comparison of two vector measures. Such a norm is defined by

$$\|\nu\|_{W^*} = \sup\left\{(\nu \,|\, f) : \|f\|_W = 1\right\}$$

where W is now an RKHS of vector fields, and we still have $\|\nu\|_{W^*}^2 = (\nu \,|\, \xi\nu)$ where ξ is the kernel of W(which is now matrix-valued). In particular, we have

$$\|\nu_\gamma\|_{W^*}^2 = \int_a^b \int_a^b \dot{\gamma}_u(u)^T \xi(\gamma(u), \gamma(v))\dot{\gamma}_u(v)dudv$$

and

$$\|\nu_{\varphi \cdot \gamma}\|_{W^*}^2 = \int_a^b \int_a^b \dot{\gamma}_u(u)^T D\varphi(\gamma(u))^T \xi(\varphi(\gamma(u)), \varphi(\gamma(v)))D\varphi(\gamma(v))\dot{\gamma}_u(v)dudv.$$

Define $E_{\gamma,\gamma'}(\varphi) = \|\nu_{\varphi \cdot \gamma} - \nu_{\gamma'}\|_{W^*}^2$. We follow the same pattern as in the previous section and define

$$\tilde{E}(\varphi) = \langle \nu_{\varphi \cdot \gamma}, \, \nu_{\gamma} - \nu_{\gamma'} \rangle_{W^*}$$

which (introducing $\varphi(\varepsilon, \cdot)$ with $\varphi(0, \cdot) = id$ and $\partial_\varepsilon \varphi(0, \cdot) = h$) is such that $\partial_\varepsilon E(\varphi(\varepsilon, \cdot)) = 2\partial_\varepsilon \tilde{E}(\varphi(\varepsilon, \cdot))$. Define

$$Z^{\tilde{\gamma}}(\cdot) = \int_{\tilde{\gamma}} \xi(\cdot, p) N^{\tilde{\gamma}}(p) dp,$$

and $\zeta = Z^{\gamma} - Z^{\gamma'}, \zeta^{\varphi} = Z^{\varphi \cdot \gamma} - Z^{\gamma'}$, so that (using $\langle T^{\varphi \cdot \gamma}, T^{\gamma'} \rangle = \langle N^{\varphi \cdot \gamma}, N^{\gamma'} \rangle$)

$$\tilde{E}(\varphi) = \int_{\varphi \cdot \gamma} \langle \zeta, \, N^{\varphi \cdot \gamma} \rangle dl.$$

We can use proposition 6.2, equation (6.4), to find

$$\partial_\varepsilon E(\varphi(\varepsilon, \cdot)) = -[\det(\zeta, h)]_0^\Delta + \int_\gamma \operatorname{div}(\zeta) \langle N^\gamma, \, h \rangle dl.$$

This yields in turn (replacing γ by $\varphi \cdot \gamma$, and letting p_0 and p_1 be the extremities of γ)

$$\frac{1}{2} \bar{\partial} E_{\gamma, \gamma'}(\varphi) = -(R_{\pi/2} \zeta^\varphi) \otimes (\delta_{\varphi(p_1)} - \delta_{\varphi(p_0)}) + \operatorname{div}(\zeta^\varphi) \nu_{\varphi \cdot \gamma}, \qquad (10.43)$$

where $R_{\pi/2}$ is a 90° rotation. This final expression is remarkably simple, especially with closed curves, for which the first term cancels. A discrete version of the matching functional can also be defined, namely, using the notation of the previous section:

$$E_{\gamma, \gamma'}(\varphi) = \sum_{i,j=1}^{N} \xi(c_i^\varphi, c_j^\varphi) \langle \tau_i^\varphi, \, \tau_j^\varphi \rangle$$

$$- 2 \sum_{i=1}^{N} \sum_{j=1}^{N} \xi(c_i^\varphi, c_j') \langle \tau_i^\varphi, \, \tau_j' \rangle + \sum_{i,j=1}^{M} \xi(c_i', c_j') \langle \tau_i', \, \tau_j' \rangle.$$

We leave the computation of the associated Eulerian differential (which is a slight variation of the one we made with measures) to the reader.

10.6.3 Surface Matching

We now extend to surfaces the matching functionals that we just studied for curves. The construction is formally very similar. If S is a surface in $\mathbb{R}^3$, one can compute a measure μ_S and a vector measure ν_S defined by

$$(\mu_S \mid f) = \int_S f(x) d\sigma_S(x) \text{ for a scalar } f$$

and

$$(\nu_S \,|\, f) = \int_S \langle f(x)\,,\, N(x)\rangle d\sigma_S(x) \text{ for a vector field } f$$

where $d\sigma$ is the area form on S and N is the unit normal (S being assumed to be oriented in the second case).

We state without proof the following result:

Proposition 10.8. *If S is a surface and φ a diffeomorphism of $\mathbb{R}^3$ that preserves the orientation (i.e., with positive Jacobian), we have*

$$\left(\mu_{\varphi(S)} \,|\, f\right) = \int_S f \circ \varphi(x)|D\varphi(x)^{-T}N|\,\det(D\varphi(x))d\sigma_S(x)$$

for a scalar f and for a vector-valued f,

$$\left(\nu_{\varphi(S)} \,|\, f\right) = \int_S f \circ \varphi(x)^T D\varphi(x)^{-T}N\,\det(D\varphi(x))d\sigma_S(x).$$

If $e_1(x), e_2(x)$ is a basis of the tangent plane to S at x, we have

$$D\varphi(x)^{-T}N\,\det(D\varphi(x)) = (D\varphi(x)e_1 \times D\varphi(x)e_2)/|e_1 \times e_2|. \qquad (10.44)$$

The last formula implies in particular that if S is parametrized by $(u,v) \mapsto m(u,v)$, then (since $N = (\partial_u m \times \partial_v m)/|\partial_u m \times \partial_v m|$ and $d\sigma_S = |\partial_u m \times \partial_v m|\,dudv$)

$$(\nu_S \,|\, f) = \int \langle f(x)\,,\, (\partial_u m \times \partial_v m)\rangle dudv$$

$$= \int \det(\partial_u m, \partial_v m, f)dudv$$

and

$$\left(\nu_{\varphi(S)} \,|\, f\right) = \int \det(D\varphi\partial_u m, D\varphi\partial_v m, f \circ \varphi)dudv.$$

If W is an RKHS of scalar functions or vector fields, we can compare two surfaces by using the norm of the difference of their associated measures on W^*. So define (in the scalar measure case)

$$E_{S,S'}(\varphi) = \|\mu_{\varphi.S} - \mu_{S'}\|_{W^*}^2$$

and the associated

$$\tilde{E}(\varphi) = \langle \mu_{\varphi.S}\,,\, \mu_S - \mu_{S'}\rangle_{W^*}$$

so that, for $\varphi(\varepsilon, \cdot)$ such that $\varphi(0, \cdot) = id$ and $\partial_\varepsilon\varphi(0, \cdot) = h$

$$\partial_\varepsilon E_{\gamma,\gamma'}(\varphi(\varepsilon, \cdot)) = 2\partial_\varepsilon\tilde{E}(\varphi(\varepsilon, \cdot)).$$

To a given surface $\tilde{S}$, associate the function

$$Z^{\tilde{S}}(\cdot) = \int_{\tilde{S}} \xi(\cdot, p)d\sigma_{\tilde{S}}(p)$$

and $\zeta = Z^S - Z^{S'}$, $\zeta^\varphi = Z^{\varphi \cdot S} - Z^{S'}$. Since

$$\tilde{E}(\varphi) = \int_{\varphi \cdot S} \zeta(p) d\sigma_{\varphi \cdot S}(p)$$

proposition 6.4 yields

$$\partial_\varepsilon \tilde{E}(\varphi(\varepsilon, \cdot)) = -\int_{\partial S} \zeta \langle n^S, h \rangle dl$$

$$+ \int_S \left(-2\zeta H^S + \langle \nabla \zeta, N^S \rangle \right) \langle N^S, h \rangle d\sigma_S$$

where H^S is the mean curvature on S. This implies

$$\frac{1}{2} \bar{\partial} E_{S,S'}(\varphi) = -\zeta^\varphi n^{\varphi \cdot S} \otimes \mu_{\varphi \cdot \partial S} + \left(-2\zeta^\varphi H^{\varphi \cdot S} + \langle \nabla \zeta^\varphi, N^{\varphi \cdot S} \rangle \right) \nu_{\varphi \cdot S}. \quad (10.45)$$

If we now use vector measures, so that

$$E_{S,S'}(\varphi) = \| \nu_{\varphi \cdot S} - \nu_{S'} \|_{W^*}^2$$

and

$$\tilde{E}(\varphi) = \langle \nu_{\varphi \cdot S}, \nu_S - \nu_{S'} \rangle_{W^*},$$

we need to define

$$Z^{\tilde{S}}(\cdot) = \int_{\tilde{S}} \xi(\cdot, p) N^{\tilde{S}} d\sigma_{\tilde{S}}(p)$$

and $\zeta = Z^S - Z^{S'}$, $\zeta^\varphi = Z^{\varphi \cdot S} - Z^{S'}$ so that

$$\tilde{E}(\varphi) = \int_{\varphi \cdot S} \langle \zeta, N^{\varphi \cdot S} \rangle d\sigma_{\varphi \cdot S}.$$

Variations derive again from Proposition 6.4, yielding

$$\partial_\varepsilon \tilde{E} = -\int_{\partial S} (\langle \zeta, N^S \rangle \langle h, n^S \rangle - \langle \zeta, n^S \rangle \langle h, N^S \rangle) dl$$

$$+ \int_S \text{div}(\zeta) \langle N^S, h \rangle d\sigma_S.$$

We therefore have

$$\frac{1}{2} \bar{\partial} E_{S,S'}(\varphi) = -(\langle \zeta^\varphi, N^{\varphi \cdot S} \rangle n^{\varphi \cdot S} - \langle \zeta^\varphi, n^{\varphi \cdot S} \rangle N^{\varphi \cdot S}) \otimes \mu_{\varphi \cdot \partial S}$$

$$+ \text{div}(\zeta^\varphi) \nu_{\varphi \cdot S}. \quad (10.46)$$

Again the expression is remarkably simple for surfaces without boundary.

Consider now the discrete case and let S be a triangulated surface [207]. Let $x_1, \ldots, x_N$ be the vertices of S, $f_1, \ldots, f_Q$ be the faces (triangles) which are ordered triplets of vertices $f_i = (x_{i1}, x_{i2}, x_{i3})$. Let c_i be the center of f_i, N_i its oriented unit normal and a_i its area. Define the discrete versions of the previous measures by

$$(\mu_S \,|\, h) = \sum_{i=1}^{Q} h(c_i) a_i, \text{ for a scalar } h$$

and

$$(\nu_S \,|\, h) = \sum_{i=1}^{Q} \langle f(c_i), N_i \rangle a_i, \text{ for a vector field } h.$$

The previous formulae can be written

$$(\mu_S \,|\, h) = \sum_{i=1}^{K} h\left(\frac{x_{i1} + x_{i2} + x_{i3}}{3}\right) |(x_{i2} - x_{i1}) \times (x_{i3} - x_{i1})|$$

and

$$(\nu_S \,|\, h) = \sum_{i=1}^{K} h\left(\frac{x_{i1} + x_{i2} + x_{i3}}{3}\right)^T (x_{i2} - x_{i1}) \times (x_{i3} - x_{i1})$$

where the last formula requires that the vertices or the triangles are ordered consistently with the orientation (see section 4.11.2). The transformed surfaces are now represented by the same expressions with x_{ik} replaced by $\varphi(x_{ik})$. If, given two triangulated surfaces, one defines $E_{S,S'}(\varphi) = \|\mu_{\varphi \cdot S} - \mu_{S'}\|_{W^*}^2$, then (leaving the computation to the reader)

$$\frac{1}{2}\bar{\partial}E_{S,S'}(id) = \sum_{k=1}^{N} \left(\sum_{i:x_k \in f_i} \left(\nabla\zeta(c_i)\frac{a_i}{3} + \zeta(c_i)e_{ik} \times N_i\right)\right) \otimes \delta_{x_k}$$

where e_{ik} is the edge opposed to x_k in f_i (oriented so that (x_k, e_{ik}) is positively ordered), and $\zeta = Z^S - Z^{S'}$, with

$$Z^{\tilde{S}}(\cdot) = \sum_{i=1}^{\tilde{K}} \xi(\cdot, \tilde{c}_i)\tilde{a}_i$$

for a triangulated surface $\tilde{S}$. The Eulerian differential at φ is obtained by replacing all x_k's by $\varphi(x_k)$.

For the vector-measure form, $E_{S,S'}(\varphi) = \|\nu_{\varphi \cdot S} - \nu_{S'}\|_{W^*}^2$, we get

$$\frac{1}{2}\bar{\partial}E_{S,S'}(id) = \sum_{k=1}^{N} \left(\sum_{i:x_k \in f_i} \left(D\zeta(c_i)N_i\right)\frac{a_i}{3} + e_{ik} \times \zeta(c_i)\right) \otimes \delta_{x_k}$$

still with $\zeta = Z^S - Z^{S'}$, but with

$$Z^{\tilde{S}}(\cdot) = \sum_{i=1}^{\tilde{K}} \xi(\cdot, \tilde{c}_i)N_i a_i.$$

10.6.4 Induced Actions and Currents

We have designed the action of diffeomorphisms on measures by $(\varphi \cdot \mu \,|\, h) = (\mu \,|\, h \circ \varphi)$. Recall that we have the usual action of diffeomorphisms on functions defined by $\varphi \cdot h = h \circ \varphi^{-1}$, so that we can write $(\varphi \cdot \mu \,|\, h) = (\mu \,|\, \varphi^{-1} \cdot h)$. In the case of curves, we have seen that this action on the induced measure did not correspond to the image of the curve by a diffeomorphism, in the sense that $\mu_{\varphi \cdot \gamma} \neq \varphi \cdot \mu_\gamma$. Here, we discuss whether the transformations $\mu_\gamma \to \mu_{\varphi \cdot \gamma}$ or $\nu_\gamma \to \nu_{\varphi \cdot \gamma}$ (and the equivalent transformations for surfaces) can be explained by a similar operation, e.g., whether one can write $(\varphi \cdot \mu \,|\, h) = (\mu \,|\, \varphi^{-1} \star h)$ where $\star$ would represent another action of diffeomorphisms on functions (or on vector fields for vector measures).

For μ_γ, the answer is negative. We have, letting $T(\gamma(u)))$ be the unit tangent to γ,

$$(\mu_{\varphi \cdot \gamma} \,|\, h) = \int_a^b h(\varphi(\gamma(u)))|D\varphi(\gamma(u))\dot{\gamma}_u(u)|du$$

$$= \int_a^b h(\varphi(\gamma(u)))|D\varphi(\gamma(u))T(u)||\dot{\gamma}_u(u)|du$$

so that $(\mu_{\varphi \cdot \gamma} \,|\, h) = (\mu_\gamma \,|\, h \circ \varphi|D\varphi T \circ \varphi^{-1}|)$, with some abuse of notation in the last formula, since T is only defined along γ. The important fact here is that the function h is transformed according to a rule which depends, not only on the diffeomorphism φ, but also on the curve γ and therefore the result cannot be put in the form $\varphi^{-1} \star h$.

The situation is different for vector measures. Indeed, we have

$$\nu_{\varphi \cdot \gamma}(h) = \int_a^b D\varphi(\gamma(u))\dot{\gamma}(u)^T h \circ \varphi(\gamma(u))du$$

$$= (\nu_\gamma \,|\, D\varphi^T h \circ \varphi).$$

So, if we define $\varphi \star h = D(\varphi^{-1})^T h \circ \varphi^{-1}$, we have $(\nu_{\varphi \cdot \gamma} \,|\, h) = (\nu_\gamma \,|\, \varphi^{-1} \star h)$. The transformation $(\varphi, h) \mapsto \varphi \star h$ is a valid action of diffeomorphisms on vector fields, since $id \star h = h$ and $\varphi \star (\psi \star h) = (\varphi \circ \psi) \star h$ as can easily be checked.

The same analysis can be made for surfaces; scalar measures do not transform in accordance to an action, but vector measures do. Let's check this last point by considering the formula in a local chart, where

$$(\nu_{\varphi(S)} \,|\, h) = \int \det(D\varphi \, \partial_u m, D\varphi \, \partial_v m, h \circ \varphi)dudv$$

$$= \int \det(D\varphi) \det(\partial_u m, \partial_v m, (D\varphi)^{-1} h \circ \varphi)dudv$$

$$= (\nu_S \,|\, \det(D\varphi)(D\varphi)^{-1} h \circ \varphi).$$

So, we need here to define

$$\varphi \star h = \det(D(\varphi^{-1}))(D\varphi^{-1})^{-1}h \circ \varphi^{-1} = (D\varphi h / \det(D\varphi)) \circ \varphi^{-1}.$$

Here again, a direct computation shows that this is an action.

We have therefore discovered that vector measures are transformed by a diffeomorphism φ according to a rule $(\varphi \cdot \mu \,|\, h) = (\mu \,|\, \varphi^{-1} \star h)$, the action $\star$ being apparently different for curves and surfaces. In fact, all these actions (including the scalar one) can be placed within a single framework if one replaces vector fields by differential forms and measures by currents [95, 207, 96].

This requires a few definitions. A p-linear form on $\mathbb{R}^d$ is a function $q : (\mathbb{R}^d)^p \to \mathbb{R}$ which is linear with respect to each of its variables. We will use the notation

$$q(e_1, \ldots, e_p) = (q \,|\, e_1, \ldots, e_p)$$

which is consistent with our notation for $p = 1$.

A p-linear form is skew-symmetric (or alternating) if $(q \,|\, e_1, \ldots, e_p) = 0$ whenever $e_i = e_j$ for some $i \neq j$. The p-linear skew-symmetric forms define a vector space denoted Λ_p (or Λ_p^d when the dimension of the underlying space needs to be specified). For $p = d$, there is, up to a multiplicative constant, only one skew-symmetric d-linear functional, the determinant: $q(e_1, \ldots, e_d) = \det(e_1, \ldots, e_d)$. In fact, for any $p \leq d$ and p by p matrix A, the skew-symmetry implies that, for any alternating p-linear form,

$$q(f_1, \ldots, f_p) = \det(A)q(e_1, \ldots, e_p) \tag{10.47}$$

when $f_i = \sum_{j=1}^p a_{ij} e_j$, $i = 1, \ldots, p$.

For $p < d$, one can associate to any family $e_1, \ldots, e_{d-p}$ of vectors, the skew-symmetric p-linear functional

$$q_{e_1, \ldots, e_{d-p}}(f_1, \ldots, f_p) = \det(e_1, \ldots, e_{d-p}, f_1, \ldots, f_p).$$

This functional is usually denoted $e_1 \wedge \ldots \wedge e_{d-p} \in \Lambda_p^d$. If $e_1, \ldots, e_d$ is a basis of $\mathbb{R}^d$, one can show that the family $(e_{j_1} \wedge \ldots \wedge e_{j_{d-p}}, 1 \leq j_1 < \cdots < j_{d-p} \leq d)$ is a basis of Λ_p which is therefore $\binom{d}{p}$-dimensional. One uniquely defines a dot product on Λ_p by first selecting an orthonormal basis $(e_1, \ldots, e_d)$ of $\mathbb{R}^d$ and deciding that $(e_{j_1} \wedge \ldots \wedge e_{j_{d-p}}, 1 \leq j_1 < \cdots < j_{d-p} \leq d)$ is an orthonormal family in Λ_p. This implicitly defines a dot product denoted $\langle \cdot, \cdot \rangle_{\Lambda_p}$.

A differential p-form on $\mathbb{R}^d$ is a function $x \mapsto q(x)$ such that, for all x, $q(x)$ is a p-linear skew symmetric form, i.e., q is a function from $\mathbb{R}^d$ to Λ_p. For example, 0-forms are ordinary functions.

The space of differential p-forms is denoted Ω_p, or Ω_p^d. We can consider spaces of smooth differential p-forms, and in particular, reproducing Hilbert spaces of such forms: a space $W \subset \Omega_p$ is a RKHS if, for every $x \in \mathbb{R}^d$ and $e_1, \ldots, e_p \in \mathbb{R}^d$, the evaluation function $q \mapsto (q(x) \,|\, e_1, \ldots, e_p)$ is continuous in W. This implies that there exists an element, denoted $\xi_x(e_1, \ldots, e_p) \in W$ such that

$$(q(x) \mid e_1, \ldots, e_p) = \langle q, \xi_x(e_1, \ldots, e_p) \rangle_W.$$

Note that $\xi_x(e_1, \ldots, e_p)$ being a differential p-form, we can compute

$$(\xi_x(e_1, \ldots, e_p)(y) \mid f_1, \ldots, f_p)$$

which will be denoted $(\xi(x, y) \mid e_1, \ldots, e_p; f_1, \ldots, f_p)$. So, $\xi(x, y)$, the kernel of W evaluated at (x, y), is $2p$-linear, and skew-symmetric with respect to its first p and its last p variables. Moreover, by construction

$$\langle \xi_x(e_1, \ldots, e_p), \xi_y(f_1, \ldots, f_p) \rangle_W = (\xi(x, y) \mid e_1, \ldots, e_p; f_1, \ldots, f_p)$$

which expresses the reproducibility of the kernel ξ. Like for vector fields, kernels for differential p-forms can be derived from scalar kernels by letting

$$(\xi(x, y) \mid e_1, \ldots, e_p; f_1, \ldots, f_p) = \xi(x, y) \langle e_1 \wedge \cdots \wedge e_p, f_1 \wedge \cdots \wedge f_p \rangle_{\Lambda_{d-p}}. \tag{10.48}$$

Elements of the dual space, W^*, to W therefore are linear forms over differential p-forms, and are special instances of p-currents [81, 152] (p-currents are continuous differential forms over C^∞ differential p-forms with compact support, which is less restrictive than being continuous on W). Important examples of currents are those associated to submanifolds of $\mathbb{R}^d$, and are defined as follows. Let M be an oriented p-dimensional submanifold of $\mathbb{R}^d$. To a differential p-form q, associate the quantity

$$(\eta_M \mid q) = \int_M (q(x) \mid e_1(x), \ldots, e_p(x)) d\sigma_M(x)$$

where $e_1, \ldots, e_p$ is, for all x, a positively oriented orthonormal basis of the tangent space to M at x (by (10.47), the result does not depend on the chosen basis).

If W is an RKHS of differential p-forms, η_M belongs to W^* and we can compute the dual norm of η_M, which is

$$\|\eta_M\|_{W^*}^2 = \int_M \int_M (\xi(x, y) \mid e_1(x), \ldots, e_p(x); e_1(y), \ldots, e_p(y)) d\sigma_M(x) d\sigma_M(y)$$

or, for a scalar kernel

$$\|\eta_M\|_{W^*}^2 = \int_M \int_M \xi(x, y) \langle e_1(x) \wedge \ldots \wedge e_p(x), e_1(y) \wedge \ldots \wedge e_p(y) \rangle_{\Lambda_{d-p}}$$
$$d\sigma_M(x) d\sigma_M(y).$$

The expressions of η_M and its norm in a local chart of M are quite simple. Indeed, if $(u_1, \ldots, u_p)$ is the parametrization in the chart and $(\partial_{u_1} m, \ldots, \partial_{u_p} m)$

the associated tangent vectors (assumed to be positively oriented), we have, for a p-form q (using (10.47))

$$(q \mid \partial_{u_1} m, \ldots, \partial_{u_p} m) = (q \mid e_1, \ldots, e_p) \det(\partial_{u_1} m, \ldots, \partial_{u_p} m)$$

which immediately yields

$$(q \mid \partial_{u_1} m, \ldots, \partial_{u_p} m) du_1 \ldots du_p = (q \mid e_1, \ldots, e_p) d\sigma_M.$$

We therefore have, in the chart,

$$\left(\eta_M \mid q\right) = \int (q \mid \partial_{u_1} m, \ldots, \partial_{u_p} m) du_1 \ldots du_p$$

and similar formulae for the norm.

Now consider the action of diffeomorphisms. If M becomes $\varphi(M)$, the formula in the chart yields

$$\left(\eta_{\varphi(M)} \mid q\right) = \int (q \circ \varphi \mid D\varphi \partial_{u_1} m, \ldots, D\varphi \partial_{u_p} m) du_1 \ldots du_p$$

so that $\left(\eta_{\varphi(M)} \mid q\right) = \left(\eta_M \mid \tilde{q}\right)$ with $(\tilde{q}(x) \mid f_1, \ldots, f_p) = (q(\varphi(x)) \mid D\varphi f_1, \ldots, D\varphi f_p)$. Like with vector measures, we can introduce the left action on p-forms (also called the push-forward of the p-form):

$$(\varphi \star q \mid f_1, \ldots, f_p) = (q \circ \varphi^{-1} \mid D(\varphi^{-1}) f_1, \ldots, D(\varphi^{-1}) f_p)$$

and the resulting action on p-currents

$$(\varphi \cdot \eta \mid q) = (\eta \mid \varphi^{-1} \star q) \tag{10.49}$$

so that we can write $\eta_{\varphi(M)} = \varphi \cdot \eta_M$.

This is reminiscent of what we have obtained for measures, and for vector measures with curves and surfaces. We now check that these examples are particular cases of the previous discussion.

Measures are linear forms on functions, which are also differential 0-forms. The definition $(\varphi \cdot \mu \mid f) = (\mu \mid f \circ \varphi)$ is exactly the same as in (10.49).

Consider now the case of curves, which are 1D submanifolds, so that $p = 1$. If γ is a curve, and T is its unit tangent, s its arc length, we have

$$\left(\eta_\gamma \mid q\right) = \int_\gamma (q(\gamma(s)) \mid T(s)) ds = \int_a^b (q(\gamma(u)) \mid \dot{\gamma}_u(u)) du.$$

To a vector field h on $\mathbb{R}^d$, we can associate the differential 1-form q_h defined by $(q_h(x) \mid v) = \langle h(x), v \rangle$. In fact all differential 1-forms can be expressed as q_h for some vector field h. Using this identification and noting that $(\nu_\gamma \mid h) = (\eta_\gamma \mid q_h)$, we can see that the vector measure for curve matching is a special case of the currents that we have considered here.

For surfaces in three dimensions, we need to take $p = 2$, and if S is a surface, we have

$$(\eta_S \,|\, q) = \int_M (q(x) \,|\, e_1(x), e_2(x)) d\sigma_S(x).$$

Again, a vector field f on $\mathbb{R}^3$ induces a 2-form q_f, defined by $(q_f \,|\, v_1, v_2) = \det(f, v_1, v_2) = f^T(v_1 \times v_2)$, and every 2-form can be obtained this way. Using the fact that, if (e_1, e_2) is a positively oriented basis of the tangent space to the surface, then $e_1 \times e_2 = N$, we retrieve $(\nu_S \,|\, f) = (\eta_S \,|\, q_f)$.

10.7 Matching Vector Fields

We now study vector fields as deformable objects, which correspond, for example, to velocity fields (that can be observed for weather data), or gradient fields that can be computed for images. Orientation fields (that can be represented by unit vector fields) are also interesting. They can correspond, for example, to fiber orientations in tissues observed in medical images.

We want to compare two vector fields f and f', i.e., two functions from Ω to $\mathbb{R}^d$. To simplify, we restrict ourselves to $E_{f,f'}(\varphi)$ being the L^2 norm between $\varphi \cdot f$ and f', and focus our discussion on the definition of the action of diffeomorphisms on vector fields.

The simplest choice is to use the same action as in image matching and use the action $\varphi \cdot f = f \circ \varphi^{-1}$ when f is a vector field on $\mathbb{R}^d$. It is, however, natural (and more consistent with applications) to combine the displacement of the points at which f is evaluated with a reorientation of f, also induced by the transformation. Several choices can be made for such an action and all may be of interest depending on the context.

For example, we can interpret a vector field as a velocity field, assuming that each point in Ω moves on to a trajectory $x(t)$ and that $f(x) = dx/dt$, say at time $t = 0$. If we make the transformation $x \mapsto \varphi(x)$, and let f' be the transformed vector field, such that $dx'/dt = f'(x')$, we get, replacing x' by $\varphi(x)$: $D\varphi(x)(dx/dt) = f' \circ \varphi(x)$ so that $f' = (D\varphi\, f) \circ \varphi^{-1}$. The transformation $f \mapsto (D\varphi\, f) \circ \varphi^{-1}$ is an important Lie group operation, called the adjoint representation $(\mathrm{Ad}_\varphi f)$. This is anecdotal here, but we will use it again later as a fundamental notion. So, our first action is

$$\varphi * f = (D\varphi\, f) \circ \varphi^{-1}.$$

To define a second action, we now consider vector fields that are obtained as gradients of a function I: $f = \nabla I$. If I becomes $\varphi \cdot I = I \circ \varphi^{-1}$, then f becomes $D(\varphi^{-1})^T \nabla I \circ \varphi^{-1}$. This defines a new action

$$\varphi \star f = D(\varphi^{-1})^T f \circ \varphi^{-1} = (D\varphi^{-T} f) \circ \varphi^{-1}.$$

This action can be applied to any vector field, not only gradients, but one can check that the set of vector fields f such that $\mathrm{curl} f = 0$ is left invariant by this action.

Sometimes, it is important that the norms of the vector fields at each point remain invariant by the transformation, when dealing, for example, with orientation fields. This can be achieved in both cases by normalizing the result, and we define the following normalized actions:

$$\varphi \bar{*} f = |f| \left(\frac{D\varphi f}{|D\varphi f|} \right) \circ \varphi^{-1}$$

$$\varphi \bar{*} f = |f| \left(\frac{D\varphi^{-T} f}{|D\varphi^{-T} f|} \right) \circ \varphi^{-1}$$

(taking, with in both cases, the right-hand side equal to 0 if $|f| = 0$).

We now evaluate the differential of $E_{f,f'}(\varphi) = \|\varphi \cdot f - f'\|_2^2$, where $\varphi \cdot f$ is one of the actions above. So let $\varphi(\varepsilon, \cdot)$ be such that $\varphi(0, \cdot) = id$ and $\partial_\varepsilon \varphi(0, \cdot) = h$, so that

$$(\bar{\partial} E_{f,f'}(id) \,|\, h) = 2 \langle \partial_\varepsilon \varphi(\varepsilon, \cdot) f_{|_{\varepsilon=0}} \,,\, f - f' \rangle_2.$$

We have, for the first action, taking the derivative at $\varepsilon = 0$,

$$\partial_\varepsilon \varphi(\varepsilon, \cdot) * f = \partial_\varepsilon (D\varphi(\varepsilon, \cdot) f) \circ \varphi(\varepsilon, \cdot)^{-1}_{|_{\varepsilon=0}}$$
$$= Dhf + Df \partial_\varepsilon \varphi(\varepsilon, \cdot)^{-1}$$
$$= Dhf - Dfh$$

by equation (10.16) applied to $\varphi = id$. This gives

$$(\bar{\partial} E_{f,f'}(id) \,|\, h) = 2 \langle Dh\, f - Df\, h \,,\, f - f' \rangle_2 = 2 \int_\Omega \langle Dh\, f - Df\, h \,,\, f - f' \rangle dx.$$

The Eulerian differential now comes by replacing f by $\varphi * f$ so that

$$(\bar{\partial} E_{f,f'}(\varphi) \,|\, h) = 2 \langle Dh\, (\varphi * f) - D(\varphi * f)\, h \,,\, \varphi * f - f' \rangle_2.$$

This expression can be combined with (10.41) to obtain the Eulerian gradient of U, namely (see equation (10.41))

$$\bar{\nabla}^V E_{f,f'}(\varphi) = 2 \int_\Omega \left((D_2 K(.,x)(\varphi * f)) - K(.,x) D(\varphi * f)^T \right) (\varphi * f - f') dx.$$

The Eulerian differential can be rewritten in another form to avoid the intervention of the differential of h. The following lemma is a consequence of the divergence theorem.

Lemma 10.9. *If Ω is a bounded open domain of $\mathbb{R}^d$ and v, w, h are smooth vector fields on $\mathbb{R}^d$, then*

$$\int_{\Omega} v^T Dhwdx = \int_{\partial\Omega} (v^T h)(w^T N)dl$$

$$- \int_{\Omega} (w^T Dv^T h + (\text{div}w)\langle v, h\rangle)dx. \quad (10.50)$$

Proof. To prove this, introduce the coordinates $h^1, \ldots, h^d$ for h and $v^1, \ldots, v^d$ for v so that

$$v^T Dhw = \sum_{i=1}^{d} v^i \langle \nabla h^i, w\rangle.$$

Now, use the fact that

$$\text{div}(\langle v, h\rangle w) = \text{div}\left(\sum_{i=1}^{d} v^i h^i w\right)$$

$$= \sum_{i=1}^{d} (v^i \langle \nabla h^i, w\rangle + h^i \langle \nabla v^i, w\rangle + h_i v^i \text{div}(w))$$

$$= v^T Dhw + h^T Dvw + \langle h, v\rangle \text{div}(w)$$

and the divergence theorem to obtain the result.   □

If, in particular h is orthogonal to the normal on $\partial\Omega$ (which is the case for variations of diffeomorphisms of Ω) equation (10.50) can be rewritten

$$\langle Dh\,w, v\rangle_2 = -\langle Dv\,w + \text{div}w\,v, h\rangle_2. \quad (10.51)$$

Using this lemma, we get

$$\left(\bar{\partial} E_{f,f'}(id) \mid h\right) = -2\langle D(f - f')\,f + \text{div}(f)\,(f - f') + Df^T(f - f'), h\rangle_2$$

which directly provides a new version of the Eulerian derivative at an arbitrary φ, with the corresponding new expression of the Eulerian gradient:

$$\overline{\nabla}^V E_{f,f'}(\varphi) =$$

$$- 2 \int_{\Omega} K(.,x)\big(D(\varphi * f - f')(\varphi * f) + \text{div}(\varphi * f)(\varphi * f - f')$$

$$+ D(\varphi * f)^T(\varphi * f - f')\big)dx.$$

Let's now consider the normalized version of this action,

$$\varphi \bar{*} f = |f| \frac{\varphi * f}{|\varphi * f|}.$$

Using the fact that the derivative of $z/|z|$ is $\dot{z}/|z| - zz^T \dot{z}/|z|^3$, we obtain (using $\varphi(\varepsilon, \cdot) * f = f$ at $\varepsilon = 0$)

$$\partial_\varepsilon(\varphi(\varepsilon,\cdot)\bar{\ast} f)_{|\varepsilon=0} = \pi_{f\perp}(\partial_\varepsilon\varphi \star f)_{|\varepsilon=0}$$

where $\pi_{f\perp}$ is the projection onto the hyperplane perpendicular to f,

$$\pi_{f\perp} = \mathrm{Id} - \frac{ff^T}{|f|^2}.$$

So we simply have to replace $(f - f')$ by $\pi_{f\perp}(f - f') = -\pi_{f\perp}f'$ in the previous formulae to compute the differential of the normalized action. This gives

$$\begin{aligned}
\left(\bar{\partial}E_{f,f'}(id) \,|\, h\right) &= -2\langle Dh\,f - Df\,h,\, \pi_{f\perp}f'\rangle_2 & (10.52)\\
&= 2\langle D(\pi_{f\perp}f')\,f + \mathrm{div}(f)\,(\pi_{f\perp}f') + Df^T(\pi_{f\perp}f'),\, h\rangle_2.
\end{aligned}$$

Replace f by $\varphi\bar{\ast} f$ to compute the Eulerian differential for $\varphi \neq id$, and apply the kernel to compute the Eulerian gradient.

The computations for $\varphi \star f = (D\varphi^{-T}f)\circ\varphi^{-1}$ and its normalized version are very similar. We find, for the unnormalized action

$$\partial_\varepsilon\varphi(\varepsilon,\cdot)\star f_{|\varepsilon=0} = -Dh^T f - Df\,h.$$

so that

$$\begin{aligned}
\left(\bar{\partial}E_{f,f'}(id) \,|\, h\right) &= -2\langle Dh(f - f'),\, f\rangle_2 - 2\langle Df^T(f - f'),\, h\rangle_2\\
&= 2\langle (Df - Df^T)(f - f') + \mathrm{div}(f - f')f,\, h\rangle
\end{aligned}$$

and $\bar{\partial}E_{f,f'}(\varphi)$ is obtained by replacing f by $\varphi \star f$. To obtain the Eulerian differential of $E_{f,f'}$ for the normalized $\star$ action, we only have to replace $f - f'$ by $-\pi_{f\perp}(f')$, yielding

$$\begin{aligned}
\left(\bar{\partial}E_{f,f'}(id) \,|\, h\right) &= 2\langle Dh(\pi_{f\perp}(f')),\, f\rangle_2 + 2\langle Df^T(\pi_{f\perp}(f')),\, h\rangle_2\\
&= -2\langle (Df - Df^T)(\pi_{f\perp}(f')) + \mathrm{div}(\pi_{f\perp}(f'))f,\, h\rangle.
\end{aligned}$$

As an example of application of vector field matching, let's consider contrast-invariant image registration [68]. If $I : \Omega \to \mathbb{R}$ is an image, a change of contrast is the transformation $I \mapsto q \circ I$ where q is a scalar diffeomorphism of the image intensity range. The *level sets* $I_\lambda = \{x, I(x) \leq \lambda\}$ are simply relabeled by a change of contrast, and one obtains a contrast-invariant representation of the image by considering the normals to these level sets, i.e., the vector field

$$f = \nabla I/|\nabla I|$$

with the convention that $f = 0$ when $\nabla I = 0$. Two images represented in this way can now be compared using vector field matching. Since we are using normalized gradients, the natural action if $(\varphi, f) \mapsto \varphi\bar{\ast} f$.

10.8 Matching Fields of Frames

We now extend vector field deformation models to define an action of diffeomorphisms on fields of positively oriented orthogonal matrices, or frames. We will restrict ourselves to dimension 3, so that the deformable objects considered in this section are mappings $x \mapsto R(x)$, with, for all $x \in \Omega$, $R(x) \in SO_3(\mathbb{R})$ (the group of rotation matrices).

The $*$ and $\star$ actions we have just defined on vector fields have the nice property to conserve the Euclidean dot product when combined, namely

$$\langle \varphi * f , \, \varphi \star g \rangle = \langle f , \, g \rangle \circ \varphi^{-1}.$$

Since $\bar{*}$ and $\bar{\star}$ also conserve the norm, we find that $(\varphi \bar{*} f, \varphi \bar{\star} g)$ is orthonormal as soon as (f, g) is.

We now define an action of diffeomorphisms on fields on frames. Writing $R(x) = (f_1(x), f_2(x), f_3(x))$, we define

$$\varphi \cdot R = (\varphi \bar{*} f_1, (\varphi \bar{*} f_3) \times (\varphi \bar{*} f_1), \varphi \bar{*} f_3). \tag{10.53}$$

That this defines an action is a straightforward consequence of $\bar{*}$ and $\bar{\star}$ being actions.

The action can be interpreted as follows. Given a local chart in $\mathbb{R}^3$, which is a diffeomorphic change of coordinates $x = m(s, t, u)$, we uniquely specify a positively oriented frame $R_m = (f_1, f_2, f_3)$ by $f_1 = \partial_s m / |\partial_s m|$ and $f_3 = (\partial_s m \times \partial_t m) / |\partial_s m \times \partial_t m|$. Then, the action we have just defined is such that $\varphi \cdot R$ is the frame associated to the change of coordinates $\varphi \circ m$, i.e.,

$$R_{\varphi \circ m} \circ \varphi = \varphi \cdot R_m.$$

The transformation $m \to R_m$ has in turn the following interpretation, which is relevant for some medical imaging modalities. Let the change of coordinates be adapted to the following stratified description of a tissue. Curves $s \mapsto m(s, t, u)$ correspond to tissue fibers, and surfaces $(s, t) \mapsto m(s, t, u)$ describe a layered organization. The cardiac muscle, for example, exhibits this kind of structure. Then f_1 in R_m represents the fiber orientation, and f_3 the normal to the layers; $\varphi \cdot R_m$ then corresponds to the tissue to which the deformation φ has been applied.

From the computations in the previous section, if $\varphi(\varepsilon, \cdot)$ is given with $\varphi(0, \cdot) = id$ and $\partial_\varepsilon \varphi(0, \cdot) = h$ at $\varepsilon = 0$, then we can write, if $R = (f_1, f_2, f_3)$,

$$\partial_\varepsilon \varphi(\varepsilon, \cdot) \cdot R_{|_{\varepsilon=0}} = (w_1, w_2, w_3)$$

with

$$w_1 = (\mathrm{Id}_d - f_1 f_1^T)(Dh f_1 - D f_1 h), \tag{10.54}$$
$$w_3 = -(\mathrm{Id}_d - f_3 f_3^T)(Dh^T f_3 + D f_3 h),$$
$$w_2 = w_3 \times f_1 + f_3 \times w_1.$$

It now remains to choose a suitable matching functional $E_{R,R'}(\varphi)$. One possible choice is to use the rotation angle, θ, from R to R', which can be computed by

$$\text{trace}(R^T R') = 1 + 2\cos\theta. \tag{10.55}$$

This can also be written $\text{trace}(\text{Id}_3 - R^T R') = 2(1 - \cos\theta)$. Since the latter quantity is always positive and vanishes only when $R = R'$, it is a good candidate for our matching functional, yielding

$$E_{R,R'}(\varphi) = \int_\Omega \text{trace}(\text{Id}_3 - (\varphi \cdot R)^T R') dx.$$

The Eulerian differential of $E_{R,R'}$ is

$$(\bar\partial E_{R,R'}(\varphi) \,|\, h) = -\int_\Omega \text{trace}(w^T.R') dx$$

where $w = (w_1, w_2, w_3)$ is given by equation (10.54) with R replaced by $\varphi \cdot R$. The differential of h can be removed using Lemma 10.9. For this, first notice that, for $\varphi = id$,

$$\begin{aligned}
\text{trace}(w^T.R') &= \langle w_1\,, f_1'\rangle + \langle w_3 \times f_1 + f_3 \times w_1\,, f_2'\rangle + \langle w_3\,, f_3'\rangle \\
&= \langle w_1\,, f_1' + f_2' \times f_3\rangle + \langle w_3\,, f_3' + f_1 \times f_2'\rangle \\
&= \langle Dh\, f_1 - Df_1\, h\,, u_{R,R'}^1\rangle + \langle -Dh^T f_3 - Df_3 h\,, u_{R,R'}^3\rangle
\end{aligned}$$

with

$$\begin{aligned}
u_{R,R'}^1 &= \pi_{f_1^\perp}(f_1' + f_2' \times f_3) \\
u_{R,R'}^3 &= \pi_{f_3^\perp}(f_3' + f_1 \times f_2').
\end{aligned}$$

With this notation and using Lemma 10.9 to eliminate Dh, we find

$$\begin{aligned}
\bar\partial E_{R,R'}(id) = \big(&Du_{R,R'}^1 f_1 + \text{div}(f_1)u_{R,R'}^1 + Df_1^T u_{R,R'}^1 \\
&- (Df_3 - Df_3)^T u_{R,R'}^3 - \text{div}(u_{R,R'}^3)f_3\big) \otimes dx. \tag{10.56}
\end{aligned}$$

10.9 Tensor Matching

The last class of deformable objects we will consider in this chapter are fields of matrices (or tensor fields). For general matrices, we can use the actions we have defined on vector fields, and apply them to each column of M, where M is a field of matrices. The computation of the differential of matching functionals is then done like in the previous two sections.

A more interesting problem is the action on symmetric matrices, for which the previous actions cannot be used, since they do not transform symmetric matrices into symmetric matrices. A simple choice is to add the transpose of the previous actions on the right, defining, for a field $x \mapsto S(x)$ of symmetric matrices

$$\varphi * S = (D\varphi S D\varphi^T) \circ \varphi^{-1}$$
$$\varphi \star S = (D\varphi^{-T} S D\varphi^{-1}) \circ \varphi^{-1}.$$

An important practical situation in which deformable objects are symmetric tensor fields is with diffusion tensor imaging (DTI), which produces, at each point x in space, a symmetric positive definite matrix $S(x)$ that measures the diffusion of water molecules in the imaged tissue. Roughly speaking, the tensor $S(x)$ is such that if a water molecule is at s at time t, the probability to be at $x + dx$ at time $t + dt$ is centered Gaussian with variance $dt^2 dx^T S(x) dx$.

If we return to the structured tissue model discussed in the last section (represented by the parametrization $x = m(s, t, u)$), we can assume that molecules travel more easily along fibers, and with most difficulty across layers. So the direction of $\partial_s m$ is the direction of largest variance, and $\partial_s m \times \partial_t m$ of smallest variance, so that the frame $R_m = (f_1, f_2, f_3)$ associated to the parametrization is such that f_1 is an eigenvector of S for the largest eigenvalue, and f_3 for the smallest eigenvalue, which implies that f_2 is an eigenvector for the intermediate eigenvalue. According to our discussion in the last section, a diffeomorphism φ should transform S so that the frame R_S formed by the eigenbasis of S transforms according to the action of diffeomorphisms on frames, namely, $R_{\varphi \cdot S} = \varphi \cdot R_S$.

So, if we express the decomposition of S in the form

$$S = \lambda_1 f_1 f_1^T + \lambda_2 f_2 f_2^T + \lambda_3 f_3 f_3^T$$

with $\lambda_1 \geq \lambda_2 \geq \lambda_3$, we need

$$\varphi \cdot S = \tilde{\lambda}_1 \tilde{f}_1 \tilde{f}_1^T + \tilde{\lambda}_2 \tilde{f}_2 \tilde{f}_2^T + \tilde{\lambda}_3 \tilde{f}_3 \tilde{f}_3^T \qquad (10.57)$$

where

$$(\tilde{f}_1, \tilde{f}_2, \tilde{f}_3) = (\varphi \bar{*} f_1, (\varphi \bar{\star} f_3) \times (\varphi \bar{*} f_1), \varphi \bar{*} f_3).$$

The new eigenvalues $\tilde{\lambda}_1, \tilde{\lambda}_2, \tilde{\lambda}_3$ must result from an action of φ on $\lambda_1, \lambda_2, \lambda_3$. The simplest choice is $\tilde{\lambda}_i = \lambda_i \circ \varphi^{-1}$, which expresses that the intrinsic tissue properties have not been affected by the deformation. If there are reasons to believe that variations in volume should affect the intensity of water diffusion, using the action of diffeomorphisms on densities may be a better option, namely $\tilde{\lambda}_i = \det D(\varphi^{-1}) \lambda_i \circ \varphi^{-1}$.

The action with $\tilde{\lambda}_i = \lambda_i \circ \varphi^{-1}$ is identical to the eigenvector-based tensor reorientation discussed in [3]. One of the important (and required) features or the construction is that, although the eigen-decomposition of S is not unique (in particular when two or three eigenvalues coincide) the transformation $S \mapsto \varphi \cdot S$ is defined without ambiguity. This will be justified below.

To compute the variations of the action, it will be convenient to introduce the three-dimensional rotation $U_S(\varphi) = (\varphi \cdot R_S) \circ \varphi R_S^T$, so that

$$\varphi \cdot S = (U_S(\varphi)SU_S(\varphi)^T) \circ \varphi^{-1}.$$

The variation $\partial_\varepsilon U_S(\varphi(\varepsilon, \cdot))$ at $\varepsilon = 0$ (with the usual assumptions on $\varphi(\varepsilon, \cdot)$) is a skew-symmetric matrix (like any infinitesimal variation of a rotation) given by

$$\omega_S(h) = \pi_{f_1^\perp} Dh f_1 f_1^T - ((\pi_{f_3^\perp} Dh^T f_3) \times f_1) f_2^T$$
$$+ (f_3 \times (\pi_{f_1^\perp} Dh f_1)) f_2^T - (\pi_{f_3^\perp} Dh^T f_3) f_3^T.$$

With this notation, we can write

$$\partial_\varepsilon \varphi(\varepsilon, \cdot) \cdot S_{|\varepsilon=0} = \omega_S(h)S - S\omega_S(h) - DSh.$$

(Here DSh is the matrix with coefficients $\langle \nabla S^{ij}, h \rangle$.)

This can be used to compute the Eulerian differential of a matching functional $E_{S,S'}(\varphi)$. Consider the simplest choice

$$E_{S,S'}(\varphi) = \int_\Omega \text{trace}((\varphi \cdot S - S')^2)dx$$

which is the sum of the squared coefficients of the difference between the matrices. We get

$$(\bar{\partial} E_{S,S'}(\varphi) \,|\, h) =$$
$$2 \int_\Omega \text{trace}((\varphi \cdot S - S')(\omega_{\varphi \cdot S}(h)(\varphi \cdot S) - (\varphi \cdot S)\omega_{\varphi \cdot S}(h) - D(\varphi \cdot S).h))dx.$$

Here again, the derivatives of h that are involved in $\omega_{\varphi \cdot f}(h)$ can be integrated by parts using the divergence theorem. Let's sketch this computation at $\varphi = id$, which leads to a vector measure form for the differential. We focus on the term

$$(\eta \,|\, h) := \int_\Omega \text{trace}((S - S')(\omega_S(h)S - S\omega_S(h)))dx = \int_\Omega \text{trace}(A\omega_S(h))dx$$

where $A = S(S - S') - (S - S')S = SS' - S'S$, and want to express η as a vector measure. We have (using the fact that A is skew symmetric and that (f_1, f_2, f_3) is orthonormal)

$$-\text{trace}(A\omega_S(h)) = \langle \omega_S(h)f_1, Af_1 \rangle + \langle \omega_S(h)f_2, Af_2 \rangle + \langle \omega_S(h)f_3, Af_3 \rangle$$
$$= \langle \pi_{f_1^\perp} Dh f_1, Af_1 \rangle - \langle (\pi_{f_3^\perp} Dh^T f_3) \times f_1, Af_2 \rangle$$
$$+ \langle f_3 \times (\pi_{f_1^\perp} Dh f_1), Af_2 \rangle - \langle \pi_{f_3^\perp} Dh^T f_3, Af_3 \rangle$$
$$= \langle Dh f_1, u_{S,S'}^1 \rangle - \langle Dh^T f_3, u_{S,S'}^3 \rangle,$$

with

$$u_{S,S'}^1 = \pi_{f_1^\perp}(Af_1 + (Af_2 \times f_3))$$
$$\text{and } u_{S,S'}^3 = \pi_{f_3^\perp}(Af_3 + (f_1 \times Af_2)).$$

It now remains to use Lemma 10.9 to identify η as

$$\eta = \left(Du_{S,S'}^1 f_1 + \text{div}(f_1)u_{S,S'}^1 - Df_3 u_{S,S'}^3 - \text{div}(u_{S,S'}^3)f_3\right) \otimes dx.$$

To write the final expression of $\bar{\partial}E_{S,S'}(id)$, define $(S - S') \odot DS$ to be the vector

$$(S - S') \odot DS = \sum_{i,j=1}^3 (S^{ij} - (S')^{ij})\nabla S^{ij}$$

so that we have

$$\bar{\partial}E_{S,S'}(id) = 2\big(Du_{S,S'}^1 f_1 + \text{div}(f_1)u_{S,S'}^1 - Df_3 u_{S,S'}^3$$
$$- \text{div}(u_{S,S'}^3)f_3 - (S - S') \odot DS\big) \otimes dx. \quad (10.58)$$

We now generalize this action to arbitrary dimensions, in a way that will provide a new interpretation of the three-dimensional case. Decompose a field of d by d symmetric matrices S in $\mathbb{R}^d$ in the form

$$S(x) = \sum_{k=1}^d \lambda_k(x)f_k(x)f_k(x)^T$$

with $\lambda_1 \geq \cdots \geq \lambda_d$ and $(f_1, \ldots, f_d)$ orthonormal. The matrices $f_k f_k^T$ represent the orthogonal projections on the one-dimensional space $\mathbb{R} f_k$ and, letting

$$W_k = \text{span}(f_1, \ldots, f_k),$$

and noting that the projection on W_k, π_{W_k} is equal to $f_1 f_1^T + \cdots + f_k f_k^T$, we can obviously write

$$S(x) = \sum_{k=1}^d \lambda_k(x)(\pi_{W_k(x)} - \pi_{W_{k-1}(x)})$$

where we have set $W_0 = \{0\}$.

Define the action $S \mapsto \varphi \cdot S$ by

$$\varphi \cdot S = \Big(\sum_{k=1}^d \lambda_k(\pi_{D\varphi W_k} - \pi_{D\varphi W_{k-1}}) \Big) \circ \varphi^{-1}.$$

First note that this coincides with the action we have just considered in the case $d = 3$. For this, it suffices to note that

$$D\varphi \operatorname{span}(f_1, f_2) = \operatorname{span}(\tilde{f}_1 \circ \varphi, \tilde{f}_2 \circ \varphi).$$

Since $\tilde{f}_1 \circ \varphi$ is proportional to $D\varphi\, f_1$, it remains to show that $D\varphi f_2$ is in $\operatorname{span}(\tilde{f}_1 \circ \varphi, \tilde{f}_2 \circ \varphi)$ which is equivalent to showing that it is orthogonal to $\tilde{f}_3 \circ \varphi$, which is true since

$$\langle D\varphi f_2, \tilde{f}_3 \circ \varphi \rangle = \langle f_2, f_3 \rangle / |D\varphi^{-T} f_3| = 0.$$

Returning to the general d-dimensional case, the definition we just gave does not depend on the choice made for the basis $e_1, \ldots, e_d$. Indeed, if we let $\mu_1 > \cdots > \mu_d$ denote the distinct eigenvalues of S, and $\Lambda_1, \ldots, \Lambda_q$ the corresponding eigenspaces, then, regrouping together the terms with identical eigenvalues in the decomposition of S and $\varphi \cdot S$, and letting

$$\Gamma_k = \Lambda_1 + \cdots + \Lambda_k, \quad \Gamma_0 = \{0\},$$

we clearly have

$$S(x) = \sum_{k=1}^{q} \mu_k(x)(\pi_{\Gamma_k(x)} - \pi_{\Gamma_{k-1}(x)})$$

and

$$\varphi \cdot S = \Big(\sum_{k=1}^{q} \mu_k (\pi_{D\varphi \Gamma_k} - \pi_{D\varphi \Gamma_{k-1}}) \Big) \circ \varphi^{-1}.$$

Since the decomposition of S in terms of its eigenspaces is uniquely defined, we obtain the fact that the definition of $\varphi \cdot S$ is non-ambiguous.

10.10 Pros and Cons of Greedy Algorithms

We have studied in this chapter a series of deformable objects, by defining the relevant action(s) that diffeomorphisms have on them and computing the variations of associated matching functionals.

This computation can be used, as we did with landmarks and images, to design "greedy" registration algorithms, which implement gradient descent to progressively minimize the functionals within the group of diffeomorphisms. These algorithms have the advantage of providing relatively simple implementations, and of requiring a relatively limited computation time.

Most of the time, however, this minimization is an ill-posed problem. Minimizers may fail to exist, for example. This has required, for image matching, the implementation of a suitable stopping rule that prevents the algorithm from running indefinitely. Even when a minimizer exists, it is generally not unique (see the example we gave with landmarks). Greedy algorithms provide the one that corresponds to the path of steepest descent from where they have been initialized (usually the identity). This solution does not have to be the "best one", and we will see that other methods can find much smoother solutions when large deformations are involved.

To design potentially well-posed problems, the matching functionals need to be combined with regularization terms, which measure the smoothness of the registration. This will be discussed in detail in the next chapter.

10.11 Summary of the Results of Chapter 10

We summarize in this section the various actions that we have considered in this chapter, with the associated matching functionals and the Eulerian differential at the identity. Eulerian differentials at $\varphi \neq id$ are formed by replacing the template by the deformed template. The Eulerian gradient is computed by applying the kernel to the differential.

We will give the differentials in vector measure form. Some of them have been computed after an integration by parts to be put in such a form. Even if these expressions are more convenient and easier to interpret, numerical implementations may benefit from using the original forms, because they involve analytically computable derivatives of the kernel instead of derivatives of the deformable object that may be known in discrete form only.

This section is for reference only and does not include any new result.

10.11.1 Labeled Points

Configurations of labeled points are $x = (x_1, \ldots, x_N)$. The action is

$$\varphi \cdot (x_1, \ldots, x_N) = (\varphi(x_1), \ldots, \varphi(x_N)).$$

For the matching functional

$$E_{x,x'}(\varphi) = \sum_{i=1}^{N} |x_i' - \varphi(x_i)|^2,$$

the Eulerian differential at the identity is

$$\bar{\partial} E(\varphi) = 2 \sum_{i=1}^{N} (x_i - x_i') \otimes \delta_{x_i}.$$

10.11.2 Images

For functions $I : \Omega \to \mathbb{R}$, the action is

$$\varphi \cdot I = I \circ \varphi^{-1}.$$

For the functional

$$E_{I,I'}(\varphi) = \|I \circ \varphi^{-1} - I'\|_2,$$

we have

$$\bar{\partial} E(id) = -2(I - I') \nabla I \otimes dx.$$

10.11.3 Measures

For a measure μ, define

$$(\varphi \cdot \mu \,|\, f) = (\mu \,|\, f \circ \varphi).$$

For densities, this gives

$$\varphi \cdot \zeta = \det(D\varphi^{-1})\zeta \circ \varphi^{-1}.$$

For the matching functional

$$E_{\zeta,\zeta'}(\varphi) = \int_\Omega (\varphi \cdot \zeta - \zeta')^2 dx$$

between densities, we have

$$\partial_\varepsilon E_{\zeta,\zeta'}(id) = 2\zeta\nabla(\zeta - \zeta') \otimes dx.$$

Let W be an RKHS of real-valued functions with kernel ξ, and

$$E_{\mu,\mu'}(\varphi) = \|\varphi \cdot \mu - \mu'\|_{W^*}^2.$$

We have

$$\bar\partial E_{\mu,\mu'}(id) = 2(\zeta^\mu - \zeta^{\mu'}) \otimes \mu$$

with

$$\zeta^{\tilde\mu} = \int \nabla_1 \xi(.,y) d\tilde\mu(y).$$

10.11.4 Plane Curves

Define

$$(\mu_\gamma \,|\, f) = \int_\gamma f\,dl,$$

$$E_{\gamma,\gamma'}(\varphi) = \|\mu_{\varphi \cdot \gamma} - \mu_{\gamma'}\|_{W^*}^2,$$

and

$$Z^{\tilde\gamma}(\cdot) = \int_{\tilde\gamma} \xi(\cdot,p)dl(p).$$

Then, letting p_0 and p_1 be the extremities of γ

$$\partial_\varepsilon \tilde E(\varphi(\varepsilon,\cdot))_{|_{\varepsilon=0}} = \zeta(p_1)\langle h(p_1)\,,\,T^\gamma(p_1)\rangle - \zeta(p_0)\langle h(p_0)\,,\,T^\gamma(p_0)\rangle$$
$$+ \int_\gamma (\langle \nabla\zeta\,,\,N^\gamma\rangle - \zeta\kappa^\gamma)\langle h\,,\,N^\gamma\rangle dl$$

with $\zeta = Z^\gamma - Z^{\gamma'}$.

With $\nu_\gamma = T^\gamma \otimes \mu_\gamma$ and W an RKHS of vector fields with kernel ξ, let $E_{\gamma,\gamma'}(\varphi) = \|\nu_{\varphi\cdot\gamma} - \nu_{\gamma'}\|^2_{W*}$. Then

$$\frac{1}{2}\bar{\partial}E_{\gamma,\gamma'}(id) = -(R_{\pi/2}\zeta) \otimes (\delta_{p_1} - \delta_{p_0}) + \mathrm{div}(\zeta)\nu_\gamma,$$

where $R_{\pi/2}$ is a 90° rotation and

$$\zeta(\cdot) = \int_\gamma \xi(\cdot,p)N^\gamma(p)dp - \int_{\gamma'} \xi(\cdot,p)N^{\gamma'}(p)dp.$$

10.11.5 Surfaces

Let

$$(\mu_S \mid f) = \int_S f(x)d\sigma_S(x) \text{ for a scalar } f$$

and

$$(\nu_S \mid f) = \int_S \langle f(x), N(x)\rangle d\sigma_S(x) \text{ for a vector field } f$$

where $d\sigma$ is the area form on S and N is the unit normal.

If W is a RKHS of scalar functions, let

$$E_{S,S'}(\varphi) = \|\mu_{\varphi\cdot S} - \mu_{S'}\|^2_{W*}.$$

Then

$$\frac{1}{2}\bar{\partial}E_{S,S'}(id) = -\zeta n^S \otimes \mu_{\partial S} + \left(-2\zeta H^S + \langle\nabla\zeta, N^S\rangle\right)\nu_S$$

with

$$\zeta(\cdot) = \int_S \xi(\cdot,p)d\sigma_S(p) - \int_{S'} \xi(\cdot,p)d\sigma_{S'}(p).$$

With

$$E_{S,S'}(\varphi) = \|\nu_{\varphi\cdot S} - \nu_{S'}\|^2_{W*},$$

then

$$\frac{1}{2}\bar{\partial}E_{S,S'}(id) = -(\langle\zeta, N^S\rangle n^S - \langle\zeta, n^S\rangle N^S) \otimes \mu_{\partial S} + \mathrm{div}(\zeta)\nu_S.$$

with

$$\zeta(\cdot) = \int_S \xi(\cdot,p)N^S d\sigma_S(p) - \int_{S'} \xi(\cdot,p)N^{S'}d\sigma_{S'}(p).$$

10.11.6 Vector Fields

Define, for a vector field f,

$$\varphi * f = (D\varphi f) \circ \varphi^{-1}$$
$$\varphi \star f = (D\varphi^{-T} f) \circ \varphi^{-1}$$
$$\varphi \bar{\ast} f = |f| \left(\frac{D\varphi f}{|D\varphi f|} \right) \circ \varphi^{-1}$$
$$\varphi \bar{\star} f = |f| \left(\frac{D\varphi^{-T} f}{|D\varphi^{-T} f|} \right) \circ \varphi^{-1}$$

and $E_{f,f'}(\varphi) = \|\varphi \cdot f - f'\|_2^2$ for one of these actions.

Then, for the $*$ action,

$$\bar{\partial} E_{f,f'}(id) = (-2D(f - f') f + \text{div}(f) (f - f') + Df^T (f - f')) \otimes dx.$$

For the $\bar{\ast}$ action,

$$\bar{\partial} E_{f,f'}(id) = 2(D(\pi_{f\perp} f') f + \text{div}(f) (\pi_{f\perp} f') + Df^T (\pi_{f\perp} f')) \otimes dx.$$

with

$$\pi_{f\perp} = \text{Id} - \frac{ff^T}{|f|^2}.$$

For the $\star$ action,

$$\left(\bar{\partial} E_{f,f'}(id) \,|\, h \right) = 2((Df - Df^T)(f - f') + \text{div}(f - f')f) \otimes dx.$$

and for the $\bar{\star}$ action,

$$\left(\bar{\partial} E_{f,f'}(id) \,|\, h \right) = -2((Df - Df^T)(\pi_{f\perp}(f')) + \text{div}(\pi_{f\perp}(f'))f) \otimes dx.$$

10.11.7 Frames

For $R(x) = (f_1(x), f_2(x), f_3(x)) \in SO_3(\mathbb{R})$, define

$$\varphi \cdot R = (\varphi \bar{\ast} f_1, (\varphi \bar{\ast} f_3) \times (\varphi \bar{\ast} f_1), \varphi \bar{\ast} f_3).$$

Consider

$$E_{R,R'}(\varphi) = \int_\Omega \text{trace}(\text{Id}_3 - (\varphi \cdot R)^T R') dx.$$

and let

$$u_{R,R'}^1 = \pi_{f_1^\perp}(f_1' + f_2' \times f_3)$$
$$u_{R,R'}^3 = \pi_{f_3^\perp}(f_3' + f_1 \times f_2').$$

We have

$$\bar{\partial} E_{R,R'}(id) = \left(Du_{R,R'}^1 f_1 + \text{div}(f_1)u_{R,R'}^1 + Df_1^T u_{R,R'}^1 \right.$$
$$\left. - (Df_3 - Df_3)^T u_{R,R'}^3 - \text{div}(u_{R,R'}^3)f_3 \right) \otimes dx.$$

10.11.8 Tensor Matching

For a symmetric tensor field decomposed as

$$S = \lambda_1 f_1 f_1^T + \lambda_2 f_2 f_2^T + \lambda_3 f_3 f_3^T$$

with $\lambda_1 \geq \lambda_2 \geq \lambda_3$, define

$$\varphi \cdot S = \tilde{\lambda}_1 \tilde{f}_1 \tilde{f}_1^T + \tilde{\lambda}_2 \tilde{f}_2 \tilde{f}_2^T + \tilde{\lambda}_3 \tilde{f}_3 \tilde{f}_3^T$$

where

$$(\tilde{f}_1, \tilde{f}_2, \tilde{f}_3) = (\varphi \bar{*} f_1, (\varphi, \bar{*} f_3) \times (\varphi \bar{*} f_1), \varphi \bar{*} f_3).$$

and $\tilde{\lambda}_i = \lambda_i \circ \varphi^{-1}$. Define

$$E_{S,S'}(\varphi) = \int_\Omega \text{trace}((\varphi \cdot S - S')^2) dx.$$

Let $A = SS' - S'S$, and

$$u_{S,S'}^1 = \pi_{f_1^\perp}(Af_1 + (Af_2 \times f_3))$$
$$\text{and } u_{S,S'}^3 = \pi_{f_3^\perp}(Af_3 + (f_1 \times Af_2)).$$

Denote

$$(S - S') \odot DS = \sum_{i,j=1}^3 (S^{ij} - (S')^{ij}) \nabla S^{ij}.$$

Then

$$\bar{\partial} E_{S,S'}(id) = 2\big(Du_{S,S'}^1 f_1 + \text{div}(f_1) u_{S,S'}^1 - Df_3 u_{S,S'}^3$$
$$- \text{div}(u_{S,S'}^3) f_3 - (S - S') \odot DS\big) \otimes dx.$$

11

Diffeomorphic Matching

11.1 Linearized Deformations

A standard way to ensure the existence of a smooth solution of a matching problem is to add a penalty term in the matching functional. This term would complete (10.1) to form

$$E_{I,I'}(\varphi) = \rho(\varphi) + D(\varphi \cdot I, I').$$ (11.1)

A large variety of methods have been designed, in approximation theory, statistics or signal processing for solving ill-posed problems. The simplest (and typical) form of penalty function is

$$\rho(\varphi) = \|\varphi - id\|_H^2$$

for some Hilbert (or Banach) space of functions. Some more complex functions of $\varphi - id$ may also be designed, related to energies of non-linear elasticity (see, among others, [17, 18, 8, 92, 67, 169, 109]). Such methods may be called "small deformation" methods because they work on the deviation of $u = \varphi - id$, and controlling the size or smoothness of u alone is most of the time not enough to guarantee that φ is a diffeomorphism (unless u is small, as we have seen in section 8.1 of chapter 8). There is, in general, no way of proving the existence of a solution of the minimization problem within some group of diffeomorphisms G, unless some restrictive assumptions are made on the objects to be matched.

Our focus here is on diffeomorphic matching. Because of this, we shall not detail much of these methods. However, it is interesting to note that these functionals also have an Eulerian gradient within an RKHS of vector fields with a smooth enough kernel, and can therefore be minimized using (10.5). Let's compute the Eulerian gradients of such functionals, as an example, in a simple situation, since the computation rapidly becomes heavy.

So, consider the function $\rho(\varphi) = \int_\Omega \|D\varphi(x)\|^2 \, dx$ where the matrix norm is

$$\|A\|^2 = \text{trace}(A^T A) = \sum_{i,j} a_{ij}^2$$

(Hilbert–Schmidt norm). We have

$$\left(\frac{\delta\rho}{\delta\varphi}(\varphi) \mid h\right) = 2\int_\Omega \text{trace}(D\varphi^T Dh)dx = -2\int_\Omega \langle \Delta\varphi, h\rangle dx$$

where $\Delta\varphi$ is the vector formed by the Laplacian of the coordinates of φ (this is under the assumption that h is tangent to $\partial\Omega$ and vanishes at infinity). This implies that

$$(\bar{\partial}\rho(\varphi) \mid h) = -2\int_\Omega \langle \Delta\varphi, h \circ \varphi\rangle dx$$

and

$$\overline{\nabla}_\varphi^V \rho(\cdot) = -2\int_\Omega K(\cdot, \varphi(x))\Delta\varphi(x)dx. \tag{11.2}$$

This provides a regularized greedy image-matching algorithm, which includes a regularization term (a similar algorithm may easily be written for point matching).

Algorithm 3 *The following procedure is an Eulerian gradient descent, on V, for the energy*

$$E_{I,I'}(\varphi) = \int_\Omega \|D\varphi(x)\|^2 \, dx + \frac{1}{\sigma^2}\int_\Omega \left|I \circ \varphi^{-1}(x) - I'(x)\right| dx.$$

Start with an initial $\varphi_0 = id$ and solve the differential equation

$$\partial_t\varphi(t,y) = -2\int_\Omega K(\varphi(t,y), \varphi(t,x))\Delta\varphi(t,x)dx \tag{11.3}$$

$$+ \frac{2}{\sigma^2}\int_\Omega (J(t,x) - I'(x))K(\varphi(t,y), x)\nabla J(t,x)dx \tag{11.4}$$

with $J(t,.) = I \circ \varphi(t)^{-1}(.)$.

This algorithm, which, like the previous greedy procedures, has the fundamental feature of providing a smooth flow of diffeomorphisms to minimize the matching functional, suffers from the same limitations as its predecessors concerning its limit behavior, which are essentially due to the fact that the variational problem itself is not well-posed; minimizers may not exist, and when they exist they are not necessarily diffeomorphisms. In order to ensure the existence of, at least, homeomorphic solutions, the energy must include terms that must not only prevent $D\varphi$ from being too large, but also from being too small (or its inverse from being too large). In [68], the following regularization is proved to ensure the existence of homeomorphic solutions:

$$\delta(\varphi) = \int_\Omega (a\|D\varphi\|^p + b\|\text{Com}(D\varphi)\|^q + c(\det \varphi)^r + d(\det \varphi)^{-s})dx \tag{11.5}$$

under some assumptions on p, q, r and s, namely $p, q > 3$, $r > 1$ and $s > 2q/(q-3)$.

11.2 The Monge–Kantorovitch Problem

We discuss in this section the mass transfer problem which is, under some assumption, a diffeomorphic method for matching probability densities, i.e., positive functions on $\mathbb{R}^d$ with integral equal to 1. Consider such a density, ζ, and a diffeomorphism φ on $\mathbb{R}^d$. If an object has density ζ, the mass included in an infinitesimal volume dx around x is $\zeta(x)dx$. Now, if each point x in the object is transported to the location $y = \varphi(x)$, the mass of a volume dy around y is the same as mass of the volume $\varphi^{-1}(dy)$ around $x = \varphi^{-1}(y)$, which is $\zeta \circ \varphi^{-1}(y)|\det(D\varphi^{-1})(y)|dy$ (this provides a physical interpretation of proposition 10.5).

Given two densities ζ and ζ', the optimal mass transfer problem consists in finding the diffeomorphism φ with minimal cost such that $\zeta' = \zeta \circ \varphi^{-1}|\det(D\varphi^{-1})|$. The cost associated to φ in this context is related to the distance along which the transfer is made. Let's measure it by a positive number, $\rho(x, \varphi(x))$. The total cost comes after summing over the transferred mass, yielding

$$E(\varphi) = \int_\Omega \rho(x, \varphi(x))\zeta(x)dx.$$

The mass transfer problem now is to minimize E over all φ's such that $\zeta' = \zeta \circ \varphi^{-1}|\det(D\varphi^{-1})|$. The problem is slightly different from the matching formulations that we discuss in the other sections of this chapter, since the minimization is associated to exact matching.

It is very interesting that this apparently very complex and highly nonlinear problem can be reduced to linear programming, albeit infinite-dimensional. Let's first consider a more general formulation. Instead of looking for a one-to-one correspondence $x \mapsto \varphi(x)$, one can decide that the mass in a small neighborhood of x is dispatched over all Ω with weights $y \mapsto q(x, y)$, where $q(x, y) \geq 0$ and $\int_\Omega q(x, y)dy = 1$. We still have the constraint that the mass density arriving at y is $\tilde{\zeta}(y)$, which gives

$$\int_\Omega \zeta(x)q(x, y)dx = \tilde{\zeta}(y).$$

The cost now has the simple expression (linear in q)

$$E = \int_{\Omega^2} \rho(x, y)\zeta(x)q(x, y)dxdy.$$

The original formulation can be retrieved by letting $q(x, y)dy \to \delta_{\varphi(x)}(y)$ (i.e., pass to the limit $\sigma = 0$ with $q(x, y) = \exp(-|y - \varphi(x)|^2/2\sigma^2)/(2\pi\sigma^2)^{d/2})$.

If we write $g(x, y) = \zeta(x)q(x, y)$, this relaxed problem is clearly equivalent to: minimize

$$E(g) = \int_{\Omega^2} \rho(x, y)g(x, y)dxdy$$

subject to the constraints $g(x, y) \geq 0$, $\int g(x, y)dy = \zeta(x)$ and $\int g(x, y)dx = \tilde{\zeta}(y)$. In fact, the natural formulation of this problem uses measures instead of densities: given two measures μ and $\tilde{\mu}$ on Ω, minimize

$$E(\nu) = \int_{\Omega^2} \rho(x, y)\nu(dx, dy)$$

subject to the constraints that the marginals of ν are μ and $\tilde{\mu}$. This provides the Wasserstein distance between the probabilities μ and $\tilde{\mu}$, associated to the transportation cost ρ. Note that this formulation generalizes the computation of the Wasserstein distance (10.22) between the measures μ and $\tilde{\mu}$.

This problem is much nicer than the original one, since it is a linear programming problem. The theory of convex optimization (which we only apply formally in this infinite-dimensional context; see [31] for rigorous proofs) implies that it has an equivalent dual formulation which is: maximize

$$F(h) = \int_{\Omega} h d\mu + \int_{\Omega} \tilde{h}\tilde{\mu}$$

subject to the constraint that, for all $x, y \in \Omega$, $h(x) + \tilde{h}(y) \leq \rho(x, y)$.

The duality equivalence means that the maximum of F coincides with the minimum of E. The solutions are, moreover, related by duality conditions (the KKT conditions) that imply that ν must be supported by the set

$$A = \left\{ (x, y) : h(x) + \tilde{h}(y) = \rho(x, y) \right\}. \tag{11.6}$$

For the dual problem, one obviously has interest in making h and $\tilde{h}$ as large as possible. Given h, one should therefore choose $\tilde{h}$ as

$$\tilde{h}(y) = \sup_x(\rho(x, y) - h(x))$$

so that the set in (11.6) is exactly the set of (y^*, y) where y^* is a point that achieves the maximum of $\rho(x, y) - h(x)$.

The situation is particularly interesting when $\rho(x, y) = |x - y|^2/2$. In this situation,

$$\tilde{h}(y) = \frac{y^2}{2} + \sup_x \left(x^T y + \frac{x^2}{2} - h(x) \right).$$

From this equation, it is natural to introduce the auxiliary functions $s(x) = h(x) - x^2/2$ and $\tilde{s}(y) = \tilde{h}(y) - y^2/2$. Using these functions, the set A in (11.6) becomes

$$A = \left\{ (x, y) : s(x) + \tilde{s}(y) = x^T y \right\}$$

with $\tilde{s}(y) = \sup_x(x^T y - s(x))$. Because the latter is a supremum of linear functions, we obtain the fact that $\tilde{s}$ is convex, as well as s by symmetry; $\tilde{s}$ is in fact what is called the complex conjugate of s, denoted $\tilde{s} = s^*$. Convex functions are almost everywhere differentiable, and, in order that $(x, y) \in A$,

x must maximize $u \mapsto u^T y - s(u)$, which implies that $y = \nabla s(x)$. So, the conclusion is that, whenever s is the solution of the dual problem, the solution of the primal problem is supported by $y = \nabla s(x)$. This shows that the relaxed mass transport problem has the same solution as the initial one, with $\varphi = \nabla s$, s being a convex function. That φ is invertible is obvious by symmetry: $\varphi^{-1} = \nabla \tilde{s}$.

This result is fundamental, since it is the basis for the construction of a numerical procedure for the solution of the mass transport problem in this case. Introduce a time-dependent vector field $v(t, .)$ and the corresponding flow of diffeomorphisms φ_{0t}^v. Let $h(t, .) = \zeta \circ \varphi_{t0}^v \det(D\varphi_{t0}^v)$. Then

$$h(t) \circ \varphi_{0t}^v \det(D\varphi_{0t}^v) = \zeta.$$

The time derivative of this equation yields

$$h_t + \mathrm{div}(hv) = 0. \tag{11.7}$$

We have the following theorem [23].

Theorem 11.1. *Consider the following energy:*

$$G(v) = \int_0^1 \int_\Omega h(t,x)|v(t,x)|^2 dx dt$$

and the variational problem: minimize G subject to the constraints $h(0) = \zeta$, $h(1) = \tilde{\zeta}$ and (11.7). If v is the solution of the above problem, then φ_{01}^v solves the optimal mass transport problem.

Proof. Indeed, in G, we can make the change of variables $x = \varphi_{0t}(y)$, which yields

$$G(v) = \int_0^1 \int_\Omega f(y) \left|v(t, \varphi_{0t}^v(y))\right|^2 dy dt$$

$$= \int_\Omega f(y) \int_0^1 |\partial_t \varphi_{0t}^v|^2 \, dt$$

$$\geq \int_\Omega f(y) \left|\varphi_{01}^v(y) - y\right|^2 dy.$$

So the minimum of G is always larger than the minimum of E. If φ solves the mass transport problem, then one can take $v(t,x)$ such that $\varphi_{0t}^v(x) = (1-t)x + t\varphi(x)$, which is a diffeomorphism [138] and achieves the minimum of G. $\square$

We refer to [23] for a numerical algorithm for the computation of the optimal φ. Note that $\rho(x,y) = |x - y|^2$ is not the only transportation cost that can be used in this context, but that others (like $|x - y|$, which is not strictly convex in the distance) may fail to provide diffeomorphic solutions. Interesting developments on this subject can be found in [34, 89, 213].

We now discuss methods that are both diffeomorphic and metric (i.e., they relate to a distance). They also rely on the representation of diffeomorphisms using flows of ordinary differential equations.

11.3 Optimizing Over Flows

We return in this section to the representation of diffeomorphisms with flows of ordinary differential equations (ODEs) and describe how this representation can be used for diffeomorphic registration. Instead of using a function norm on the difference between φ and the identity mapping, we now consider, as a regularizing term, the distance d_V, which has been defined in Section 8.2.4. More precisely, we set

$$\rho(\varphi) = d_V(id, \varphi)^2$$

and henceforth restrict the matching to diffeomorphisms belonging to G_V.

In this context, we have the following important theorem:

Theorem 11.2. *Let V be a Hilbert space which is embedded in $C_0^{p+1}(\Omega, \mathbb{R}^d)$. Assume that the functional $U : G_V \mapsto \mathbb{R}$ is bounded from below and continuous with respect to uniform convergence on compact sets of derivatives up to order p. Then, there exists a minimizer of*

$$E(\varphi) = d_V(id, \varphi)^2 + U(\varphi) \tag{11.8}$$

over G_V.

Continuity for uniform convergence on compact sets means that, if $\varphi_n \to \varphi$ uniformly on compact subsets of Ω, and the same holds for all partial derivatives up to order p, then $U(\varphi_n) \to U(\varphi)$.

Proof. E has an infimum E_{min} over G_V, since it is bounded from below. We first use the following lemma (recall that we have denoted by $\mathcal{X}_V^1(\Omega)$ (resp. $\mathcal{X}_V^2(\Omega)$) the set of time-dependent vector fields on Ω with integrable (resp. square integrable) V norm over $[0, 1]$):

Lemma 11.3. *Minimizing $E(\varphi) = d(id, \varphi)^2 + U(\varphi)$ over G_V is equivalent to minimizing the function*

$$\tilde{E}(v) = \int_0^1 \|v(t)\|_V^2 \, dt + U(\varphi_{01}^v)$$

over $\mathcal{X}_V^2(\Omega)$.

Let's prove this lemma. For $v \in \mathcal{X}_V^2(\Omega)$, we have, by definition of the distance

$$d_V(id, \varphi_{01}^v)^2 \leq \int_0^1 \|v(t)\|_V^2 \, dt$$

which implies $E(\varphi_{01}^v) \leq \tilde{E}(v)$. This obviously implies $\inf_{G_V} E(\varphi) \leq \tilde{E}(v)$, and since this is true for all $v \in \mathcal{X}_V^2(\Omega)$, we have $\inf E \leq \inf \tilde{E}$. Now, assume that φ is such that $E(\varphi) \leq \inf E + \varepsilon$. Then, by definition of the distance, there exists v such that

$$\int_0^1 \|v(t)\|_V^2 \, dt \le d_V(id, \varphi)^2 + \varepsilon,$$

which implies that

$$\tilde{E}(v) \le E(\varphi) + \varepsilon \le \inf E + 2\varepsilon.$$

Since the left-hand side is larger than $\inf \tilde{E}$, we have $\inf \tilde{E} \le \inf E + 2\varepsilon$ for all $\varepsilon > 0$, which implies that $\inf E \le \inf \tilde{E}$.

We therefore have $\inf E = \inf \tilde{E}$. Moreover, if there exists v such that $\tilde{E}(v) = \min \tilde{E} = \inf E$, then, since we know that $E(\varphi_{01}^v) \le \tilde{E}$, we must have $E(\varphi_{01}^v) = \min E$. Conversely, if $E(\varphi) = \min E(\varphi)$, by Theorem 8.20, $E(\varphi) = E(\varphi_{01}^v)$ for some v and this v must achieve the infimum of $\tilde{E}$, which proves the lemma.

This lemma shows that it suffices to study the minima of $\tilde{E}$. Now, like in Theorem 8.20, one can find, by taking a subsequence of a minimizing sequence, a sequence v^n in $\mathcal{X}_V^2(\Omega)$ which converges weakly to some $v \in \mathcal{X}_V^2(\Omega)$ and $\tilde{E}(v^n)$ tends to E_{min}. Since

$$\liminf \int_0^1 \|v^n(t)\|_V^2 \, dt \ge \int_0^1 \|v(t)\|_V^2 \, dt$$

and since weak convergence in $\mathcal{X}_V^2(\Omega)$ implies convergence of the flow and its derivatives on compact sets (cf. Theorem 8.11 and the remarks following it) we also have $U(\varphi_{01}^{v^n}) \to U(\varphi_{01}^v)$ so that $\tilde{E}(v) = E_{min}$ and v is a minimizer. $\qquad\square$

11.4 Euler–Lagrange Equations and Gradient

We now detail the computation of the gradient for energies like (11.8). As remarked in the proof of Theorem 11.2, the variational problem which has to be solved is conveniently expressed as a problem over $\mathcal{X}_V^2(\Omega)$. The function which is minimized over this space takes the form

$$E(v) = \int_0^1 \|v(t)\|_V^2 \, dt + U(\varphi_{01}^v).$$

We want to obtain the expression of the gradient of E for the Hilbert structure of $\mathcal{X}_V^2(\Omega)$, which is the natural inner-product space for the problem. So we want to identify a function, denoted $\nabla^V E : v \mapsto \nabla^V E(v) \in \mathcal{X}_V^2(\Omega)$, such that for all v, h in $\mathcal{X}_V^2(\Omega)$

$$\partial_\varepsilon E(v + \varepsilon h)_{|_{\varepsilon=0}} = \left\langle \nabla^V E(v), h \right\rangle_{\mathcal{X}_V^2} = \int_0^1 \left\langle (\nabla^V E(v))(t), h(t) \right\rangle_V dt.$$

Since the set V is fixed in this section, we will drop the exponent from the notation, and simply refer to the gradient $\nabla E(v)$. Note that this is different from the Eulerian gradient we have dealt with before; ∇E now represents the usual gradient of a function defined over a Hilbert space. One important thing to keep in mind is that the gradient we define here is an element of $\mathcal{X}_V^2(\Omega)$, henceforth a time-dependent vector field, whereas the Eulerian gradient simply was an element of V (a vector field on Ω). Theorem 11.5 relates the two (and allows us to reuse the computations we have made in Chapter 10). For this, we need to introduce the following operation of diffeomorphisms acting on vector fields.

Definition 11.4. *Let φ be a diffeomorphism of Ω and v a vector field on Ω. We denote by $\mathrm{Ad}_\varphi v$ the vector field on Ω defined by*

$$\mathrm{Ad}_\varphi v(x) = (D\varphi\, v) \circ \varphi^{-1}(x). \tag{11.9}$$

$\mathrm{Ad}_\varphi v$ is called the adjoint representation of φ in the group, evaluated at v.

We want to define the conjugate of this adjoint operator. However, even if φ is smooth and $v \in V$, the vector field $\mathrm{Ad}_\varphi v$ does not necessarily belong to V, and we need to consider larger spaces that would contain it. The admissibility of V leads naturally to the following assumption. We will assume that there exists $p \geq 1$ such that V is continuously embedded in $C_0^p(\Omega, \mathbb{R}^d)$. If, in addition, $\varphi - id \in C_0^q(\Omega, \mathbb{R}^d)$, then $Ad_\varphi v \in C^r(\Omega, \mathbb{R}^d)$ for $r = \min(p, q-1)$. Under these conditions, we can define, for $\rho \in C^r(\Omega, \mathbb{R}^d)^*$, the linear form $\mathrm{Ad}_\varphi^* \rho$ by

$$\left(\mathrm{Ad}_\varphi^* \rho \,|\, v\right) = \left(\rho \,|\, \mathrm{Ad}_\varphi v\right) \tag{11.10}$$

for $v \in C^p(\Omega, \mathbb{R}^d)$.

Now, still under the same assumptions, we consider the set $V^{(r)}$ of vector fields $v \in V$ such that $Lv \in C^r(\Omega, \mathbb{R}^d)^*$. Then, for $v \in V^{(r)}$, we define

$$Ad_\varphi^T v = K(Ad_\varphi^* Lv). \tag{11.11}$$

This is well-defined, because, by construction, $Ad_\varphi^* Lv \in C^p(\Omega, \mathbb{R}^d)^* \subset V^*$. We have in particular, for $v \in V^{(r)}$ and $w \in V$,

$$\langle Ad_\varphi^T v,\, w \rangle_V = \left(Ad_\varphi^* Lv \,|\, w\right)$$
$$= \left(Lv \,|\, Ad_\varphi w\right).$$

With this notation, we have the following theorem.

Theorem 11.5. *Let U be a function defined on G_V. Assume that for any $\varphi \in C_0^p(\Omega, \mathbb{R}^d)$, the Eulerian differential of U at φ, $\bar{\partial}U(\varphi)$, exists and belongs to $C_0^{p-1}(\Omega, \mathbb{R}^d)^*$. Then, the $\mathcal{X}_V^2$ gradient of U is given by the formula*

$$\nabla U(v)(t) = K Ad_{\varphi_{t1}^v}^* \bar{\partial}U(\varphi_{01}^v) = Ad_{\varphi_{t1}^v}^T \overline{\nabla}^V U(\varphi_{01}^v). \tag{11.12}$$

Proof. We have, using Theorem 8.10,

$$\frac{d}{d\varepsilon}\varphi_{01}^{v+\varepsilon h}(x)_{|\varepsilon=0} = \int_0^1 D\varphi_{u1}^v(\varphi_{0u}^v(x))h(u)\circ\varphi_{0u}^v(x)du$$

$$= \int_0^1 D\varphi_{u1}^v \circ \varphi_{1u}^v\, h(u) \circ \varphi_{1u}^v \circ \varphi_{01}(x)du$$

$$= \int_0^1 (\mathrm{Ad}_{\varphi_{u1}^v}h(u)) \circ \varphi_{01}(x)du$$

Letting $w = \int_0^1 \mathrm{Ad}_{\varphi_{u1}^v} h(u)du$, we have, noting that, by Theorem 8.9, $\varphi_{1u}^v - id \in C_0^p(\Omega, \mathbb{R}^d)$ (so that Ad_φ^* is well-defined on $C_0^{p-1}(\Omega, \mathbb{R}^d)^*$)

$$\frac{d}{d\varepsilon}U(\varphi_{01}^{v+\varepsilon h})_{|\varepsilon=0} = \frac{d}{d\varepsilon}U((id + \varepsilon w) \circ \varphi_{01}^v + o(\varepsilon))_{|\varepsilon=0}$$

$$= (\bar{\partial}U(\varphi_{01}^v)\,|\,w)$$

$$= \int_0^1 (\bar{\partial}U(\varphi_{01}^v)\,|\,\mathrm{Ad}_{\varphi_{u1}^v}h(u))du$$

$$= \int_0^1 (\mathrm{Ad}_{\varphi_{u1}^v}^*\bar{\partial}U(\varphi_{01}^v)\,|\,h(u))du$$

$$= \int_0^1 \langle \mathrm{Ad}_{\varphi_{u1}^v}^T \overline{\nabla}U(\varphi_{01}^v)\,,\,h_u\rangle_V du,$$

which proves (11.12).   □

This is a very important result, which has the following simple consequences.

Proposition 11.6. *Let U satisfy the assumptions of Theorem 11.5. If $v \in \mathcal{X}_V^2(\Omega)$ is a minimizer of*

$$\tilde{E}(v) = \int_0^1 \|v_t\|_V^2 dt + U(\varphi_{01}^v), \tag{11.13}$$

Then, for all t

$$v_t = \mathrm{Ad}_{\varphi_{t1}^v}^T v_1 \tag{11.14}$$

with $v_1 = -\frac{1}{2}\overline{\nabla}^V U(\varphi_{01}^v)(x)$.

Corollary 11.7. *Under the same conditions on U, if $v \in \mathcal{X}_V^2(\Omega)$ is a minimizer of*

$$\tilde{E}(v) = \int_0^1 \|v_t\|_V^2 dt + U(\varphi_{01}^v)$$

then, for all t,

$$v_t = \mathrm{Ad}_{\varphi_{t0}^v}^T v_0. \tag{11.15}$$

Proposition 11.6 is a direct consequence of Theorem 11.5. For the corollary, we need to use the fact that $\mathrm{Ad}_\varphi \mathrm{Ad}_\psi = \mathrm{Ad}_{\varphi \circ \psi}$, which can be checked by direct computation, and write

$$v_t = \mathrm{Ad}^T_{\varphi^v_{t1}} v_1 = \mathrm{Ad}^T_{\varphi^v_{t1}} \mathrm{Ad}^T_{\varphi^v_{10}} v_0 = (\mathrm{Ad}_{\varphi^v_{10}} \mathrm{Ad}_{\varphi^v_{t1}})^T v_0 = \mathrm{Ad}^T_{\varphi^v_{t0}} v_0.$$

Proposition 11.6 is an expression of what is called the Pontryagin principle in optimal control. Equations $v_t = \mathrm{Ad}^T_{\varphi^v_{t0}} v_0$ and $v_1 = -\frac{1}{2}\overline{\nabla}^V U(\varphi^v_{01})(x)$ together are equivalent to the Euler–Lagrange equations for $\tilde{E}$ and will lead to interesting numerical procedures. Equation (11.15) is a cornerstone of the theory. It results in a general mechanical property called the conservation of momentum, on which we will return later.

The conjugate of the adjoint can be put into a more explicit form. Before this, we introduce a notation that will be used throughout this chapter. If ρ is a linear form on function spaces, we have been denoting by $(\rho \,|\, v)$ the result of ρ applied to V. In the formulae that will come, we will need to emphasize the variable on which v depends, and we will use the alternate notation $(\rho \,|\, v(x))_x$ to denote the same quantity. Thus,

$$\rho(v) = (\rho \,|\, v) = (\rho \,|\, v(x))_x.$$

In particular, when v depends on two variables, the notation $(\rho \,|\, v(x,y))_x$ will represent ρ applied to the function $x \mapsto v(x,y)$ with constant y.

We still assume that V is continuously embedded in $C^p(\Omega, \mathbb{R}^d)$. Then, the following theorem holds.

Theorem 11.8. *Assume that $\varphi \in C^q_0(\Omega, \mathbb{R}^r)$ and $\rho \in C^r_0(|\Omega, \mathbb{R}^d)^*$, with $r = \min(p, q-1)$. Let $v = K\rho$ and $(e_1, \ldots, e_d)$ be an orthonormal basis of $\mathbb{R}^d$. Then, for $y \in \Omega$, we have*

$$\mathrm{Ad}^T_\varphi v(y) = \sum_{i=1}^d \big(\rho \,\big|\, Ad_\varphi(K(x,y)e_i)\big)_x e_i. \tag{11.16}$$

Proof. For $b \in \mathbb{R}^d$, we have

$$\begin{aligned}
b^T \mathrm{Ad}^T_\varphi v(y) &= \big\langle Ad^T_\varphi v,\, K(.,y)b \big\rangle_V \\
&= \big\langle v,\, Ad_\varphi(K(.,y)b) \big\rangle_V \\
&= \big(\rho \,\big|\, Ad_\varphi(K(x,y)b)\big)_x.
\end{aligned}$$

Theorem 11.8 is now a consequence of the decomposition

$$\mathrm{Ad}^T_\varphi v(y) = \sum_{i=1}^d e_i^T \mathrm{Ad}^T_\varphi v(y) e_i.$$

$\square$

Recall that $K(.,.)$ is a matrix, so that $K(.,y)e_i$ is the ith column of $K(.,y)$, which we can denote by K^i. Equation (11.16) states that the ith coordinate of $\mathrm{Ad}_\varphi^T v$ is $\left(\rho\,|\,\mathrm{Ad}_\varphi K^i(x,y)\right)_x$.

Using Proposition 11.6 and Theorem 11.8, we obtain the expression of the V-gradient of E:

Corollary 11.9. *Under the hypotheses of Proposition 11.6, the V gradient of*

$$\tilde{E}(v) = \int_0^1 \|v(t)\|_V^2\,dt + U(\varphi_{01}^v)$$

is equal to

$$\nabla E(v)(y) = 2v(t,y) + \sum_{i=1}^d \left(\rho(1)\,|\,D\varphi_{t1}(\varphi_{1t}^v(x))K^i(\varphi_{1t}^v(x),y)\right)_x e_i \quad (11.17)$$

with $\rho(1) = \bar{\partial}U(\varphi_{01}^v)(x)$.

11.5 Conservation of Momentum

We rapidly justify why we are referring to (11.15) as an expression for the conservation of a momentum. The justification of the term momentum comes from the analogy of $E_k = (1/2)\|v(t)\|_V^2$ with the total kinetic energy at time t of a dynamical system. In fluid mechanics, this energy can be defined as

$$E_k = \frac{1}{2}\int z(t,y)|v(t,y)|^2 dy$$

the momentum here being $\rho(t) = z(t,y)v(t,y) \otimes dy$ with $E_k = (1/2)(\rho\,|\,v)$. In our case, taking $\rho(t) = Lv(t)$, we still have $E_k = (1/2)(\rho\,|\,v)$, so that $\rho(t)$ is also interpreted as a momentum.

The interpretation of (11.15) as a conservation equation is related to the effect of a change of coordinate system. Indeed, interpret $v(t,y)$ as the velocity of a particle located at coordinates y, so $v = dy/dt$. Now assume that we want to use a new coordinate system, and replace y by $x = \varphi(y)$. In the new coordinates, the same particle moves with velocity

$$\partial_t x = D\varphi(y)\frac{dy}{dt} = D\varphi(y)v(t,y) = (D\varphi v(t)) \circ \varphi^{-1}(x)$$

so that the translation from the old to the new expression of the velocity is precisely given by the adjoint operator: $v(y) \to \tilde{v}(x) = \mathrm{Ad}_\varphi v(x)$ if $x = \varphi(y)$. To obtain the correct transformation of the momentum, it suffices to notice that the energy of the system must remain the same if we just change the coordinates, so that, if ρ and $\tilde{\rho}$ are the momenta before and after the change of coordinates, we must have

$$(\tilde{\rho} \,|\, \tilde{v}) = (\rho \,|\, v)$$

which yields $Ad_\varphi^* \tilde{\rho} = \rho$ or $\tilde{\rho} = Ad_{\varphi^{-1}}^* \rho$.

Now, return to equation (11.15). Here, $v(t, y)$ is the velocity at time t of the particle that was at $x = \varphi_{t0}^v(y)$ at time 0. So it is the expression of the velocity in a coordinate system that evolves with the flow, and $Lv(t)$ is the momentum in the same system. By the previous argument, the expression of the momentum in the fixed coordinate system, taken at time $t = 0$, is $Ad_{\varphi_{0t}^v}^* Lv(t)$. Equation (11.15) simply states that this expression remains constant over time, i.e., the momentum is conserved when measured in a fixed coordinate system.

The conservation of momentum equation, described in Corollary 11.7 is a fundamental equation in *Geometric Mechanics* [135, 112], which appears in a wide variety of contexts. It has been described in abstract form by Arnold [11, 12] in his analysis of invariant metrics on Lie groups. This equation also derives from an application of the Euler–Poincaré principle, as described in [113, 136, 112]. Combined with a volume-preserving constraint, this equation is equivalent to the Euler equation for incompressible fluids, in the case when $\|v(t)\|_V = \|v(t)\|_2$, the L^2 norm. Another type of norm on V (called the H_α^1 norm) relates to models of waves in shallow water, and provides the Camassa–Holm equation [35, 86, 112]. A discussion of (11.15) in the particular case of template matching is provided in [149], and a parallel with the soliton emerging from the Camassa–Holm equation is discussed in [114].

11.5.1 Properties of the Momentum Conservation Equation

Combining equation (11.16) and the fact that $\partial_t \varphi_{0t}^v = v(t, \varphi_{0t}^v)$, we get, for the optimal v (letting $v_0 = K\rho_0$)

$$\partial_t \varphi(t, y) = \sum_{i=1}^d \left(\rho_0 \,|\, (D\varphi(t, x))^{-1} K^{\cdot i}(\varphi(t, x), \varphi(t, y)) \right)_x e_i.$$

This implies that the evolution of the diffeomorphism at any point depends on the diffeomorphism and its derivative everywhere (this is a global equation). Since it depends on the differential of φ, it is natural to combine it with the evolution of $D\varphi$, which here takes a particularly simple form. Indeed, formally taking the differential with respect to y, we get, for $a \in \mathbb{R}^d$

$$\partial_t D\varphi(t, y)a = \sum_{i=1}^d \left(\rho_0 \,|\, (D\varphi(t, x))^{-1} D_2 K^i(\varphi(t, x), \varphi(t, y))D\varphi(t, y)a \right)_x e_i.$$

$$(11.18)$$

The derivation "under" ρ_0 requires some care. As we have seen, the expression $(\rho_0 \,|\, u)$ may depend on derivatives of u and will typically satisfy

$$(\rho_0 \,|\, u) \le C\|u\|_{r,\infty}$$

for some r, meaning that $\rho_0 \in C_0^r(\Omega, \mathbb{R}^d)^*$. (The examples we have seen so far corresponded to either $r = 0$ of $r = 1$.) Let u depend on an additional parameter, say θ. We want to discuss the validity of the equation

$$\partial_\theta(\rho_0 \mid u(\theta)) = (\rho_0 \mid \partial_\theta u),$$

which corresponds to

$$\lim_{\varepsilon \to 0} (\rho_0 \mid (u(\theta + \varepsilon) - u(\theta))/\varepsilon - \partial_\theta u) = 0.$$

A direct consequence of our hypotheses is

$$(\rho_0 \mid (u(\theta + \varepsilon) - u(\theta))/\varepsilon - \partial_\theta u) \leq C \left\| (u(\theta + \varepsilon) - u(\theta))/\varepsilon - \partial_\theta u \right\|_{r,\infty};$$

we see that we need to ensure that all derivatives up to order r of u (with respect to the variable x) are differentiable in the parameter θ, with a uniform convergence of the increment rate in θ. We summarize this in the following lemma.

Lemma 11.10. *Assume that* $\theta \mapsto u(\theta)$ *is defined on some open real interval containing* θ_0 *with values in* $C_0^r(\Omega, \mathbb{R}^d)$. *Assume that, for some* $u_{\theta_0} \in C_0^r(\Omega, \mathbb{R}^d)$, *one has*

$$\lim_{\varepsilon \to 0} \left\| (u(\theta_0 + \varepsilon) - u(\theta_0))/\varepsilon - u_{\theta_0} \right\|_{r,\infty} = 0.$$

Then, for $\rho_0 \in C_0^r(\Omega, \mathbb{R}^d)^*$, *the function* $\theta \mapsto (\rho_0 \mid u(\theta))$ *is differentiable at* θ_0 *with derivative* $(\rho_0 \mid u_{\theta_0})$.

The hypothesis of the lemma, which applies to a generic context, can be relaxed in particular cases. For example, if ρ_0 is a measure, standard theorems for differentiation of integrals can be applied.

In the case of (11.18), $D\varphi(t, y)a$ is the derivative of $\varphi(t, y + \theta a)$ so that the function $u(\theta)$ is given by (fixing the coordinate i)

$$u(\theta)(x) = (D\varphi(t, x))^{-1} K^i(\varphi(t, x), \varphi(t, y + \theta a)).$$

We see that the existence of the rth derivative of $u(\theta)$ requires the existence of the rth derivative of K and of the $(r + 1)$th derivative of φ. From Theorem 8.9, we see that it suffices to assume that V is continuously embedded in $C_0^{r+1}(\Omega, \mathbb{R}^d)$. Moreover, once this is assumed, we can iterate the argument to higher derivatives (up to $r + 1$) of φ_{st}^v. This is summarized in the following result.

Theorem 11.11. *Assume that* v *satisfies equation (11.15) with* $v_0 = K\rho_0$. *Assume that*

$$(\rho_0 \mid u) \leq C \|u\|_{p-1,\infty}$$

for some constant C *and that* V *is continuously embedded in* $C_0^p(\Omega, \mathbb{R}^d)$. *Then the diffeomorphism and its first* p *differentials evolve according to the following* *differential system (recall that a differential of order q is a symmetric q-form, acting on q vectors $a_1, \ldots, a_q$):*

$$
\begin{cases}
\partial_t \varphi(t, y) = \displaystyle\sum_{i=1}^{d} \left(\rho_0 \,|\, (D\varphi(t, x))^{-1} K^i(\varphi(t, x), \varphi(t, y)) \right)_x e_i \\[2em]
\partial_t D\varphi(t, y) a = \displaystyle\sum_{i=1}^{d} \left(\rho_0 \,|\, (D\varphi(t, x))^{-1} \partial_y K^i(\varphi(t, x), \varphi(t, y))(D\varphi(t, y) a) \right)_x e_i \\[2em]
\vdots \\[1em]
\partial_t D^p \varphi(t, y)(a_1, \ldots, a_p) \\[1.5em]
\qquad = \displaystyle\sum_{i=1}^{d} \left(\rho_0 \,|\, (D\varphi(t, x))^{-1} \partial_y^p K^i(\varphi(t, x), \varphi(t, y))(a_1, \ldots, a_p) \right)_x e_i
\end{cases}
\tag{11.19}
$$

The interest of this theorem lies in the fact that it expresses the family $(\varphi_{0t}, D\varphi_{0t}, \ldots D^{(p)}\varphi_{0t})$ as the solution of an ordinary differential equation in function space, that takes the form

$$
(d/dt)(\varphi(t), M^{(1)}(t), \ldots, M^{(p)}(t)) = F(\varphi(t), M^{(1)}(t) \ldots, M^{(p)}(t)).
$$

This fact is used in the next proposition.

11.5.2 Existence of Solutions

Proposition 11.12. *Under the hypotheses of Theorem 11.11, the momentum conservation equation admits a unique solution in small time.*

Proof. We want to apply the uniqueness and existence theorem for differential equations in Banach spaces. Write $\varphi = id + u$ so that system (11.19) can be seen as a differential equation in

$$
U = (u, Du, \ldots, D^{(p)}u) = (U^{(0)}, \ldots, U^{(p)}),
$$

of the form $dU/dt = G(U)$. Define $\|U\| = \sum_{k=1}^{p} \|U^{(k)}\|_{p-k,\infty}$. Then, for $\max(\|U\|, \|U'\|)$ bounded, there exists a constant C such that

$$
\|G(U) - G(\tilde{U})\| \leq C\|U - \tilde{U}\|.
$$

This fact (which implies the existence and uniqueness of the solutions) comes from direct, but lengthy, estimates of the terms in system (11.19) that we will not detail here, but only take the first equation as an example. We need to estimate

$$\left\| \sum_{i=1}^{d} \left(\rho_0 \,|\, (\mathrm{Id} + U^{(1)})^{-1} K^i(x + U^{(0)}(x), y + U^{(0)}(y)) \right)_x e_i \right.$$
$$\left. - \left(\rho_0 \,|\, (\mathrm{Id} + \tilde{U}^{(1)}(x))^{-1} K^i(x + \tilde{U}^{(0)}(x), y + \tilde{U}^{(0)}(y)) \right)_x e_i \right\|_{p,\infty}$$

for which it suffices to estimate each coordinate separately, i.e.,

$$\left\| \left(\rho_0 \,|\, (\mathrm{Id} + U^{(1)}(x))^{-1} K^i(x + U^{(0)}, y + U^{(0)}(y)) \right)_x \right.$$
$$\left. - \left(\rho_0 \,|\, (\mathrm{Id} + \tilde{U}^{(1)}(x))^{-1} K^i(x + \tilde{U}^{(0)}, y + \tilde{U}^{(0)}(y)) \right)_x \right\|_{p,\infty}.$$

The important remark is that the p derivatives in the norm are computed in the variable y, and will therefore only involve p derivatives in $U^{(0)}$. The computation ρ_0 is done with respect to the other variable, x, and its estimate requires using $p - 1$ derivatives in $U^{(0)}$ and $U^{(1)}$. The variation will finally be controlled by estimates that are linear in $\|U^{(0)} - \tilde{U}^{(0)}\|_{p,\infty}$ and $\|U^{(1)} - \tilde{U}^{(1)}\|_{p-1,\infty}$. $\qquad\square$

The result is, in fact, stronger, because the solution can be integrated over all times (not only short times). The time length during which a solution can be shown to exist using the Cauchy–Lipschitz theorem can be directly related to the Lipschitz constant of the function G in the previous proof. If unique solutions have been obtained up to $t < t_0$, we can try to restart the equation at $t \simeq t_0$. If the Lipschitz constant there is bounded, then an extension exists.

In the case of (11.15), the analysis is even simpler, because, after solving the equation up to time t_0, we can consider the evolution of $\varphi_{t_0 t}$ and its derivatives. Because of the properties of the adjoint, it is easy to check that the evolution is exactly the same, simply with ρ_0 replaced by $\rho(t_0)$. So being able to push the solution forward after time t_0 only requires that, if C_t is such that $(\rho(t) \,|\, u) \leq C_t \|u\|_{p,\infty}$, then C_t does not go to infinity in finite time.

Since $(\rho(t) \,|\, u) = (\rho_0 \,|\, \mathrm{Ad}_{\varphi(t)} u)$ it suffices to check that the first p derivatives of $\varphi(t)$ remain bounded. From Theorem 8.9, it suffices in turn to prove that $\int_0^t \|v(t)\|_{p,\infty}$ remains finite. Finally, our hypotheses imply that $\|v(t)\|_{p,\infty} = O(\|v(t)\|_V)$ so that it suffices to show that $\|v(t)\|_V$ remains bounded.

In fact, $\|v(t)\|_V$ remains constant during the evolution, so that long-time existence is ensured. The constancy of $\|v(t)\|_V$ will be proved in the next section, after a discussion of the EPdiff equation, which is the time-derivative of the momentum conservation equation.

11.5.3 Time Variation of the Eulerian Momentum

We want here to compute the variation $d\rho(t)/dt$ when ρ satisfies (11.15). We have

$$(\rho(t) \,|\, w) = (\rho_0 \,|\, \mathrm{Ad}_{\varphi(t)^{-1}} w).$$

We assume here that $\varphi(t)$ is a solution of system (11.19) (we have already proved that solutions exist over small time intervals), with v the associated velocity.

Fix t and $\varepsilon > 0$ and denote $\psi_\varepsilon = \varphi(t)\varphi(t+\varepsilon)^{-1}$. From the definition of Ad, one directly checks that

$$\mathrm{Ad}_{\varphi(t+\varepsilon)^{-1}}w = \mathrm{Ad}_{\varphi(t)^{-1}}\mathrm{Ad}_{\psi_\varepsilon}w,$$

with

$$\partial_\varepsilon \mathrm{Ad}_{\psi_\varepsilon}w = -Dv(t)w + Dwv(t).$$

The term in right-hand side in the adjoint representation of $v(t)$, as expressed in the following definition.

Definition 11.13. *If v is a differentiable vector field on Ω, we denote by ad_v the mapping that transform a differentiable vector field w into*

$$\mathrm{ad}_v w = Dv\, w - Dw\, v. \tag{11.20}$$

We therefore have:

$$\partial_t \mathrm{Ad}_{\varphi(t)^{-1}}w = -\mathrm{Ad}_{\varphi(t)^{-1}}\mathrm{ad}_{v(t)}w.$$

Because φ is a solution of (11.19), its first p derivatives are bounded in the supremum norm, so that

$$\begin{aligned}
\partial_t\big(\rho(0)\,|\,\mathrm{Ad}_{\varphi(t)^{-1}}w\big) &= \big(\rho(0)\,|\,\partial_t \mathrm{Ad}_{\varphi(t)^{-1}}w\big)\\
&= -\big(\rho(0)\,|\,\mathrm{Ad}_{\varphi(t)^{-1}}\mathrm{ad}_{v(t)}w\big)\\
&= -\big(\rho(t)\,|\,\mathrm{ad}_{v(t)}w\big).
\end{aligned}$$

This yields

$$\partial_t\big(\rho(t)\,|\,w\big) + \big(\rho(t)\,|\,\mathrm{ad}_{v(t)}w\big) = 0$$

or

$$\partial_t\rho(t) + \mathrm{ad}^*_{v(t)}\rho(t) = 0. \tag{11.21}$$

This equation is often referred to as the EPDiff equation, for Euler-Poincaré on groups of diffeomorphisms (cf. [113, 136]). By extension, we will also refer to momentum conservation as EPDiff in the following. Equation (11.21) can be used to prove the following proposition.

Proposition 11.14. $\|v(t)\|_V$ *is constant along solutions of* (11.19).

Proof. Indeed, we have, for $\varepsilon > 0$,

$$\frac{1}{\varepsilon}\big(\|v(t+\varepsilon)\|^2 - \|v(t)\|_V^2\big) = \frac{2}{\varepsilon}\big(\rho(t+\varepsilon) - \rho(t)\,|\,v(t)\big) + \frac{1}{\varepsilon}\|v(t+\varepsilon) - v(t)\|_V^2$$

The first term on the right-hand side converges, from (11.21) to

$$-2\big(\rho(t)\,|\,\mathrm{ad}_{v(t)}v(t)\big).$$

For the second term, we have

$$\|v(t+\varepsilon) - v(t)\|_V = \sup_{\|w\|_V \leq 1} \left(\rho(t+\varepsilon) - \rho(t) \,|\, w\right)$$

$$= \sup_{\|w\|_V \leq 1} \int_0^\varepsilon \left(\rho(t+s) \,|\, \mathrm{ad}_{v(t+s)} w\right) ds$$

$$= O(\varepsilon)$$

$\square$

As discussed after Proposition 11.12, this implies that the unique solution of the EPDiff equation exists over arbitrary time intervals.

Theorem 11.15. *Under the hypotheses of Proposition 11.12, the EPDiff equation has solutions over all times, uniquely specified by its initial conditions.*

11.5.4 Hamiltonian Form of EPDiff

We now provide an alternate form of (11.19). Introduce the linear form

$$(\mu(t) \,|\, w) = (\rho_0 \,|\, D\varphi(t)^{-1}w) = (\rho(t) \,|\, w \circ \varphi(t)^{-1}). \tag{11.22}$$

Denote $M(t) = (D\varphi(t))^{-1}$. We have, using $M(t)D\varphi(t) = \mathrm{Id}_d$

$$\partial_t M = -M\left(\partial_t D\varphi\right) M.$$

Using the second equation in (11.19), this yields

$$\partial_t M(t)a = -\sum_{i=1}^d \left(\rho_0 \,|\, M(t,x)D_2 K^i(\varphi(t,x), \varphi(t,y))a\right)_x M(t,y)e_i.$$

This implies that, for any $w \in V$

$$\left(\partial_t \mu(t) \,|\, w\right) = -\left(\rho_0 \,|\, \partial_t M w\right)$$

$$= \sum_{i=1}^d \left(\rho_0 \,|\, (\rho_0 \,|\, M(t,x)D_2 K^i(\varphi(t,x), \varphi(t,y))w)_x M(t,y)e_i\right)_y$$

$$= -\sum_{i=1}^d \left(\mu(t) \,|\, (\mu(t) \,|\, D_2 K^i(\varphi(t,x), \varphi(t,y))w)_x e_i\right)_y.$$

We therefore have the system

$$\partial_t \varphi(t,y) = \sum_{i=1}^d \left(\mu(t) \,|\, K^i(\varphi(t,x), \varphi(t,y))\right)_x e_i \tag{11.23}$$

$$\left(\partial_t \mu(t) \,|\, w\right) = -\sum_{i=1}^d \left(\mu(t) \,|\, (\mu(t) \,|\, D_2 K^i(\varphi(t,x), \varphi(t,y))w)_x e_i\right)_y.$$

We will refer to this (for reasons that will be clarified later) as the Hamiltonian form of EPDiff. When $(\rho_0 \mid w)$ does not depend on the derivatives of w (more precisely, $\rho_0 \in C_0^0(\Omega, \mathbb{R}^d)^*$), this provides an ordinary differential equation in the variables (φ, μ) (of the form $(d/dt)(\varphi, \mu) = F(\varphi, \mu)$.) The initial conditions are $\varphi_0 = id$ and $\mu_0 = \rho_0$.

11.5.5 Case of Measure Momenta

An interesting feature of (11.23) is that it can easily be reduced to a smaller number of dimensions when ρ_0 takes specific forms. As a typical example, we perform the computation in the case

$$\rho_0 = \sum_{k=1}^{N} z_k(0, .) \otimes \omega_k \tag{11.24}$$

where ω_k is an arbitrary measure on Ω and $z_k(0)$ a vector field. (We recall the notation $(z \otimes \omega \mid w) = \int z(x)^T w(x) \omega(dx)$.) Most of the Eulerian differentials that we have computed have been reduced to this form. From the definition of $\mu(t)$, we have $\mu(t) = \sum_{k=1}^{N} z_k(t, .) \otimes \omega_k$ (where $z_k(t, x) = D\varphi_{0t}(x)^{-T} z_k(0, x)$). The first equation in (11.23) is

$$\partial_t \varphi(t, y) = \sum_{i=1}^{d} \sum_{k=1}^{N} \int_{\Omega} z_k(t, x)^T K^i(\varphi(t, x), \varphi(t, y)) e_i d\omega_k(x).$$

For a matrix A with ith column vector A^i, and a vector z, $z^T A^i$ is the ith coordinate of $A^T z$. Applying this to the previous equation yields

$$\partial_t \varphi(t, y) = \sum_{k=1}^{N} \int_{\Omega} K(\varphi(t, y), \varphi(t, x)) z_k(t, x) d\omega_k(x), \tag{11.25}$$

where we have used the fact that $K(\varphi(t, x), \varphi(t, y))^T = K(\varphi(t, y), \varphi(t, x))$. The second equation in (11.23) becomes

$$\left(\partial_t \mu(t) \mid w \right)$$

$$= -\sum_{i=1}^{d} \left(\mu(t) \mid \left(\mu(t) \mid D_2 K^i(\varphi(t, x), \varphi(t, y)) w(y) \right)_x e_i \right)_y$$

$$= -\sum_{k,l=1}^{N} \int_{\Omega} \int_{\Omega} \sum_{i=1}^{d} z_l^T(t, x) D_2 K^i(\varphi(t, x), \varphi(t, y)) w(y) z_k(t, y)^T e_i d\omega_l(x) d\omega_k(y)$$

$$= -\sum_{k=1}^{N} \int_{\Omega} \left(\int_{\Omega} \sum_{l=1}^{N} \sum_{i=1}^{d} z_k^i(t, y) z_l(t, x)^T D_2 K^i(\varphi(t, x), \varphi(t, y)) d\omega_l(x) \right) w(y) d\omega_k(y)$$

where z_k^i is the ith coordinate of z_k. From the expression of $\mu(t)$, we also have

$$\partial_t \mu = \sum_{k=1}^{N} (\partial_t z_k) \otimes \omega_k.$$

Letting K^{ij} denote the entries of K, we can identify $\partial_t z_k$ as

$$\partial_t z_k(t, y) =$$

$$\int_\Omega \sum_{l=1}^{N} \sum_{i,j=1}^{d} z_k^i(t, y) z_l^i(t, x) \nabla_2 K^{ij}(\varphi(t, x), \varphi(t, y)) d\omega_l(x)$$

$$= - \int_\Omega \sum_{l=1}^{N} \sum_{i,j=1}^{d} z_l^i(t, y) z_j^k(t, x) \nabla_1 K^{ij}(\varphi(t, y), \varphi(t, x)) d\omega_l(x). \quad (11.26)$$

This equation is somewhat simpler when K is a scalar kernel, in which case $K^{ij}(x, y) = \Gamma(x, y)$ if $i = j$ and 0 otherwise, where Γ takes real values. We get, in this case

$$\partial_t z_k(t, y) = - \sum_{l=1}^{N} \int_\Omega \nabla_2 \Gamma(\varphi(t, x), \varphi(t, y)) z_k(t, y)^T z_l(t, x) d\omega_l(x)$$

$$= - \sum_{l=1}^{N} \int_\Omega \nabla_1 \Gamma(\varphi(t, y), \varphi(t, x)) z_k(t, y)^T z_l(t, x) d\omega_l(x).$$

In all cases, we see that the evolution of μ can be completely described using the evolution of $z_1, \ldots, z_N$. In the particular case when the z_k's are constant vectors (which corresponds to most of the point-matching problems), this provides a finite-dimensional system on the μ part.

11.6 Optimization Strategies for Flow-Based Matching

We have formulated flow-based matching as an optimization problem over time-dependent vector fields. We discuss here other possible optimization strategies that take advantage of the different formulations that we obtained for the EPDiff equation. They will correspond to taking different control variables with respect to which the minimization is performed, and we will in each case provide the expression of the gradient of E with respect to a suitable metric. Optimization can then be performed by gradient descent, conjugate gradient or any feasible optimization algorithm (see Appendix D or [158]).

After discussing the general formulation of each of these strategies, we will provide the specific expression of the gradients for point-matching problems, in the following form: minimize

$$E(\varphi) = d_V(id, \varphi)^2 + F(\varphi(x_1), \ldots, \varphi(x_N)) \quad (11.27)$$

with respect to φ, where $x_1, \ldots, x_N$ are fixed points in Ω. These problems are important because, in addition to the labeled and unlabeled point matching problems we have discussed, other problems, like curve and surface matching end up being discretized in this form (we will discuss algorithms for image matching in the next section). The following discussion describes (and often extends) several algorithms that have been proposed in the literature, in [119, 21, 207, 147, 148, 223] among other references.

11.6.1 Gradient Descent in $\mathcal{X}_V^2$

The original problem having been expressed in this form, Corollary 11.9 directly provides the expression of the gradient of E considered as a function defined over $\mathcal{X}_V^2$, with respect to the metric in this space. Using $t \mapsto v(t, \cdot)$ as an optimization variable has some disadvantages, however. The most obvious is that it results in solving a huge dimensional problem (over a $d+1$-dimensional variable) even if the original objects are, say, N landmarks in $\mathbb{R}^d$.

When the matching functional U is only function of the deformation of a template object, i.e.,

$$U = F(\varphi_{01} \cdot I),$$

then we can write, when $\tilde{E}$ is given by (11.13)

$$\min_{v(t, \cdot)} \tilde{E}(v) = \min_{J(t, \cdot), J(0) = I} \left(\min_{v: \partial_t J = v(t) \cdot J(t)} \tilde{E}(v) \right)$$

where the notation $v \cdot J$ refers to the infinitesimal action defined as in Appendix B (Section B.5.3) by

$$\partial_\varepsilon \psi(\varepsilon, \cdot) \cdot I = h \cdot I$$

whenever $\psi(0, \cdot) = id$ and $\partial_\varepsilon \psi(0, \cdot) = h$. The equation $\partial_t J = v(t) \cdot J(t)$ combined with $J(0) = I$ is equivalent to $J(t) = \varphi^v(t, \cdot) \cdot I$, so that the iterated minimization first minimizes with respect to v for fixed object trajectory, then minimizes with respect to object trajectories.

When $J(t, \cdot)$ is given, the inner minimization is

$$\min_{v: \partial_t J = v(t) \cdot J(t)} \tilde{E}(v) = \min_{v: \partial_t J = v(t) \cdot J(t)} \left(\int_0^1 \|v(t)\|_V^2 + F(J(1)) \right)$$

$$= \int_0^1 \left(\min_{w: \partial_t J = w \cdot J(t)} \|w\|_V^2 \right) dt + F(J(1)) \quad (11.28)$$

since the constraints apply separately to each $v(t)$. This expression only depends on the trajectory $J(t)$. One can therefore try to compute its gradient with respect to this object trajectory and run a minimization algorithm accordingly. One difficulty in this approach is that, given an object trajectory $J(t)$, there may exist no $w \in V$ such that $\partial_t J = w \cdot J(t)$ (which result in the minimum in the integral being infinite), so that the possibility of expressing the trajectory as evolving according to a flow is a constraint of the problem. This may be intractable in the general case, but always satisfied with point-matching problems as long as the points remain distinct. We will discuss this in the next section.

However, what (11.28) tells us is that, if a time-dependent vector field $\tilde{v}(t\cdot)$ is given, one always reduces the value of $\bar{E}(\tilde{v})$ by replacing $\tilde{v}(t,\cdot)$ by

$$v(t,\cdot) = \text{argmin}_{w:w\cdot J(t)=\tilde{v}\cdot J(t)}\|w\|_V^2 \tag{11.29}$$

with $J(t) = \varphi_{0t}^{\tilde{v}} \cdot I$. If we write $w = u + \tilde{v}$, then the solution v is equal to $\hat{u} + \tilde{v}$, where $\hat{u}$ is the orthogonal projection of $(-\tilde{v})$ on the space

$$V_J = \{u : u \cdot J = 0\}.$$

This orthogonal projection is well-defined when V_J is closed, which requires the transformation $u \mapsto u \cdot J$ to be continuous on V in the sense that, if $\|w_n - w\|_V \to 0$, then $w_n \cdot J$ converges to $w \cdot J$ (in some appropriate sense depending on the space of deformable objects).

The numerical computation of this orthogonal projection is not always easy, but when it is, it generally has a form which is more specific than a generic time-dependent vector field, and provides an improved gradient descent algorithm in $\mathcal{X}_V^2$ as follows. Assume that, at time τ in the algorithm, the current vector field v^τ in the minimization of E is such that $v^\tau(t) \in V_{J^\tau(t)}$ at all times t. Then define a vector field at the next step $\tau + \delta\tau$ by

$$\tilde{v}^{\tau+\delta\tau}(t,y) = v^\tau(t,y) - \delta\tau\Big(2v(t,y) + \sum_{i=1}^{d}\big(\rho(1) \mid D\varphi_{t1}(\varphi_{1t}^v(x))K^i(\varphi_{1t}^v(x),y)\big)_x e_i\Big)$$

which corresponds to one step of gradient descent, as specified in (11.17), then compute $J(t) = \varphi_{0t}^{\tilde{v}^{\tau+\delta\tau}} \cdot I$ and define

$$v^{\tau+\delta\tau}(t) = \pi_{V_{J(t)}}\tilde{v}^{\tau+\delta\tau}$$

at all times t.

Application to Point Matching

Consider the point-matching energy. In this case, letting

$$U(\varphi) = F(\varphi(x_1), \ldots, \varphi(x_N)),$$

we have

$$\rho(1) = \bar{\partial}U(\varphi_{01}^v) = \sum_{i=1}^{N} \partial_i F(\varphi_{01}^v(x_k)) \otimes \delta_{\varphi_{01}^v(x_k)}$$

where $\partial_i F(\varphi_{01}^v(x))$ is the partial derivative of F with respect to its ith variable computed at $\varphi_{01}^v(x_1), \ldots, \varphi_{01}^v(x_N)$. We therefore have, by corollary 11.9,

$$\nabla^V U(v)(t,y) = \sum_{i=1}^{d} \left(\rho(1) \mid D\varphi_{t1}^v(\varphi_{1t}^v(x)) K(\varphi_{1t}^v(x),y)e_i \right)_x e_i$$

$$= \sum_{i=1}^{d} \sum_{q=1}^{N} \left(\partial_q F(\varphi_{01}^v(x))^T D\varphi_{t1}^v(\varphi_{0t}^v(x_q)) K(\varphi_{0t}^v(x_q),y)e_i \right) e_i$$

$$= \sum_{q=1}^{N} K(y,\varphi_{0t}^v(x_q)) D\varphi_{t1}^v(\varphi_{0t}^v(x_q))^T \partial_q F(\varphi_{01}^v(x_q))$$

so that

$$\nabla^V E(v)(t,y) = 2v(t,y) + 2\sum_{q=1}^{N} K(y,\varphi_{0t}^v(x_q)) D\varphi_{t1}^v(\varphi_{0t}^v(x_q))^T \partial_q F(\varphi_{01}^v(x_q)).$$

$$(11.30)$$

So, a basic gradient descent algorithm in $\mathcal{X}_V^2$ would implement the evolution (letting τ denote the algorithm time)

$$\partial_\tau v^\tau(t,y) = -2\gamma\left(v^\tau(t,y) + \sum_{q=1}^{N} K(y,\varphi_{0t}^{v^\tau}(x_q)) D\varphi_{t1}^{v^\tau}(\varphi_{0t}^{v^\tau}(x_q))^T \partial_q F(\varphi_{01}^{v^\tau}(x_q)) \right).$$

$$(11.31)$$

The two-step algorithm defined in the previous section is especially efficient with point sets. It is because, with $x = (x_1,\dots,x_N)$, $v \cdot x = (v(x_1),\dots,v(x_N))$ and the projection on

$$V_x = \{v : v \cdot x\} = \{v : v(x_1) = \cdots = v(x_N) = 0\}$$

is given by spline interpolation with the kernel, as described in Theorem 9.6.

More precisely, define $x_i^v(t) = \varphi_{0t}^v(x_i)$. We assume that, at time τ, we have a time-dependent vector field v^τ which takes the form

$$v^\tau(t,y) = \sum_{i=1}^{N} K(y,x_i^{v^\tau}(t))\alpha_i^\tau(t). \qquad (11.32)$$

Using (11.31), we define

$$\tilde{v}(t,y) = v^\tau - \delta\tau\left(v^\tau(t,y) + \sum_{q=1}^{N} K(y,\varphi_{0t}^{v^\tau}(x_q)) D\varphi_{t1}^{v^\tau}(\varphi_{0t}^{v^\tau}(x_q))^T \partial_q F(\varphi_{01}^{v^\tau}(x_q)) \right)$$

The values of $\tilde{v}(t,.)$ are in fact only needed at the points $x_i^{\tilde{v}}(t) = \varphi_{0t}^{\tilde{v}}(x_i)$. These points are obtained by solving the differential equation

$$\partial_t x = v^\tau(t,x) - 2\delta\tau(v^\tau(t,x) + \sum_{i=1}^{N} K(x_i^{v^\tau}(t),x) D\varphi_{t1}^{v^\tau}(x_i^{v^\tau}(t))^T F_i(x^{v^\tau}(1)))$$

$$(11.33)$$

with $x(0) = x_i$. Solving this equation simultaneously provides $x_i^{\tilde{v}}(t)$ and $\tilde{v}(x_i^{\tilde{v}}(t))$ for $t \in [0,1]$.

Once this is done, define $v^{\tau+\delta\tau}(t,.)$ to be the solution of the approximation problem $\inf_w \|w\|_V$ with $w(x_i^{\tilde{v}}(t)) = v(x_i^{\tilde{v}}(t))$ which will therefore take the form

$$v^{\tau+\delta\tau}(t,y) = \sum_{i=1}^{N} K(y, x_i^{v^{\tau+\delta\tau}}(t))\alpha_i^{\tau+\delta\tau}(t).$$

Solving (11.33) requires evaluating the expression of v^τ, which can be done exactly using (11.32). It also requires computing the expression of $D\varphi_{t1}^{v^\tau}(x_i^{v^\tau}(t))$. This is provided by Proposition 8.8 and by the fact that

$$\partial_t D\varphi_{t1}^v \circ \varphi_{1t} = \partial_t(D\varphi_{1t})^{-1} = -(D\varphi_{1t})^{-1}(\partial_t(D\varphi_{1t}))(D\varphi_{1t})^{-1}$$

which yield:

$$\partial_t D\varphi_{t1}^v(x_i^v(t)) = -D\varphi_{t1}^v(x_i^v(t))Dv(t, x_i^v(t)).$$

Thus, $D\varphi_{t1}^v(x_i^v(t))$ is solution of $\partial_t M = -MDv(x_i^v(t))$ with initial condition $M = \text{Id}$. The matrix $Dv(t, x_i^v(t))$ can be computed explicitly as a function of the point trajectories $x_j^v(t), j = 1, \ldots, N$, using the explicit expression (11.32). This algorithm has been introduced in [20].

11.6.2 Gradient in the Hamiltonian Form

Because the Hamiltonian form of EPDiff directly incorporates the dimension reduction that arises from the projection on V_J, it can be more efficient to use this representation instead of the two-step algorithm discussed above. Since we have, from (11.23),

$$\partial_t \varphi(t,y) = \sum_{i=1}^{d} \left(\mu(t) \mid K^i(\varphi(t,x), \varphi(t,y))\right)_x e_i,$$

the evolving diffeomorphism is uniquely specified if we are given the linear form $\mu(t)$ at all times. Moreover, since

$$v(t,y) = \sum_{i=1}^{d} \left(\mu(t) \mid K^i(\varphi(t,x), y)\right)_x e_i,$$

we can write, using the fact that $\left(\rho(t) \mid w\right) = \left(\mu(t) \mid w \circ \varphi(t,x)\right)_x$,

$$\begin{aligned}
\|v(t)\|_V^2 &= \left(\rho(t) \mid v(t)\right) \\
&= \left(\mu(t) \mid v(t) \circ \varphi(t,x)\right)_x \\
&= \sum_{i=1}^{d} \left(\mu(t) \mid \left(\mu(t) \mid K^i(\varphi(t,x), \varphi(t,y))\right)_x e_i\right)_y.
\end{aligned}$$

This implies that we can formulate the minimization problem in terms of μ as: minimize

$$E(\mu) = \sum_{i=1}^{d} \int_0^1 \left(\mu(t) \,\middle|\, \left(\mu(t) \,\middle|\, K(\varphi(t,x), \varphi(t,y)) e_i \right)_x e_i \right)_y dt + U(\varphi(1))$$

subject to the constraint

$$\partial_t \varphi(t,y) = \sum_{i=1}^{d} \left(\mu(t) \,\middle|\, K^i(\varphi(t,x), \varphi(t,y)) \right)_x e_i.$$

Although this is apparently more complex than the original problem in terms of v, using μ takes advantage of the fact that this linear form is often simpler (sparser) than the vector field. We now discuss the computation of the gradient of E with respect to μ.

Given a diffeomorphism φ, introduce the operator $H_\varphi : \rho \mapsto H_\varphi \rho$ which associates to a linear form ρ the vector field defined by

$$(H_\varphi \rho)(y) = \sum_{i=1}^{d} \left(\rho \,\middle|\, K^{(i)}(\varphi(x), \varphi(y)) \right)_x e_i. \tag{11.34}$$

We have the following result.

Lemma 11.16. *The operator H_φ defined in (11.34) is self-adjoint: for all φ, $H_\varphi = H_\varphi^*$.*

(Recall that if $A : B \to \tilde{B}$ is an operator, its conjugate $A^* : \tilde{B}^* \to B^*$ is defined by $(A^* \eta \,|\, w) = (\eta \,|\, Aw)$. In this case, assuming that V is included in C^p and that $\varphi \in C^p$, H_φ is well-defined for $\rho \in (C^p)^*$ and takes values in C^p, so that H_φ^* is defined on $(C^p)^*$ and takes values in $(C^p)^{**}$ which can be identified with C^p.)

Proof. Indeed, it is a direct consequence of the symmetry of the kernel that, for all $a, b \in \mathbb{R}^d$ and all $x, y \in \Omega$,

$$\left(a \otimes \delta_x \,\middle|\, H_\varphi(b \otimes \delta_y) \right) = b^T K(\varphi(y), \varphi(x)) a = a^T K(\varphi(x), \varphi(y)) b$$
$$= \left(b \otimes \delta_y \,\middle|\, H_\varphi(a \otimes \delta_x) \right) = \left(a \otimes \delta_x \,\middle|\, H_\varphi^*(b \otimes \delta_y) \right),$$

which implies that for all b, y, the vector fields $H_\varphi(b \otimes \delta_y)$ and $H_\varphi^*(b \otimes \delta_Y)$ coincide. Thus, for any ρ,

$$\left(\rho \,\middle|\, H_\varphi(b \otimes \delta_y) \right) = \left(\rho \,\middle|\, H_\varphi^*(b \otimes \delta_y) \right),$$

which is the same as $\left(b \otimes \delta_y \,\middle|\, H_\varphi \rho \right) = \left(b \otimes \delta_y \,\middle|\, H_\varphi^* \rho \right)$, which implies $H_\varphi \rho = H_\varphi^* \rho$. $\qquad \square$

Returning to the optimization problem, we must minimize

$$E(\mu) = \int_0^1 \big(\mu(t) \mid H_{\varphi(t)}\mu(t)\big)dt + U(\varphi(1))$$

subject to the constraint

$$\partial_t\varphi(t) = H_{\varphi(t)}\mu(t).$$

Consider a variation $\mu \mapsto \mu + \varepsilon\delta\mu$, which induces a variation $\varphi \mapsto \varphi + \varepsilon\delta\varphi$. The variation of the energy is

$$\partial_\varepsilon E = 2\sum_{i=1}^d \int_0^1 \big(\delta\mu(t) \mid H_{\varphi(t)}\mu(t)\big)dt$$

$$+ \sum_{i=1}^d \int_0^1 \big(\mu(t) \mid \big(\mu(t) \mid D_1K^i(\varphi(t,x),\varphi(t,y))\delta\varphi(t,x)\big)_x e_i\big)_y dt$$

$$+ \sum_{i=1}^d \int_0^1 \big(\mu(t) \mid \big(\mu(t) \mid D_2K^i(\varphi(t,x),\varphi(t,y))\delta\varphi(t,y)e_i\big)_x e_i\big)_y dt$$

$$+ \big(\frac{\delta U}{\delta\varphi} \mid \delta\varphi(1)\big).$$

Let $\beta(t)$ denote the linear form

$$\big(\beta(t) \mid h\big) = \sum_{i=1}^d \big(\mu(t) \mid \big(\mu(t) \mid D_1K^i(\varphi(t,x),\varphi(t,y))h(x)\big)_x e_i\big)_y$$

$$+ \sum_{i=1}^d \big(\mu(t) \mid \big(\mu(t) \mid D_2K^i(\varphi(t,x),\varphi(t,y))h(y)e_i\big)_x e_i\big)_y. \quad (11.35)$$

Then we can write

$$\partial_\varepsilon E = 2\int_0^1 \big(\delta\mu(t) \mid H_{\varphi(t)}\mu(t)\big)dt + \int_0^1 \big(\beta(t) \mid \delta\varphi(t)\big)dt + \big(\frac{\delta U}{\delta\varphi} \mid \delta\varphi(1)\big). \quad (11.36)$$

We would like to write

$$\big(\beta(t) \mid \delta\varphi(t)\big) = \big(\beta(t) \mid \frac{\delta\varphi(t)}{\delta\mu}\delta\mu\big)$$

$$= \big(\delta\mu \mid \big(\frac{\delta\varphi(t)}{\delta\mu}\big)^* \beta(t)\big)$$

to obtain an expression that only depends on $\delta\mu$. The variation of φ, however, cannot be directly computed as a function of $\delta\mu$, but as a solution of the differential equation

$$\partial_t \delta\varphi(t,y) = H_{\varphi(t)}\delta\mu(t)(y)$$

$$+ \sum_{i=1}^{d} \left(\mu(t) \,|\, D_1 K^i(\varphi(t,x), \varphi(t,y))\delta\varphi(t,x) \right)_x e_i$$

$$+ \sum_{i=1}^{d} \left(\mu(t) \,|\, D_2 K^i(\varphi(t,x), \varphi(t,y))\delta\varphi(t,y) \right)_x e_i.$$

Letting

$$J_{\varphi,\rho}w(y) = \sum_{i=1}^{d} \left(\rho \,|\, D_1 K^i(\varphi(x), \varphi(y))w(x) \right)_x e_i \qquad (11.37)$$

$$+ \sum_{i=1}^{d} \left(\rho \,|\, D_2 K^i(\varphi(x), \varphi(y))w(y) \right)_x e_i,$$

this takes the form

$$\partial_t \delta\varphi(t) = H_{\varphi(t)}\delta\mu(t) + J_{\varphi(t),\mu(t)}\delta\varphi(t), \qquad (11.38)$$

the initial condition being $\delta\varphi(0) = 0$. The following lemma provides the necessary tool to infer the dependency of $\delta\varphi(t)$ on $\delta\mu$ in (11.36).

Lemma 11.17. *Let B be a Banach space and consider a time-dependent linear operator $J(t) : B \to B$, $t \in [0,1]$. We assume that $\sup_t \|J(t)\|$ is bounded, so that the linear equation*

$$\partial_t x = J(t)x \qquad (11.39)$$

has a unique solution over $[0,1]$ for any initial condition. Denote by $M_{st}h$ the solution at time t of this equation with condition $x(s) = h$ at time s. Then, for any t and any $\eta \in B^*$, $M_{0t}^*\eta$ is given by the solution at time 0 of the equation

$$\partial_s \xi(s) = -J(s)^*\xi(s) \qquad (11.40)$$

with condition $\xi(t) = \eta$.

Thus $M_{0t}^*\eta$ is computed by solving (11.40) backwards in time from times t to 0.

Proof. We first remark that, from the uniqueness of the solution, we have

$$M_{st} = M_{0t}M_{0s}^{-1},$$

which implies

$$\partial_s M_{st} = -M_{0t}M_{0s}^{-1}(\partial_s M_{0s})M_{0s}^{-1} = -M_{st}J(s).$$

Therefore, for $\eta \in B^*$,

$$\partial_s M_{st}^*\eta = -J(s)^* M_{st}^*\eta$$

so that $\partial_s \xi = M_{st}^*\eta$ satisfies (11.40). $\square$

This lemma can be applied to (11.38) as follows. First, note that the general solution of (11.38) is given by

$$\delta\varphi(t) = \int_0^t M_{st}H_{\varphi(s)}\delta\mu(s)ds,$$

where $M_{st}h(s)$ is the solution of $\partial_t h(t) = J_{\varphi(t),\mu(t)}h(t)$. This is obtained by variation of the constant on the homogeneous equation (11.39), and can be directly checked to satisfy (11.38). We can therefore rewrite (11.36) in the form

$$\delta E = 2\int_0^1 \left(\delta\mu(t) \,|\, H_{\varphi(t)}\mu(t)\right)dt + \int_0^1 \int_0^t \left(\beta(t) \,|\, M_{st}H_{\varphi(s)}\delta\mu(s)\right)dsdt$$

$$+ \int_0^1 \left(\frac{\delta U}{\delta\varphi}(\varphi(1)) \,|\, M_{t1}H_{\varphi(t)}\delta\mu(t)\right)dt$$

$$= 2\int_0^1 \left(\delta\mu(t) \,|\, H_{\varphi(t)}\mu(t)\right)dt + \int_0^1 \int_s^1 \left(\delta\mu(s) \,|\, H_{\varphi(s)}M_{st}^*\beta(t)\right)dsdt$$

$$+ \int_0^1 \left(\delta\mu(t) \,|\, H_{\varphi(t)}M_{t1}^*\frac{\delta U}{\delta\varphi}(\varphi(1))\right)dt$$

$$= \int_0^1 \left(\delta\mu(t) \,|\, z(t)\right)dt$$

with

$$z(t) = H_{\varphi(t)}\left(2\mu(t) + \int_t^1 M_{ts}^*\beta(s)ds + M_{t1}^*\frac{\delta U}{\delta\varphi}(\varphi(1))\right).$$

Let $\zeta(t) = \int_t^1 M_{ts}^*\beta(s)ds + M_{t1}^*\frac{\delta U}{\delta\varphi}(\varphi(1))$. Using Lemma 11.17, we have

$$\frac{d\zeta(t)}{dt} = -J_{\varphi(t),\mu(t)}^*\zeta(t) - \beta(t) \text{ with } \zeta(1) = \frac{\delta U}{\delta\varphi}(\varphi(1)).$$

We now can summarize the computation of $\delta E/\delta\mu$, which requires the following steps:

Algorithm 4 1. *Compute, for the specific problem at hand, the expressions of $\beta(t)$, $J_{\varphi,\mu}$ and H_φ.*
2. *Compute the conjugate map $J_{\varphi,\mu}^*$.*
3. *Solve the following ordinary differential equation backwards in time:*

$$\frac{d\zeta(t)}{dt} = -J_{\varphi(t),\mu(t)}^*\zeta(t) - \beta(t)$$

with initial condition $\zeta(1) = (\delta U/\delta\varphi)(\varphi(1))$.
4. *Let*

$$z(t) = 2\mu(t) + \zeta(t)$$

so that $\partial_\varepsilon E = \int_0^1 \left(\delta\mu(t) \,|\, H_{\varphi(t)}z(t)\right)dt$.

An interesting feature of this approach is that $z(t)$ is the gradient of E for the following metric, over time-dependent linear forms, indexed by time-dependent diffeomorphisms:

$$\langle \mu \,,\, \tilde{\mu} \rangle_{\varphi} = \int_0^1 \big(\mu(t) \,|\, H_{\varphi(t)} \tilde{\mu}(t) \big) dt.$$

This is a very reasonable choice of a Riemannian metric over linear forms, which corresponds to displacing the kernel along the time-dependent diffeomorphism, since $\mu(t)$ applies to vector fields composed with $\varphi(t)$.

Application to Point Matching

The minimization in terms of $t \mapsto \mu(t)$ directly captures the finite-dimensional nature of the point-matching problem. We repeat some of the steps of the general computation in this less abstract situation. Since we found

$$\rho(1) = \sum_{k=1}^{N} \partial_k F(\varphi_{01}^v(x_k)) \otimes \delta_{\varphi_{01}^v(x_k)},$$

we know that μ also takes this form, so that

$$\mu(t) = \sum_{k=1}^{N} \xi_k(t) \otimes \delta_{x_k}$$

at all times, for some coefficients $\xi_1, \ldots, \xi_N$. In this case, the evaluation of $E(\mu)$ expressed in terms of ξ gives

$$E(\xi) = \sum_{k,l=1}^{d} \int_0^1 \xi_k(t)^T K(\varphi(t, x_k), \varphi(t, x_l)) \xi_l(t) dt + U(\varphi(1))$$

so that, letting $y_k(t) = \varphi(t, x_k)$ and making a variation $\xi \to \xi + \varepsilon \delta \xi$, we get

$$\partial_\varepsilon E = 2 \sum_{k,l=1}^{N} \int_0^1 \delta \xi_k(t)^T K(y_k(t), y_l(t)) \xi_l(t) dt$$

$$+ \sum_{k,l=1}^{N} \int_0^1 \xi_k(t)^T (D_1 K(y_k(t), y_l(t)) \delta y_k(t)) \xi_l(t) dt$$

$$+ \sum_{k,l=1}^{N} \int_0^1 \xi_k(t)^T (D_2 K(y_k(t), y_l(t)) \delta y_l(t)) \xi_l(t) dt$$

$$+ \nabla U(z)^T \delta y(1).$$

This takes the form

$$\partial_\varepsilon E = 2 \int_0^1 \xi(t)^T H_{y(t)} \delta\xi(t) dt + \int_0^1 \beta(t)^T \delta y(t) dt + \left(\frac{\delta U}{\delta y} \mid \delta y(1) \right)$$

with

$$H_y = (K(y_k, y_l), k, l = 1, \ldots, N) \text{ and}$$

$$\beta_k(t) = \sum_{l=1}^N \nabla_1 (\xi_k(t)^T K(y_k(t), y_l(t)) \xi_l(t))$$

$$+ \sum_{l=1}^N \nabla_2 (\xi_l(t)^T K(y_l(t), y_k(t)) \xi_k(t)).$$

Let's write

$$\partial_t \delta y_k(t) = \sum_{l=1}^N K(y_k(t), y_l(t)) \delta\xi_l(t)$$

$$+ \sum_{l=1}^N D_1(K(y_k(t), y_l(t)) \xi_l(t)) \delta y_k(t)$$

$$+ \sum_{l=1}^N D_2(K(y_k(t), y_l(t)) \xi_l(t)) \delta y_l(t).$$

This takes the form (cf. (11.38))

$$\partial_t y(t) = H_{y(t)} \delta\xi(t) + J_{y(t), \xi(t)} \delta y(t),$$

where $J_{y, \xi}$ is the block matrix with (k, l) block given by

$$\sum_{l'=1}^N \left(D_1(K(y_k, y_{l'}) \xi_{l'}) \delta_{kl} + D_2(K(y_k, y_l) \xi_l) \right).$$

The general method can now be applied to this problem, namely: Solve backwards in time

$$\partial_t \zeta = -J_{y(t), \xi(t)}^T \zeta(t) - \beta(t)$$

with initial condition $\zeta(1) = (\delta U / \delta y)(y(1))$. Then let

$$z(t) = 2\xi(t) + \zeta(t)$$

so that $\partial_\varepsilon E = \int_0^1 (\delta\xi(t))^T H(t) z(t) dt$. This algorithm has been introduced in [207] for surface matching.

11.6.3 Gradient in the Initial Momentum

We now use the fact the equation (11.15) implies that the optimal $v(t)$ is uniquely constrained by its value at $t = 0$ for formulating the variations of the objective function in terms of these initial conditions. We therefore optimize with respect to v_0, or equivalently with respect to $\mu_0 = \rho_0$. This requires finding ρ_0 such that

$$\int_0^1 \|v(t)\|^2 dt + \lambda U(\varphi(1))$$

is minimal under the constraints $\partial \varphi(t) = v(t) \circ \varphi(t)$, with

$$v(t) = \sum_{i=1}^d \left(\rho_0 \mid D\varphi(t) K^{(i)}(x, y)\right)_x e_i.$$

Proposition 11.14 helps simplify this expression, since it implies that $\int_0^1 \|v(t)\|^2 dt = (\rho_0 \mid K\rho_0)$ and the minimization problem therefore is to find ρ_0 such that

$$E(\rho_0) = (\rho_0 \mid K\rho_0) + \lambda U(\varphi(1))$$

is minimal, with

$$\partial_t \varphi(t, y) = \sum_{i=1}^d \left(\rho_0 \mid D\varphi(t, x)^{-1} K^{(i)}(\varphi(t, x), \varphi(t, y))\right)_x e_i.$$

Here also, the relation between ρ_0 and $\varphi(1)$ comes as the solution of a differential equation in which ρ_0 is a parameter. The resulting relation between a small variation $\rho_0 \to \rho_0 + \varepsilon \delta \rho_0$ and $\varphi \to \varphi + \varepsilon \delta \varphi$ can be derived as follows (at first order in ε). We still have, like in Section 11.6.2,

$$\partial_t \delta \varphi(t) = J_{\varphi(t)\mu(t)} \delta \varphi(t) + H_{\varphi(t)} \delta \mu(t), \tag{11.41}$$

with H_φ and $J_{\varphi,\mu}$ respectively given by (11.34) and (11.37). Here, we also need to compute the variation of μ, which is, using (11.23),

$$\left(\partial_t \delta \mu(t) \mid w\right) = -\sum_{i=1}^d \left(\delta \mu(t) \mid (\mu(t) \mid D_2 K^i(\varphi(t, x), \varphi(t, y)) w(y))_x e_i\right)_y \tag{11.42}$$

$$-\sum_{i=1}^d \left(\mu(t) \mid (\delta \mu(t) \mid D_2 K^i(\varphi(t, x), \varphi(t, y)) w(y))_x e_i\right)_y$$

$$-\sum_{i=1}^d \left(\mu(t) \mid (\mu(t) \mid D_1(D_2 K^i(\varphi(t, x), \varphi(t, y)) w(y)) \delta \varphi(t, x))_x e_i\right)_y$$

$$-\sum_{i=1}^d \left(\mu(t) \mid (\mu(t) \mid D_2(D_2 K^i(\varphi(t, x), \varphi(t, y)) w(y)) \delta \varphi(t, y))_x e_i\right)_y.$$

Thus, $\partial_t \delta \mu(t)$ takes the form

$$\partial_t \delta \mu(t) = \tilde{H}_{\varphi(t), \mu(t)} \delta \varphi(t) + \tilde{J}_{\varphi(t), \mu(t)} \delta \mu(t) \tag{11.43}$$

with

$$\big(\tilde{H}_{\varphi,\mu}\tilde{w}\,|\,w\big) = -\sum_{i=1}^{d}\big(\mu\,|\,(\mu\,|\,D_1(D_2K^i(\varphi(x),\varphi(y))w(y))\tilde{w}(x))_x e_i\big)_y$$
$$-\sum_{i=1}^{d}\big(\mu(t)\,|\,(\mu(t)\,|\,D_2(D_2K^i(\varphi(x),\varphi(y))w(y))\tilde{w}(y))_x e_i\big)_y$$

and

$$\big(\tilde{J}_{\varphi\mu}\nu\,|\,w\big) = -\sum_{i=1}^{d}\big(\nu\,|\,(\mu\,|\,D_2K^i(\varphi(x),\varphi(y))w(y))_x e_i\big)_y$$
$$-\sum_{i=1}^{d}\big(\mu\,|\,(\nu\,|\,D_2K^i(\varphi(x),\varphi(y))w(y))_x e_i\big)_y.$$

Equations (11.41) and (11.43) form a differential system of equations that provide the evolution of $\delta\varphi(t)$ and $\delta\mu(t)$, the initial condition being $\delta\varphi(0) = 0$ and $\delta\mu(0) = \delta\rho_0$.

The variation of the energy is (using the same notation as in the previous section)

$$\partial_\varepsilon E = 2\big(\delta\rho_0\,|\,K\rho_0\big) + \Big(\frac{\delta U}{\delta\varphi}(\varphi(1))\,|\,\delta\varphi(1)\Big)$$

and we need to compute the dual of the transformation $(\delta\varphi(1)/\delta\rho_0)$ to express $\partial_\varepsilon E$ in the form

$$\partial_\varepsilon E = \Big(\delta\rho_0\,|\,2K\rho_0 + \frac{\delta\varphi(1)}{\delta\rho_0}^* \frac{\delta U}{\delta\varphi}\Big).$$

Since $\delta\varphi$ evolves jointly with $\delta\mu$, we need in fact to compute the dual of the joint variation $\delta(\varphi(t),\mu(t))/\delta\rho_0$. This is again an application of Lemma 11.17 since the joint evolution takes the form

$$\partial_t\begin{pmatrix}\delta\varphi(t)\\\delta\mu(t)\end{pmatrix} = \mathcal{J}_{\varphi(t),\mu(t)}\begin{pmatrix}\delta\varphi(t)\\\delta\mu(t)\end{pmatrix}$$

with

$$\mathcal{J}_{\varphi,\mu} = \begin{pmatrix}J_{\varphi,\mu} & H_\mu\\\tilde{H}_{\varphi,\mu} & \tilde{J}_{\varphi,\mu}\end{pmatrix}.$$

We can now compute the gradient of E at a given ρ_0, using the metric that derives from V: $\langle\rho,\,\rho'\rangle_{V^*} = (\rho\,|\,K\rho')$.

Algorithm 5 *1. Compute the variation $(\delta U/\delta\varphi)(\varphi(1))$.*
2. Solve the equation

$$\partial_t\begin{pmatrix}\zeta_\varphi\\\zeta_\mu\end{pmatrix} = -\mathcal{J}^*_{\varphi(t),\mu(t)}\begin{pmatrix}\zeta_\varphi\\\zeta_\mu\end{pmatrix}$$

backwards in time with $\zeta_\varphi(1) = (\delta U/\delta\varphi)(\varphi(1))$ and $\zeta_\mu(1) = 0$.
3. The $\nabla E(\rho_0) = 2\rho_0 + \lambda K^{-1}\zeta_\mu(0))$.

Application to Point Matching

We now apply this approach to point-matching problems. Since ρ_0 takes the form

$$\rho_0 = \sum_{k=1}^{N} \xi_k(0) \otimes \delta_{x_k(0)}$$

we are in the situation studied for (11.24). Since the equations get somewhat messy, we will assume that we work with a scalar radial kernel $K(x, y) = \gamma(|x - y|^2)\mathrm{Id}_d$, which implies (with the same notation as in Section 11.6.2)

$$\partial_t y_k(t) = \sum_{l=1}^{N} \gamma(|y_k(t) - y_l(t)|^2)\xi_l(t)$$

and

$$\partial_t \xi_k(t) = -2\sum_{l=1}^{N} \dot{\gamma}(|y_l(t) - y_k(t)|^2)\xi_k(t)^T \xi_l(t)(y_k(t) - y_l(t)).$$

Let's consider the effect of a variation $\xi_k(0) \to \xi_k(0) + \varepsilon\delta\xi_k(0)$. Denote $\gamma_{kl}(t) = \gamma(|y_k(t) - y_l(t)|^2)$ and similarly for the derivatives. All quantities in the following equations depend on time, the t variable being dropped to simplify the expressions. We have

$$\partial_t \delta y_k = \sum_{l=1}^{N} \gamma_{kl}\delta\xi_l + 2\sum_{l=1}^{N} \dot{\gamma}_{kl}(y_k - y_l)^T(\delta y_k - \delta y_l)\xi_l$$

and

$$\partial_t \delta\xi_k = -2\sum_{l=1}^{N} \dot{\gamma}_{kl}(\delta\xi_k^T\xi_l + \delta\xi_l^T\xi_k)(y_k - y_l)$$

$$- 2\sum_{l=1}^{N} \dot{\gamma}_{kl}\xi_k^T\xi_l(\delta y_k - \delta y_l)$$

$$- 4\sum_{l=1}^{N} \ddot{\gamma}_{kl}\xi_k^T\xi_l(y_k - y_l)^T(\delta y_k - \delta y_l)(y_k - y_l).$$

By reordering the terms, we can rewrite this system in the form

$$\partial_t \begin{pmatrix} \delta x \\ \delta\xi \end{pmatrix} = \mathcal{J}_{y\xi} \begin{pmatrix} \delta x \\ \delta\xi \end{pmatrix}$$

where $t \mapsto \mathcal{J}_{y\xi}(t)$ is a $2dN$ by $2dN$ matrix that depends on $y(t)$ and $\xi(t)$. Given this, the gradient of the objective function at a given ξ_0 can be computed as follows.

1. Compute $\delta U/\delta\varphi$. For example, if $U(\varphi) = \sum_{i=1}^{N}|\varphi(x_i)-y_i|^2$, take $\delta U/\delta\varphi = 2\sum_{i=1}^{N}(\varphi(x_i) - y_i)$.
2. Solve backwards in time the equation

$$\frac{d}{dt}\begin{pmatrix}\zeta_y\\\zeta_\xi\end{pmatrix} = -\mathcal{J}_{y\xi}^T\begin{pmatrix}\zeta_y\\\zeta_\xi\end{pmatrix},$$

with $\zeta_y(1) = \delta U/\delta\varphi(\varphi(1))$ and $\zeta_\xi(1) = 0$.
3. Then $\nabla E(\xi_0) = 2\xi_0 + K^{-1}\zeta_{\xi_0}(0)$.

Finally, we provide the explicit expression of $\begin{pmatrix}\delta^*\zeta_y\\\delta^*\zeta_\xi\end{pmatrix} = \mathcal{J}_{yx}^T\begin{pmatrix}\zeta_y\\\zeta_\xi\end{pmatrix}$, skipping the straightforward computation. We have

$$(\delta^*\zeta_y)_k = 2\sum_{l=1}^{N}\dot{\gamma}_{kl}(\xi_k^T(\zeta_y)_l + \xi_l^T(\zeta_y)_k)(y_k - y_l)$$

$$- 2\sum_{l=1}^{N}\dot{\gamma}_{kl}(\xi_k^T\xi_l)((\zeta_\xi)_k - (\zeta_\xi)_l)$$

$$- 4\sum_{l=1}^{N}\ddot{\gamma}_{kl}(\xi_k^T\xi_l)(y_k - y_l)^T((\zeta_\xi)_k - (\zeta_\xi)_l)(y_k - y_l)$$

$$(\delta^*\zeta_\xi)_k = \sum_{l=1}^{N}\gamma_{kl}(\zeta_y)_l - 2\sum_{l=1}^{N}\dot{\gamma}_{kl}(y_k - y_l)^T((\zeta_\xi)_k - (\zeta_\xi)_l)\xi_l$$

This algorithm is illustrated in Figure 11.1. In the same figure, we also provide (for comparison purposes) the results provided by spline interpolation, which computes $\varphi(x) = x + v(x)$, where v is computed (using Theorem 9.7) in order to minimize

$$\|v\|_V^2 + C\sum_{i=1}^{N}|v(x_i) - (y_i - x_i)|^2.$$

Although this is a widely spread registration method [29], Figure 11.1 shows that it is far from being diffeomorphic for large deformations.

11.6.4 Shooting

The Euler–Lagrange equations for our problem are $\rho(1) = -\frac{1}{2}\bar{\partial}U(\varphi(1))$ with $(\rho(1)\,|\,w\circ\varphi(1)) = (\mu(1)\,|\,w)$ and $\mu(t)$ given by (11.23). Note that

$$(\bar{\partial}U(\varphi)\,|\,w\circ\varphi^{-1}) = (\frac{\delta U}{\delta\varphi}\,|\,w),$$

so that $m(1) = -\frac{1}{2}\bar{\partial}U(\varphi(1))$ is equivalent to

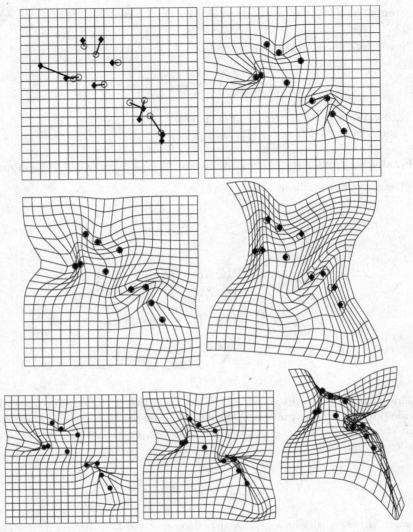

Fig. 11.1. Metric point matching. The first two rows provide results obtained with gradient descent in the initial momentum for point matching, with the same input as in Figure 10.1, using Gaussian kernels $K(x, y) = \exp(-|x - y|^2 / 2\sigma^2)$ with $\sigma = 1, 2, 4$ in grid units. The impact of the diffeomorphic regularization on the quality of the result is particularly obvious in the last experiment. The last row provides the output of Gaussian spline registration with the same kernels, exhibiting singularities and ambiguities in the registration.

$$\mu(1) = -\frac{1}{2}\frac{\delta U}{\delta \varphi}(\varphi(1)). \tag{11.44}$$

Since $\varphi(1)$ and $\mu(1)$ can be considered as functions of $\mu(0) = \rho_0$, one can try to solve (11.44) directly as a function of this initial value using a root-finding algorithm. Newton's method can in principle be used for this purpose: given a current value of ρ_0, find a variation $\rho_0 \to \rho_0 + \delta\rho_0$ such that (11.44) is true at first order, i.e.,

$$\mu(1) + \frac{\delta \mu(1)}{\delta \rho_0}\delta\rho_0 = -\frac{1}{2}\frac{\delta U}{\delta \varphi}(\varphi(1)) - \frac{1}{2}\frac{\delta^2 U}{\delta \varphi^2}(\varphi(1))\frac{\delta \varphi(1)}{\delta \rho_0}\delta\rho_0. \tag{11.45}$$

This linear equation in $\delta\rho_0$ needs to be solved to update the current ρ_0. We have already described the variations $(\delta\mu(1)/\delta\rho_0)\delta\rho_0$ and $(\delta\varphi(1)/\delta\rho_0)\delta\rho_0$ as solutions of the system (11.41) and (11.43). However, inverting this system cannot be done as easily as computing its conjugate. In fact, Newton's method can mainly be applied for systems of small dimension, which is the case, in our context, for some point-matching problems.

Application to Point Matching

Assuming that U is a function $U(y_1, \ldots, y_N)$ where $y_j = \varphi(x_j)$, $j = 1, \ldots, N$, the Euler equation can be expressed as

$$\xi(1) + \nabla U(x(1)) = 0.$$

For a variation $\xi(0) \to \xi(0) + \delta\xi(0)$, we have

$$\begin{pmatrix} \delta y(1) \\ \delta \xi(1) \end{pmatrix} = M(1)\begin{pmatrix} 0 \\ \delta\xi(0) \end{pmatrix}$$

where $M(1)$ is obtained by solving the differential equation (using the matrix $\mathcal{J}$ computed in the previous section) $dM/dt = \mathcal{J}(t)M$. Denoting by $M_{yy}(1)$, $M_{y\xi}(1)$, $M_{\xi y}(1)$ and $M_{\xi\xi}(1)$ the four dN by dN blocks of $M(1)$, the variation of the Euler equation yields, for the current $\xi(1)$ and $y(1)$

$$\xi(1) + M_{\xi\xi}(1)\delta\xi(0) + \nabla U(y(1)) + D^2 U(y(1))M_{y\xi}(1)\delta\xi(0) \simeq 0,$$

which yields the increment

$$\delta\xi(0) = -(M_{\xi\xi}(1) + D^2 U(y(1))M_{y\xi}(1))^{-1}(\xi(1) + \nabla U(y(1))).$$

This algorithm has been used in [208] for labeled point matching. It is a very efficient procedure for small- to average-dimensional problems (a small number of points) but cannot scale easily. Another issue is that root finding algorithms are not guaranteed to converge. Usually, a good initial solution must be found, using, for example, a few preliminary steps of gradient descent.

11.6.5 Gradient in the Deformable Object

Finally, we consider the option of using the deformable object as a control variable, using the fact that, by (11.28), the objective function can be reduced to

$$E(J) = \int_0^1 L(\partial_t J(t), J(t))dt + F(J(1))$$

with $L(\eta, J) = \min_{w:\eta = w \cdot J(t)} \|w\|_V^2$. This formulation is limited, in that $L(\eta, J)$ is not always defined for all (η, J), resulting in constraints in the minimization that are not always easy to handle. Even if well-defined, the computation of L may be numerically demanding. To illustrate this, consider the image-matching case, in which $v \cdot J = -\langle \nabla J, v \rangle$. An obvious constraint is that, in order for

$$\langle \nabla J, w \rangle = -\eta$$

to have at least one solution, the variation η must be supported by the set $\nabla J \neq 0$. To compute this solution when it exists, one can write, for $x \in \Omega$,

$$\langle \nabla J(x), w(x) \rangle = \langle K(\cdot, x)\nabla J(x), w \rangle_V$$

and it is possible to look for a solution in the form

$$w(y) = \int_\Omega \lambda(x) K(y, x) \nabla J(x) dx$$

where $\lambda(x)$ can be interpreted as a continuous family of Lagrange multipliers. This results in a linear equation in λ, namely

$$\int_\Omega \lambda(x) K(y, x) \langle \nabla J(y), \nabla J(x) \rangle dx = -\eta(y)$$

which is numerically challenging.

For point sets, however, the approach is feasible [119] because L can be made explicit. Given a point-set trajectory $x(t) = (x^{(1)}(t), \ldots, x^{(N)}(t))$, let $S(x(t))$ denote the block matrix with (i, j) block given by $K(x^{(i)}(t), x^{(j)}(t))$. The constraints are $x_t = S(x(t))\xi(t)$ so that $\xi(t) = S(x(t))^{-1}\dot{x}_t$ and the minimization reduces to

$$E(x) = \int_0^1 \dot{x}_t(t)^T S(x(t))^{-1} \dot{x}_t(t) dt + U(x(1)).$$

Minimizing this function with respect to x by gradient descent is possible, and has been described in [118, 119] for labeled landmark matching. The basic computation is as follows: if $s_{pq} = \partial_{x_r} s_{pq}$, we can write (using the fact that $\partial_{x_r}(S^{-1}) = -S^{-1}(\partial_{x_r} S)S^{-1}$)

$$\frac{d}{d\varepsilon}E(x + \varepsilon h)_{|\varepsilon=0} = 2 \int_0^1 \dot{x}_t(t)^T S(x(t))^{-1} \dot{h}_t(t) dt$$

$$- \int_0^1 \sum_{p,q,r} \xi^{(p)}(t)\xi^{q)}(t) s_{pq,r}(x(t))h^{(r)}(t) dt + \nabla U(x(1))^T h(1).$$

After an integration by parts in the first integral, we obtain a differential given by

$$-2\partial_t\left(S(x(t))^{-1}x_t\right) - z(t) + \left(S(x(1))^{-1}\dot{x}_t(1) + \nabla U(x(1))\right)\delta_1(t)$$

where $z_r(t) = \sum_{p,q} \xi_p(t)\xi_q(t)s_{pq,r}(x(t))$ and δ_1 is the Dirac measure at $t = 1$ (the differential is singular).

This singular part can be dealt with by computing the gradient in a Hilbert space in which the evaluation function $x(.) \mapsto x(1)$ is continuous. This method has been suggested, in particular, in [98, 120]. Let H be the space of all landmark trajectories $x : t \mapsto x(t) = (x^{(1)}(t), \ldots, x^{(N)}(t))$, with fixed starting point $x(0)$, free end-point $x(1)$ and square integrable time derivative. This is a space of the form $x(0) + H$ where H is the Hilbert space of time-dependent functions $t \mapsto h(t)$, considered as column vectors of size Nk, with $h(0) = 0$ and

$$\langle h, \tilde{h}\rangle_H = \int_0^1 h_t^T \tilde{h}_t dt + h(1)^T \tilde{h}(1).$$

To compute the gradient for this inner product, we need to express $\partial_\varepsilon E(x + \varepsilon h)|_{\varepsilon=0}$ in the form $\langle \nabla^H E(x), h\rangle_H$. We will make the assumption that

$$\int_0^1 \left|S(x(t))^{-1}\dot{x}_t(t)\right|^2 dt < \infty,$$

which implies that

$$\int_0^1 \dot{x}_t(t)^T S(x(t))^{-1}h_t(t)dt \leq \sqrt{\int_0^1 \left|S(x(t))^{-1}\dot{x}_t(t)\right|^2 dt \int_0^1 \left|h_t(t)\right|^2 dt}$$

is continuous in h. Similarly, the linear form $\xi \mapsto \nabla U(x(1))^T h(1)$ is continuous since

$$\nabla U(x(1))^T h(1) \leq |\nabla U(x(1))|\,|h(1)|.$$

Finally, $h \mapsto \int_0^1 z(t)^T h_t dt$ is continuous provided that we assume that

$$\eta(t) = \int_0^t z(s)dt$$

is square integrable over $[0, 1]$, since this yields

$$\int_0^1 z(t)^T h(t)dt = \eta(1)h(1) - \int_0^1 \eta(t)\dot{h}_t(t)dt,$$

which is continuous in h for the H norm.

Thus, under these assumptions,

$$h \mapsto \frac{d}{d\varepsilon}E(x + \varepsilon h)|_{\varepsilon=0}$$

is continuous over H, and the Riesz representation theorem implies that $\nabla^H E(x)$ exists as an element of H. We now proceed to its computation. Letting

$$\mu(t) = 2 \int_0^t S(x(s))^{-1} \dot{h}_t(s) ds$$

and $a = \nabla U(x(1))$, the problem is to find $\zeta \in H$ such that, for all $h \in H$,

$$\langle \zeta, h \rangle_H = \int_0^1 \dot{\mu}_t^T h_t dt + \int_0^1 z(t)^T h(t) dt + a^T \xi(1).$$

This expression can also be written

$$\int_0^1 \left(\dot{\zeta}_t + \zeta(1) \right)^T h_t dt = \int_0^1 (\dot{\mu}_t + \eta(1) - \eta(t) + a)^T h_t dt.$$

This suggests selecting ζ such that $\zeta(0) = 0$ and

$$\dot{\zeta}_t + \zeta(1) = \dot{\mu}_t + \eta(1) - \eta(t) + a,$$

which implies

$$\zeta(t) + t\zeta(1) = \mu(t) - \int_0^t \eta(s)ds + t(\eta(1) + a).$$

At $t = 1$, this yields

$$2\zeta(1) = \mu(1) - \int_0^1 \eta(s)ds + \eta(1) + a$$

and we finally obtain

$$\zeta(t) = \mu(t) - \int_0^t \eta(s)ds + \frac{t}{2} \left(\int_0^1 \eta(s)ds - \mu(1) + \eta(1) + a \right).$$

We summarize this in an algorithm, in which τ is again the computation time.

Algorithm 6 (Gradient descent algorithm for landmark matching)
Start with initial landmark trajectories $x(t, \tau) = (x^{(1)}(t, \tau), \ldots, x^{(N)}(t, \tau))$.
Solve

$$\partial_\tau x(t, \tau) = -\gamma \Big(\mu(t, \tau) - \int_0^t \eta(s, \tau) ds$$

$$+ \frac{t}{2} \Big(\int_0^1 \eta(s, \tau) ds - \mu(1, \tau) + \eta(1, \tau) + a(\tau) \Big) \Big)$$

with $a(\tau) = \nabla U(x(1, \tau))$, $\mu(t, \tau) = 2 \int_0^t \xi(s, \tau) ds$, $\eta(t, \tau) = \int_0^t z(s, \tau) dt$ and

$$\xi(t, \tau) = S(x(t, \tau))^{-1} \dot{x}_t(t, \tau)$$

$$z^{(q)}(t, \tau) = \sum_{p,r} \xi^{(p)}(t, \tau) \xi^{(r)}(t, \tau) s_{pq,r}(x(t, \tau)).$$

11.6.6 Image Matching

We now take an infinite-dimensional example to illustrate some of the previously discussed methods and focus on the image-matching problem. We therefore consider

$$U(\varphi) = \int_\Omega (I \circ \varphi^{-1} - \tilde{I})^2 dx,$$

where $I, \tilde{I}$ are functions $\Omega \to \mathbb{R}$, I being differentiable. The Eulerian differential of U is given by (10.19):

$$\bar\partial U(\varphi) = -2(I \circ \varphi^{-1} - I')\nabla(I \circ \varphi^{-1}) \otimes dx.$$

So, according to (11.16),

$$\nabla U(v)(t,y) = \sum_{i=1}^d \left(\bar\partial U(\varphi_{01}^v) \mid D\varphi_{t1}^v(\varphi_{1t}^v(.))K(y,\varphi_{1t}^v(.))e_i \right) e_i$$

$$= -2\sum_{i=1}^d e_i \int_\Omega (I \circ \varphi_{10}^v(x) - \tilde{I}(x))$$
$$\nabla(I \circ \varphi_{10}^v)(x)^T D\varphi_{t1}^v(\varphi_{1t}^v(x))K(y,\varphi_{1t}^v(x))e_i dx$$

$$= -2\sum_{i=1}^d e_i \int_\Omega (I \circ \varphi_{10}^v(x) - \tilde{I}(x))$$
$$\nabla I \circ \varphi_{10}^v(x)^T D\varphi_{10}^v(x)D\varphi_{t1}^v(\varphi_{1t}^v(x))K(y,\varphi_{1t}^v(x))e_i dx$$

$$= -2\sum_{i=1}^d e_i \int_\Omega (I \circ \varphi_{10}^v(x) - \tilde{I}(x))$$
$$\nabla I \circ \varphi_{10}^v(x)^T D\varphi_{t0}^v(x)(\varphi_{1t}^v(x))K(y,\varphi_{1t}^v(x))e_i dx$$

$$= -2\sum_{i=1}^d e_i \int_\Omega (I \circ \varphi_{10}^v(x) - \tilde{I}(x))\nabla(I \circ \varphi_{t0}^v)(\varphi_{1t}^v(x))^T K(y,\varphi_{1t}^v(x))e_i dx$$

$$= -2\sum_{i=1}^d e_i \int_\Omega (I \circ \varphi_{10}^v(x) - \tilde{I}(x))e_i^T K(\varphi_{1t}^v(x),y)\nabla(I \circ \varphi_{t0}^v)(\varphi_{1t}^v(x)) dx$$

$$= -2\int_\Omega (I \circ \varphi_{10}^v(x) - \tilde{I}(x))K(\varphi_{1t}^v(x),y)\nabla(I \circ \varphi_{t0}^v)(\varphi_{1t}^v(x)) dx.$$

Note that in this formula, the transpose of the kernel appears $(K(x,y) = K(y,x)^T)$, the two being the same in the case of a scalar kernel.

This provides the expression of the V-gradient of $\tilde{E}$ for image matching, namely

$$(\nabla^V E(v))(t,y) = 2v(t,y)$$
$$- 2\int_\Omega (I \circ \varphi_{10}^v(x) - \tilde{I}(x))K(\varphi_{1t}^v(x),y)\nabla(I \circ \varphi_{t0}^v)(\varphi_{1t}^v(x)) dx. \quad (11.46)$$

Using a change of variable in the integral, the gradient may also be written

$$(\nabla^V E)(t, y) = 2v(t, y)$$
$$- 2 \int_\Omega (I \circ \varphi_{t0}^v(x) - \tilde{I} \circ \varphi_{t1}^v(x)) K(x, y) \nabla (I \circ \varphi_{t0}^v)(x) \det(D\varphi_{t1}^v(x)) dx,$$

$$(11.47)$$

the associated gradient descent algorithm having been proposed in [21].

Let's now consider an optimization with respect to the initial ρ_0. First note that, by (10.18), $\rho(1) = 2(I \circ \varphi(1)^{-1} - I') \nabla (I \circ \varphi(1)^{-1}) \otimes dx$ so that

$$(\rho(1) \,|\, w) = 2 \int_\Omega (I \circ \varphi(1)^{-1} - I') \nabla (I \circ \varphi(1)^{-1]})^T w(x) dx.$$

We have, letting $\varphi = \varphi(1)$ and $\psi = \varphi(1)^{-1}$,

$$(\rho_0 \,|\, w) = (\rho(1) \,|\, D\psi^{-1} w \circ \psi)$$
$$= 2 \int_\Omega (I \circ \psi - I') \nabla (I \circ \psi)^T D\psi^{-1} w \circ \psi dx$$
$$= 2 \int_\Omega (I \circ \psi - I') (\nabla I \circ \psi)^T w \circ \psi dx$$
$$= 2 \int_\Omega (I - I' \circ \varphi)(\nabla I)^T w \det(D\varphi) dx,$$

which takes the form $\rho_0 = Z(0) \otimes dx$ for a vector-valued function $Z(0)$. We can therefore use the evolution equations (11.25) and (11.26) with $N = 1$ and $\omega_1 = dx$, which yield (assuming a scalar kernel $K(x, y) = \Gamma(x, y)I$ to simplify the expressions)

$$\partial_t \varphi(t, y) = \int_\Omega \Gamma(\varphi(t, y), \varphi(t, x)) Z(t, x) dx$$

and

$$\partial_t Z(t, y) = - \int_\Omega Z(t, y)^T Z(t, x) \nabla_1 \Gamma(\varphi(t, y), \varphi(t, x)) dx.$$

Still to avoid over-complicated expressions, let's assume that Γ is radial and takes the form $\Gamma(x, y) = \gamma(|x - y|^2)$. This implies $\nabla_1 \Gamma(x, y) = 2\dot{\gamma}(|x - y|^2)(x - y)$ and

$$D_{11} \Gamma(x, y) = 4\ddot{\gamma}(|x - y|^2)(x - y)(x - y)^T + 2\dot{\gamma}(|x - y|^2)\mathrm{Id}_d = -D_{21} \Gamma(x, y).$$

A variation $\rho_0 \to \rho_0 + \delta\rho_0$ yields variations $\varphi \to \varphi + \delta\varphi$ and $Z \to Z + \delta Z$ that satisfy the following equations, starting with $\delta\varphi$. We do not write the dependency on t in the following computation.

$$\partial_t \delta\varphi(y) = \int_\Omega \gamma(|\varphi(y) - \varphi(x)|^2) \delta Z(x) dx$$
$$+ 2 \int_\Omega \dot{\gamma}(|\varphi(y) - \varphi(x)|^2)(\varphi(y) - \varphi(x))^T (\delta\varphi(y) - \delta\varphi(x)) Z(x) dx$$

For δZ, we have

$$\partial_t \delta Z(y) = -2 \int_\Omega (\delta Z(y)^T Z(x) + \delta Z(x)^T Z(y))$$
$$\dot\gamma(|\varphi(y) - \varphi(x)|^2)(\varphi(y) - \varphi(x))dx$$
$$- 2 \int_\Omega Z(y)^T Z(x)\dot\gamma(|\varphi(y) - \varphi(x)|^2)(\delta\varphi(y) - \delta\varphi(x))dx$$
$$- 4 \int_\Omega Z(y)^T Z(x)\ddot\gamma(|\varphi(y) - \varphi(x)|^2)$$
$$(\varphi(y) - \varphi(x))(\varphi(y) - \varphi(x))^T(\delta\varphi(y) - \delta\varphi(x))dx$$

This provides an expression of the form $(d/dt)(\delta\varphi, \delta Z) = J.(\delta\varphi, \delta Z)$. We now compute its transpose, i.e., letting J_φ and J_Z be the two components of J, we solve

$$\int_\Omega (\eta_\varphi^T \delta\varphi + \eta_Z^T \delta Z)dx = \int_\Omega (w_\varphi^T J_\varphi(\delta\varphi, \delta Z) + w_Z^T J_Z(\delta\varphi, \delta Z))dy$$

to obtain $\eta_\varphi = \eta\varphi(w)$ and $\eta_Z = \eta_Z(w)$ as a function of $w = (w_\varphi, w_Z)$. Let's detail the computation. To shorten the formulas we will denote $\gamma_{xy}^\varphi = \gamma(|\varphi(x) - \varphi(y)|^2)$ and the same for the derivatives, yielding notation $\dot\gamma_{xy}^\varphi$ and $\ddot\gamma_{xy}^\varphi$, remembering that they all are symmetric functions of x and y. First, we have

$$\int_\Omega w_\varphi^T J_\varphi(\delta\varphi, \delta Z)dy$$
$$= \int_\Omega w_\varphi(y)^T \int_\Omega \gamma_{xy}^\varphi \delta Z(x)dxdy$$
$$+ 2 \int_\Omega w_\varphi(y)^T \int_\Omega \dot\gamma_{xy}^\varphi(\varphi(y) - \varphi(x))^T(\delta\varphi(y) - \delta\varphi(x))Z(x)dxdy$$
$$= \int_\Omega \delta Z(x)^T \int_\Omega \gamma_{xy}^\varphi w_\varphi(y)dydx$$
$$+ 2 \int_\Omega \delta\varphi(x)^T \int_\Omega \dot\gamma_{xy}^\varphi(\varphi(x) - \varphi(y))(w_\varphi(y)^T Z(x) + w_\varphi(x)^T Z(y))dydx.$$

For the second term, we write

$$\int_\Omega w_Z^T J_Z(\delta\varphi, \delta Z)dy$$

$$= -2\int_\Omega w_Z(y)^T \int_\Omega \dot\gamma_{xy}^\varphi(\delta Z(y)^T Z(x) + \delta Z(x)^T Z(y))(\varphi(y) - \varphi(x))dxdy$$

$$\quad -2\int_\Omega w_Z(y)^T \int_\Omega \dot\gamma_{xy}^\varphi Z(y)^T Z(x)(\delta\varphi(y) - \delta\varphi(x))dxdy$$

$$\quad -4\int_\Omega w_Z(y)^T \int_\Omega \ddot\gamma_{xy}^\varphi Z(y)^T Z(x)(\varphi(y) - \varphi(x))(\varphi(y) - \varphi(x))^T(\delta\varphi(y) - \delta\varphi(x))dxdy$$

$$= -2\int_\Omega \delta Z(x)^T \int_\Omega \dot\gamma_{xy}^\varphi Z(y)(w_Z(y) - w_Z(x))^T(\varphi(y) - \varphi(x))dydx$$

$$\quad -2\int_\Omega \delta\varphi(x)^T \int_\Omega \dot\gamma_{xy}^\varphi Z(y)^T Z(x)(w_Z(x) - w_Z(y))dydx$$

$$\quad -4\int_\Omega \delta\varphi(x)^T \int_\Omega \ddot\gamma_{xy}^\varphi Z(y)^T Z(x)(\varphi(y) - \varphi(x))(\varphi(y) - \varphi(x))^T(w_Z(x) - w_Z(y))dydx$$

This yields

$$\eta_\varphi(w) = 2\int_\Omega \dot\gamma_{xy}^\varphi(w_\varphi(y)^T Z(x) + w_\varphi(x)^T Z(y))(\varphi(x) - \varphi(y))dy$$

$$\quad -2\int_\Omega \dot\gamma_{xy}^\varphi Z(y)^T Z(x)(w_Z(x) - w_Z(y))dy$$

$$\quad -4\int_\Omega \ddot\gamma_{xy}^\varphi Z(y)^T Z(x)(\varphi(y) - \varphi(x))(\varphi(y) - \varphi(x))^T(w_Z(x) - w_Z(y))dy$$

and

$$\eta_Z(w) = \int_\Omega \gamma_{xy}^\varphi w_\varphi(y)dy - 2\int_\Omega \dot\gamma_{xy}^\varphi(w_Z(y) - w_Z(x))^T(\varphi(y) - \varphi(x))Z(y)dy.$$

We summarize the computation of the gradient of the image-matching functional with respect to Z_0 such that $\rho_0 = Z_0 \otimes dx$:

Algorithm 7   1. *Compute* $\partial U/\partial\varphi = -2(I - I' \circ \varphi)\det(D\varphi)D\varphi^{-T}\nabla I$.
2. *Solve, backwards in time, until time* $t = 0$ *the equations* $\partial_t w_\varphi = -\eta_\varphi.w$ *and* $\partial_t w_Z = -\eta_Z.w$ *with initial condition* $w_\varphi(1) = \partial U/\partial\varphi$ *and* $w_Z(1) = 0$.
3. *Then* $\nabla E(Z_0) = 2Z_0 + \lambda K^{-1}w_Z(0)$.

The gradient is computed with the metric $\langle Z, Z'\rangle = \int_\Omega Z(y)^T KZ(y)dy$. Results obtained with this algorithm are presented in Figure 11.2.

One can also use the fact that $Z_0 = z_0\nabla I$ for a scalar-valued z_0. Since we have

$$\left(\frac{\delta E}{\delta Z_0} \,\middle|\, \delta Z_0\right) = \int_\Omega (2KZ_0 + \lambda w_Z(0))^T \delta Z_0 dy,$$

we can write, with $\delta z_0 \nabla I = \delta Z_0$:

$$\left(\frac{\delta E}{\delta z_0} \,\middle|\, \delta z_0\right) = \int_\Omega (2K(z_0\nabla I) + \lambda w_Z(0))^T \nabla I \delta z_0 dy,$$

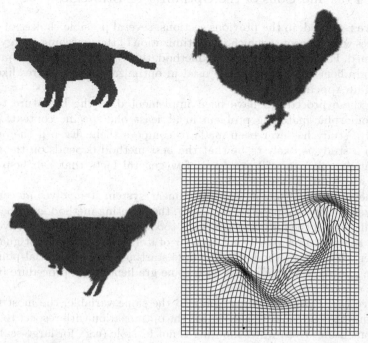

Fig. 11.2. Metric image matching . Output of Algorithm 7 when estimating a deformation of the first image to match the second one (compare to figure 10.2). The third image is the obtained deformation of the first one and the last provides the deformation applied to a grid.

which leads to replacing the last step in Algorithm 7 by

$$\delta z_0 = -\nabla I^T (2K(z_0 \nabla I) + \lambda w_Z(0)),$$

which corresponds to using the L^2 metric in z_0 for gradient descent. However, a more natural metric, in this case, is the one induced by the kernel, i.e.,

$$\langle z, z' \rangle_I = \int_\Omega \int_\Omega K(z \nabla I)(y)(z'(y) \nabla I(y)) dy = \int_\Omega k_I(x,y) z(x) z'(y) dx dy$$

with $k_I(x,y) = \nabla I(x)^T K(x,y) \nabla I(y)$. With this metric, z_0 is updated with

$$\delta z_0 = 2 z_0 + \lambda k_I^{-1} \nabla I^T w_Z(0).$$

Although this metric is more satisfactory, the inversion of k_I might be difficult, numerically.

11.6.7 Pros and Cons of the Optimization Strategies

We have reviewed, in the previous sections, several possible choices of control variables with respect to which the optimization of the matching energy can be performed. For all but the shooting method, this results in specific expressions of the gradient that can then be used in optimization procedures like those defined in Appendix D.

All these procedures have been implemented in the literature to solve a diffeomorphic-matching problem in at least one specific context, but no extensive study has ever been made to compare them. Even if the outcome of such a study is likely to be that the best method depends on the specific application, one can still provide a few general facts that can help a user decide which one to use.

When feasible (that is, when the linear system it involves at each step can be efficiently computed and solved), the shooting method is probably the most efficient. If the initialization is not too far from the solution, convergence can be achieved in a very small number of iterations. One cannot guarantee, however, that the method will converge starting from any initial point, and shooting needs to be combined with some gradient-based procedure in order to find a good starting position.

Since they optimize with respect to the same variable, the most natural procedure to combine with shooting is the optimization with respect to the initial momentum. Even when shooting is not feasible (e.g., for large-scale problems), this specific choice of control variable is important, because it makes sure that the final solution satisfies the EPDiff equation, which guarantees the consistency of the momentum representation, which wil be discussed in Section 12.5.2. The limitation is that, with large and complex deformations, the sensibility of the solution to small changes in the control variable can be large, which may result in an unstable optimization procedure.

The other methods, which optimize with respect to time-dependent quantities, are generally more able to compute very large deformations. Beside the obvious additional burden in computer memory that they require, one must be aware that the discrete solution can sometimes be far from satisfying the EPDiff equation unless the time discretization is fine enough (which may be impossible to achieve within a feasible implementation for large-scale problems). Therefore, these methods do not constitute the best choice if obtaining a reliable momentum representation is important. Among the three time-dependent control variables that we have studied (velocity, momentum and deformable object), one may have a slight preference for the representation using the time-dependent momenta, even if the computation it involves is slightly more complex than the others. There are at least two reasons for this. Firstly, the momenta generally are more parcimonious in the space variables, because they incorporate the normality constraints to transformations that leave the deformable objects invariant. Secondly, because the forward and backward equations solved at each iteration immediately provide a gradient with respect to the correct metric, so that the implementation does not have to include the solution of a possibly large-dimensional linear system which is required by other representations.

11.7 Numerical Aspects

11.7.1 Discretization

The implementation of the diffeomorphic-matching algorithm must combine a number of basic numerical procedures. In this section, we discuss some of these procedures, which are involved in the computation of the gradients defined in the previous sections.

To start with, one must decide how to discretize the problem. There are two aspects in this regard, which are time and space discretization. Let's start with point-set matching problems, since they are already discrete in space. The discussion depends on which kind of control variable is used, i.e., if one optimizes with respect to time-dependent velocities (with projection) or momenta, or with respect to the initial momentum.

In the third case, the computation of the gradient requires solving forward the EPDiff equation for point sets, in order to compute the trajectories, and its conjugate variation to finally obtain the gradient. Both are finite-dimensional ODEs, and a possible solution is to use standard ODE solvers, like Runge–Kutta methods, which are available in many software and library packages (landmark experiments in this book use this approach). One can be a little more accurate by ensuring that the backward conjugate equation indeed provides the gradient of the forward one. For this to be feasible, the method used for the forward equation cannot be too complex. The following proposition provides the correct backward variation for a generic ODE that depends on some parameter ε when the forward one uses either basic Euler iterations, or Euler's midpoint method, with fixed step, the latter being a good compromise between simplicity and accuracy.

Proposition 11.18. *Consider the ordinary differential equation on $\mathbb{R}^Q$*

$$\partial_t y = w(t, y, \varepsilon)$$

where ε is a parameter. Let h be a time discretization step and $t_k = kh$. Define $w_k(y, \varepsilon) = w(t_k, y, \varepsilon)$ and the Euler iteration

$$y_{k+1}(\varepsilon) = y_k(\varepsilon) + h w_k(y_k, \varepsilon)$$

initialized at $y_0(\varepsilon)$ which also depends on the parameter. Denote for short $y_k = y_k(0)$, $w_k(y) = w_k(y, 0)$, $\delta y_k = \partial_\varepsilon y_k(0)$ and $\delta w_k(y) = \partial_\varepsilon w_k(y, 0)$. Then, for any $N \geq 0$, any $\eta_N \in \mathbb{R}^Q$, we have, for $k \leq N$:

$$\langle \delta y_N \,, \eta_N \rangle = \langle \delta y_k \,, \eta_k \rangle + h \sum_{l=k}^{N-1} \langle \delta w_l(y_l) \,, \eta_{l+1} \rangle \qquad (11.48)$$

where the sequence $\eta_l, l \leq N$ satisfies

$$\eta_l = \eta_{l+1} + h D w_l(y_l)^T \eta_{l+1}.$$

For the Euler midpoint scheme

$$y_{k+1}(\varepsilon) = y_k(\varepsilon) + h w'_k(y'_k, \varepsilon)$$
$$y'_k(\varepsilon) = y_k + \frac{h}{2} w_k(y_k, \varepsilon)$$

and the notation $w'_k(y, \varepsilon) = w(t_k + h/2, y, \varepsilon)$. We have

$$\langle \delta y_N \,, \eta_N \rangle = \langle \delta y_k \,, \eta_k \rangle + h \sum_{l=k}^{N-1} \langle \delta w'_l(y_l) + \frac{h}{2} D w'_l(y'_l) \delta w_l(y_l) \,, \eta_{l+1} \rangle \quad (11.49)$$

where the sequence $\eta_l, l \leq N$ satisfies

$$\eta_l = \eta_{l+1} + h D w'_l(y'_l)^T \eta_{l+1} + \frac{h}{2} D w_l(y_l)^T D w'_l(y'_l)^T \eta_{l+1}.$$

Proof. Start with the Euler case, and let's prove (11.48) by induction on N, since it is obviously true for $N = k$. Now, differentiating the iterations yields

$$\delta y_{N+1} = \delta y_N + h \delta w_N(y_N) + h D w_N(y_N) \delta y_N.$$

This implies

$$\langle \delta y_{N+1} \,, \eta_{N+1} \rangle = \langle \delta y_N \,, \eta_{N+1} + h D w_N(y_N)^T \rangle + h \langle \delta w_N(y_N) \,, \eta_{N+1} \rangle,$$

which immediately gives (11.48) for $N + 1$, assuming that it is true for N.

The argument for (11.49) is similar. Write

$$\delta y_{N+1} = \delta y_N + h \delta w'_N(y'_N) + h D w'_N(y'_N) \delta y'_N$$
$$\delta y'_N = \delta y_N + \frac{h}{2} \delta w_N(y_N) + \frac{h}{2} D w_N(y_N) \delta y_N$$

so that

$$\delta y_{N+1} = \delta y_N + h \delta w'_N(y'_N) + h D w'_N(y'_N) \delta y_N$$
$$+ \frac{h^2}{2} D w'_N(y'_N) \delta w_N(y_N) + \frac{h^2}{2} D w'_N(y'_N) D w_N(y_N) \delta y_N.$$

This gives

$$\langle \delta y_{N+1} \,, \eta_{N+1} \rangle = \langle (\mathrm{Id} + h D w'_N(y'_N) + \frac{h^2}{2} D w'_N(y'_N) D w_N(y_N)) \delta y_N \,, \eta_{N+1} \rangle$$
$$+ h \langle \delta w'_N(y'_N) + \frac{h}{2} D w'_N(y'_N) \delta w_N(y_N) \,, \eta_{N+1} \rangle,$$

which enables the induction argument for (11.49). $\qquad \square$

We now consider time-dependent control variables (like $v(t, \cdot)$, $\xi(t)$ or $x(t)$ in the examples above), still restricting ourselves to point-matching algorithms. In this case, the variables need to be discretized in time. Consider, to fix the ideas, an optimization in the time-dependent momentum, i.e., minimize

$$\int_0^1 \xi(t)^T S(x(t))\xi(t)dt + F(x(1))$$

with $\partial_t x = S(x)\xi(t)$ and $S(x)$ is the matrix formed from the kernel at $x_1, \ldots, x_N$. Using a time discretization $h = 1/Q$ (Q being an integer) and letting $t_k = hk$, we can discretize the problem by looking for $\xi(t)$ constant over intervals $[t_k, t_{k+1})$. The resulting problem then minimizes, letting $\xi^k = \xi(t_k)$,

$$\sum_{k=1}^{Q-1} (\xi^k)^T \left(\int_{t_k}^{t_{k+1}} S(x(t))dt \right) \xi^k + F(x(1))$$

still with $\partial_t x = S(x)\xi(t)$. Without getting into details, we can see that the variation with respect to the discrete ξ can be computed as in Section 11.6.2. A further approximation is to assume that h is so small that $x(t)$ can be considered as constant over $[t_k, t_{k+1})$. This leads to the minimization of

$$h \sum_{k=1}^{Q-1} (\xi^k)^T S(x^k)\xi^k + F(x^Q)$$

with $x^{k+1} = x^k + hS(x^k)\xi^k$. This corresponds to discretizing the evolution of x using an Euler scheme. Using the Euler midpoint, which is more accurate, would require minimizing

$$h \sum_{k=1}^{Q-1} (\xi^k)^T S(\tilde{x}^k)\xi^k + F(x^Q)$$

with

$$\tilde{x}^k = x^k + \frac{h}{2} S(x^k)\xi^k$$
$$x^{k+1} = x^k + hS(\tilde{x}^k)\xi^k.$$

The resulting expression is fully discrete. Its gradient can be computed with the discrete analog of the method discussed in Section 11.6.2, using Proposition 11.18. The accuracy of the approximation can be shown to be controlled by the size of the products $h\xi^k$, $k = 0, \ldots Q-1$. Since the size of ξ^k determines how large the deformation is, we can see that a given accuracy requires smaller time steps for larger deformations, which is natural. One issue with this is that one cannot know the size of the deformation before having computed it, so that one can only assess the accuracy of the solution after the algorithm has run (and possibly rerun it with a smaller time step if the deformation is found to be too large for the one that has been used).

Now consider the situation in which the deformed objects are continuous. In some cases (like with curves and surfaces), one can first discretize the energy, and reduce to point-matching problems, for which the previous discussion applies. There is still the question of whether the solution of the discrete problem (even accurately computed) is close to the solution of the continuous problem. Once again, this is related to the size of the deformation. Indeed, in the curve and surface examples we have considered, the discretized deformation cost cannot be a good approximation of the original uniformly in the size of the deformation, since these examples involve approximations of integrals over arcs of curves or surface patches by evaluations over line segments or triangles, and these segments or triangles can become large after deformation and deviate from the true deformed structure. Like with time discretization, it is therefore necessary to revise a solution that results in large deformed faces, by refining the discretization of the template.

Let's now discuss image matching. First consider the computation of the gradient in (11.47). Assume that time is discretized at $t_k = kh$ for $h = 1/Q$ and that $v_k(\cdot) = v(t_k, \cdot)$ is discretized over a regular grid $\mathcal{G}$, with nodes given by $(z_j, j \in \mathcal{G})$.

It will be convenient to introduce the momentum and express v_k in the form

$$v_k(y) = \sum_{z_j \in \mathcal{G}} K(y, z_j) \rho_k^j. \tag{11.50}$$

We can consider $(\rho_k^j, j \in \mathcal{G})$ as new control variables, and note that (11.47) directly provides the gradient of the energy in V^*, namely

$$(\nabla^{V^*} E)(t, y) = 2\rho(t) - 2\det(D\varphi_{t1}^v)(I \circ \varphi_{t0}^v - \tilde{I} \circ \varphi_{t1}^v)\nabla(I \circ \varphi_{t0}^v) \otimes dx.$$

From this continuous expression, we see that we must interpret the family $(\rho_k^j, j \in \mathcal{G})$ as discretizing a measure, namely

$$\rho_k = \sum_{j \in \mathcal{G}} \rho_k^j \otimes \delta_{z_k}.$$

Given this, the gradient in V^* can be discretized as

$$\xi_k^j = 2\rho_k^j - 2\det(D\varphi_{t_k 1}^v(z_j))(I \circ \varphi_{t_k 0}^v(z_j) - \tilde{I} \circ \varphi_{t_k 1}^v(z_j))\nabla(I \circ \varphi_{t_k 0}^v(z_j)) \otimes dz_j$$

which can be used to update ρ_k^j.

The last requirement in order to obtain a fully discrete procedure is to select interpolation schemes for the computation of the diffeomorphisms φ^v and for the compositions of I and I' with them. Interpolation algorithms (linear, or cubic, for example) are standard procedures that are included in many software packages [167]. In mathematical representation, they are linear operators that take a discrete signal f on a grid $\mathcal{G}$, that is $f \in \mathbb{R}^{\mathcal{G}}$, and return a function defined everywhere. Let's denote this function by $\mathcal{R}f$, which must, by linearity, decompose as

$$(\mathcal{R}f)(y) = \sum_{j \in \mathcal{G}} r_j(y)f_j.$$

Linear interpolation, for example, corresponds, in one dimension, to $r_j(y) = (z_{j+1} - y)/\eta$ if $z_j \leq y < z_{j+1}$, $r_j(y) = (y - z_{j-1})/\eta$ is $z_{j-1} \leq y < z_j$ and 0 otherwise, where η is the space discretization step, $\eta = z_{j+1} - z_j$.

So, assuming that I and I' are discretized as $(I_j, j \in \mathcal{G})$ and $(I'_j, j \in \mathcal{G})$, we want to replace, say, $I \circ \varphi_{t_k 0}(z_j)$ in the expression of the gradient by

$$(\mathcal{R}I)(\varphi_{t_k 0}(z_j)) = \sum_{j \in \mathcal{G}} r_j(\varphi_{t_k 0}(z_j))I_j.$$

For computational purposes, it is also convenient to replace the definition of v_k in (11.50) by an interpolated form

$$v_k(y) = \sum_{j \in \mathcal{G}} r_j(y) \sum_{i \in \mathcal{G}} K(z_j, z_i)\rho_k^i \tag{11.51}$$

because the inner sum can be computed very efficiently using Fourier transforms (see next section).

So the only quantities that need to be discretized are the $\varphi_{t_k 0}$, $\varphi_{1 t_k}$ and its Jacobian determinant, which are involved in the expression of the gradient. The evolution equations

$$\partial_t \varphi_{1t}(z_j) = v(t, \varphi_{1t}(z_j))$$
$$\partial_t(\det D\varphi_{1t}(z_j)) = \operatorname{div} v(t, \varphi_{1t}(z_j))$$
$$\partial_t \varphi_{t0}(z_j) = -D\varphi_{t0}(z_j)v(t, z_j)$$

(where the first two need to be solved backward in time starting from $\varphi_{11} = id$) can be discretized (at first-order accuracy in h) as

$$\varphi_{1t_k}(z_j) = (id - hv_k(\varphi_{1t_k}(z_j))) \circ \varphi_{1t_{k+1}}(z_j)$$
$$\det D\varphi_{1t_k}(z_j) = (1 - h\operatorname{div} v_k(\varphi_{1t_k}(z_j))) \det D\varphi_{1t_{k+1}}(z_j)$$
$$\varphi_{t_{k+1}0}(z_j) = = \varphi_{t_k 0}(z_j - hv_k(\varphi_{1t_k}(z_j)))$$

with $\varphi_{11} = \varphi_{00} = id$.

So, we can build, at this point, a comprehensive discrete procedure for the computation of the gradient in (11.47). Of course, it is a discretization of the gradient of a continuous cost function, not the gradient of a discretized one. In extreme cases (very large deformation), this approximation may be inaccurate and result in failing optimization procedures. For this reason, it may be interesting to analyze how the whole process can be revised to provide an exact gradient of a discrete cost function.

For this, define v_k as before as a function of ρ_k and

$$\psi_{lk} = (id - hv_l) \circ \cdots \circ (id - hv_{k-1})$$

where an empty product of compositions is equal to the identity. From the iteration formula above, ψ_{0k} is an approximation of $\varphi_{t_k 0}$ and ψ_{lk} of $\varphi_{t_k t_l}$. Define the cost function, which is explicitly computable as a function of $\rho_0, \ldots, \rho_{Q-1}$:

$$E(\rho) = \sum_{k=0}^{Q-1} \sum_{i,j \in \mathcal{G}} K(z_i, z_j) \langle \rho_k^i, \rho_k^j \rangle + \sum_{j \in \mathcal{G}} ((\mathcal{R}I)(\psi_{0Q-1}(z_j)) - I_j')^2.$$

If we make a variation $\rho \mapsto \varepsilon\delta\rho$, then $v \mapsto v + \varepsilon\delta v$ with (using the interpolated expression of v)

$$\delta v_k(y) = \sum_{i \in \mathcal{G}} r_i(y) \sum_{j \in \mathcal{G}} K(z_i, z_j)\delta\rho_j$$

and letting $\delta\varphi_{lk} = \partial_\varepsilon \psi_{lk}$, we have, by direct computation

$$\delta\psi_{lk} = -h \sum_{q=l}^{k-1} D\psi_{lq} \circ \psi_{qk} \, \delta v_q \circ \psi_{q+1k}.$$

Using this, we can compute the variation of the energy, yielding

$$\partial_\varepsilon E = 2 \sum_{k=0}^{Q-1} \sum_{i,j, \in CG} K(z_i, z_j) \langle \rho_k^i, \delta\rho_k^j \rangle$$

$$- h \sum_{k=0}^{Q-1} \sum_{i,i',j \in \mathcal{G}} K(z_i, z_j) r_i(\psi_{k+1\,Q-1}(z_{i'}))((\mathcal{R}I)(\psi_{0Q-1}(z_{i'})) - I_{i'}')$$

$$\langle \nabla(\mathcal{R}I)(\psi_{0Q-1}(z_{i'})), D\psi_{0q} \circ \psi_{q\,Q-1}(z_{i'})\delta\rho_k^j \rangle$$

This provides the expression of the gradient of the discretized E in V^*, namely

$$(\nabla^{V^*} E(\rho))_k^i = 2\rho_k^i$$

$$- h \sum_{i' \in \mathcal{G}} r_i(\psi_{k+1\,Q-1}(z_{i'}))((\mathcal{R}I)(\psi_{0Q-1}(z_{i'})) - I_{i'}')\nabla(\mathcal{R}I \circ \psi_{0q})(\psi_{qQ-1}(z_{i'})).$$

11.7.2 Kernel-Related Numerics

Most of the previously discussed methods included repeated computations of linear combination of the kernel. A basic such step is to compute, given points $y_1, \ldots, y_M, x_1, \ldots, x_N$ and vectors (or scalars) $\alpha_1, \ldots, \alpha_N$, the sums

$$\sum_{k=1}^{N} K(y_j, x_k)\alpha_k, \quad j = 1, \ldots, M.$$

Such sums are involved when deriving velocities from momenta, for example, or when evaluating dual RKHS norms in curve or surface matching.

Computing these sums explicitly requires NM evaluations of the kernel (and this probably several times per iteration of an optimization algorithm). When N or M are reasonably small (say, less than 1,000), such a direct evaluation is not a problem. But for large-scale methods, like for triangulated surface matching, where the surface may have tens of thousands of nodes, or for image matching, where a three-dimensional grid typically has millions of nodes, this becomes unfeasible.

If $x = y$ is supported by a regular grid $\mathcal{G}$, and K is translation invariant, i.e., $K(x, y) = \Gamma(x - y)$, then, letting $x_k = hk$ where k is a multi-index ($k = (k_1, \ldots, k_d)$) and h the discretization step, we see that

$$\sum_{k \in \mathcal{G}} \Gamma(h(k - l))\alpha_l$$

is a convolution that can be implemented with $O(N \log N)$ operations, using fast Fourier transforms (with $N = |\mathcal{G}|$). The same conclusion holds if K takes the form $K(x, y) = A(x)^T \Gamma(x - y)A(y)$ for some matrix A (which can be used to censor the kernel at the boundary of a domain), since the resulting operation is

$$A(x_k)^T \Big(\sum_{k \in \mathcal{G}} \Gamma(h(k - l))(A(x_l)\alpha_l) \Big),$$

which can still be implemented in $O(N \log N)$ operations.

The situation is less favorable when x and y are not regularly spaced. In such situations, feasibility must come with some approximation.

Still assuming a translation-invariant kernel $K(x, y) = \Gamma(x - y)$, we associate to a grid $\mathcal{G}$ in $\mathbb{R}^d$, with nodes $(z_j, j \in \mathcal{G})$, the interpolated kernel

$$K_{\mathcal{G}}(x, y) = \sum_{j, j' \in \mathcal{G}} r_j(x)\Gamma(h(j - j'))r_{j'}(y)$$

where the r_j's are interpolants adapted to the grid. This approximation provides a non-negative kernel, with null space equal to the space of functions with vanishing interpolation on $\mathcal{G}$. With such a kernel, we have

$$\sum_{k=1}^{N} K(y_j, x_k)\alpha_k = \sum_{i \in \mathcal{G}} r_i(y_j) \sum_{i' \in \mathcal{G}} \Gamma(h(i - i')) \sum_{k=1}^{N} r_{i'}(x_k)\alpha_k.$$

The computation therefore requires using the following sequence of operations:

1. Compute, for all $i' \in \mathcal{G}$ the quantity

$$a_{i'} = \sum_{k=1}^{N} r_{i'}(x_k)\alpha_k.$$

Because, for each x_k, only a fixed number of $r_{i'}(x_k)$ are non-vanishing, this requires an $O(N)$ number of operations.

2. Compute, for all $i \in \mathcal{G}$,

$$b_i = \sum_{i' \in \mathcal{G}} \Gamma(h(i - i'))a_{i'},$$

which is a convolution requiring $O(K \log K)$ operations, with $K = |\mathcal{G}|$.
3. Compute, for all $j = 1, \ldots, M$, the interpolation

$$\sum_{i \in \mathcal{G}} r_i(y_j)b_i,$$

which requires $O(M)$ operations.

So the resulting cost is $O(M + N + K \log K)$, which must be compared to the original $O(MN)$, the comparison being favorable essentially when MN is larger than the number of nodes in the grid, K. This formulation has the advantage that the resulting algorithm is quite simple, and that the resulting $K_{\mathcal{G}}$ remains a non-negative kernel, which is important.

Another class of methods, called fast multipole, compute sums like

$$\sum_{k=1}^{N} K(y, x_k)\alpha_k,$$

by taking advantage of the fact that $K(y, x)$ varies slowly when x varies in a region which is remote from y. By grouping the x_k's in clusters, assigning centers to these clusters and approximating the kernel using asymptotic expansions valid at large enough distance from the clusters, false multipole methods can organize the computation of the sums with a resulting cost of order $M + N$ when M sums over N terms are computed. Even if it is smaller than a constant times $(M + N)$, the total number of operations increases (via the size of the constant) with the required accuracy. The interested reader may refer to [19, 105] for more details.

Another important operation involving the kernel is the inversion of the system of equations (say, with a scalar kernel)

$$\sum_{k=1}^{N} K(x_k, x_l)\alpha_l = u_k, \quad k = 1, \ldots, N. \tag{11.52}$$

This is the spline interpolation problem, but it is also part of several of the algorithms that we have discussed, including for example the projection steps that have been introduced to obtain a gradient in the correct metric.

Such a problem is governed by an uncertainty principle [186], between accuracy of the approximation, which is given by the distance between a smooth function $x \mapsto u(x)$ and its interpolation

$$x \mapsto \sum_{k=1}^{N} K(x, x_l)\alpha_l,$$

where $\alpha_1, \ldots, \alpha_N$ are given by (11.52) with $u_k = u(x_k)$, and the stability of the system (11.52), measured by the condition number (the ratio of the largest by the smallest eigenvalue) of the matrix $S(x) = (K(x_i, x_j), i, j = 1, \ldots, N)$.

When $K(x, y) = \Gamma(x - y)$, the trade-off is measured by how fast $\xi \mapsto \hat{\Gamma}(\xi)$ (the Fourier transform of Γ) decreases at infinity. One extreme is given by the Gaussian kernel, for which $\hat{\Gamma}$ decreases like $e^{-c|\xi|^2}$, which is highly accurate and highly numerically unstable. On the other side of the range are Laplacian kernels, which decrease polynomially in the Fourier domain. The loss of stability is measured by the smallest distance between two distinct x_k's ($S(x)$ is singular if two x_k's coincide). In this dilemma, one possible rule is to prefer accuracy for small values of N, therefore using a kernel like the Gaussian, and go for stability for large-scale problems (using a Laplacian kernel with high enough degree).

For the numerical inversion of system (11.52), iterative methods should be used (especially for large N), conjugate gradient probably being the best choice. Recent advances in this domain [79, 106] use preconditioned conjugate gradient. The interested reader may refer to these references for more detail.

12

Distances and Group Actions

12.1 General Principles

We start, in this chapter, a description of how metric comparison between deformable objects can be performed, and how it interacts with the registration methods that we have studied in the previous chapters. We start with a general discussion on the interaction between distances on a set and transformation groups acting on it.

12.1.1 Distance Induced by a Group Action

Transformation groups acting on sets can help in defining or altering distances on these sets. We will first give a generic construction, based on a least action principle. We will then develop the related differential point of view, when a Lie group acts on a manifold.

We start with some terminology. A distance on a set M is a mapping $d : M^2 \mapsto \mathbb{R}_+$ such that: for all $m, m', m'' \in M$,

D1. $d(m, m') = 0 \Leftrightarrow m = m'$.
D2. $d(m.m') = d(m', m)$.
D3. $d(m, m'') \leq d(m, m') + d(m', m'')$.

If D1 is not satisfied, but only the fact that $d(m, m) = 0$ for all m, we will say (still assuming D2 and D3) that d is a *pseudo-distance*.

If G is a group acting on M, we will say that a distance d on M is G-equivariant if and only if for all $g \in G$, for all $m, m' \in M$, $d(g \cdot m, g \cdot m') = d(m, m')$.

A mapping $d : M^2 \mapsto \mathbb{R}_+$ is a G-invariant distance if and only if it is a pseudo-distance such that $d(m, m') = 0 \Leftrightarrow \exists g \in G, g \cdot m = m'$. This is equivalent to stating that d is a distance on the coset space M/G, composed with cosets, or orbits,

$$[m] = \{g \cdot m, g \in G\},$$

with the identification $d([m], [m']) = d(m, m')$.

The next proposition shows how a G-equivariant distance induces a G-invariant pseudo-distance.

Proposition 12.1. *Let d be equivariant for the left action of G on M. The function $\tilde{d}$, defined by*

$$\tilde{d}([m], [m']) = \inf\{d(g \cdot m, g' \cdot m'), g, g' \in G\}$$

is a pseudo-distance on M/G.

If, in addition, the orbits $[m]$ are closed subsets of M (for the topology associated to d), then $\tilde{d}$ is a distance.

Since d is G-equivariant, $\tilde{d}$ in the previous proposition is also given by

$$\tilde{d}([m], [m']) = \inf\{d(g \cdot m, m'), g \in G\}.$$

Proof. The symmetry of $\tilde{d}$ is obvious, as well as the fact that $\tilde{d}((m], [m]) = 0$ for all m. For the triangle inequality, D3, it suffices to show that, for all $g_1, g_1', g_2', g_1'' \in G$, there exists $g_2, g_2'' \in G$ such that

$$d(g_2 \cdot m, g_2'' \cdot m'') \leq d(g_1 \cdot m, g_1' \cdot m') + d(g_2' \cdot m', g_1'' \cdot m''). \tag{12.1}$$

Indeed, if this is true, the minimum of the right-hand term in g_1, g_1', g_2', g_1'', which is $\tilde{d}([m], [m']) + \tilde{d}([m'], [m''])$ is larger than the minimum of the left-hand term in g_2, g_2'', which is $\tilde{d}([m], [m''])$. This is D3. To prove (12.1), write $d(g_2' \cdot m', g_1'' \cdot m'') = d(g_1' \cdot m', g_1'(g_2')^{-1}g_1'' \cdot m'')$, take $g_2 = g_1$ and $g_2'' = g_1'(g_2')^{-1}g_1''$; (12.1) is then a consequence of the triangle inequality for d.

We now make the additional assumption that the orbits are closed and prove that D1 is true. Since $\tilde{d}([m], [m]) = 0$ is an obvious fact, we only need to show that $\tilde{d}([m], [m']) = 0$ is only possible if $[m] = [m']$. So, take $m, m' \in M$ such that $\tilde{d}([m], [m']) = 0$. This implies that there exists a sequence $(g_n, n \geq 0)$ in G such that $d(g_n \cdot m, m') \to 0$ when $n \to \infty$, so that m' belongs to the closure of the orbit of m. Since the latter is assumed to be closed, this yields $m' \in [m]$ which is equivalent to $[m] = [m']$. □

The same statement can clearly be made with G acting on the right on m, writing $m \mapsto m \cdot g$. We state it without proof.

Proposition 12.2. *Let d be equivariant for the right action of G on M. The function $\tilde{d}$, defined by*

$$\tilde{d}([m], [m']) = \inf\{d(m \cdot g, m' \cdot g'), g, g' \in G\}$$

is a pseudo-distance on $G\backslash M$.

If, in addition, the orbits $[m]$ are closed subsets of M (for the topology associated to d), then $\tilde{d}$ is a distance.

Here $G\backslash M$ denotes the coset space for the right action of G.

12.1.2 Distance Altered by a Group Action

In this section, G is still a group acting on the left on M, but we consider the product space $\mathcal{M} = G \times M$ and project on M a distance defined on $\mathcal{M}$. The result of this analysis will be to allow a distance on M to incorporate a component that accounts for possible group transformation being part of the difference between the compared objects.

The left action of G on M induces a right action of G on $\mathcal{M}$, defined, for $k \in G$, $z = (h, m) \in \mathcal{M}$, by

$$z \cdot k = (hk, k^{-1} \cdot m)$$

For $z = (h, m) \in \mathcal{M}$, we define the projection $\pi(z) = h \cdot m$. This projection is constant on the orbits $z.G$ for a given z, i.e., for all $k \in G$, $\pi(z \cdot k) = \pi(z)$.

Let $d_{\mathcal{M}}$ be a distance on $\mathcal{M}$. We let, for $m, m' \in M$

$$d(m, m') = \inf\{d_{\mathcal{M}}(z, z'), z, z' \in \mathcal{M}, \pi(z) = m, \pi(z') = m'\}. \qquad (12.2)$$

We have the following proposition:

Proposition 12.3. *If $d_{\mathcal{M}}$ is equivariant by the right action of G, then, the function d defined by 12.2 is a pseudo-distance on M.*

If, in addition, the orbits $[z] = \{z \cdot k, k \in G\}$ are closed in $\mathcal{M}$ for the topology associated to $d_{\mathcal{M}}$, then d is a distance.

This is in fact a corollary of Proposition 12.2. One only has to remark that the quotient space $G \backslash \mathcal{M}$ can be identified to M via the projection π, and that the distance in (12.2) then becomes the projection distance introduced in Proposition 12.2.

12.1.3 Transitive Action

Induced Distance

In this section, we assume that $\mathcal{G}$ is a group that acts transitively on M. The action being transitive means that for any m, m' in M, there exists an element $z \in \mathcal{G}$ such that $m' = z \cdot m$.

We fix a reference element m_0 in M, and define the group G by

$$G = \mathrm{Iso}_{m_0}(\mathcal{G}) = \{z \in \mathcal{G}, z \cdot m_0 = m_0\}.$$

This group is the isotropy group, or stabilizer, of m_0 in $\mathcal{O}$. We can define a distance in M by projecting a distance on $\mathcal{G}$ as in Section 12.1.2, because, as we see now, $\mathcal{G}$ can be identified with $\mathcal{M} := G \times M$.

To see this, assume that a function $\rho : M \to \mathcal{G}$ has been defined, such that for all $m \in M$, $m = \rho(m) \cdot m_0$. This is possible, since the action is transitive (using the axiom of choice). Define

$$\Psi : G \times M \to \mathcal{G}$$
$$(h, m) \mapsto \rho(h \cdot m)h \cdot$$

Ψ is a bijection: if $z \in \mathcal{G}$, we can compute a unique (h, m) such that $z = \Psi(h, m)$; this (h, m) must satisfy

$$z \cdot m_0 = \rho(h \cdot m)h \cdot m_0 = \rho(h \cdot m)m_0 = h \cdot m,$$

which implies that $\rho(h \cdot m) = \rho(z \cdot m_0)$ and therefore $h = \rho(z \cdot m_0)^{-1}z$, which is uniquely specified; but this also specifies $m = h^{-1}z \cdot m_0$. This proves that Ψ is one-to-one and onto and provides the identification we were looking for.

The right action of G on $\mathcal{M}$, which is $(h, m) \cdot k = (hk, k^{-1}.m)$, translates to $\mathcal{G}$ via Ψ with

$$\Psi((h, m) \cdot k) = \rho(hkk^{-1} \cdot m)hk = \Psi(h, m) \cdot k$$

so that the right actions (of G on $\mathcal{M}$ and of G on $\mathcal{G}$) "commute" with Ψ. Finally, the constraint $\pi(h, m_1) = m$ in Proposition 12.3 becomes $z \cdot m_0 = m$ via the identification. All this provides a new version of Proposition 12.3 for transitive actions, given by:

Corollary 12.4. *Let $d_{\mathcal{G}}$ be a distance on $\mathcal{G}$, which is equivariant for the right action of the isotropy group of $m_0 \in M$. Define, for all $m, m' \in M$*

$$d(m, m') = \inf\{d_{\mathcal{G}}(z, z'), z \cdot m_0 = m, z' \cdot m_0 = m'\}. \qquad (12.3)$$

Then d is a pseudo-distance on M.

Note that, if $d_{\mathcal{G}}$ is right equivariant for the action of $\mathrm{Iso}_{m_0}(\mathcal{G})$, the distance

$$\tilde{d}_{\mathcal{G}}(z, z') = d_{\mathcal{G}}(z^{-1}, (z')^{-1})$$

is left equivariant, which yields the symmetric version of the previous corollary.

Corollary 12.5. *Let $d_{\mathcal{G}}$ be a distance on $\mathcal{G}$, which is equivariant for the left action of the isotropy group of $m_0 \in M$. Define, for all $m, m' \in M$*

$$d(m, m') = \inf\{d_{\mathcal{G}}(z, z'), z \cdot m = m_0, z' \cdot m' = m_0\}. \qquad (12.4)$$

Then d is a pseudo-distance on M.

From Propositions 12.1 and 12.2, d in Propositions 12.4 and 12.5 is a distance as soon as the orbits $g \cdot \mathrm{Iso}_{m_0}(\mathcal{G})$ (assuming, for example, a left action) are closed for $d_{\mathcal{G}}$. If the left translations $h \mapsto g \cdot h$ are continuous, this is true as soon as $\mathrm{Iso}_{m_0}(\mathcal{G})$ is closed. This last property is itself true as soon as the action $g \mapsto g \cdot m_0$ is continuous, from G to M, given some topology on M.

Finally, if $d_{\mathcal{G}}$ is left or right invariant by the action of the whole group, $\mathcal{G}$, on itself, then the distances in (12.3) and (12.4) both reduce to

$$d(m, m') = \inf\{d_{\mathcal{G}}(id, z), z \cdot m = m'\}.$$

Indeed, assume right invariance (the left-invariant case is similar): then, if $z \cdot m_0 = m$ and $z' \cdot m_0 = m'$, then $z'z^{-1} \cdot m = m'$ and $d_{\mathcal{G}}(id, z'z^{-1}) = d_{\mathcal{G}}(z, z')$. Conversely, assume that $\zeta \cdot m = m'$. Since the action is transitive, we know that there exists z such that $z \cdot m_0 = m$, in which case $\zeta z \cdot m_0 = m'$ and $d_{\mathcal{G}}(id, \zeta) = d_{\mathcal{G}}(z, \zeta z)$. We summarize this in the following, in which we take $\mathcal{G} = G$:

Corollary 12.6. *Assume that G acts transitively on M. Let d_G be a distance on G, which is left or right equivariant. Define, for all $m, m' \in M$*

$$d(m, m') = \inf\{d_G(id, g), g \cdot m = m'\}. \tag{12.5}$$

Then d is a pseudo-distance on M.

Effort Functional

As formalized in [102], one can build a distance on M on which a group acts transitively using the notion of effort functionals. The definition we give here is slightly more general than in [102], to take into account a possible influence of the deformed object on the effort. We also make a connection with the previous, distance based, formulations.

We let $\mathcal{G}$ be a group acting transitively on M. Assume that a cost $\Gamma(z, m)$ is assigned to a transformation $m \rightarrow z \cdot m$. If m and m' are two objects, we define $d(m, m')$ as the minimal cost (effort) required to transform m in m', i.e.,

$$d(m, m') = \inf\{\Gamma(z, m), z \in \mathcal{G}, z \cdot m = m'\}. \tag{12.6}$$

The proof of the following proposition is almost obvious.

Proposition 12.7. *If Γ satisfies:*

C1. $\Gamma(z, m) = 0 \Leftrightarrow z = id_{\mathcal{G}}$,
C2. $\Gamma(z, m) = \Gamma(z^{-1}, z \cdot m)$.
C3. $\Gamma(zz', m) \leq \Gamma(z, m) + \Gamma(z', om)$, then, d defined by (12.6) is a pseudo-distance on M.

In fact, this is equivalent to the construction of Corollary 12.5. To see this, let G be the isotropy group of m_0 for the action of $\mathcal{G}$ on M. We have

Proposition 12.8. *If Γ satisfies C1, C2 and C3, then, for all $m_0 \in M$, the function $d_{\mathcal{G}}$ defined by*

$$d_{\mathcal{G}}(z, z') = \Gamma(z'z^{-1}, z \cdot m_0) \tag{12.7}$$

is a distance on $\mathcal{O}$ which is equivariant for the right action of G. Conversely, given such a distance $d_{\mathcal{O}}$, one builds an effort functional Γ satisfying C1, C2, C3 letting

$$\Gamma(h, m) = d_{\mathcal{G}}(z, h \cdot z)$$

where z is any element of $\mathcal{G}$ with the property $z \cdot m = m_0$.

The proof of this proposition is straightforward and left to the reader.

12.1.4 Infinitesimal Approach

The previous sections have demonstrated the usefulness of building distances on a space $\mathcal{M}$ that are equivariant to the actions of a group G. The easiest way, probably, to construct such a distance (at least when $\mathcal{M}$ is a differential manifold and G is a Lie group) is to design a right-invariant Riemannian metric in $\mathcal{M}$, so that the computation of the right-invariant distance requires finding minimizing geodesics (Appendix B).

Recall that a Riemannian metric on $\mathcal{M}$ requires, for all $z \in \mathcal{M}$, an inner product $\langle . , . \rangle_z$ on the tangent space, $T_z\mathcal{M}$, to $\mathcal{M}$ at z, which depends smoothly on z. With such a metric, one defines the energy of a differentiable path $z(.)$ in $\mathcal{M}$ by

$$E(z(.)) = \int_0^1 \|\dot{z}_t\|_{z(t)}^2 \, dt. \tag{12.8}$$

The associated Riemannian distance on $\mathcal{M}$ is

$$d_{\mathcal{M}}(z_0, z_1) = \inf\{\sqrt{E(z(.))}, z(0) = z_0, z(1) = z_1\}. \tag{12.9}$$

To obtain a right-invariant distance, it suffices to ensure that the metric has this property. For $h \in G$, let R_h denote the right action of h on $\mathcal{M}$: $R_h : z \mapsto z \cdot h$. Let $DR_h(z)$ be its differential at $z \in \mathcal{M}$. It is a linear map, defined on $T_z\mathcal{M}$, with values in $T_{z \cdot h}\mathcal{M}$, such that, for all differentiable path $\varepsilon \mapsto z(\varepsilon)$ on $\mathcal{M}$ such that $z(0) = z_0$, we have

$$\partial_\varepsilon(z \cdot h)(0) = DR_h(z_0).\partial_\varepsilon z(0). \tag{12.10}$$

The right invariance of the metric is expressed by the identity, true for all $z \in \mathcal{M}$, $A \in T_z\mathcal{M}$ and $h \in G$,

$$\|A\|_z = \|DR_h(z).A\|_{z \cdot h}. \tag{12.11}$$

When $\mathcal{M} = G \times M$, condition (12.11) implies that it suffices to define $\langle . , . \rangle_z$ at elements $z \in \mathcal{M}$ of the form $z = (id, m)$ with $m \in M$. The metric at a generic point (h, m) can then be computed, by right invariance, from the metric at $(h, m) \cdot h^{-1} = (id, h^{-1} \cdot m)$. Since the metric at (id, m) can be interpreted as a way to attribute a cost to a deformation $(id, h(t) \cdot m)$ with $h(0) = id$ and small t, defining it corresponds to an analysis of the cost of an infinitesimal perturbation of m by elements of G.

Of course, an identical construction could be made with left actions and left-invariant distances.

12.2 Invariant Distances Between Point Sets

12.2.1 Introduction

The purpose of this section is to present the construction provided by Kendall [123] on distances between landmarks, taking the infinitesimal point of view that we have just outlined. Here, configurations of landmarks are considered up to similitude (translation, rotation, scaling). Since its introduction, this space has led to a rich literature that specially focuses on statistical data analysis on landmark data. The reader interested by further developments can refer to [193, 69, 124] and the references therein.

We only consider the two-dimensional case, which is also the simplest. For a fixed integer $N > 0$ let $\mathcal{P}_N$ denote the set of configurations of N points $(z^{(1)}, \ldots, z^{(N)}) \in (\mathbb{R}^2)^N$. We assume that the order in which the points are listed matters, which means that we consider labeled landmarks. The set $\mathcal{P}_N$ can therefore be identified to $\mathbb{R}^{2N}$.

Two configurations $(z^{(1)}, \ldots, z^{(N)})$ and $(\tilde{z}^{(1)}, \ldots, \tilde{z}^{(N)})$ will be identified if one can be deduced from the other by the composition, say g, of a translation and a plane similitude, i.e., $\tilde{z}^{(k)} = g.z^{(k)}$ for $k = 1, \ldots, N$. Objects of interest therefore are equivalent classes of landmark configurations, which will be referred to as N-shapes.

It will be convenient to identify the plane $\mathbb{R}^2$ with the set of complex numbers $\mathbb{C}$, a point $z = (x, y)$ being represented as $x + iy$. A plane similitude composed with a translation can then be written in the form $z \mapsto a.z + b$ with $a, b \in \mathbb{C}$, $a \neq 0$.

For $Z = (z^{(1)}, \ldots, z^{(N)}) \in \mathcal{P}_N$, we let $\overline{z}$ be the center of inertia

$$\overline{z} = (z^{(1)} + \cdots + z^{(N)})/N.$$

We also let $\|Z\|^2 = \sum_{k=1}^{N} |z^{(k)} - \overline{z}|^2$.

12.2.2 Space of Planar N-Shapes

Construction of a Distance

Let Σ_N be the quotient space of $\mathcal{P}_N$ by the equivalent relation $Z \sim Z'$ if there exist $a, b \in \mathbb{C}$ such that $Z' = aZ + b$. We denote by $[Z]$ the equivalence class of Z for this relation. We want to define a distance between two equivalence classes $[Z]$ and $[Z']$.

We can apply the results of Section 12.1.1, which ensure that, in order to define a distance in the quotient space, it suffices to start with a distance on $\mathcal{P}_N$ which is equivariant for the considered operations. So we need a distance D such that for all $a, b \in \mathbb{C}$ and all $W, Y \in \mathcal{P}_N$,

$$D(aY + b, aW + b) = D(Y, W).$$

To obtain such a distance, we define a Riemannian metric on $\mathcal{P}_N = \mathbb{R}^{2N}$ which is invariant by the action. We therefore must define, for all $Z \in \mathcal{P}_N$, a norm $\|A\|_Z$ over all $A = (a_1, \ldots, a_N) \in \mathbb{C}^N$ such that for all $a, b \in \mathbb{C}$:

$$\|A\|_Z = \|a.A\|_{a.Z+b}.$$

This formula implies that it suffices to define such a norm for Z such that $\|Z\| = 1$ and $\bar{z} = 0$, since we have, for all Z,

$$\|A\|_Z = \left\| \frac{1}{\|Z\|} . A \right\|_{\frac{Z-\bar{z}}{\|Z\|}} . \tag{12.12}$$

Once the metric has been chosen, the distance $D(W, Y)$ is defined by

$$D(W, Y)^2 = \inf \int_0^1 \left\| \dot{Z}_t \right\|_{Z(t)}^2 dt, \tag{12.13}$$

the infimum being taken over all paths $Z(.)$ such that $Z(0) = W$ and $Z(1) = Y$.

Let's make the following choice. When $\bar{z} = 0$ and $\|Z\| = 1$, take

$$\|A\|_Z = \sum_{k=1}^N |a^{(k)}|^2.$$

From (12.12) and (12.13), we need to minimize, among all paths between W and Y,

$$\int_0^1 \frac{\sum_{k=1}^N |\dot{Z}_t^{(k)}|^2}{\sum_{k=1}^N |Z^{(k)}(t) - \bar{z}(t)|^2} dt.$$

Denote $v^{(k)}(t) = (Z^{(k)}(t) - \bar{z}(t))/\|Z(t)\|$ and $\rho(t) = \|Z(t)\|$. The path $Z(.)$ is uniquely characterized by $(v(.), \rho(.), \bar{z}(.))$. Moreover, we have

$$\dot{Z}_t^{(k)} = \dot{\bar{z}}_t + \rho \dot{v}_t + v \dot{\rho}_t$$

so that we need to minimize

$$\int_0^1 \sum_{k=1}^N \left| \frac{\dot{\bar{z}}_t}{\rho} + \frac{\dot{\rho}_t}{\rho} . v^{(k)} + \dot{v}_t^{(k)} \right|^2 dt.$$

This is equal (using $\sum_i \mathbf{v}_i = 0$ and $\sum_i |\mathbf{v}_i|^2 = 1$, together with the differentials of these expressions) to

$$N \int_0^1 \left(\frac{\dot{\bar{z}}_t}{\rho} \right)^2 dt + \int_0^1 \left(\frac{\dot{\rho}_t}{\rho} \right)^2 dt + \int_0^1 \sum_{k=1}^N \left| \dot{v}_t^{(k)} \right|^2 dt. \tag{12.14}$$

The end-point conditions are directly deduced from those initially given in terms of W and Y.

The last term in (12.14), which only depends on v, can be minimized explicitly, under the constraints $\sum_k v^{(k)} = 0$ and $\sum_k |v^{(k)}|^2 = 1$, which imply that v varies on a $(2N - 3)$-dimensional real sphere. The geodesic distance is therefore given by the length of great circles, which yields the expression of the minimum: $\arccos(\langle v(0), v(1) \rangle)^2$.

Using this, we have

$$D(W,Y)^2 = \inf \left(N \int_0^1 \left(\frac{\dot{\bar{z}}_t}{\rho} \right)^2 + \int_0^1 \left(\frac{\dot{\rho}_t}{\rho} \right)^2 dt \right) \qquad (12.15)$$

$$+ \arccos \left(\left\langle \frac{W - \overline{w}}{\|W\|}, \frac{Y - \overline{z}}{\|Y\|} \right\rangle \right)^2,$$

where the first infimum is over functions $t \mapsto (\overline{z}(t), \rho(t)) \in \mathbb{C} \times [0, +\infty[$, such that $\overline{z}(0) = \overline{w}$, $\overline{z}(1) = \overline{y}$, $\rho(0) = |W|$, $\rho(1) = |Y|$.

The induced distance on Σ_N is then given by

$$d([Y],[W]) = \inf\{D(Y, aW + b), a, b \in \mathbb{C}\}.$$

Clearly, if $\overline{y} = \overline{w}$ and $|Y| = |W|$, the infimum in the right-hand side of (12.15) is zero. Since we can always select $|a|$ and b such that $|Y| = |aW + b|$ and $\overline{y} = a\overline{w} + b$, we have

$$d([Y],[W]) = \inf_a \left[\arccos \left(\left\langle \frac{a}{|a|} \frac{W - \overline{w}}{\|W\|}, \frac{Y - \overline{y}}{\|Y\|} \right\rangle \right) \right].$$

Finally, optimizing this over the unit vector $a/|a|$, we get

$$d([Y],[W]) = \arccos \left| \left\langle \frac{W - \overline{w}}{\|W\|}, \frac{Y - \overline{y}}{\|Y\|} \right\rangle \right|. \qquad (12.16)$$

We now make additional remarks on the structure of the obtained shape space.

Denote by S^{2N-3} the set of $v = (v_1, \ldots, v_{N-1}) \in \mathbb{C}^{N-1}$ such that $\sum_i |v_i|^2 = 1$ (this can be identified to a real sphere of dimension $2N - 3$). The complex projective space, denoted $\mathbb{C}P^{N-2}$, is defined as the space S^{2N-3} quotiented by the equivalence relation: $v\mathcal{R}v'$ if and only if $\exists \nu \in \mathbb{C}$ such that $v' = \nu v$; in other words, $\mathbb{C}P^{N-2}$ contains all sets

$$S^1.v = \{\nu.v, \nu \in \mathbb{C}, |\nu| = 1\|$$

when v varies in S^{2N-3}. This set has the structure of an $(N-2)$-dimensional complex manifold, which means that it can be covered with an atlas of open sets which are in bijection with open subsets of $\mathbb{C}^{N-2}$ (with analytic changes of coordinates). Such an atlas is provided, for example, by the family $(\mathcal{O}_k, \Psi_k)$, $k = 1, \ldots, N$ where $\mathcal{O}_k$ is the set of $S^1.v \in \mathbb{C}P^{N-2}$ with $v^{(k)} \neq 0$, and

$$\Psi_k(S^1 v) = (v^{(1)}/v^{(k)}, \ldots, v^{(k-1)}/v^{(k)}, v^{(k+1)}/v^{(k)}, \ldots, v^{(N-1)}/v^{(k)}) \in \mathbb{C}^{N-2}$$

for all $S^1 v \in \mathcal{O}_i$. This provides $\mathbb{C}P^{N-2}$ with a structure of analytic complex manifold. In fact, Σ_N is also an analytic complex manifold that can be identified with $\mathbb{C}P^{N-2}$.

Let us be more explicit with this identification [123]. Associate to $Z = (z^{(1)}, \ldots, z^{(N)})$ the family $(\zeta^{(1)}, \ldots, \zeta^{(N-1)})$ defined by

$$\zeta^{(k)} = (kz^{(k+1)} - (z^{(1)} + \cdots + z^{(k)}))/\sqrt{k^2 + k}.$$

One can verify that $\sum_{k=1}^{N-1} |\zeta^{(k)}|^2 = \|Z\|^2$ (similar decompositions are used, for example, for the analysis of large-dimensional systems of particles [220]). Denote by $F(Z)$ the element $S^1.(\zeta/\|Z\|)$ in $\mathbb{C}P^{N-2}$. One can check that $F(Z)$ only depends on $[Z]$ and that $[Z] \mapsto F(Z)$ is an isometry between Σ_N and $\mathbb{C}P^{N-2}$.

The Space of Triangles

This construction, applied to the case $N = 3$ (which corresponds to triangles with labeled vertices), yields a quite interesting result. For a triangle $Z = (z^{(1)}, z^{(2)}, z^{(3)})$, the previous function $F(Z)$ can be written

$$F(Z) = S^1 \cdot \left(\frac{\left[\frac{z^{(2)} - z^{(1)}}{\sqrt{2}}, \frac{2z^{(3)} - z^{(1)} - z^{(2)}}{\sqrt{6}} \right]}{\sqrt{|z^{(2)} - z^{(1)}|^2/2 + |2z^{(3)} - z^{(1)} - z^{(2)}|^2/6}} \right)$$

$$= S^1.[v^{(1)}, v^{(2)}].$$

On the set $v^{(1)} \neq 0$ (i.e., the set $z^{(1)} \neq z^{(2)}$) we have the local chart

$$Z \mapsto v^{(2)}/v^{(1)} = \frac{1}{\sqrt{3}} \left(\frac{2z^{(3)} - z^{(2)} - z^{(1)}}{z^{(2)} - z^{(1)}} \right) \in \mathbb{C}.$$

If we let $v^{(2)}/v^{(1)} = \tan \frac{\theta}{2} e^{i\varphi}$, and $M(Z) = (\sin\theta \cos\varphi, \sin\theta \sin\psi, \cos\theta) \in \mathbb{R}^3$, we obtain a correspondence between the triangles and the unit sphere S^2. This correspondence is isometric: the distance between two triangles $[Z]$ and $[\tilde{Z}]$, which has been defined above by

$$d([Z], [\tilde{Z}]) = \arccos \left| \frac{\sum_{k=1}^{3} z^{(k)} (\tilde{z}^{(k)})^*}{\|Z\| . \|\tilde{Z}\|} \right|$$

gives, after passing to coordinates θ and φ, exactly the length of the great circle between the images $M(Z)$ and $M(\tilde{Z})$. We therefore obtain a representation of labeled triangular shapes as points on the sphere S^2, with the possibility of comparing them using the standard metric on S^2.

12.2.3 Extension to Plane Curves

It is interesting to analyze the limit case when the landmarks correspond to increasingly refined discretizations of plane curves. Consider for this two plane curves m and $\tilde{m}$, parametrized on $[0, 1]$. Discretize the interval $[0, 1]$ into $N+1$ points $t^{(k)} = k/N, k = 0, \ldots, N$. Denote $Y^{(k)} = m(t^{(k)})$ and $W^{(k)} = \tilde{m}(t^{(k)})$. Let also m^N and $\tilde{m}^N$ be the piecewise constant approximations of m and $\tilde{m}$ given by $m^N(t) = Y^{(k)}$ (resp. $\tilde{m}^N(t) = W_k$) on $(t^{(k-1)}, t^{(k)}]$.

We have $\bar{y} = (1/N) \sum_k m(t^{(k)}) \simeq \int_0^1 m(t)dt$. Similarly

$$\|Y\|^2 \simeq N \left[\int_0^1 |m(t)|^2 dt - \left| \int_0^1 m(t)dt \right|^2 \right]$$

and

$$\langle Y, W \rangle \simeq N \left[\int_0^1 m(t)\tilde{m}(t)^* dt - \int_0^1 m(t)dt \int_0^1 \tilde{m}(t)^* dt \right]$$

which gives, passing to the limit,

$$d(m, \tilde{m}) = \arccos \left\{ \frac{\left| \int_0^1 m(t)^T \tilde{m}(t)dt - \left(\int_0^1 m(t)dt \right)^T \int_0^1 \tilde{m}(t)dt \right|}{[V_m]^{1/2} [V_{\tilde{m}}]^{1/2}} \right\}$$

with $V_m = \int_0^1 |m(t)|^2 dt - \left| \int_0^1 m(t)dt \right|^2$ (and a similar definition for $V_{\tilde{m}}$).

We can get rid of the translation component by comparing the differences $Y_k = m(t_k) - m(t_{k-1})$ instead of the points $m(t_k)$. Doing so, we obtain the same formula, replacing $m(t)$ by $\dot{m}_t$. Since

$$\int_0^1 \dot{m}_t dt = m(1) - m(0),$$

vanishes if m and $\tilde{m}$ are closed, we have, making this assumption,

$$d(m, \tilde{m}) = \arccos \left\{ \frac{\left| \int_0^1 \dot{m}_t^T \dot{\tilde{m}}_t dt \right|}{\left[\int_0^1 |\dot{m}_t|^2 dt \right]^{1/2} \left[\int_0^1 |\dot{\tilde{m}}_t|^2 dt \right]^{1/2}} \right\}.$$

(If the curves are open, the obtained formulae correspond to comparing them after applying the closing operation $m(t) \mapsto m(t) - t(m(1) - m(0))$.)

We now introduce the curve parametrization in this formula. Assume that the curves have length one (which is no loss of generality, since the distance is scale-invariant). We let $s = \varphi(t)$ for curve m and $s = \psi(t)$ for curve $\tilde{m}$, where s is the arc length. We then have $\partial_t m(t) = \dot{\varphi}_t(t) T(\varphi(t))$ where T is the unit tangent (expressed as a function of the arc length). To keep with complex notation, we let $T = e^{i\theta}$. We then have

$$\int_0^1 |\dot{m}_t|^2 \, dt = \int_0^1 \dot{\varphi}_t^2 dt$$

and

$$\int_0^1 \dot{m}_t^T \dot{\tilde{m}}_t dt = \int_0^1 \dot{\varphi}_t \dot{\psi}_t e^{i(\theta(\varphi(t)) - \tilde{\theta}(\psi(t)))} dt.$$

This yields the formula

$$d(m, \tilde{m}) = \arccos \left\{ \frac{\left| \int_0^1 \dot{\varphi}_t \dot{\psi}_t e^{i(\theta(\varphi(t)) - \tilde{\theta}(\psi(t)))} dt \right|}{\sqrt{\int_0^1 \dot{\varphi}_t^2 dt} \sqrt{\int_0^1 \dot{\psi}_t^2 dt}} \right\}.$$

This provides a metric between *parametrized curves*, which are continuous generalizations of labeled landmarks. This metric, however, is not equivariant for the right action of diffeomorphisms on m and $\tilde{m}$ (i.e., a common change or parameter) so that it cannot be quotiented out to obtain a distance modulo parametrization. Such distances are addressed in the next section.

12.3 Parametrization-Invariant Distances Between Plane Curves

We now describe distances between 2D shapes when they are defined as plane curves modulo changes of parameter. Such distances have been the subject of extensive and detailed mathematical studies, [141, 146], but here we only give an overview of the ideas and results.

Simple parametrization-free distances can be defined, based on the images of the curves. For example, it is possible to use standard norms applied to arc-length parametrizations of the curves, like L^p or Sobolev norms of the difference. With simple closed curves, we can measure the area of the symmetric difference between the interiors of the curves. A more advanced notion, the Hausdorff distance, is defined by

$$d(m, \tilde{m}) = \inf \{ \varepsilon > 0, m \subset \tilde{m}^\varepsilon \text{ and } \tilde{m} \subset m^\varepsilon \}$$

where m^ε is the set of points at a distance less than ε from m (and similarly for $\tilde{m}^\varepsilon$). The same distance can be used with the interiors for simple closed curves. In fact, the Hausdorff distance is a distance between closed sets as stated in the following proposition.

Proposition 12.9. *For $\varepsilon > 0$ and a subset A of $\mathbb{R}^d$, let A^ε be the set of points $x \in \mathbb{R}^d$ such that there exists $a \in A$ with $|a - x| < \varepsilon$. Let*

$$d_H(A, B) = \inf \{\varepsilon > 0 : A \subset B^\varepsilon \text{ and } B \subset A^\varepsilon\}.$$

Then d_H is a distance on the set of closed subsets of $\mathbb{R}^d$.

Proof. Symmetry is obvious, and we leave to the reader the proof of the triangular inequality, which is a direct consequence of the fact that $(A^\varepsilon)^{\varepsilon'} \subset A^{\varepsilon + \varepsilon'}$.

Assume that $d_H(A, B) = 0$. Then $A \subset B^\varepsilon$ for all $\varepsilon > 0$. But $\bigcap_\varepsilon B^\varepsilon = \bar{B}$, the closure of B. We therefore have

$$d_H(A, B) = 0 \Rightarrow A \subset \bar{B} \text{ and } B \subset \bar{A},$$

which implies that $A = B$ if both sets are closed.

One can also proceed like in Section 12.1.1. First define equivariant distances over parametrized curves, then optimize them with respect to changes of parameters. Let $\mathcal{C}$ be a set of parametrized curves, defined as functions $m : [0, 1] \mapsto \mathbb{R}^2$, subject to additional properties (like smoothness, closedness, etc.), and G a group of changes of parameter over $[0, 1]$ (including changes of offset for closed curves). Then the space of shapes must at least be a quotient space $\mathcal{S} = \mathcal{C}/G$ for the action $\varphi \cdot m = m \circ \varphi^{-1}$, which are curves modulo change of parameter (one may also want to quotient out rotation, scaling). So, based on our discussion in Section 12.1.1, a pseudo-distance on $\mathcal{S}$ can be defined from a distance on $\mathcal{C}$ provided it is equivariant by changes of parameter. L^p norms between parametrized curves are not equivariant, except for $p = \infty$, with

$$d_\infty(m, \tilde{m}) = \sup_u |m(u) - \tilde{m}(u)|.$$

The distance obtained after reduction by diffeomorphism is called the Fréchet distance, and is therefore defined by

$$d_F(m, \tilde{m}) = \inf_\varphi d_\infty(m \circ \varphi, \tilde{m}).$$

Note that, if for some diffeomorphism φ, $d_\infty(m \circ \varphi, \tilde{m}) \leq \varepsilon$, then $m \subset \tilde{m}^\varepsilon$ and $\tilde{m} \subset m^\varepsilon$. So we get the relation

$$\varepsilon > d_F(m, \tilde{m}) \Rightarrow \varepsilon > d_H(m, \tilde{m})$$

which implies $d_H \leq d_F$. This and Proposition 12.9 prove that d_F is a distance between curves (we already knew that it was a pseudo-distance).

We can also design equivariant distances on $\mathcal{C}$ based on Riemannian metrics associated to invariant norms on the tangent space. We only give here a formal discussion, ignoring the complications that arise from the infinite dimension of the space of curves (see [144, 145] for a rigorous presentation). Tangent vectors to $\mathcal{C}$ are derivatives of paths in $\mathcal{C}$, which are time-dependent parametrized curves $t \mapsto m(t, .)$. Tangent vectors therefore take the form $v = m_t(t, .)$, which are functions $v : [0, 1] \to \mathbb{R}^2$. Since a change of parameter in a time-dependent curve induces the same change of parameter on the time derivative, a norm on the tangent space to $\mathbb{C}$ is equivariant for the action of changes of parameter, if, for any m, v, φ,

$$\|v \circ \varphi^{-1}\|_{m \circ \varphi^{-1}} = \|v\|_m. \tag{12.17}$$

It is therefore sufficient to define $\|v\|_m$ for curves parametrized with arc length, since (12.17) then defines the metric for any parametrized curve.

We now want to define tangent vectors to plane curves modulo change of parameters. We know that we can modify the tangential component of the time derivative of a time-dependent parametrized curve $t \mapsto m(t, .)$ without changing the geometry of the evolving curve. It follows from this that tangent vectors to $\mathcal{S}$ at a curve m are equivalent classes of vector fields along m that share the same normal component to m, and can therefore be identified to this normal component itself, $a := v^T N$. The induced metric on $\mathcal{S}$ is

$$\|a\|_m := \inf \left\{ \|v\|_m : v^T N = a \right\}.$$

The associated pseudo-distance on $\mathcal{S}$ is

$$d(m, \tilde{m})^2 = \inf \left\{ \int_0^1 \left\| \dot{\mu}_t^T N \right\|_{\mu(t)}^2 dt, \mu(0, .) = m, \mu(1, .) = \tilde{m} \right\}. \tag{12.18}$$

The fact that we only get a pseudo-distance in general is interestingly illustrated by the following simple example. Define

$$\|v\|_m^2 = \int_0^{L_m} |v(s)|^2 ds. \tag{12.19}$$

This is the L^2 norm in the arc-length parametrization. Then we have $d(m, \tilde{m}) = 0$:

Theorem 12.10 (Mumford–Michor). *The distance defined in* (12.18) *with the norm given by* (12.19) *vanishes between any smooth curves m and $\tilde{m}$.*

A proof of this result can be found in [145, 144]. It relies on the remark that one can grow thin protrusions ("teeth") on the curve at a cost which is negligible compared to the size of the tooth. It is an easy exercise to compute the geodesic length of a path that starts with a horizontal segment and progressively grows an isosceles triangle of width ε and height t (at time t) on the segment until $t = 1$. This length is $o(\varepsilon)$ (in fact, $O(\varepsilon^2 \ln \varepsilon)$). This implies that one can turn a curve into $O(1/\varepsilon)$ thin non-overlapping teeth at almost no cost. A repeated application of this concept is the basic idea in the construction made in [145] to create almost-zero-length paths between two arbitrary curves.

To prevent the distance from vanishing, one possibility is to penalize the curve length more than (12.19) does. For example, the distance associated with the metric

$$\|v\|_m^2 = L_m \int_0^{L_m} |v(s)|^2 ds, \tag{12.20}$$

introduced in [187, 141], does not degenerate. It corresponds in fact to the area swept by the path relating the compared curves [187].

A second way to control degeneracy is to penalize high-curvature points, using for example

$$\|v\|_m^2 = \int_0^{L_m} (1 + a\kappa_m(s)^2)|v(s)|^2 ds. \tag{12.21}$$

This metric has been studied in [145], where it is shown (among other results) that the distance between distinct curves is positive. Finally, one can add derivatives of v (with respect to arc length) in the definition of the metric; this provides *Sobolev metrics* [146, 141] that we have already used for curve evolution.

12.4 Invariant Metrics on Diffeomorphisms

We discuss here the construction of a right-invariant distance between diffeomorphisms. We will see, in particular, that it coincides with the direct construction made in Chapter 8.

Like with curves, we only make a formal (non-rigorous) discussion. We consider a group G of diffeomorphisms of Ω, and define (to fix the ideas) the tangent space to G at $\varphi \in G$ by the set of $u : \Omega \to \mathbb{R}^d$ such that $id + tu \circ \varphi^{-1} \in G$ for small enough t. Since the group product on G is the composition, $\varphi\psi = \varphi \circ \psi$, the right translation $R_\varphi : \psi \mapsto \psi \circ \varphi$ is linear, and therefore "equal" to its differential: for $u \in T_\psi G$,

$$DR_\varphi(\psi)u = u \circ \varphi.$$

A metric on G is right invariant if, for all $\varphi, \psi \in G$ and for all $u \in T_\psi G$,

$$\|DR_\varphi(\psi)u\|_{\psi \circ \varphi} = \|u\|_\psi$$

which yields, taking $\varphi = \psi^{-1}$:

$$\|u\|_\psi = \|u \circ \psi^{-1}\|_{id}.$$

This implies that the energy of a path $(t \mapsto \varphi(t,.))$ in G must be defined by

$$E(\varphi(.)) = \int_0^1 \left\|(\partial_t\varphi)(t, \varphi^{-1}(t, .))\right\|_{id}^2 dt.$$

If we let

$$v(t, x) = (\partial_t\varphi)(t, \varphi^{-1}(t, x)),$$

the energy can be written

$$E(\varphi) = \int_0^1 \|v(t, .)\|_{id}^2 dt$$

with the identity

$$\partial_t\varphi(t, x) = v(t, \varphi(t, x)).$$

This implies that φ is the flow associated to the velocity field $v(t, .)$. We therefore retrieve the construction given in Chapter 8, with $\|.\|_{id} = \|.\|_V$. Thus, V in Chapter 8 has a role similar to the tangent space to G at id here. Because of this, we let $V = T_{id}G$ and $\|.\|_{id} = \|.\|_V$ in the remaining discussion to homogenize the notation.

Assume that V is admissible, according to Definition 8.12. The right-invariant distance on G is

$$d(\varphi_0, \varphi_1) = \inf \sqrt{\int_0^1 \|v(t, .)\|_V^2 \, dt} \qquad (12.22)$$

where the minimum is taken over all v such that, for all $x \in \Omega$, the solution of the ordinary differential equation

$$\partial_t y = v(t, y)$$

with initial conditions $y(0) = \varphi_0(x)$ is such that $y(1) = \varphi_1(x)$, consistently with Section 8.2.4.

12.4.1 Geodesic Equation

The geodesic equation on G is equivalent to the Euler–Lagrange equations associated to the variational problem (12.22). This is similar to what we have computed in Section 11.4, except that here we have a fixed end-point condition. One may address this with a method called the Euler–Poincaré reduction [113, 136], and the presentation we make here is related to it. The energy

$$E(v) = \int_0^1 \|v(t)\|_V^2 dt,$$

is minimized over all v such that $\varphi_{01}^v = \varphi_1$ (without loss of generality, because the distance is right invariant, we can assume that $\varphi_0 = id$).

If v is optimal, we have

$$\int_0^1 \langle v(t), h(t) \rangle_V \, dt = 0$$

for all "admissible" h, i.e., all $h \in \mathcal{X}_V^2$ such that there exists a variation $\epsilon \mapsto v^\epsilon$ such that $\partial_\epsilon v^\epsilon_{|\epsilon=0} = h$ and $\varphi^{v^\epsilon}_{01} = \varphi_1$ for all ϵ in a neighborhood of 0.

Let $\mathcal{X}_{\varphi_1}$ denote this set. If $h \in \mathcal{X}_{\varphi_1}$, we have, using $\partial_\epsilon \varphi^{v^\epsilon}_{01} = 0$ and applying Theorem 8.10 and equation (11.9):

$$0 = \int_0^1 \mathrm{Ad}_{\varphi^v_{t0}} h(t) dt.$$

Letting $w(t) = \mathrm{Ad}_{\varphi^v_{t0}} h(t)$, we have, whenever $h \in \mathcal{X}_{\varphi_1}$

$$\int_0^1 w(t) dt = 0 \text{ and } \int_0^1 \langle v(t), \mathrm{Ad}_{\varphi^v_{0t}} w(t) \rangle_V \, dt = 0.$$

Assume that the family of w's that can be built this way is large enough (i.e., it is dense in $\mathcal{X}_V^2$). How to ensure that this is true will not be addressed here, but this will essentially be true for some spaces V if some regularity assumptions are made on the optimal trajectory $v(t)$. Then this will ensure that, for $u \in V$, and any function f such that $\int_0^1 f(t) dt = 0$, taking $w(t) = uf(t)$:

$$\int_0^1 f(t) \langle v(t), \mathrm{Ad}_{\varphi^v_{0t}} u \rangle_V \, dt = 0.$$

which is only possible if $\langle v(t), \mathrm{Ad}_{\varphi^v_{0t}} u \rangle_V$ is constant. This is the conservation of momentum property already discussed in Section 11.5.

It should not be surprising that we retrieve the same optimality property as in Section 11.4. In that section, the minimized energy had the form

$$E(v) = \int_0^1 \|v(t)\|_V^2 \, dt + U(\varphi^v_{01}).$$

The optimal v for this problem must minimize $\int_0^1 \|v(t)\|_V^2 \, dt$ given the endpoint φ^v_{01} and therefore satisfy the optimality condition found in the present section.

We can also (still formally) compute the Hamiltonian form of the geodesic equation. Following section B.6.6, define the linear form $\mu(t)$ so that

$$\left(\mu(t) \mid u \right) = \left\langle \dot{\varphi}_t , u \right\rangle_{\varphi(t)} = \left(L(\dot{\varphi}_t \circ \varphi(t)^{-1}) \mid u \circ \varphi(t)^{-1} \right).$$

Then, with $v(t) = \dot{\varphi}_t \circ \varphi(t)^{-1}$ and $\rho(t) = Lv(t)$, we have

$$\left(\mu(t) \mid u \right) = \left(\rho(t) \mid u \circ \varphi(t)^{-1} \right)$$

so that $\mu(t)$ coincides with the one introduced in equation (11.22). Its evolution in system (11.23) therefore provides the Hamiltonian form of the geodesics in the group.

12.4.2 A Simple Example

An explicit computation of the distance is generally impossible, but here is an exception, in one dimension. Take $\Omega = [0, 1]$ and

$$\|u\|_{id}^2 = \int_0^1 |u_x|^2 \, dx.$$

Note that this norm is not admissible, because it cannot be used to control the supremum norm of $u_x = \partial u / \partial x$. The associated energy of a path of diffeomorphisms $\varphi(t, .)$ is

$$U(\varphi(.)) = \int_0^1 \int_0^1 \left| \frac{\partial}{\partial x} \left(\varphi_t \circ \varphi^{-1}(t, .) \right) \right|^2 \, dx dt.$$

This gives, after expanding the derivative and making the change of variables $x = \varphi(t, y)$:

$$U(\varphi(.)) = \int_0^1 \int_0^1 |\varphi_{tx}|^2 \, |\varphi_x|^{-1} \, dy dt.$$

Define $q(t, y) = \sqrt{\varphi_x(t, y)}$: we have

$$U(\varphi(.)) = 4 \int_0^1 \int_0^1 |q_t|^2 \, dy dt,$$

which yields

$$U(\varphi(.)) = 4 \int_0^1 \|q_t(t, .)\|_2^2 \, dt.$$

If the problem were to minimize this energy under the constraints $q(0, .) = \sqrt{\varphi_x(0, .)}$ and $q(1, .) = \sqrt{\varphi_x(1, .)}$, the solution q would be given by the line segment

$$q(t, x) = tq(1, x) + (1 - t)q(0, x).$$

There is, however, an additional constraint that comes from the fact that $q(t, .)$ must provide a homeomorphism of $[0, 1]$ for all t, which implies $\varphi(t, 1) = 1$, or, in terms of q

$$\|q(t, .)\|_2^2 = \int_0^1 q(t, x)^2 \, dx = 1.$$

We therefore need to minimize the length of the path q under the constraint that it remains on a Hilbert (L^2) sphere. Like in finite dimension, geodesics on Hilbert spheres are great circles. This implies that the optimal q is given by

$$q(t, .) = \frac{1}{\sin \alpha} (\sin(\alpha(1 - t))q_0 + \sin(\alpha t)q_1)$$

with $\alpha = \arccos \langle q_0, q_1 \rangle_2$. The length of the geodesic is precisely given by α, which provides a closed-form expression of the distance on G

$$d(\varphi, \tilde{\varphi}) = 2 \arccos \int_0^1 \sqrt{\varphi_x \tilde{\varphi}_x} \, dx.$$

12.4.3 Gradient Descent

Assume that a function $\varphi \mapsto E(\varphi)$ is defined over diffeomorphisms. Take C^1 h and ε_0 small enough so that $\varphi + \varepsilon h$ is a diffeomorphism is $|\varepsilon| \leq \varepsilon_0$, and assume that the Gâteaux derivative $\partial_\varepsilon U(\varphi + \varepsilon h)$ exists at $\varepsilon = 0$, denoting it, as in Section 10.2, by

$$\partial_\varepsilon U(\varphi + \varepsilon h) = \left(\frac{\partial U}{\partial \varphi}(\varphi) \,|\, h \right).$$

If a right-invariant metric is given, in the form

$$\langle h, h' \rangle_\varphi = \langle h \circ \varphi^{-1}, h' \circ \varphi^{-1} \rangle_V$$

as above, the gradient of U at φ is computed by identifying

$$\begin{aligned} \left(\frac{\partial U}{\partial \varphi}(\varphi) \,|\, h \right) &= \langle \nabla U(\varphi), h \rangle_\varphi \\ &= \langle \nabla U(\varphi) \circ \varphi^{-1}, h \circ \varphi^{-1} \rangle_V \\ &= \left(L(\nabla U(\varphi) \circ \varphi^{-1}) \,|\, h \circ \varphi^{-1} \right) \end{aligned}$$

where $L = K^{-1}$ is the inverse of the kernel operator on V. Since (with the notation of Section 10.2)

$$\left(\frac{\partial U}{\partial \varphi}(\varphi) \,|\, h \right) = \left(\bar{\partial} U(\varphi) \,|\, h \circ \varphi^{-1} \right)$$

we see that, using $\overline{\nabla}^V U = K \partial U$,

$$\nabla U(\varphi) = \overline{\nabla}^V U(\varphi) \circ \varphi$$

and the evolution equation introduced in (10.5) is nothing but a Riemannian gradient descent for U for the considered metric.

12.4.4 Diffeomorphic Active Contours

As a new example of an application of this formalism, we provide a Riemannian version of the active contours algorithm discussed in Section 6.4. Indeed, letting

$$E(m) = \int_m F(p)dl + \int_{\Omega_m} \tilde{F}(x)dx, \tag{12.23}$$

we can, fixing a template curve m_0 (a similar argument can be made with surfaces), define the functional

$$U(\varphi) = E(\varphi(m_0)).$$

Since, with a straightforward computation, and letting $m = \varphi(m_0)$

$$\left(\bar{\partial} U(\varphi) \,|\, v\right) = -\int_m (\kappa F - \langle F, N \rangle + \tilde{F}) \langle v, N \rangle dl$$

we find

$$\nabla U(\varphi)(x) = -\int_m (\kappa F - \langle F, N \rangle + \tilde{F}) K(\varphi(x), \cdot) N dl$$

defining the gradient descent algorithm,

$$\partial_t \varphi(t, x) = \int_{m(t)} (\kappa F - \langle F, N \rangle + \tilde{F}) K(\varphi(t, x), \cdot) N dl$$

with $m(t) = \varphi(t, \cdot) \circ m_0$.

This can also be expressed as an evolution equation in terms of $m(t)$ only, yielding the *diffeomorphic active contours evolution equation* [224, 14]

$$\partial_t m(t, u) = \int_{m(t)} (\kappa F - \langle F, N \rangle + \tilde{F}) K(m(t, u), \cdot) N dl. \tag{12.24}$$

Examples of segmentation using this equation are provided in Figure 12.1.

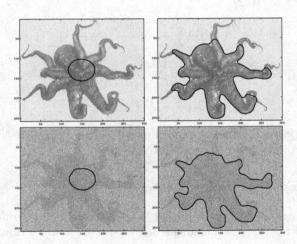

Fig. 12.1. Diffeomorphic active contours (compare with Figure 6.4). On each row, the left image is the initial contour, and the right one is the solution obtained with diffeomorphic active contours. The first row presents a clean image and the second a noisy one.

12.5 Horizontality

12.5.1 Riemannian Projection

We temporarily switch to general (but finite-dimensional) Lie groups before returning to diffeomorphisms. Let G be a Lie group acting transitively on a manifold, M. We know, from Corollary 12.6, that if d_G is a right-invariant distance on G, the least-cost measure given by (12.5) provides a pseudo-distance on M. We now discuss the infinitesimal version of this result.

So, we assume that G is equipped with a right-invariant metric such that, for any $w \in T_g G$,

$$\|w\|_g = \|DR_g^{-1}w\|_{id}.$$

Since the infinitesimal counterpart of $d_G(id, g)$ is $\|v\|_{id}$ for $v \in T_{id_G}$, and since the action $m' = g \cdot m$ becomes the infinitesimal action $\xi = v \cdot m \in T_m M$ (cf. Appendix, Section B.5.3), one defines the projected metric, for $m \in M$ and $\xi \in T_m M$:

$$\|\xi\|_m = \inf \{\|v\|_{id} : v \cdot m = \xi\}. \tag{12.25}$$

This is a Riemannian metric on M, as stated in the following theorem.

Theorem 12.11. *Let G be a Lie group with a right-invariant Riemannian metric. Assume that G acts transitively on M, and take $m \in M$ such that $v \mapsto v \cdot m$ has constant rank in a neighborhood of 0 in $T_{id}G$. Then, the right-hand side of (12.25) is finite, and defines a Riemannian metric on M.*

Proof. We start with a lemma, which we state without proof.

Lemma 12.12. *If G acts transitively on M, and $v \mapsto v \cdot m$ has constant rank in a neighborhood of 0, then, any smooth curve $t \mapsto m(t) \in M$ with $m(0) = m_0$ can be represented (for small enough t) as $t \mapsto g(t) \cdot m_0$ for a smooth curve $t \mapsto g(t) \in G$.*

A direct consequence of this is the fact that any tangent vector, ξ, to M at m can be written in the form $\xi = v \cdot m$ with $v \in T_{id}G$, so that the set

$$B_m^\xi = \{v \in T_{id}G : v \cdot m = \xi\}.$$

is not empty. Now, if $v_0^\xi \in B_m^\xi$, we can write

$$B_m^\xi = \left\{v \in T_{id}G : (v - v_0^\xi) \cdot m = 0\right\} = v_0^\xi + B_m^0.$$

So, minimizing $\|v\|_{id}$ over B_m^ξ is the same as minimizing $\|v_0^\xi + h\|_{id}$ over $h \in B_m^0$: the optimal h is therefore

$$h^* = -\pi_{B_m^0}(v_0^\xi)$$

where $\pi_{B_m^0}$ is the orthogonal projection on B_m^0 for $\|\cdot\|_{id}$. The optimal v is then

$$v^\xi = v_0^\xi - \pi_{B_m^0}(v_0^\xi) = \pi_{(B_m^0)^\perp}(v_0^\xi).$$

This does not depend on the choice of $v_0^\xi \in B_m^\xi$.

From this formula, we see directly that $\xi \mapsto v^\xi$ is linear because, given ξ and η, and any choice for v_0^ξ and v_0^η, we can always take $v_0^{\alpha\xi+\beta\eta} = \alpha v_0^\xi + \beta v_0^\eta$, which implies

$$v^{\alpha\xi+\beta\eta} = \alpha v^\xi + \beta v^\eta.$$

So this implies that $\langle \xi, \eta \rangle_m := \langle v^\xi, v^\eta \rangle_{id}$ defines a positive bilinear form, and that $\|\xi\|_m$ derives from a dot product. Finally, if $\|\xi\|_m = 0$, then $v^\xi = 0$. This implies that $\xi = v^\xi \cdot m = 0$ (in infinite dimension, we need the fact that B_m^0 is closed, which is true as soon as $v \mapsto v \cdot m$ is continuous).

$\square$

Important concepts and notation have been introduced in the previous proof. The first one is the space B_m^0, which is the set of vectors $v \in T_{id}G$ such that $v \cdot m = 0$. This is the set of *vertical vectors at m*, and will henceforth be denoted $\mathcal{V}_m$, so

$$\mathcal{V}_m = \{v \in T_{id}G : v \cdot m = 0\}. \tag{12.26}$$

The vectors that are perpendicular to $\mathcal{V}_m$ are quite naturally called horizontal vectors. The set of horizontal vectors at m will be denoted $\mathcal{H}_m$, with

$$\mathcal{H}_m = \mathcal{V}_m^\perp = \left\{w \in T_{id}G : \langle v, w \rangle_{id} = 0 \text{ if } v \cdot m = 0\right\}. \tag{12.27}$$

We will also speak of *horizontal linear forms*, or *horizontal covectors* which are linear forms $z \in T_{id}G^*$ such that $(z \mid v) = 0$ for all $v \in \mathcal{V}_m$.

If $\xi \in T_mM$, we have defined v^ξ as the vector in $T_{id}G$ that minimizes $\|v\|_{id}$ among all v such that $v \cdot m = \xi$. We have identified it as the orthogonal projection on $\mathcal{H}_m$ of any v_0^ξ such that $v_0^\xi \cdot m = \xi$. This leads to the following definitions, in which we let $v^\xi = h_m(\xi)$:

Definition 12.13. *Let G and M be as in Proposition 12.17. If $m \in M$ and $\xi \in T_mM$, the horizontal lift of ξ is the vector $h_m(\xi) \in \mathcal{H}_m$ such that $h_m(\xi) \cdot m = \xi$.*

If $v \in T_{id}G$, we let $\pi_{\mathcal{H}_m}(v)$ be the horizontal part of v at m and $v - \pi_{\mathcal{H}_m}(v)$ its vertical part at m. With the above notation

$$\pi_{\mathcal{H}_m}(v) = h_m(v \cdot m). \tag{12.28}$$

The projection on M of the Riemannian metric on G is defined by

$$\langle \xi, \eta \rangle_m = \langle h_m(\xi), h_m(\eta) \rangle_{id}. \tag{12.29}$$

Geodesics for the projected metric are immediately deduced from those on G, as implied by the following proposition.

Proposition 12.14. *The geodesic on M starting at m in the direction ξ is deduced from horizontal ones on G by*

$$\text{Exp}_m(t\xi) = \text{Exp}_{id}(th_m(\xi)) \cdot m. \qquad (12.30)$$

Proof. We will use the fact that geodesics are locally energy-minimizing [65], so that, if $t \mapsto \gamma(t)$ is a geodesic on M, there exists a neighborhood $(s-\eta, s+\eta)$ of each given s such that γ is energy-minimizing between $\gamma(s-\eta)$ and $\gamma(s+\eta)$. Obviously without loss of generality, we take $s = 0$ in the rest of the proof.

Write $\gamma(\varepsilon) = g(\varepsilon) \cdot \gamma(0)$ for small ε. This implies that $\partial_\varepsilon \gamma = v(\varepsilon) \cdot \gamma(\varepsilon)$ with $v(\varepsilon) = DR_g(id)^{-1} \dot{\partial}_\varepsilon g$. By definition of h_m, we have $\partial_\varepsilon \gamma = h_{\gamma(\varepsilon)}(v(\varepsilon)) \cdot \gamma(\varepsilon)$, so that, if we define $\tilde{g}(\varepsilon)$ by $\tilde{g}(0) = id$ and $\partial_\varepsilon \tilde{g} = h_{\gamma(\varepsilon)}(v(\varepsilon)) \cdot \tilde{g}(\varepsilon)$, then $\tilde{\gamma}(\varepsilon) := \tilde{g}(\varepsilon) \cdot \gamma(0)$ satisfies $\tilde{\gamma}(0) = \gamma(0)$ and

$$\partial_\varepsilon \tilde{\gamma} = h_{\tilde{\gamma}(\varepsilon)}(v(\varepsilon)) \cdot \tilde{\gamma}(\varepsilon).$$

The latter is an ordinary differential equation in $\tilde{\gamma}$ which is satisfied also by γ with the same initial condition, so that γ and $\tilde{\gamma}$ coincide. All this proves that there is no loss of generality in assuming that $v(\varepsilon)$ is horizontal along γ (and this is true for any curve).

Now, assume that γ is energy-minimizing over $(s-\eta, s+\eta)$, with horizontal $v(\varepsilon)$. Then g is also energy-minimizing between $g(s - \eta)$ and $g(s + \eta)$: if not, take a curve $\tilde{g}$ with lower energy and replace γ by $\tilde{g} \cdot \gamma(0)$, which reduces the energy of γ. So $g(\varepsilon) = \text{Exp}_{id}(\varepsilon v(0))$, with, as we have seen, $v(0)$ horizontal.

Since the same construction can be done everywhere along γ, we see that any geodesic in M can be expressed as $\text{Exp}_{id}(tv) \cdot \gamma(0)$.

Conversely, any curve of the form

$$\gamma(t) = \text{Exp}_{id}(tv) \cdot \gamma(0)$$

with v horizontal at $\gamma(0)$ is a geodesic on M, since such a geodesic is uniquely specified by $\gamma(0)$ and $\dot{\gamma}_t(0) = v \cdot \gamma(0)$, which in turn uniquely specifies the horizontal vector, v. $\qquad \square$

12.5.2 The Momentum Representation

We now apply these general results to groups of diffeomorphisms. The previous discussion, however, relied on being able to prove Lemma 12.12, which itself relies on the constant rank assumption and may not hold in infinite dimensions. To make things simple, we will only consider curves in object spaces that can be written as $\varphi(t) \cdot m$ and optimize over these. Then the rest of the discussion above is valid.

Our main purpose in this section is to discuss exponential charts for the projected metric, in object spaces acted upon transitively by diffeomorphisms, which, by Proposition 12.14, directly derive from the exponential charts in groups of diffeomorphisms. (We use the term chart loosely here, and do not discuss basic properties, like being one-to-one, neither will we try to describe its range.)

As we have seen, the geodesic equation over diffeomorphisms is (under some regularity conditions) the EPDiff equation. Namely, it is given by

$$\begin{cases} \partial_t \varphi(t,.) = v(t, \varphi(t,.)) \\ \partial_t(Lv(t)) + ad^*_{v(t)}(Lv(t)) = 0. \end{cases}$$

The second equation describes, as we have remarked, the conservation of the momentum, and can also be written

$$Lv(t) = Ad_{\varphi(t)^{-1}}(Lv(0)).$$

The main part of describing exponential charts for object spaces is to characterize horizontal vector fields for the specific actions. As we will see, the simplest description will be the one of the momenta associated to horizontal vector fields, which we will call horizontal momenta.

Definition 12.15. *Given a deformable object I, let*

$$\mathcal{V}_I = \{v \in V : v \cdot I = 0\} \tag{12.31}$$

be the set of vertical vector fields at I. Then, a momentum $\rho \in V^*$ is horizontal at I if and only if $(\rho \,|\, v) = 0$ for any $v \in \mathcal{V}_I$.

Note that horizontal momenta span the set $L\mathcal{H}_I$, where $\mathcal{H}_I$ is the space of horizontal vectors considered above.

We introduce the momentum version of the exponential charts, namely the map $R : \rho_0 \mapsto \varphi(1)$ where $\varphi(1)$ is the solution, at time 1, of the EPDiff equation, starting at $\varphi(0) = id$ and $Lv(0) = \rho_0$. This map is therefore defined on V^* (more precisely, on a subset of V^* on which Theorem 11.15 is valid) and takes values in the group of diffeomorphisms.

Now, let a reference object, or template, be given, denoted I_{temp}. We restrict the object space to all possible deformations of I_{temp} (its orbit), over which the action is obviously transitive. On this space, the momentum representation must be restricted to horizontal momenta at I_{temp}. We summarize this in the following definition.

Definition 12.16. *The momentum representation of a deformable template I_{temp} is the map*

$$M_{I_{temp}} : L\mathcal{H}_{I_{temp}} \to \text{Diff}.I_{temp},$$

which associates to a horizontal momentum ρ_0 the object $R(\rho_0) \cdot I_{temp}$ where $R(\rho_0)$ is the position at time 1 of the diffeomorphism, solution of EPDiff with initial condition $Lv(0) = \rho_0$.

(More precisely, we define the momentum representation over the set of momenta in $H_{I_{temp}}$ that satisfy the hypotheses of Theorem 11.15.)

From the general discussion in the finite-dimensional case, we know that horizontality is preserved along geodesics. We retrieve this fact directly here, as a consequence of the conservation of momentum. We assume that the map

$$v \mapsto v \cdot I$$

is continuous on V.

Proposition 12.17. *Let I_0 be a deformable object and $\rho_0 \in L\mathcal{H}_{I_0}$ satisfying the hypotheses of Theorem 11.15. Let $(\rho(t), \varphi(t))$ be the evolving momentum and diffeomorphism provided by EPDiff initialized with $\rho(0) = \rho_0$. Let $I(t) = \varphi(t) \cdot I_0$ be the evolution of the deformable object. Then, at all times t, $\rho(t) \in L\mathcal{H}_{I(t)}$.*

Proof. The proof relies on the remark that if w is vertical at I and φ a diffeomorphism, then $\mathrm{Ad}_\varphi w$ is vertical at $\varphi \cdot I$.

Before proving this fact, let's see how it implies the proposition. We want to prove that $w \in \mathcal{V}_{I(t)} \Rightarrow (\rho(t) \,|\, w) = 0$. But, from the conservation of momentum we have

$$(\rho(t) \,|\, w) = (\rho_0 \,|\, \mathrm{Ad}_{\varphi(t)^{-1}} w)$$

and $\mathrm{Ad}_{\varphi(t)^{-1}} w \in \mathcal{V}_{I_0}$ if $w \in \mathcal{V}_{I(t)}$, which implies that $(\rho(t) \,|\, w) = 0$.

We now prove our claim. Let $\psi(\varepsilon)$ be such that $d\psi/d\varepsilon = w$, and $(d/d\varepsilon)\psi(\varepsilon) \cdot I = 0$ at $\varepsilon = 0$. By definition, $\mathrm{Ad}_\varphi w = (d/d\varepsilon)\varphi \circ \psi(\varepsilon) \circ \varphi^{-1}$ at $\varepsilon = 0$. Now, apply the chain rule,

$$\frac{d}{d\varepsilon}(\varphi \circ \psi(\varepsilon) \circ \varphi^{-1}) \cdot (\varphi.I) = \frac{d}{d\varepsilon}\varphi \cdot (\psi_\varepsilon \cdot I) = 0.$$

This implies that $\mathrm{Ad}_\varphi w \in V_{\varphi.I}$. $\qquad\square$

We now describe horizontal momenta in a few special cases. First assume that deformable objects are point sets. If $I = (x_1, \ldots, x_N)$, we have

$$\mathcal{V}_I = \{v \in V : v(x_1) = \cdots = v(x_N) = 0\}.$$

Letting $e_1, \ldots, e_d$ be the canonical basis of $\mathbb{R}^d$, $\mathcal{V}_I$ is therefore defined as the set of v's such that $(e_j \otimes \delta_{x_k} \,|\, v) = 0$ for all $j = 1, \ldots, d$ and $k = 1, \ldots N$. So $\mathcal{V}_I = W^\perp$, where W is the vector space generated by the $d \times N$ vectors fields $K(., x_k)e_j$. Since W is finite-dimensional, it is closed and $\mathcal{H}_I = \mathcal{V}_I^\perp = (W^\perp)^\perp = W$. Switching to momenta, we obtain the fact that, for point sets $I = (x_1, \ldots, x_N)$

$$L\mathcal{H}_I = \left\{ \sum_{k=1}^N z_k \otimes \delta_{x_k}, z_1, \ldots, z_N \in \mathbb{R}^d \right\}.$$

In particular, we see that the momentum representation is parametrized by the Nd-dimensional set $(z_1, \ldots, z_N)$ and therefore has the same dimension as the considered objects.

The description of $\mathcal{V}_I$ is still valid when I is a general parametrized subset of $\mathbb{R}^d$: $I : u \mapsto I(u) = x_u \in \mathbb{R}^d$, defined for u in a, so far, arbitrary set U. Then

$$\mathcal{V}_I = \{v \in V : v(x_u) = 0, u \in U\}$$

and we still have $\mathcal{V}_I = W^\perp$ where W is the vector space generated by the vector fields $K(., x_u)e_j$, $j = 1, \ldots, d$, $u \in U$. The difference is now that W is not finite-dimensional if I is infinite, and not a closed subspace of V so that

$$\mathcal{H}_I = (W^\perp)^\perp = \overline{W},$$

the closure of W in V. Turning to the momenta, this says that

$$L\mathcal{H}_I = \overline{\left\{\sum_{k=1}^{n} z_k \otimes \delta_{x_{u_k}}, n \geq 0, z_1, \ldots, z_n \in \mathbb{R}^d, u_1, \ldots, u_n \in U\right\}}$$

where the closure is now in V^*.

This applies to parametrized curves and surfaces, but must be adapted for geometric objects, that is, curves and surfaces seen modulo a change of parametrization. In this case, deformable objects are equivalent classes of parametrized manifolds. One way to address this is to use infinite-dimensional local charts that describe the equivalence classes in a neighborhood of a given object I. We will not detail this rigorously here, but the interested reader can refer to [146] for such a construction with plane curves.

Intuitively, however, the resulting description of $\mathcal{V}_I$ is clear. In contrast to the parametrized case, for which vector fields in $\mathcal{V}_I$ were not allowed to move any point in I, it is now possible to do so, provided the motion happens within I, i.e., the vector fields are tangent to I. This leads to the following set:

$$\mathcal{V}_I = \{v \in V, v(x) \text{ is tangent to } I, x \in I\}.$$

Since $v(x)$ tangent to I is equivalent to $N^T v(x) = 0$ for all N normal to I at x, we see that $\mathcal{V}_I = W^\perp$ where W is the vector space generated by vector fields $K(., x)N$, with $x \in I$ and N normal to I at x. Again, this implies that $\mathcal{H}_I = \overline{W}$ and that

$$L\mathcal{H}_I = \overline{\left\{\sum_{k=1}^{n} z_k \otimes \delta_{x_k}, n \geq 0, x_1, \ldots, x_n \in I, z_1, \ldots, z_n \in N_{x_k} I\right\}}$$

where $N_x I$ is the set of vectors that are normal to I at x.

Now, consider the example of smooth scalar functions (or images): $I : \mathbb{R}^d \to \mathbb{R}$. In this case, the action being $\varphi \cdot I = I \circ \varphi^{-1}$, the set $\mathcal{V}_I$ is

$$\mathcal{V}_I = \{v \in V : \nabla I^T v = 0\}$$

which directly implies that $\mathcal{V}_I = W^\perp$ where W is the vector space generated by $K(.,x)\nabla I(x)$ for $x \in \mathbb{R}^d$. Horizontal momenta therefore span the set

$$L\mathcal{H}_I = \overline{\left\{ \sum_{k=1}^{n} \nabla I(x_k) \otimes \delta_{x_k}, n \geq 0, x_1, \dots, x_n \in \mathbb{R}^d \right\}}.$$

We conclude with the action of diffeomorphisms on measures, for which:

$$(\varphi \cdot I \,|\, f) = (I \,|\, f \circ \varphi)$$

so that $v \in \mathcal{V}_I$ if and only if $(I \,|\, \nabla f^T v) = 0$ for all smooth f. So $\mathcal{V}_I = W^\perp$ where

$$W = \left\{ K(\nabla f \otimes I), f \in C^1(\Omega, \mathbb{R}) \right\}$$

so that

$$L\mathcal{H}_I = \overline{\left\{ \nabla f \otimes I, f \in C^1(\Omega.\mathbb{R}) \right\}}.$$

The momentum representation provides a diffeomorphic version of the deformable template approach described for polygons in section 7.3. As we have seen, it can be applied to a wide class of deformable objects. Applications to datasets three-dimensional medical images can be found in [208, 108, 216, 168].

13

Metamorphosis

13.1 Definitions

The infinitesimal version of the construction of Section 12.1.2 provides a metric based on transformations in which the objects can change under the action of diffeomorphisms but also under independent variations. We shall refer to such metrics as *metamorphoses* [150, 205, 206, 115]. They will result in formulations that enable both object registration and metric comparison.

We first provide an abstract description of the construction. We consider the setting in which deformations belong to a Lie group G with Lie algebra denoted V, acting on a Riemannian manifold M. We assume that V is a Hilbert space with norm $\|\cdot\|_V$; the metric on M at a given point $a \in M$ is denoted $\langle \cdot, \cdot \rangle_a$ and the corresponding norm $|\cdot|_a$.

Consider the following maps associated with the group structure and with the action. For $\varphi \in G$ and $a \in M$, define

$$
\begin{array}{ccc}
A_\varphi : M \to M & R_a : G \to M & R_\varphi : G \to G \\
b \mapsto \varphi \cdot b, & \varphi \mapsto \varphi \cdot a, & \psi \mapsto \psi\varphi.
\end{array}
\tag{13.1}
$$

The first two maps are the components of the action, and the third is the right translation on G. To simplify the expressions, we will also use the usual notation for infinitesimal actions,

$$
v \cdot \varphi = DR_\varphi(id)v \text{ and } v \cdot a = DR_a(id)v.
$$

If $(\varphi(t), t \in [0,1])$ is a differentiable curve on G, we define its velocity $v(t)$ (which is a curve in V) by the relation:

$$
\partial_t\varphi = DR_{\varphi(t)}(id)v(t) = v(t) \cdot \varphi(t).
\tag{13.2}
$$

Definition 13.1. *A* metamorphosis *is a pair of curves $(\varphi(t), \alpha(t))$ respectively on G and M, with $\varphi(0) = id$. Its image is the curve $a(t)$ on M defined by $a(t) = \varphi(t) \cdot \alpha(t)$. We will say that $\varphi(t)$ is the deformation part of the metamorphosis, and that $\alpha(t)$ is the residual part. When $\alpha(t)$ is constant, the metamorphosis is a pure deformation.*

13.2 A New Metric on M

Metamorphoses, by the evolution of their images, provide a convenient representation of combinations of a group action and of a variation on M. Let a metamorphosis $((\varphi(t), \alpha(t)), t \in [0,1])$ be given and $a(t) = \varphi(t) \cdot \alpha(t)$ be its image. Then, we can write

$$\partial_t a(t) = DA_{\varphi(t)}(\alpha(t))\partial_t\alpha(t) + DR_{\alpha(t)}(\varphi(t))\varphi_t$$
$$= DA_{\varphi(t)}(\alpha(t))\partial_t\alpha(t) + DR_{\alpha(t)}(\varphi(t))DR_{\varphi(t)}(id)v(t).$$

Since $R_\alpha \circ R_\varphi = R_{\varphi\alpha}$, we get

$$\partial_t a(t) = DA_{\varphi(t)}(\alpha(t))\partial_t\alpha(t) + DR_{a(t)}(id)v(t). \qquad (13.3)$$

In particular, when $t = 0$:

$$\partial_t a(0) = \partial_t\alpha(0) + v_0 \cdot a_0. \qquad (13.4)$$

This expression provides a decomposition of a generic element $\eta \in T_aM$ in terms of an *infinitesimal metamorphosis*, represented by an element of $V \times T_mM$. Indeed, for $a \in M$, introduce the map

$$\Phi^{(a)} : V \times T_aM \to T_aM$$
$$(v, \rho) \mapsto \rho + v \cdot a$$

so that (13.4) can be written

$$\partial_t a(0) = \Phi^{(a(0))}\left(\dot\alpha_t(0), v_0\right).$$

We now introduce the Riemannian metric associated to metamorphoses.

Proposition 13.2. *Assume that $v \mapsto v \cdot a$ is continuous on V. With $\sigma^2 > 0$, the norm*

$$\|\eta\|_a^2 = \inf\left\{|v|_V^2 + \frac{1}{\sigma^2}|\rho|_a^2 : \eta = \Phi^{(a)}(v, \rho)\right\}. \qquad (13.5)$$

defines a new Riemannian metric on M. (Note that we are using double bars instead of single ones to distinguish between the new metric and the initial one.)

Proof. Since $\Phi^{(a)}$ is onto ($\Phi^{(a)}(0, \rho) = \rho$), $\|\eta\|_a$ is finite (and bounded by $|\eta|/\sigma^2$). Define $V_a = (\Phi^{(a)})^{-1}(0)$. It is a linear subspace of $V \times T_aM$ and $\|\eta\|_a^2$ is the norm of the linear projection of $(0, \eta)$ on V_a, for the Hilbert structure on $V \times T_aM$ defined by

$$\|(v, \rho)\|_{e,a}^2 = \|v\|_V^2 + \frac{1}{\sigma^2}|\rho|_a^2.$$

Thus $\|.\|_a = \|\pi_{V_a}(0, \eta)\|_{e,a}$ is associated to an inner product. Since it is a projection on a closed subspace ($\Phi^{(a)}$ is continuous), the infimum is attained and by definition, cannot vanish unless $\eta = 0$. This therefore provides a new Riemannian metric on M. $\qquad \square$

With this metric, the energy of a curve is

$$E(a(t)) = \int_0^1 \|\partial_t a(t)\|_{a(t)}^2 \, dt$$

$$= \inf \left(\int_0^1 \|v(t)\|_V^2 \, dt + \frac{1}{\sigma^2} \int_0^1 |\partial_t a(t) - v(t).a(t)|_a^2 \, dt \right), \quad (13.6)$$

the infimum being over all curves $t \mapsto v(t)$ on V. This can also be written

$$E(a(t)) = \inf_{\varphi} \left(\int_0^1 \|v(t)\|_V^2 \, dt + \frac{1}{\sigma^2} \int_0^1 \left| DA_{\varphi(t)}(\alpha(t)) \partial_t \alpha(t) \right|_{a(t)}^2 \, dt \right). \quad (13.7)$$

The distance between two elements a and a' in M can therefore be computed by minimizing

$$U(v, a) = \int_0^1 \|v(t)\|_V^2 \, dt + \frac{1}{\sigma^2} \int_0^1 |\partial_t a(t) - v(t) \cdot a(t)|_{a(t)}^2 \, dt \quad (13.8)$$

over all curves $((v(t), a(t)), t \in [0, 1])$ on $V \times M$, with boundary conditions $a(0) = a$ and $a(1) = a'$ (no condition on v). From (13.7), this may also be seen as finding an optimal metamorphosis, by minimizing

$$\tilde{U}(\varphi, \alpha) = \int_0^1 \left\| (DR_{\varphi(t)}(id))^{-1} \partial_t \varphi(t) \right\|_V^2 \, dt + \frac{1}{\sigma^2} \int_0^1 \left| DA_{\varphi(t)}(\alpha(t)) \partial_t \alpha(t) \right|_{\alpha(t)}^2 \, dt$$

with boundary conditions $\varphi(0) = id_G$, $\alpha(0) = a$, $\varphi(1) \cdot \alpha(1) = a'$.

13.3 Euler–Lagrange Equations

Assume, to simplify, that M is, as in the examples below, a vector space (otherwise, consider that the computations that follow are done in a local chart). The problem considered in (13.8) has the general form: minimize

$$U(v, a) = \int_0^1 F\big(v(t), a(t), \dot{a}_t - v(t) \cdot a(t)\big) dt \quad (13.9)$$

for some function F. Let

$$z(t) = \partial_t a - v(t) \cdot a(t). \quad (13.10)$$

We will denote by $\partial F/\partial v$, $\partial F/\partial a$ and $\partial F/\partial z$ the partial differentials of F with respect to each of its variables. Computing the variation with respect to v, we get, for all $t \mapsto h(t) \in V$

$$\int_0^1 \left(\frac{\partial F}{\partial v} \,\middle|\, h(t) \right) dt - \int_0^1 \left(\frac{\partial F}{\partial z} \,\middle|\, h(t) \cdot a(t) \right) dt = 0.$$

Denoting by $Q_a(\partial F/\partial z)$ the linear form

$$\left(Q_a \frac{\partial F}{\partial z} \mid h(t)\right) := \left(\frac{\partial F}{\partial z} \mid h(t) \cdot a\right),$$

we obtain the optimality equation for v

$$\frac{\partial F}{\partial v} - Q_{a(t)} \frac{\partial F}{\partial z} = 0. \tag{13.11}$$

If we now make the variation with respect to a, the result is, for every $t \mapsto \alpha(t) \in M$,

$$\int_0^1 \left(\frac{\partial F}{\partial a} \mid \alpha(t)\right) dt + \int_0^1 \left(\frac{\partial F}{\partial z} \mid \dot{\alpha}_t - v(t) \cdot \alpha(t)\right) dt = 0.$$

Integrating by parts and using the notation

$$\left(\tilde{Q}_v \frac{\partial F}{\partial z} \mid \alpha(t)\right) := \left(\frac{\partial F}{\partial z} \mid v \cdot \alpha(t)\right),$$

we obtain the optimality equation

$$-\partial_t \frac{\partial F}{\partial z} - \tilde{Q}_v \frac{\partial F}{\partial z} + \frac{\partial F}{\partial a} = 0. \tag{13.12}$$

Equations (13.10), (13.11) and (13.12) provide the Euler–Lagrange equations for metamorphosis. They also provide the differentials of the energy with respect to v and a and can be used to design minimization algorithms.

13.4 Applications

We now specialize this construction to landmark and image-matching problems. This essentially means working out the expressions of $v \cdot a$ and $d_m A_\varphi$, Q_a and $\tilde{Q}_v$ in these situations.

13.4.1 Labeled Point Sets

We consider here diffeomorphisms acting on collections of points $a = (y^{(1)}, \dots, y^{(N)})$, with $y^{(k)} \in \Omega$. We therefore have $M = \Omega^N$. We consider the function [36, 137]

$$F(v, a, z) = \|v\|_V^2 + \frac{1}{\sigma^2} \sum_{k=1}^N |z^{(k)}|^2.$$

(Note that F does not depend on a.) Here, since $\varphi(a) = (\varphi(y^{(1)}), \dots, \varphi(y^{(N)}))$, we have

$$v \cdot a = (v(y^{(1)}), \dots, v(y^{(N)}))$$

so that $z^{(k)} = \partial_t y^{(k)} - v(y^{(k)})$. We can therefore write

$$U(v, a) = \int_0^1 \|v(t)\|_V^2 \, dt + \frac{1}{\sigma^2} \sum_{k=1}^N \int_0^1 \left| \dot{y}_t^{(k)} - v(t, y^{(k)}(t)) \right|^2 dt.$$

We have $\partial F / \partial v = 2Lv$ and $\partial F / \partial z = (2/\sigma^2)(z^{(1)}, \ldots, z^{(N)})$. Moreover,

$$\left(Q_a \frac{\partial F}{\partial z} \,\middle|\, h \right) = \frac{2}{\sigma^2} \sum_{k=1}^N (z^{(k)})^T h(y^{(k)})$$

so that the first Euler–Lagrange equation is

$$Lv - \frac{1}{\sigma^2} \sum_{k=1}^N z^{(k)} \otimes \delta_{y^{(k)}} = 0.$$

For the second equation, we write

$$\left(\tilde{Q}_v \frac{\partial F}{\partial z} \,\middle|\, \alpha \right) = \frac{2}{\sigma^2} \sum_{k=1}^N (z^{(k)})^T Dv(y^{(k)}) \alpha_k$$

yielding

$$-\partial_t z^{(k)} - Dv(y^{(k)})^T z^{(k)} = 0.$$

This provides the system of Euler–Lagrange equations for labeled point-set metamorphosis:

$$\begin{cases} Lv = \dfrac{1}{\sigma^2} \displaystyle\sum_{k=1}^N z^{(k)} \otimes \delta_{y^{(k)}}, \\[2ex] \partial_t z^{(k)} + Dv(y^{(k)})^T z^{(k)} = 0, \\[2ex] \partial_t y^{(k)} - v(y^{(k)}) = 0. \end{cases} \qquad (13.13)$$

Note that, introducing the reproducing kernel of V, the first equation is equivalent to

$$v(t, x) = \sum_{k=1}^N K(x, y^{(k)}(t)) z^{(k)}(t).$$

This implies that minimizing E is equivalent to minimizing

$$\tilde{E}(a, z) = \sum_{k,l=1}^N \int_0^1 z^{(k)}(t)^T K(y^{(k)}(t), y^{(l)}(t)) z^{(l)}(t) dt$$

$$+ \frac{1}{\sigma^2} \sum_{k=1}^N \int_0^1 \left| \partial_t y^{(k)}(t) - \sum_{l=1}^n K(y^{(k)}(t), y^{(l)}(t)) z^{(l)}(t) \right|^2 dt.$$

The vectors $z^{(1)}, \ldots, z^{(N)}$ can be computed explicitly given the trajectories $y^{(1)}, \ldots, y^{(N)}$, namely

$$z = (S(y) + \lambda I)^{-1} \partial_t y.$$

This provides an expression of the energy in terms of $y^{(1)}, \ldots, y^{(N)}$ that can be minimized directly, as proposed in [36]. Most of the methods developed for point sets in Chapter 11 can in fact be adapted to this new framework. Figure 13.1 provides examples of deformations computed with this method.

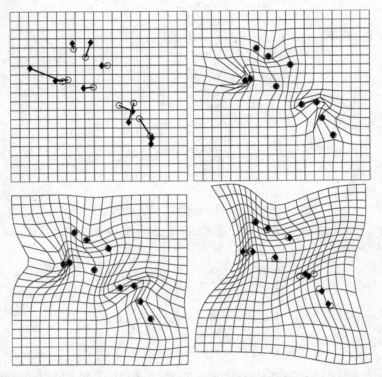

Fig. 13.1. Point-set matching using metamorphosis, with the same input as in Figures 10.1 and 11.1, using Gaussian kernels $K(x,y) = \exp(-|x-y|^2/2\sigma^2)$ with $\sigma = 1, 2, 4$ in grid units.

13.4.2 Deformable Images

Let M be a set of square integrable and differentiable functions $a : \Omega \to \mathbb{R}$. We let G act on M by $ga = a \circ g^{-1}$. We use the L^2 norm as initial metric on M. We have $v \cdot a = -\nabla a^T v$. Using this, we define

$$F(v, a, z) = \|v\|_V^2 + \frac{1}{\sigma^2}\|z\|_2^2.$$

Here again, F does not depend on a.

Equation (13.10) is $z(t) = \partial_t a(t) + \nabla a(t)^T v(t)$. We have

$$\left(Q_a \frac{\partial F}{\partial z}\,|\,h\right) = -\frac{2}{\sigma^2}\int_\Omega z\nabla a^T h\, dx$$

so that (13.11) is $Lv = -z\nabla a \otimes dx$. Also,

$$\left(\tilde{Q}_v \frac{\partial F}{\partial z}\,|\,\alpha\right) = -\frac{2}{\sigma^2}\int_\Omega z^T v^T \nabla a\, dx$$

so that, using the divergence theorem, $\tilde{Q}_v \frac{\partial F}{\partial z} = \mathrm{div}(zv)$ and (13.12) is

$$-\dot{z}_t - \mathrm{div}(zv) = 0.$$

So the evolution equations for image metamorphosis are

$$\begin{cases} \partial_t a + \nabla a^T v = z, \\[2mm] \partial_t z + \mathrm{div}(zv) = 0, \\[2mm] Lv = -(z\nabla a) \otimes dx. \end{cases}$$

Figures 13.2, 13.3 and 13.4 provide examples of images matched using this energy. In these examples, the first and last images are given in input and two interpolated images are provided. The numerical scheme is described in the next section.

13.4.3 Discretization of Image Metamorphosis

Images in Figures 13.2 and 13.3 have been obtained after minimization of a discretized version of the image metamorphosis energy,

$$E(a, v) = \int_0^1 \|v(t)\|_V^2 dt + \frac{1}{\sigma^2}\int_0^1 \|\partial_t a(t) + \nabla a(t)^T v(t)\|_2^2 dt.$$

The second integrand must be discretized with care. It represents the total derivative of a along the flow. Stable results can be obtained using the following scheme.

Introduce the auxiliary variable $w = L^{1/2}v$, so that $v = K^{1/2}w$ where $K = L^{-1}$ (the numerical implementation can in fact explicitly specify $\tilde{K} = K^{1/2}$ and use $K = \tilde{K}^2$). The following discretized energy has been used in the experiments of Figures 13.2 and 13.3:

$$E = \sum_{t=1}^T \sum_{x\in\tilde{\Omega}} |w(x)|^2 + \lambda\delta_t^{-2}\sum_{t=1}^T \sum_{x\in\tilde{\Omega}} \left| a(t+1, x + \delta t(K^{1/2}w)(t, x)) - a(t, x)\right|^2$$

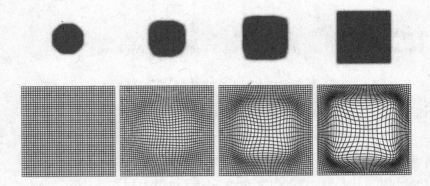

Fig. 13.2. Image metamorphosis. Estimation of a geodesic between a disc and a square. *First row:* image evolution in time. *Second row:* evolution of the diffeomorphism.

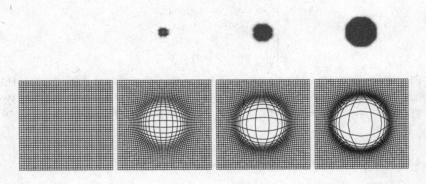

Fig. 13.3. Creation of new patterns with image metamorphosis. Geodesic between an empty image and a disc. *First row:* image evolution in time. *Second row:* evolution of the diffeomorphism.

where x and t now are discrete variables, $\tilde{\Omega}$ a discrete grid on Ω and δt the time discretization step (the space discretization is 1). The optimization algorithm alternates a few steps of nonlinear conjugate gradient in v, and a few steps of linear conjugate gradient in a [90, 91].

13.4.4 Application: Comparison of Plane Curves

Comparison Based on Orientation

We consider the issue of comparing plane curves based on the orientation of their tangents [226, 222, 127]. If m is a plane curve parametrized with arc

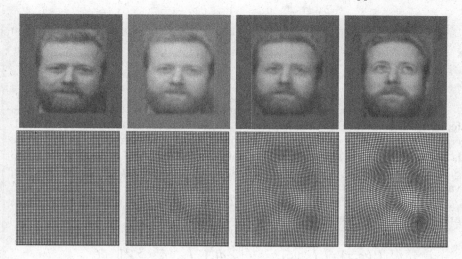

Fig. 13.4. Gray-level image metamorphosis. Geodesic between two gray-level face images. *First row:* image evolution in time. *Second row:* evolution of the diffeomorphism.

length, and L is it length, we define the normalized tangent T^m by

$$T^m : [0,1] \to [0,1]$$
$$s \mapsto \dot{m}_s(Ls).$$

The function T^m characterizes m up to translation and scaling, and the pair (L, T^m) characterizes m up to translation.

We identify, as we have often done before, the sets $\mathbb{R}^2$ and $\mathbb{C}$. We consider the group of diffeomorphisms G of $\Omega = (0,1)$, which acts on functions $T :$ $[0,1] \to S^1$ by $\varphi T \doteq T \circ \varphi^{-1}$. We also consider the Hilbert space V of functions $v : [0,1] \to \mathbb{R}$ such that $v(0) = v(1) = 0$ and

$$\|v\|_V^2 = \int_0^1 \dot{v}_s^2 ds. \tag{13.14}$$

Like for image matching, we consider the function (with $a = T$)

$$F(v, T, z) = \|v\|_V^2 + \|z\|_2^2. \tag{13.15}$$

Here, the independence on T is only apparent. Indeed, this function is defined for $z = \dot{T}_t + v\dot{T}_s$ which is orthogonal to T since T has unit norm, so F must be restricted to v, T, z such that $|T| = 1$ and $\langle T, z \rangle = 0$.

The associated Euler–Lagrange equations are

$$\begin{cases} \partial_t T + \partial_s T v = z, \\[2mm] -\partial_s^2 v_{ss} = -z \partial_s T, \\[2mm] \partial_t z + \partial_s(vz) = 0. \end{cases}$$

There is, however, a characterization of the optimal solutions with the diffeomorphism φ related to v by $\partial_t \varphi = v \circ \varphi$. (Note that this ordinary differential equation does not necessarily have a unique solution for a generic $v \in V$. This is because the norm selected on V is not admissible in the sense of Definition 8.12, since V is only included in $C^0([0,1])$. The following computations are valid for smooth v.)

Define $\tau = T \circ \varphi$ so that $\dot{\tau}_t = (\dot{T}_t + \dot{T}_s v) \circ \varphi$. Using the relation between φ and v, the cost of the path $(\varphi(t,.), \tau(t,.))$ is then

$$\tilde{U}(\varphi, \tau) = \int_0^1 \left\| \dot{\varphi}_t \circ \varphi^{-1} \right\|_V^2 dt + \int_0^1 \left\| \dot{\tau}_t \circ \varphi^{-1} \right\|_2^2 dt.$$

Expand the derivative and make the change of variables $x = \varphi(t, y)$ to write

$$\tilde{U}(\varphi, \tau) = \int_0^1 \int_0^1 \left[(\ddot{\varphi}_{ts})^2 (\dot{\varphi}_s)^{-1} + \dot{\varphi}_s \left| \dot{\tau}_t \right|^2 \right] ds\, dt.$$

Now, assume that $q(t,.)$ is differentiable in t and is such that

$$q(t, s)^2 = \dot{\varphi}_s \tau$$

(τ being considered as a unit complex number). Such a q always exists: for all s, there exists (since $|\tau(.,s)| = 1$) a unique lifting $\tau(t, s) = \exp(i\theta(t, s))$ with $\theta(t, s)$ differentiable in t and $\theta(0, s) \in [0, 2\pi)$. One can then define

$$q(t, s) = \sqrt{\dot{\varphi}_s} \exp(i\theta/2). \tag{13.16}$$

This path is not uniquely defined, since one can always change the sign of $q(t, s)$, uniformly in t (for q to remain differentiable in t), separately for each s.

We have

$$\frac{\partial q^2}{\partial t} = \ddot{\varphi}_{ts} \tau + \dot{\varphi}_s \dot{\tau}_t.$$

Taking the complex modulus and using the fact that τ is orthogonal to its derivative, we obtain

$$\left| \partial_t(q^2) \right|^2 = \ddot{\varphi}_{ts}^2 + \dot{\varphi}_s^2 \left| \dot{\tau}_t \right|^2.$$

Moreover

$$\left| \partial_t(q^2) \right|^2 = 4|q|^2 \left| \dot{q}_t \right|^2 = 4\dot{\varphi}_s \left| \dot{q}_t \right|^2$$

which implies

$$|\dot{q}_t|^2 = \frac{1}{4}\left[\ddot{\varphi}_{ts}^2\dot{\varphi}_s^{-1} + \dot{\varphi}_s\,|\dot{\tau}_t|^2\right].$$

Therefore

$$\tilde{U}(\varphi,\tau) = 4\int_0^1\int_0^1 |\dot{q}_t|^2\,dsdt. \tag{13.17}$$

We have the following theorem.

Theorem 13.3. *Let two pairs (φ^0,τ^0) and (φ^1,τ^1) be given, with $\dot{\varphi}_s^0$ and $\dot{\varphi}_s^1$ positive almost everywhere on $[0,1]$. The minimum of $\tilde{U}(\varphi,\tau)$ over all paths $(\varphi(t,.),\tau(t,))$ linking (φ^0,τ^0) and (φ^1,τ^1) and consisting of diffeomorphisms with square integrable derivative and unit vector, is equal to the minimum of*

$$\bar{U}(q) = 4\int_0^1\int_0^1 |\dot{q}_t|^2\,ds$$

over all paths of square integrable functions q such that $q(0,.)^2 = \dot{\varphi}_s^0\tau^0$ and $q(1,.)^2 = \dot{\varphi}_s^1\tau^1$.

Proof. We can already state that the minimum of $\tilde{U}(\varphi,\tau)$ is larger than the minimum of $\bar{U}$, since we have shown that we can always build q from (φ,τ). So, we need to prove the converse inequality, which requires building a valid path (φ,τ) from an optimal q.

First, note that, the computation of the minimum of the right-hand term of (13.17) with fixed boundary conditions q^0 and q^1 can be handled similarly to the distance for one-dimensional diffeomorphisms that we have computed at the end of Section 12.4. The only difference here is that q is a complex-valued function. Since we still have the constraint that the integral of the squared modulus of q is 1, we are still computing geodesics on an infinite-dimensional sphere. The optimal path is therefore given by

$$q(t,s) = (\sin((1-t)\alpha)q^0(s) + \sin(t\alpha)q^1(s))/\sin\alpha$$

with $\alpha = \arccos\langle q^0,q^1\rangle_2$ and

$$\langle q^0,q^1\rangle_2 = \int_0^1 \Re(q^0(s)^*q^1(s))ds$$

where q^* is the complex conjugate of q. For any choice of q^0 and q^1 satisfying $(q^0)^2 = \dot{\varphi}_s^0\tau^0$ and $(q^1)^2 = \dot{\varphi}_s^1\tau^1$, the optimal path q uniquely specifies the path $(\varphi(t,.),\tau(t,.))$ as long as q does not vanish, which must now be checked.

Let us choose q^0 and q^1 to ensure that α is minimal. This is achieved by choosing, for all s, $q^0(s)$ and $q^1(s)$ such that $\Re(q^0(s)^*q^1(s)) \geq 0$ (since arccos is decreasing); we then have

$$|q(t,s)|^2 = \frac{1}{\sin^2\alpha}(\sin^2((1-t)\alpha)|q^0(s)|^2 + \sin^2(t\alpha)|q^1(s)|^2$$
$$+ 2\sin((1-t)\alpha)\sin(t\alpha)\Re(q^0(s)\bar{q}^1(s)))$$
$$\geq \frac{1}{\sin^2\alpha}(\sin^2((1-t)\alpha)|q^0(s)|^2 + \sin^2(t\alpha)|q^1(s)|^2).$$

The last inequality shows that the function $q(t,s)$ can vanish (for $t \in (0,1)$) only if both $q_0(s)$ and $q_1(s)$ vanish. We are assuming that this can happen only when s belongs to a set of Lebesgue's measure 0 in $[0,1]$. On this set, the function $t \mapsto \tau(t,s)$ is undefined, but we can make an arbitrary choice without changing the value of α.   □

We have computed the minimum of $\tilde{U}$ with fixed boundary conditions (φ^0, τ^0) and (φ^1, τ^1). Returning to the original problem, the boundary conditions are in fact only constrained by $\tau^0 = T^0 \circ \varphi^0$ and $\tau^1 = T^1 \circ \varphi^1$. So the distance between T^0 and T^1 is therefore obtained by minimizing the above with respect to φ^0 and φ^1. Since, by construction, the metric is invariant by right composition with diffeomorphisms, there is no loss of generality in taking $\varphi^0 = id$ so that

$$d_0(T^0, T^1) = 2 \arccos \sup_{\varphi} \int_0^1 \sqrt{\dot{\varphi}_s} \left| \cos((\theta^0 - \theta^1 \circ \varphi)/2) \right| dx \qquad (13.18)$$

with $T^0 = e^{i\theta^0}$ and $T^1 = e^{i\theta^1}$. The existence (under some conditions on θ and θ') of an optimal φ in the previous formula is proved in [204].

We can add a rotation parameter to the minimization. The distance is equivariant for the action of rotations, so that optimization with respect to it can also be included in the definition to yield a translation, scale, parametrization and rotation-invariant distance. The formula is

$$\tilde{d}_0(T^0, T^1)$$
$$= 2 \inf_{\varphi, \alpha} \arccos \int_0^1 \sqrt{\dot{\varphi}_s(x)} \left| \cos((\theta^0(x) - \theta^1 \circ \varphi(x) - \alpha)/2) \right| dx \qquad (13.19)$$

Case of Closed Curves

Note that the previous developments have been implicitly made for open curves, since we did not ensure that curves were periodic. For closed curves, the boundary condition on v should be $v(0) = v(1)$, but non-necessarily 0. From a practical point of view, in the construction of the distance, this can be taken into account by introducing an additional variable in the transformations that can act on a curve parametrization, under the form of an offset, s_{off}. In fact, a change of offset on the parametrization of a curve can be seen as an action, namely $s_{\text{off}} \cdot m(s) = m(s + s_{\text{off}})$ where the addition is modulo 1. Because the distance is equivariant for this action, the offset parameter can be be included in the optimization process for closed curves, yielding

$$\bar{d}_0(T^0, T^1)$$
$$= 2 \inf_{\varphi, s_{\text{off}}, \alpha} \arccos \int_0^1 \sqrt{\dot{\varphi}_s(x)} \left| \cos((\theta^0(x + s_{\text{off}}) - \theta^1 \circ \varphi(x) - \alpha)/2) \right| dx.$$
$$(13.20)$$

This simple way of dealing with closed curves is not completely satisfactory, however, because it fails to enforce the closedness constraint for the evolving curves $m(t,.)$. Although this issue can be addressed [225], describing this would push us too far away from the scope of this book.

Figure 13.5 provides examples of curve matching that can be obtained using this method.

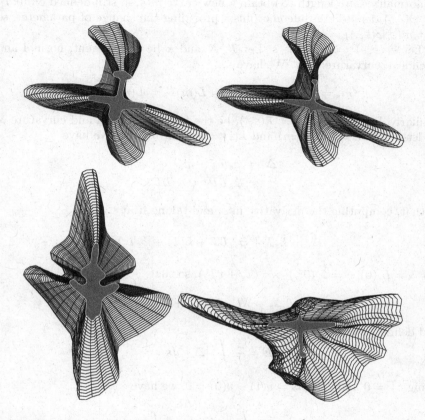

Fig. 13.5. Diffeomorphic matching between curves. In each image, the geodesic is computed between the outer curve and the inner curve. Scaling is for visualization only (the computation is scale invariant).

Infinitesimal Analysis

We now provide an infinitesimal justification for the choice of the metric (13.15). This can also provide a way to define alternative curve metrics.

We consider infinitesimal variations of plane curves. Let m be a parametrized curve, defined on $[0,1]$, such that, for all s

$$|\dot{m}_s| \equiv L(m)$$

where $L(m)$ is the length of m (this corresponds to the arc length normalized by the length). Consider a small perturbation along m, denoted $s \mapsto \varepsilon\Delta(s) \in \mathbb{R}^2$. We assume that ε is small and only keep terms of order one in the following computation. The deformed curve is $s \mapsto m(s) + \varepsilon\Delta(s)$, which we parametrize by normalized arc length to obtain a new curve $\tilde{m}(\varepsilon, .)$, still defined on $[0, 1]$. Let $\varphi(\varepsilon, .)$ denote the diffeomorphism providing the change of parameter, so that $\tilde{m}(\varepsilon, \varphi(\varepsilon, .)) = m + \varepsilon\Delta$.

Let $\varepsilon v(\varepsilon, s) = \varphi(\varepsilon, s) - s$. Let T, N and κ be the tangent, normal and Euclidean curvature of m. We have

$$\dot{m}_s = L(m)T \quad \text{and} \quad \ddot{m}_{ss} = L(m)\dot{T}_s = L(m)^2\kappa N.$$

Similarly, let $\tilde{T}(\varepsilon, .)$, $\tilde{N}(\varepsilon, .)$, $\tilde{\kappa}(\varepsilon, .)$, be the tangent, normal and curvature of $\tilde{m}$; let finally $L = \text{length}(m)$ and $\tilde{L}(\varepsilon) = \text{length}(\tilde{m}(\varepsilon, .))$. We have

$$\varepsilon\dot{\Delta}_s = \dot{\varphi}_s\dot{\tilde{m}}_s - \dot{m}_s$$
$$= \dot{\varphi}_s\tilde{L}\tilde{T} \circ \varphi - LT$$

so that, computing the derivative in ε, and taking it at $\varepsilon = 0$:

$$\dot{\Delta}_s = \dot{\tilde{L}}_\varepsilon T + \ddot{\varphi}_{s\varepsilon}LT + L(\dot{\tilde{T}}_\varepsilon + \dot{\varphi}_\varepsilon\dot{T}_s).$$

Let $\lambda = \dot{\tilde{L}}_\varepsilon(0), v = \dot{\varphi}_\varepsilon(0, .), z = (\dot{\tilde{T}}_\varepsilon + v\dot{T}_s)$, so that

$$\Delta_s = \lambda T + L\dot{v}_s T + Lz,$$

and define

$$E = \frac{1}{L}\int_0^1 |\dot{\Delta}_s|^2 ds.$$

Using $z_T^T = 0$ and $\int_0^1 \dot{v}_s ds = v(1) - v(0) = 0$, we have

$$E = \frac{\lambda^2}{L} + L\int_0^1 |\dot{v}_s|^2\,ds + L\int_0^1 |z|^2 dx$$
$$= \frac{\lambda^2}{L} + L(\|v\|_V^2 + \|z\|_2^2)$$

where $\|v\|_V$ is given by (13.14). This therefore provides the functional F we have used for scale-invariant curve metamorphoses combined with length terms. In fact, this can be interpreted as metamorphoses for pairs of the form $o = (L, T)$, still under the action of diffeomorphisms of $[0, 1]$, with the extended functional

$$\tilde{F}(v, (L, T), (\lambda, Z)) = \frac{\lambda^2}{L} + L(\|v\|_V^2 + \|z\|_2^2).$$

(Given $L > 0$ and $T : [0,1] \mapsto \mathbb{R}^2$ such that $|T(s)| = 1$ for all s, there exists a unique curve m (modulo translation) with length L and having T as a unit tangent.)

The corresponding problem can be analyzed with an approach similar to the computation made before, this time with $q = \sqrt{L\dot{\varphi}_s} \exp(i\theta/2)$ instead of (13.16). The isometric space is even simpler, since it is $L^2([0,1],\mathbb{C})$ instead of a unit sphere. This is detailed in [222]. The resulting distance is (compare to (13.18))

$$d_1(o^0, o^1) =$$

$$\inf_{\varphi} \left(L^0 + L^1 - \sqrt{2}L^0L^1 \int_0^1 \sqrt{\dot{\varphi}_s}\, \left|\cos((\theta^0 - \theta^1 \circ \varphi)/2)\right| ds \right)^{1/2}, \quad (13.21)$$

so that $d_1(o^0, o^1) = (L^0 + L^1 - 2L^0L^1 \cos^2 d_0(o^0, o^1))^{1/2}$.

Instead of using the energy E, one can choose a scale-invariant formulation and take $\bar{E} = E/L$, which yields

$$\tilde{F}(v, (L, T), (\lambda, Z)) = \frac{\lambda^2}{L^2} + \|v\|_V^2 + \|z\|_2^2.$$

The resulting distance is now

$$d_2(o^0, o^1) =$$

$$\inf_{\varphi} \left(\log^2(L^0/L^1) + 4a\cos^2 \int_0^1 \sqrt{\dot{\varphi}_s}\, \left|\cos((\theta^0 - \theta^1 \circ \varphi)/2)\right| ds \right)^{1/2} \quad (13.22)$$

or $d_2(o^0, o^1) = (\log^2(L^0/L^1) + d_0(o^0, o^1)^2)^{1/2}$.

A

Elements from Hilbert Space Theory

A.1 Definition and Notation

A set H is a (real) Hilbert space if:

(i) H is a vector space on $\mathbb{R}$ (addition and scalar multiplication are defined on H).

(ii) H has an inner product denoted $(h, h') \mapsto \langle h, h' \rangle_H$, for $h, h' \in H$. This inner product is a bilinear form, which is positive definite and the associated norm is $\|h\|_H = \sqrt{\langle h, h \rangle_H}$.

(iii) H is a complete space for the topology associated to the norm.

Converging sequences for the norm topology are sequences h_n for which there exists $h \in H$ such that $\|h - h_n\|_H \to 0$. Property (iii) means that if a sequence $(h_n, n \geq 0)$ in H is a Cauchy sequence, i.e., it collapses in the sense that, for every positive ε there exists $n_0 > 0$ such that $\|h_n - h_{n_0}\|_H \leq \varepsilon$ for $n \geq n_0$, then it necessarily has a limit: there exists $h \in H$ such that $\|h_n - h\|_H \to 0$ when n tends to infinity.

If condition (ii) is weakened to the fact that $\|.\|_H$ is a norm (not necessarily induced by an inner product), one says that H is a Banach space.

If H satisfies (i) and (ii), it is called a pre-Hilbert space. On pre-Hilbert spaces, the Schwartz inequality holds:

Proposition A.1 (Schwartz inequality). *If H is pre-Hilbert, and $h, h' \in H$, then*

$$\langle h, h' \rangle_H \leq \|h\|_H \|h'\|_H.$$

The first consequence of this property is:

Proposition A.2. *The inner product on H is continuous for the norm topology.*

Proof. The inner product is a function $H \times H \to \mathbb{R}$. Letting $\varphi(h, h') = \langle h, h' \rangle_H$, we have, by Schwartz inequality, and introducing a sequence (h_n) which converges to h

$$|\varphi(h, h') - \varphi(h_n, h')| \leq \|h - h_n\|_H \|h'\|_H \to 0,$$

which proves the continuity with respect to the first coordinate, and also with respect to the second coordinate by symmetry.   □

Working with a complete normed vector space is essential when dealing with infinite sums of elements: if (h_n) is a sequence in H, and if $\|h_{n+1} + \cdots + h_{n+k}\|$ can be made arbitrarily small for large n and any $k > 0$, then completeness implies that the series $\sum_{n=0}^{\infty} h_n$ has a limit in h. In particular, absolutely converging series in H converge:

$$\sum_{n \geq 0} \|h_n\|_H < \infty \Rightarrow \sum_{n \geq 0} h_n \text{ converges.} \tag{A.1}$$

(In fact, (A.1) is equivalent to (iii) in normed spaces.)

We shall add a fourth condition to our definition of a Hilbert space:

(iv)H is separable for the norm topology: there exists a countable subset S in H such that, for any $h \in H$ and any $\varepsilon > 0$, there exists $h' \in S$ such that $\|h - h'\| \leq \varepsilon$.

In the following, Hilbert spaces will always be separable without further mention.

A Hilbert space isometry between two Hilbert spaces H and H' is an invertible linear map $\varphi : H \to H'$ such that, for all $h, h' \in H$, $\langle \varphi(h), \varphi(h') \rangle_{H'} = \langle h, h' \rangle_H$.

A Hilbert subspace of H is a subspace H' of H (i.e., a non-empty subset, invariant under linear combination) which is closed for the norm topology. Closedness implies that H' is itself a Hilbert space, since Cauchy sequences in H' also are Cauchy sequences in H, hence converge in H, hence in H' since H' is closed. The next proposition shows that every finite-dimensional subspace is a Hilbert subspace.

Proposition A.3. *If K is a finite-dimensional subspace of H, then K is closed in H.*

Proof. let $e_1, \ldots, e_p$ be a basis of K. Let (h_n) be a sequence in K which converges to some $h \in H$; we need to show that $h \in K$. We have, for some coefficients (a_{kn}), $h_n = \sum_{k=1}^{p} a_{kn} e_k$ and for all $l = 1, \ldots, p$:

$$\langle h_n, e_l \rangle_H = \sum_{k=1}^{p} \langle e_k, e_l \rangle_H a_{kn}.$$

Let a_n be the vector in $\mathbb{R}^p$ with coordinates $(a_{kn}, k = 1, \ldots, p)$ and $u_n \in \mathbb{R}^p$ with coordinates $(\langle h_n, e_k \rangle_H, k = 1, \ldots, p)$. Let also S be the matrix with coefficients $s_{kl} = \langle e_k, e_l \rangle_H$, so that the previous system may be written: $u_n = Sa_n$. The matrix S is invertible: if b belongs to the null space of S, a quick computation shows that

$$b^T S b = \left\| \sum_{i=1}^p b_i e_i \right\|_H^2 = 0,$$

which is only possible when $b = 0$, since $(e_1, \ldots, e_n)$ is a basis of K. We therefore have $a_n = S^{-1} u_n$. Since u_n converges to u with coordinates $(\langle h, e_k \rangle_H, k = 1, \ldots, p)$ (by continuity of the inner product), we obtain the fact that a_n converges to $a = S^{-1} u$. But this implies that $\sum_{k=1}^p a_{kn} e_k \to \sum_{k=1}^p a_k e_k$ and since the limit is unique, we have $h = \sum_{k=1}^p a_k e_k \in K$. $\qquad \square$

A.2 Examples

A.2.1 Finite-Dimensional Euclidean spaces

$H = \mathbb{R}^n$ with $\langle h, h' \rangle_{\mathbb{R}^n} = \sum_{j=1}^n h_i h_i'$ is the standard example of a finite-dimensional Hilbert space.

A.2.2 ℓ^2 Space of Real Sequences

Let H be the set of real sequences $h = (h_1, h_2, \ldots)$ such that $\sum_{i=1}^\infty h_i^2 < \infty$. Then H is a Hilbert space, with dot product

$$\langle h, h' \rangle = \sum_{i=1}^\infty h_i h_i'.$$

A.2.3 L^2 Space of Functions

Let k and d be two integers. Let Ω be an open subset of $\mathbb{R}^k$. We define $L^2(\Omega, \mathbb{R}^d)$ as the set of all square integrable functions $h : \Omega \to \mathbb{R}^d$, with inner product

$$\langle h, h' \rangle_{L^2} = \int_\Omega h(x) h'(x) dx.$$

Integrals are taken with respect to the Lebesgue measure on Ω, and two functions which coincide everywhere except on a set of null Lebesgue measure are identified. The fact that $L^2(\Omega, \mathbb{R}^d)$ is a Hilbert space is a standard result in integration theory.

A.3 Orthogonal Spaces and Projection

Let O be a subset of H. The orthogonal space to O is defined by

$$O^{\perp} = \left\{ h : h \in H, \forall o \in O, \langle h, o \rangle_H = 0 \right\}.$$

Theorem A.4. $O^{\perp}$ *is a sub-Hilbert space of* H.

Proof. Stability by linear combination is obvious, and closedness is a consequence of the continuity of the inner product. $\square$

When K is a subspace of H and $h \in H$, one can define the variational problem:

$$(\mathcal{P}_K(h)) : \text{ find } k \in K \text{ such that } \|k - h\|_H = \min\left\{ \|h - k'\|_H : k' \in K \right\}$$

The following theorem is fundamental:

Theorem A.5. *If* K *is a closed subspace of* H *and* $h \in H$, $(\mathcal{P}_K(h))$ *has a unique solution, characterized by the property* $k \in K$ *and* $h - k \in K^{\perp}$.

Definition A.6. *The solution of problem* $(\mathcal{P}_K(h))$ *in the previous theorem is called the orthogonal projection of* h *on* K *and will be denoted* $\pi_K(h)$.

Proposition A.7. $\pi_K : H \to K$ *is a linear, continuous transformation and* $\pi_{K^{\perp}} = id - \pi_K$.

Proof. We prove the theorem and the proposition together. Let

$$d = \inf\left\{ \|h - k'\|_H : k' \in K \right\}.$$

The proof relies on the following construction: if $k, k' \in K$, a direct computation shows that

$$\left\| h - \frac{k + k'}{2} \right\|_H^2 + \|k - k'\|_H^2 / 4 = \left(\|h - k\|_H^2 + \|h - k'\|_H^2 \right) / 2.$$

The fact that $(k + k')/2 \in K$ implies $\left\| h - \frac{k+k'}{2} \right\|_H^2 \geq d$ so that

$$\|k - k'\|_H^2 \leq \frac{1}{2} \left(\|h - k\|_H^2 + \|h - k'\|_H^2 \right) - d^2.$$

Now, from the definition of the infimum, one can find a sequence k_n in K such that $\|k_n - h\|_H^2 \leq d^2 + 2^{-n}$ for each n. The previous inequality implies that

$$\|k_n - k_m\|_H^2 \leq \left(2^{-n} + 2^{-m} \right) / 2,$$

which implies that k_n is a Cauchy sequence, and therefore converges to a limit k which belongs to K, since K is closed, and such that $\|k - h\|_H = d$. If the minimum is attained for another k' in K, we have, by the same inequality

$$\|k - k'\|_H^2 \leq \frac{1}{2}\left(d^2 + d^2\right) - d^2 = 0,$$

so that $k = k'$ and uniqueness is proved, so that π_K is well-defined.

Let $k = \pi_K(h)$ and $k' \in K$ and consider the function $f(t) = \|h - k - tk'\|_H^2$, which is by construction minimal for $t = 0$. We have $f(t) = \|h - k\|_H^2 - 2t\langle h - k,\, k'\rangle_H + t^2\|k'\|_H^2$ and this can be minimal at 0 only if $\langle h - k,\, k'\rangle_H = 0$. Since this has to be true for every $k' \in K$, we obtain the fact that $h - k \in K^\perp$. Conversely, if $k \in K$ and $h - k \in K^\perp$, we have for any $k' \in K$

$$\|h - k'\|_H^2 = \|h - k\|_H^2 + \|k - k'\|_H^2 \geq \|h - k\|_H^2$$

so that $k = \pi_K(h)$. This proves Theorem A.5.

We now prove the proposition: let $h, h' \in H$ and $\alpha, \alpha' \in \mathbb{R}$. Let $k = \pi_K(h)$, $k' = \pi_K(h')$; we want to show that $\pi_K(\alpha h + \alpha' h') = \alpha k + \alpha' k'$ for which it suffices to prove (since $\alpha k + \alpha' k' \in K$) that $\alpha h + \alpha' h' - \alpha k - \alpha' k' \in K^\perp$. This is true since $\alpha h - \alpha' h' - \alpha k - \alpha' k' = \alpha(h - k) + \alpha'(h' - k')$, $h - k \in K^\perp$, $h' - k' \in K^\perp$, and $K^\perp$ is a vector space. Continuity comes from

$$\|h\|_H^2 = \|h - \pi_K(h)\|_H^2 + \|\pi_K(h)\|_H^2$$

so that $\|\pi_K(h)\|_H \leq \|h\|_H$.

Finally, if $h \in H$ and $k = \pi_K(h)$, then $k' = \pi_{K^\perp}(h)$ is characterized by $k' \in K^\perp$ and $h - k' \in (K^\perp)^\perp$. The first property is certainly true for $h - k$, and for the second, we need to show that $K \subset (K^\perp)^\perp$, which is a direct consequence of the definition of the orthogonal. $\qquad\square$

We have the interesting property:

Corollary A.8. *K is a Hilbert subspace of H if and only if $(K^\perp)^\perp = K$.*

Proof. The $\Leftarrow$ implication is a consequence of theorem A.4. Now assume that K is a Hilbert subspace. The fact that $K \subset (K^\perp)^\perp$ is obviously true for any subset $K \subset H$, so that it suffices to show that every element of $(K^\perp)^\perp$ belongs to K. Assume that $h \in (K^\perp)^\perp$: this implies that $\pi_{K^\perp}(h) = 0$ but since $\pi_{K^\perp}(h) = h - \pi_K(h)$, this implies that $h = \pi_K(h) \in K$. $\qquad\square$

Let's compute the orthogonal projection on a finite-dimensional subspace of H,

$$K = \left\{ \sum_{i=1}^n \alpha_i f_i, \alpha_1, \ldots, \alpha_n \in \mathbb{R} \right\}$$

where $f_1, \ldots, f_n$ are linearly independent elements of H. If $h \in H$, its orthogonal projection $\hat{h} = \pi_K(h)$ is characterized by $\hat{h} \in K$ and $\langle \hat{h},\, f_i\rangle_H = \langle h,\, f_i\rangle_H$ for $i = 1, \ldots, N$. This implies that

$$\hat{h} = \sum_{i=1}^{N} \alpha_i f_i$$

where $\alpha_1, \ldots, \alpha_n$ are obtained by solving the system

$$\sum_{j=1} \langle f_i, f_j \rangle_H \alpha_i = \langle f_i, h \rangle_H.$$

A.4 Orthonormal Sequences

A sequence $(e_1, e_2, \ldots)$ in a Hilbert space H is orthonormal if and only if $\langle e_i, e_j \rangle_H = 1$ if $i = j$ and 0 otherwise. In such a case, if $\alpha = (\alpha_1, \alpha_2, \ldots) \in \ell^2$ (the space defined is Section A.2.2), the series $\sum_{i=1}^{\infty} \alpha_i e_i$ converges in H (its partial sums form a Cauchy sequence) and if h is the limit, then $\alpha_i = \langle h, e_i \rangle_H$.

Conversely, if $h \in H$ then the sequence $(\langle h, e_1 \rangle_H, \langle h, e_2 \rangle_H, \ldots)$ belongs to ℓ^2. Indeed, letting $h_n = \sum_{i=1}^{n} \langle h, e_i \rangle_H e_i$, one has $\langle h_n, h \rangle_H = \sum_{i=1}^{n} \langle h, e_i \rangle_H^2 = \|h_n\|_H^2$. On the other hand, one has, by Schwartz's inequality $\langle h_n ; h \rangle_H \leq \|h_n\|_H \|h\|_H$ which implies that $\|h_n\|_H \leq \|h\|_H$: therefore

$$\sum_{i=1}^{\infty} \langle h, e_i \rangle^2 < \infty.$$

Denoting by $K = \mathrm{Hilb}(e_1, e_2, \ldots)$ the smallest Hilbert subspace of H containing this sequence, one has the identity

$$K = \left\{ \sum_{n=1}^{\infty} \alpha_k e_k : (\alpha_1, \alpha_2, \ldots) \in l^2 \right\}. \tag{A.2}$$

The proof is left to the reader. As a consequence of this, we see that $h \mapsto (\langle h, e_1 \rangle_H, \langle h, e_2 \rangle_H, \ldots)$ is an isometry between $\mathrm{Hilb}(e_1, e_2, \ldots)$ and ℓ^2. Moreover, we have, for $h \in H$

$$\pi_K(h) = \sum_{i=1}^{\infty} \langle h, e_i \rangle_H e_i$$

(because $h - \pi_K(h)$ is orthogonal to every e_i).

An orthonormal set $(e_1, e_2, \ldots)$ is *complete* in H if $H = \mathrm{Hilb}(e_1, e_2, \ldots)$. In this case, we see that H is itself isometric to ℓ^2, and the interesting point is that (in the separable case) this is always true.

Theorem A.9. *Every (separable) Hilbert space has a complete orthonormal sequence.*

A complete orthonormal sequence in H is also called an orthonormal basis of H.

Proof. The proof relies on the important Schmidt orthonormalization procedure. Let $f_1, f_2, \ldots$ be a dense sequence in H. We let $e_1 = f_{k_1} / \|f_{k_1}\|_H$ where f_{k_1} is the first non-vanishing element in the sequence.

Assume that an orthonormal sequence $e_1, \ldots, e_n$ have been constructed with a sequence $k_1, \ldots, k_n$ such that $e_i \in V_n = \mathrm{vect}(f_1, \ldots, f_{k_i})$ for each i. First assume that $f_k \in \mathrm{vect}(e_1, \ldots, e_n)$ for all $k > k_n$: then $H = V_n$ is finite-dimensional. Indeed, H is equal to the closure of $(f_1, f_2, \ldots)$ which is included in $\mathrm{vect}(e_1, \ldots, e_n)$ which is closed, as a finite-dimensional vector subspace of H.

Assume now that there exists k_{n+1} such that $f_{k_{n+1}} \notin V_n$. Then, we may set

$$e_{n+1} = \lambda \left(f_{k_{n+1}} - \pi_{V_n}(f_{k_{n+1}}) \right)$$

where λ is selected so that e_{n+1} has unit norm, which is always possible.

So there are two cases: either the previous construction stops at some point, and H is finite-dimensional and the theorem is true, or the process carries on indefinitely, yielding an orthonormal sequence $(e_1, e_2, \ldots)$ and an increasing sequence of integers $(k_1, k_2, \ldots)$. But $\mathrm{Hilb}(e_1, e_2, \ldots)$ contains (f_n) which is dense in H and therefore is equal to H so that the orthonormal sequence is complete. $\qquad\square$

A.5 The Riesz Representation Theorem

The dual space of a normed vector space H is the space containing all continuous linear functionals $\varphi : H \to \mathbb{R}$. It is denoted H^*, and we will use the notation, for $\varphi \in H^*$ and $h \in H$:

$$\varphi(h) = (\varphi \,|\, h). \tag{A.3}$$

Thus, parentheses indicate linear forms, angles indicate inner products.

H being a normed space, H^* also has a normed space structure defined by:

$$\|\varphi\|_{H^*} = \max \left\{ (\varphi \,|\, h) : h \in H, \|h\|_H = 1 \right\}.$$

Continuity of φ is in fact equivalent to the finiteness of this norm.

When H is Hilbert, the function $\varphi_h : h' \mapsto \langle h, h' \rangle_H$ belongs to H^*, and by the Schwartz inequality $\|\varphi_h\|_{H^*} = \|h\|_H$. The Riesz representation theorem states that there exists no other continuous linear form on H.

Theorem A.10 (Riesz). *Let H be a Hilbert space. If $\varphi \in H^*$, there exists a unique $h \in H$ such that $\varphi = \varphi_h$.*

Proof. For uniqueness, it suffices to prove that $\varphi_h = 0 \Rightarrow h = 0$, which is obvious since $\varphi_h(h) = \|h\|_H^2$. To prove existence, we introduce the orthogonal of the null space of φ. So, let $K = \left\{ h' \in H, (\varphi \,|\, h') = 0 \right\}$; K is a closed linear subspace of H (because φ is linear and continuous). If $K^\perp = \{0\}$, then $K =$

$(K^{\perp})^{\perp} = H$ and $\varphi = 0 = \varphi_h$ for $h = 0$, which proves the theorem in this case. So, we assume that $K^{\perp} \neq 0$.

Now, let h_1 and h_2 be two non-zero elements in $K^{\perp}$. We have

$$\big(\varphi \,|\, \alpha_1 h_1 + \alpha_2 h_2\big) = \alpha_1\big(\varphi \,|\, h_1\big) + \alpha_2\big(\varphi \,|\, h_2\big),$$

so that it is always possible to find non-vanishing α_1 and α_2 such that $\big(\varphi \,|\, \alpha_1 h_1 + \alpha_2 h_2\big) = 0$ since $K^{\perp}$ is a vector space. But this implies that $\alpha_1 h_1 + \alpha_2 h_2$ also belongs to K, which is only possible when $\alpha_1 h_1 + \alpha_2 h_2 = 0$. Thus for any $h_1, h_2 \in K^{\perp}$, there exists a $(\alpha_1, \alpha_2) \neq (0,0)$ such that $\alpha_1 h_1 + \alpha_2 h_2 = 0$: $K^{\perp}$ has dimension 1. Fix $h_1 \in K^{\perp}$ such that $\|h_1\|_H = 1$. Using orthogonal projections, and the fact that K is closed, any vector in H can be written: $h = \langle h, h_1 \rangle h_1 + k$ with $k \in K$. This implies

$$\big(\varphi \,|\, h\big) = \langle h, h_1 \rangle \big(\varphi \,|\, h_1\big)$$

so that $\varphi = \varphi_{\alpha h_1}$ with $\alpha = \big(\varphi \,|\, h_1\big)_H$.

$\square$

A.6 Embeddings and the Duality Paradox

A.6.1 Definition

Assume that H and H_0 are two Banach spaces. An embedding of H in H_0 is a continuous, one-to-one, linear map from H to H_0, i.e., a map $\iota : H \to H_0$ such that, for all $h \in H$,

$$\|\iota(h)\|_{H_0} \leq C \, \|h\|_H . \tag{A.4}$$

The smallest C for which this is true is the operator norm of ι, denoted $\|\iota\|_{H \to H_0}$.

One says that the embedding is *compact* when the set $\{\iota(h), \|h\|_H \leq 1\}$ is compact in H_0. In the separable case (to which we restrict ourselves), this means that for any bounded sequence $(h_n, n > 0)$ in H, there exists a subsequence of $(\iota(h_n))$ which converges in H_0. One says that the embedding is *dense* if $\iota(H)$ is dense in H_0.

In all the applications we will be interested in, H and H_0 will be function spaces, and we will have a set inclusion $H \subset H_0$. For example H may be a set of smooth functions and H_0 a set of less smooth functions (see examples of embeddings below). Then, one says that H is embedded (resp. compactly embedded) in H_0 if the canonical inclusion map: $\iota : H \to H_0$ is continuous (resp. compact).

If φ is a continuous linear form on H_0, and $\iota : H \to H_0$ is an embedding, then one can define the form $\iota^*(\varphi)$ on H by $\big(\iota^*(\varphi) \,|\, h\big) = \big(\varphi \,|\, \iota(h)\big)$, and $\iota^*(\varphi)$ is continuous on H. Indeed, we have, for all $h \in H$:

$$\left|\left(\iota^*(\varphi)\,|\,h\right)\right| = \left|\left(\varphi\,|\,\iota(h)\right)\right| \le \|\varphi\|_{H_0^*}\,\|\iota(h)\|_{H_0} \le \|\iota\|_{H \to H_0}\,\|\varphi\|_{H_0^*}\,\|h\|_H$$

where the first inequality comes from the continuity of φ and the last one from (A.4). This proves the continuity of $\iota^*(\varphi)$ as a linear form on H together with the inequality:

$$\|\iota^*(\varphi)\|_{H^*} \le \|\iota\|_{H \to H_0}\,\|\varphi\|_{H_0^*}\,.$$

This in fact proves the first statement of the theorem:

Theorem A.11. *If $\iota : H \to H_0$ is a Banach space embedding, then the map $\iota^* : H_0^* \to H^*$ is a continuous linear map.*
If ι is dense, then ι^* is also a dense embedding.

Proof. We now prove the second statement, i.e., that ι^* is a dense embedding if ι is dense. What remains to be shown is that ι^* is one-to-one and that $\iota^*(H_0^*)$ is dense in H^*. First assume that $\iota^*(\varphi) = 0$. This implies that $\left(\varphi\,|\,\iota(h)\right) = 0$ for all $h \in H$ so that φ vanishes on $\iota(H)$, which implies that $\varphi = 0$ since $\iota(H)$ is dense in H_0 and φ is continuous.

To prove that ι^* is dense, it suffices to show that no non-zero linear form in H^* is orthogonal to $\iota^*(H_0^*)$, which, by Theorem A.10, is equivalent to showing that no non-zero vector in h is such that $\left(\iota^*(\varphi)\,|\,h\right) = 0$ for all $\varphi \in H_0^*$. But since $\left(\iota^*(\varphi)\,|\,h\right) = \left(\varphi\,|\,\iota(h)\right)$, this is only possible when $\iota(h) = 0$, which yields $h = 0$ since ι is one-to-one.   □

A.6.2 Examples

Banach Spaces of Continuous Functions

Let Ω be an open subset of $\mathbb{R}^d$. The following results apply to a large class of sets Ω, including $\Omega = \mathbb{R}^d$, or an open, bounded, subset of $\mathbb{R}^d$ with a smooth (C^1) and bounded boundary, or a half-space of $\mathbb{R}^d$. The space of continuous functions on Ω with at least p continuous derivatives will be denoted $C^p(\Omega, \mathbb{R})$. If Ω is bounded, and thus $\overline{\Omega}$ is compact, $C^p(\overline{\Omega}, \mathbb{R})$ is the set of functions on Ω which are p times differentiable on Ω, each partial derivative being extended to a continuous function on $\overline{\Omega}$; $C^p(\overline{\Omega}, \mathbb{R})$ has a Banach space structure when equipped with the norm

$$\|f\|_{p,\infty} = \max_h \|h\|_\infty$$

where h varies in the set of partial derivatives of f or order lower or equal to p.

Still under the assumption that Ω is bounded, and denoting by $\partial\Omega$ the boundary of Ω we let $C_0^p(\Omega, \mathbb{R})$ be the set of functions f in $C^p(\overline{\Omega}, \mathbb{R})$ with vanishing derivatives up to order p on $\partial\Omega$ (in the following sense: $f \in C_0^p(\Omega, \mathbb{R})$ if it can be extended in a C^p function on $\mathbb{R}$ which vanishes on Ω^c).

We obviously have, almost by definition, the fact that $C^p(\overline{\Omega}, \mathbb{R})$ is embedded in $C^q(\overline{\Omega}, \mathbb{R})$ as soon as $p \leq q$. In fact the embedding is compact when $p < q$.

Compact sets in $C^0(\overline{\Omega}, \mathbb{R})$ are exactly described by Ascoli's theorem: they are bounded subsets $M \subset C^0(\overline{\Omega}, \mathbb{R})$ (for the supremum norm) which are uniformly continuous, meaning that, for any $x \in \overline{\Omega}$, for any $\varepsilon > 0$ there exists $\eta > 0$ such that,

$$\sup_{y \in \Omega, |x-y| < \eta} \sup_{h \in M} |h(x) - h(y)| < \varepsilon.$$

Compact sets in $C^p(\overline{\Omega}, \mathbb{R})$ are bounded subsets of $C^p(\overline{\Omega}, \mathbb{R})$ over which all the pth partial derivatives are uniformly continuous.

Hilbert Sobolev Spaces

We now define the space $H^1(\Omega, \mathbb{R})$ of functions with square integrable generalized derivatives. A function u belongs to this set if and only if $u \in L^2(\Omega, \mathbb{R})$, and for each $i = 1, \ldots, k$, there exists a function $u_i \in L^2(\Omega, \mathbb{R})$ such that for any C^∞ function φ with compact support in Ω, one has

$$\int_\Omega u(x) \dot{\varphi}_{x_i}(x) dx = - \int_\Omega u_i(x) \varphi(x) dx.$$

The function u_i is called the (generalized) directional derivative of u, and will be denoted $\dot{u}_{x_i}$. The integration by parts formula shows that this coincides with the standard partial derivative when it exists.

$H^1(\Omega, \mathbb{R}^d)$ has a Hilbert space structure with the inner product:

$$\langle u, v \rangle_{H^1} = \langle u, v \rangle_{L^2} + \sum_{i=1}^k \langle \dot{u}_{x_i}, \dot{v}_{x_i} \rangle_{L^2}.$$

The space $H^m(\Omega, \mathbb{R})$ can be defined by induction as the set of functions $f \in H^1(\Omega, \mathbb{R})$ with all partial derivatives belonging to $H^{m-1}(\Omega, \mathbb{R})$. Partial derivatives of increasing order are defined by induction, and the inner product on H^m is the sum of the L^2 inner products of all partial derivatives up to order m.

These spaces can be embedded in classic spaces of functions. We will be concerned with a special case of Morrey's theorem, which is stated below:

Theorem A.12. *Assume that* $m - d/2 > 0$. *Then, for any* $j \geq 0$, $H^{j+m}(\Omega, \mathbb{R})$ *is embedded in* $C^j(\overline{\Omega}, \mathbb{R})$. *If* Ω *is bounded, the embedding is compact.*

Moreover, if $\theta \in]0, m - d/2]$, *and* $u \in H^{j+m}(\Omega, \mathbb{R})$, *then every partial derivative, h, of order j has Hölder regularity θ: for all $x, y \in \Omega$*

$$|h(x) - h(y)| \leq C \|u\|_{H^{m+j}} |x - y|^\theta.$$

Let Ω be bounded. As a final definition, we let $H_0^m(\Omega, \mathbb{R})$ be the completion in $H_0^m(\Omega, \mathbb{R})$ of the set of C^∞ functions with compact support in Ω: u belongs to $H_0^m(\Omega, \mathbb{R})$ if and only if $u \in H^m(\Omega, \mathbb{R})$ and there exists a sequence of functions u_n, C^∞ with compact support in Ω such that $\|u - u_n\|_{H^m}$ tends to 0. A direct application of Morrey's theorem shows that, if $m - d/2 > 0$, then, for any $j \geq 0$, $H_0^{j+m}(\Omega, \mathbb{R})$ is embedded in $C_0^j(\overline{\Omega}, \mathbb{R})$.

General Sobolev spaces $(W^{p,k})$ are defined like the spaces H^p, using L^k norms of derivative instead of the L^2 one. The interested reader can refer to classical textbooks (e.g., [221, 2, 32]) on functional analysis for more information.

A.6.3 The Duality Paradox

The Riesz representation theorem allows one to identify a Hilbert space H and its dual H^*. However, when H is densely embedded in another Hilbert space H_0, every continuous linear form on H_0 is also continuous on H, and H_0^* is densely embedded in H^*. We therefore have the sequence of embeddings

$$H \to H_0 \simeq H_0^* \to H^*$$

but this sequence loops since $H^* \simeq H$. This indicates that H_0 is also embedded in H. This is arguably a strange result. For example, let $H = H^1(\Omega, \mathbb{R})$, and $H_0 = L^2(\Omega, \mathbb{R})$: the embedding of H in H_0 is clear from their definition (it is dense because C^∞ functions form a dense subset for both of them); but the converse inclusion does not seem natural, since there are more constraints in belonging to H than to H_0. To understand this reversed embedding, we must think in terms of linear forms.

If $u \in L^2(\Omega, \mathbb{R})$, we may consider the linear form φ_u defined by

$$(\varphi_u \,|\, v) = \langle u, v \rangle_{L^2} = \int_\Omega u(x)v(x)dx.$$

When $v \in H^1(\Omega, \mathbb{R})$, we have

$$(\varphi_u \,|\, v) \leq \|u\|_{L^2} \|v\|_{L^2} \leq \|u\|_{L^2} \|v\|_{H^1}$$

so that φ_u is continuous when seen as a linear form on $H^1(\Omega, \mathbb{R})$. The Riesz representation theorem implies that there exists a $\tilde{u} \in H^1(\Omega, \mathbb{R})$ such that, for all $v \in H^1(\Omega, \mathbb{R})$, $\langle u, v \rangle_{L^2} = \langle \tilde{u}, v \rangle_{H^1}$: the relation $u \mapsto \tilde{u}$ provides the embedding of $L^2(\Omega, \mathbb{R})$ into $H^1(\Omega, \mathbb{R})$. Let us be more specific and take $\Omega =]0, 1[$. The relation states that, for any $v \in H^1(\Omega, \mathbb{R})$,

$$\int_0^1 \dot{\tilde{u}}(t)\dot{v}(t)dt + \int_0^1 \tilde{u}(t)v(t)dt = \int_0^1 \tilde{u}(t)v(t)dt.$$

Let us make the simplifying assumption that $\tilde{u}$ has two derivatives, in order to integrate by part the first integral and obtain

$$\dot{u}(1)v(1) - \dot{u}(0)v(0) - \int_0^1 \ddot{\tilde{u}}(t)v(t) + \int_0^1 \tilde{u}(t)v(t)dt = \int_0^1 \tilde{u}(t)v(t)dt. \quad (A.5)$$

Such an identity can be true for every v in $H^1(\Omega, \mathbb{R})$ if and only if $\dot{u}(0) = \dot{u}(1) = 0$ and $-\ddot{\tilde{u}} + \tilde{u} = u$: $\tilde{u}$ is thus a solution of a second-order differential equation, with first-order boundary conditions, and the embedding of $L^2(]0,1[, \mathbb{R})$ into $H^1(]0,1[, \mathbb{R})$ just shows that a unique solution exists, at least in the generalized sense of equation (A.5).

As seen in these examples, even if, from an abstract point of view, we have an identification between the two Hilbert spaces, the associated embeddings are of very different nature, the first one corresponding to a set inclusion (it is canonical), the second to the solution of a differential equation in one dimension, and in fact to a partial differential equation in the general case.

A.7 Weak Convergence in a Hilbert Space

Let us start with the definition:

Definition A.13. *When V is a Banach space, a sequence (v_n) in V is said to weakly converge to some $v \in V$ if and only if, for all continuous linear form $\alpha : V \to \mathbb{R}$, one has $\alpha(v_n) \to \alpha(v)$ when n tends to infinity.*

Recall that $h_n \to h$ "strongly" if $\|h_n - h\|_H \to 0$. The following proposition describes how convergence is related with inclusion.

Proposition A.14. *Assume that V and W are Banach spaces and that $W \subset V$ with a continuous inclusion. If a sequence (w_n) in W weakly converges (in W) to some $w \in W$, then w_n also weakly converges to w in V.*

This just says that if $\alpha(w_n) \to \alpha(w)$ for all continuous linear functional on W, then the convergence holds for all continuous linear functional on V, which is in fact obvious because the restriction to W of any continuous linear functional on V is a continuous linear functional on W.

In the case of a Hilbert space, the Riesz representation theorem immediately provides the following proposition:

Proposition A.15. *Let V be a Hilbert space. A sequence v_n in V weakly converges to an element $v \in V$ if and only if, for all $w \in V$,*

$$\lim_{n \to \infty} \langle w, v_n \rangle_V = \langle w, v \rangle_V.$$

Moreover, if v_n weakly converges to v, then $\|v\| \leq \liminf \|v_n\|$.

The last statement comes from the inequality: $\langle v_n, v \rangle_V \leq \|v_n\|_V \|v\|_V$ which provides at the limit

$$\|v\|_V^2 \leq \|v\|_V \liminf_{n \to \infty} \|v_n\|_V.$$

Finally, the following result is about the weak compactness of bounded sets [221, 228].

Theorem A.16. *If V is a Hilbert space and (v_n) is a bounded sequence in V (there exists a constant C such that $\|v_n\| \leq C$ for all n), then one can extract a subsequence from v_n which weakly converges to some $v \in V$.*

A.8 The Fourier Transform

We provide here a brief overview of Fourier transforms with their definition and basic properties. We let $H = L^2(\mathbb{R}^d, \mathbb{C})$. For a function $f \in H$, one defines

$$\mathcal{FT}(f)(\xi) = \hat{f}(\xi) = \int_{\mathbb{R}^d} f(x) e^{-2\iota\pi\xi^T x} dx.$$

This is well-defined as soon as f is absolutely integrable, i.e.,

$$\int_{\mathbb{R}^d} |f(x)| \, dx < \infty$$

but can be extended (based on isometry properties, see below) to the whole space H.

We have the following inversion formula:

Theorem A.17. *Assume that f and $\hat{f}$ are both absolutely integrable. Then, we have*

$$f(x) = \int_{\mathbb{R}^d} \hat{f}(\xi) e^{2\iota\pi x^T \xi} d\xi$$

The inversion formula illustrates the bijective nature of the Fourier transform. Another important property is that, when both f and $\hat{f}$ are in $L^1 \cap L^2$, $\|\hat{f}\|_2 = \|f\|_2$. This implies that there is a unique isometric extension of the Fourier transform to L^2. In particular, we have, for this extension:

$$\langle \hat{f}, \hat{g} \rangle_2 = \langle f, g \rangle_2.$$

The Fourier transform has some other useful properties.

Proposition A.18.

(1) $\mathcal{FT}(f(x - x_0))(\xi) = e^{-2\iota\pi\xi^T x_0} \hat{f}(\xi)$;

(2) $\mathcal{FT}(e^{2\iota\pi x^T \xi_0} f(x))(\xi) = \hat{f}(\xi - \xi_0)$;

(3) $\mathcal{FT}(f(ax))(\xi) = \dfrac{1}{|a|^d} \hat{f}\left(\dfrac{\xi}{a}\right)$;

(4) $\mathcal{FT}(\alpha_1 f_1 + \alpha_2 f_2) = \alpha_1 \hat{f}_1 + \alpha_2 \hat{f}_2$;

(5) $\mathcal{FT}(e^{-\pi x^T x})(\xi) = e^{-\pi\xi^T \xi}$;

(6) $\mathcal{FT}(\dfrac{\partial f}{\partial x_i})(\xi) = (2\iota\pi\xi_i)\hat{f}$.

The Fourier transform also interacts nicely with convolutions.

Proposition A.19. *If h and f are absolutely integrable, then*

$$\mathcal{FT}(h * f) = \hat{h}\hat{f}$$

with

$$(h * f)(x) = \int_{\mathbb{R}^d} h(x - y)f(y)dy.$$

B

Elements from Differential Geometry

B.1 Introduction

This appendix describes a few notions of differential geometry that are relevant for this book. One of the primary goals of differential geometry is to provide mathematical tools that make possible the use of infinitesimal calculus for sets that are more general than Euclidean spaces. In this setting, we will not go further than the notion of tangent spaces and Riemannian metrics, leaving aside some fundamental features (like the curvature), for which we refer the reader to general textbooks on the subject, like [65, 30, 1, 107, 121, 143].

B.2 Differential Manifolds

B.2.1 Definition

A differential manifold is a set within which points may be described by coordinate systems, which must satisfy some compatibility constraints from which one can develop intrinsic differential operations. We start with the definition:

Definition B.1. *Let M be a topological Hausdorff space. A d-dimensional local chart on M is a pair (U, Φ) where U is an open subset of M and Φ a homeomorphism between U and some open subset of $\mathbb{R}^d$.*

Two d-dimensional local charts, (U_1, Φ_1) and (U_2, Φ_2) are C^∞-compatible if either U_1 and U_2 do not overlap, or the function $\Phi_1 \circ \Phi_2^{-1}$ is a C^∞-diffeomorphism between $\Phi_2(U_2 \cap U_1)$ and $\Phi_1(U_2 \cap U_1)$.

A d-dimensional atlas on M is a family of pairwise compatible local charts $((U_i, \Phi_i), i \in I)$, such that $M = \bigcup_I U_i$. Two atlases on M are equivalents if their union is also an atlas, i.e., if every local chart of the first one is compatible with every local chart of the second one.

A Hausdorff space with an d-dimensional atlas is called an d-dimensional (C^∞) differential manifold.

If M is a manifold, a local chart on M will always be assumed to be compatible with the atlas on M. If M and N are two manifolds, their product $M \times N$ is also a manifold; if (U, Φ) is a chart on M, (V, Ψ) a chart on N, $(U \times V, (\Phi, \Psi))$ is a chart on $M \times N$, and one shows easily that one can form an atlas for $M \times N$ from cross products of two atlases of M and N.

When a local chart (U, Φ) is given, the coordinate functions $x_1, \ldots, x_d$ are defined by $\Phi(p) = (x_1(p), \ldots, x_d(p))$ for $p \in U$. Formally, x_i is a function from U to $\mathbb{R}$. However, when a point p is given, one generally refers to $x_i = x_i(p) \in \mathbb{R}$ as the ith coordinate of p in the chart (U, Φ).

According to these definitions, $\mathbb{R}^d$ is a differential manifold, and so are open sets in $\mathbb{R}^d$. Another example is given by the d-dimensional sphere, S^d, defined as the set of points $p \in \mathbb{R}^{d+1}$ with $|p| = 1$. The sphere can be equipped with a $2(d+1)$-chart atlas (U_i, Φ_i), $i = 1, \ldots, 2(d+1)$, letting

$$U_1 = \{p = (p_1, \ldots, p_{d+1}) : p_1 > 0\} \cap S^d$$

and $\Phi_1(p) = (p_2, \ldots, p_{d+1}) \in \mathbb{R}^n$, (U_2, Φ_2) being the same, with $p_1 < 0$ instead, and so on with the other coordinates $p_2, \ldots, p_{d+1}$.

We now consider functions on manifolds.

Definition B.2. *A function $\psi : M \to \mathbb{R}$ is C^∞ if, for every local chart (U, Φ) on M, the function $\psi \circ \Phi^{-1} : \Phi(U) \subset \mathbb{R}^d \to \mathbb{R}$, is C^∞ in the usual sense. The function $\psi \circ \Phi^{-1}$ is called the interpretation of ψ in (U, Φ).*

From the compatibility condition, if this property is true for an atlas, it is true for all charts compatible with it. The set of C^∞ functions on M is denoted $C^\infty(M)$. If U is open in M, the set $C^\infty(U)$ contains functions defined on U which can be interpreted as C^∞ functions of the coordinates for all local charts of M which are contained in U. The first example of C^∞ functions are the coordinates: if (U, Φ) is a chart, the ith coordinate $(x_i(p), p \in U)$ belongs to $C^\infty(U)$, since, when interpreted in (U, Φ), it reduces to $(x_1, \ldots, x_d) \mapsto x_i$.

B.2.2 Vector Fields, Tangent Spaces

We fix, in this section, a differential manifold, denoted M, of dimension d. We define vector fields and tangent vectors via their actions on functions, and later provide alternative interpretations.

Definition B.3. *A vector field on M is a function $X : C^\infty(M) \to C^\infty(M)$, such that: $\forall \alpha, \beta \in \mathbb{R}$, $\forall \varphi, \psi \in C^\infty(M)$:*

$$X(\alpha.\varphi + \beta.\psi) = \alpha.X(\varphi) + \beta.X(\psi),$$

$$X(\varphi\psi) = X(\varphi)\psi + \varphi X(\psi).$$

The set of vector fields on M is denoted by $\mathcal{X}(M)$.

Definition B.4. *If $p \in M$, a tangent vector to M at p is a function ξ : $C^\infty(M) \to \mathbb{R}$ such that: $\forall \alpha, \beta \in \mathbb{R}$, $\forall \varphi, \psi \in C^\infty(M)$:*

$$\xi(\alpha.\varphi + \beta.\psi) = \alpha.\xi(\varphi) + \beta.\xi(\psi),$$

$$\xi(\varphi\psi) = \xi(\varphi)\psi(p) + \varphi(p)\xi(\psi).$$

The set of tangent vectors to M at p is denoted T_pM.

So vector fields assign C^∞ functions to C^∞ functions and tangent vectors assign real numbers to C^∞ functions.

One can go from vector fields to tangent vectors and vice versa as follows. If $X \in \mathcal{X}(M)$ is a vector field on M, and if $p \in M$, we may define X_p : $C^\infty(M) \to \mathbb{R}$ by

$$X_p(\varphi) = (X(\varphi))(p)$$

to obtain a tangent vector at p. Conversely, if a collection $(X_p \in T_pM, p \in M)$ is given, we can define $(X(\varphi))(p) = X_p(\varphi)$ for $\varphi \in C^\infty(M)$ and $p \in M$; X will be a vector field on M if and only if, for all $\varphi \in C^\infty(M)$ the function $p \mapsto X_p(\varphi)$ is C^∞. Finally, one can show that, for all $\xi \in T_pM$, there exists a vector field X such that $\xi = X_p$ [107].

The linear nature of Definitions B.3 and B.4 is clarified in the next proposition:

Proposition B.5. *For all $p \in M$, the tangent space T_pM is an d-dimensional vector space.*

Proof. Let $C = (U, \Phi)$ be a local chart with $p \in U$. Denote $x^0 = \Phi(p)$, $x^0 \in \mathbb{R}^d$. If $\varphi \in C^\infty(M)$, then, by definition,

$$\begin{aligned} \varphi_C : \varphi(U) \subset \mathbb{R}^d &\to \mathbb{R} \\ x &\mapsto \varphi \circ \Phi^{-1}(x) \end{aligned}$$

is C^∞. Define

$$\partial_{x_i,p}(\varphi) := \frac{\partial \varphi_C}{\partial x_i}(x_0).$$

It is easy to check that $\partial_{x_i,p}$ satisfies the conditions given in Definition B.4, so that $\partial_{x_i,p} \in T_pM$. We show that every $\xi \in T_pM$ may be uniquely written in the form

$$\xi = \sum_{i=1}^d \lambda_i \partial_{x_i,p}. \tag{B.1}$$

Indeed,

$$\varphi_C(x) = \varphi_C(x^0) + \sum_{i=1}^d (x_i - x_i^0)\psi_i^*(x)$$

with $\psi_i^*(x) = \int_0^1 \dfrac{\partial \varphi_C}{\partial x_i}(x^0 + t(x - x^0))dt$. Thus, if $p' \in U$,

$$\varphi(p') = \varphi_0 + \sum_{i=1}^d (x_i(p') - x_i(p))\psi_i(p')$$

with $\psi_i(p') = \psi_i^*(\Phi(p'))$, and $\varphi_0 = \varphi(p)$. If $\xi \in T_pM$ and f is constant, we have $\xi(f) = f\xi(1) = f\xi(1^2) = 2f\xi(1)$ so that $\xi(f) = 0$. Thus, for all $\xi \in T_pM$,

$$\xi(\varphi) = 0 + \sum_{i=1}^d \psi_i(p)\xi(x_i)\,.$$

But $\psi_i(p) = (\partial_{x_i,m})(\varphi)$, which yields (B.1) with $\lambda_i = \xi(x_i)$.   □

This also implies that, *within a local chart*, a vector field can always be interpreted in the form

$$X = \sum_{i=1}^d \varphi_i \partial_{x_i}$$

with $\varphi_i \in C^\infty(M)$ and $[\partial_{x_i}]_p = \partial_{x_i,p}$.

There is another standard definition of tangent vectors on M, in relation with differentiable curves on M. This starts with the following definitions.

Definition B.6. *Let $t \mapsto \mu(t) \in M$ be a continuous curve, $\mu : [0, T] \to M$. One says that this curve is C^∞ if, for any local chart $C = (U, \Phi)$, the curve $\mu^C : s \mapsto \Phi \circ \mu(s)$, defined on $\{t \in [0, T] : \mu(t) \in U\}$ is C^∞.*

Let $p \in M$. One says that two C^∞ curves, μ and ν, starting at p (i.e., $\mu(0) = \nu(0) = p$) have the same tangent at p, if and only if, for all charts $C = (U, \Phi)$, the curves μ^C and ν^C have identical derivatives at $t = 0$.

Proposition B.7. *The tangential identity at p is an equivalence relation. The tangent space to M at p can be identified with the set of equivalence classes for this relation.*

Proof. We sketch the argument. If a curve μ is given, with $\mu(0) = p$, define, for $\varphi \in C^\infty(M)$,

$$\xi^\mu(\varphi) = \frac{d}{dt}_{|t=0} \varphi \circ \mu.$$

One can check that $\xi^\mu \in T_pM$, and that $\xi^\mu = \xi^\nu$ if μ and ν have the same tangent at p. Conversely, if $\xi \in T_pM$ there exists a curve μ such that $\xi = \xi^\mu$, and the equivalence class of μ is uniquely specified by ξ. To show this, consider a chart (U, Φ). We must have

$$\xi^\mu(x_i) = \xi(x_i) = \frac{d}{dt}_{|t=0} x_i \circ \mu(.),$$

which indeed shows that the tangent at $\Phi(p)$ is uniquely defined. To determine μ, start from a line segment in $\Phi(U)$, passing through $\Phi(p)$, with direction given by $(\xi(x_i), i = 1, \ldots, d)$ and apply Φ^{-1} to it to obtain a curve on M.   □

When $X \in \mathcal{X}(M)$ is given, one can consider the differential equation

$$\partial_t \mu(t) = X_{\mu(t)}.$$

Such a differential equation always admits a unique solution, with some initial condition $\mu(0) = p$, at least for t small enough. This can be proved by translating the problem in a local chart and applying the results of Appendix C.

Finally, we can define the differential of a scalar-valued function.

Definition B.8. *If $\varphi \in C^\infty(M)$, one defines a linear form on T_pM by $\xi \mapsto \xi(\varphi)$. It will be denoted $D\varphi(p)$, and called the differential of φ at p.*

B.2.3 Maps Between Two Manifolds

Definition B.9. *Let M and M' be two differential manifolds. A map $\Phi : M \to M'$ has class C^∞ if and only if, for all $\varphi \in C^\infty(M')$, one has $\varphi \circ \Phi \in C^\infty(M)$.*

Definition B.10. *If $p \in M$ and $p' = \Phi(p)$, define the tangent map of Φ at p,*

$$D\Phi(p) : T_pM \to T_{p'}M'$$

by: for all $\xi \in T_pM$, $\varphi \in C^\infty(M')$,

$$(D\Phi(p)\xi)(\varphi) = \xi(\varphi \circ \Phi).$$

The tangent map $D\Phi(p)$ is also called the differential of Φ at p.

The chain rule is true for tangent maps: if $\Phi : M \to M'$ and $\Psi : M' \to M''$ are differentiable, then $\Psi \circ \Phi : M \to M''$ is also differentiable and

$$D(\Psi \circ \Phi)(p) = D\Psi(\Phi(p)) \circ D\Phi(p).$$

This almost directly follows from the definition.

B.3 Submanifolds

One efficient way to build manifolds is to characterize them as submanifolds of simple manifolds like $\mathbb{R}^d$. If M is a manifold, a submanifold of M is a subset of M, that is itself a manifold, but also inherits the manifold structure of M.

Definition B.11. *Let M be a d-dimensional differential manifold. We say that P is a d'-dimensional submanifold of M (with $d' \leq d$) if, for all $m_0 \in P$, there exists a local chart (U, Φ) of M such that $m_0 \in U$, with local coordinates $(x_1, \ldots, x_d)$, such that*

$$U \cap P = \{m \in M : x_i = 0, i = d' + 1, \ldots, d\}.$$

The next theorem is one of the main tools for defining manifolds:

Theorem B.12. *Let M be a d-dimensional differential manifold, Φ a differentiable map from M to $\mathbb{R}^k$. Let $a \in \Phi(M)$ and*

$$P = \Phi^{-1}(a) = \{p \in M : \Phi(p) = a\}.$$

If there exists an integer l, such that, for all $p \in P$, the linear map $D\Phi(p) : T_pM \to \mathbb{R}^k$ has constant rank l (independent ón p), then P is a submanifold of M, with dimension $d' = d - l$.

This applies to the sphere S^d, defined by $x_1^2 + \cdots + x_{d+1}^2 = 1$, which is a submanifold of $\mathbb{R}^{d+1}$ of dimension d.

If $P \subset M$ is a submanifold of M defined as in Theorem B.12, the tangent space to P at p can be identified with the null space of $D\Phi(p)$ in T_pM:

$$T_pP = \{\xi \in T_pM, D\Phi(p)\xi = 0\}.$$

Another way to define submanifolds is via immersions, as outlined below.

Definition B.13. *Let M and P be two differential manifolds. An immersion of M into P is a C^∞ map $\Phi : M \to P$, such that:*
(i) For all $p \in M$, the tangent map, $D\Phi(p)$, is one-to-one, from T_pM to $T_{\Phi(p)}P$.
(ii) Φ is a homeomorphism between M and $\Phi(M)$ (this last set being considered with the topology induced by P).

The second condition means that Φ is one-to-one, and, for all open subsets U in M, there exists an open subset V in P such that $\Phi(U) = V \cap \Phi(M)$.

We then have:

Proposition B.14. *If $\Phi : M \to P$ is an immersion, then $\Phi(M)$ is a submanifold of P, with same dimension as M.*

B.4 Lie Groups

B.4.1 Definitions

A group is a set G with a composition rule $(g, h) \mapsto gh$ which is associative, has an identity element (denoted id_G, or id if there is no risk of confusion) and such that every element in G has an inverse in G. A Lie group is both a group and a differential manifold, such that the operations $(g, h) \mapsto gh$ and $g \mapsto g^{-1}$, respectively from $G \times G$ to G and from G to G are C^∞.

B.4.2 Lie Algebra of a Lie Group

If G is a Lie group, $g \in G$ and $\varphi \in C^\infty(G)$, one defines $\varphi \cdot g \in C^\infty(G)$ by $(\varphi \cdot g)(g') = \varphi(g'g)$. A vector field on G is right-invariant if, for all $g \in G, \varphi \in C^\infty(G)$, one has $X(\varphi \cdot g) = (X(\varphi))g$. Denoting by R_g the right translation on G (defined by $R_g(g') = g'g$), right invariance is equivalent to the identity, true for all g: $X = DR_g X$. The set of right-invariant vector fields is called the Lie algebra of the group G, and denoted $\mathfrak{g}$.

Since $(X(\varphi.g))(id) = ((X(\varphi)).g)(id) = (X(\varphi))(g)$ whenever $X \in \mathfrak{g}$, an element X of $\mathfrak{g}$ is entirely specified by the values of $X(\varphi)(id)$ for $\varphi \in C^\infty(G)$. This implies that the Lie algebra $\mathfrak{g}$ may be identified to the tangent space to G at id, $T_{id}G$. If $\xi \in T_{id}G$, its associated right-invariant vector field is

$$X^\xi : g \mapsto X_g = DR_g(id)\xi \in T_g G.$$

The operation which provides a structure of algebra on $\mathfrak{g}$ is the Lie bracket. Recall that a vector field on a manifold M is a function $X : C^\infty(M) \to C^\infty(M)$, which satisfies the conditions of definition B.3. When X and Y are two vector fields, it is possible to combine them and compute $(XY)(\varphi) = X(Y(\varphi))$; XY also transforms C^∞ functions into C^∞ functions, but will not satisfy the conditions of definition B.3, essentially because it involves second derivatives. However, it is easy to check that the second derivatives cancel in the difference $XY - YX$ which is a vector field on M, denoted $[X, Y]$, and called the bracket of X and Y. A few important properties of Lie brackets are listed (without proof) in the next proposition.

Proposition B.15.

i. $[X, Y] = -[Y, X]$.
ii. $[[X, Y], Z] = [X, [Y, Z]]$.
iii. $[[X, Y], Z] + [[Z, X], Y] + [[Y, Z], X] = 0$.
iv. If $\Phi \in C^\infty(M, N)$, $D\Phi[X, Y] = [D\Phi X, D\Phi Y]$.

The last property is important for Lie groups, since, applied with $\Phi = R_g : G \mapsto G$, and $X, Y \in \mathfrak{g}$, it yields

$$DR_g[X, Y] = [DR_g X, DR_g Y] = [X, Y]$$

so that $[X, Y] \in \mathfrak{g}$. The Lie algebra of G is closed under the Lie bracket operation (which induces the term Lie algebra). Because of the identification of $\mathfrak{g}$ with $T_{id}G$, the bracket notation is also used for tangent vectors at the identity, letting $[\xi, \eta] = [X^\xi, X^\eta]_{id}$.

There is another possible definition of the Lie bracket on $\mathfrak{g}$. For $g \in G$, one can define the group isomorphism $I_g : h \mapsto ghg^{-1}$. It is differentiable, and the differential of I_g at $h = id$ is called the adjoint map, denoted $Ad_g : T_{id}G \to T_{id}G$. We therefore have, for $\eta \in T_{id}G$,

$$Ad_g(\eta) = DI_g(id)\eta.$$

Now consider the application $U_\eta : g \mapsto Ad_g(\eta)$ which is defined on G and takes values in $T_{id}G$. One can show [107] that

$$d_{id}U_\eta\xi = [\xi, \eta]. \tag{B.2}$$

The notation $ad_\xi\eta = [\xi, \eta]$ is commonly used to represent the Lie bracket.

When a vector field $X \in \mathfrak{g}$ is given, the solution of the associated differential equation

$$\partial_t\mu(t) = X_{\mu(t)} \tag{B.3}$$

with initial condition $\mu(0) = id$ always exists, not only for small time, but for arbitrary times. The small time existence comes from the general theory, and the existence for arbitrary time comes from the fact that, wherever it is defined, $\mu(t)$ satisfies the semi-group property $\mu(t + s) = \mu(t)\mu(s)$: this implies that if $\mu(t)$ is defined on some interval $[0, T]$, one can always extend it to $[0, 2T]$ by letting $\mu(t) = \mu(\frac{t}{2})^2$. The semi-group property can be proved to be true as follows: if $X = X^\xi$, for $\xi \in T_{id}G$, the ordinary differential equation can be written

$$\partial_t\mu(t) = DR_{\mu(t)}(id)\xi$$

Consider now $\nu : s \mapsto \mu(t + s)$. It is solution of the same equation with initial condition $\nu(0) = \mu(t)$. If $\tilde{\nu}(s) = \mu(s)\mu(t)$, we have

$$\begin{aligned}\partial_s\tilde{\nu}(s) &= DR_{\mu(t)}(\mu(s))(\partial_s\mu(s)) = DR_{\mu(t)}(\mu(s))DR_{\mu(t)}(id)\xi \\ &= D(R_{\mu(t)} \circ R_{\mu(s)})(id)\xi \\ &= DR_{\nu(s)}(id)\xi.\end{aligned}$$

Thus, ν and $\tilde{\nu}$ satisfy the same differential equation, with the same value, $\mu(t)$, at $s = 0$, and therefore coincide, which is the semi-group property.

The solution of (B.3) with initial condition id is called the exponential map on G, and is denoted $\exp(tX)$ or $\exp(t\xi)$ if $X = X^\xi$. The semi-group property becomes $\exp((t + s)X) = \exp(tX)\exp(sX)$. Equation (B.2) can be written, using the exponential map

$$\partial_t\left[\partial_s \exp(t\xi)\exp(s\eta)\exp(-t\xi)_{|s=0}\right]_{|t=0} = [\xi, \eta].$$

We finally quote the last important property of the exponential map [107, 65]:

Theorem B.16. *There exists a neighborhood V of 0 in $\mathfrak{g}$ and a neighborhood U of id in G such that $\exp$ is a diffeomorphism between V and U.*

B.4.3 Finite-Dimensional Transformation Groups

Finite-dimensional transformation groups, and in particular matrix groups, are fundamental examples of Lie groups. They also provide important transformation in the analysis of shapes. Denote by $\mathcal{M}_n(\mathbb{R})$ the n^2-dimensional space of real n by n matrices. For $i, j \in \{1, \ldots, n\}$, denote by ∂x_{ij} the matrix with (i, j) coefficient equal to 1, and all others to 0. Let $\text{Id}_{\mathbb{R}^n}$ denote the identity matrix of size n.

Linear Group

$GL_n(\mathbb{R})$ is the group of invertible matrices in $\mathcal{M}_n(\mathbb{R})$. It is open in $\mathcal{M}_n(\mathbb{R})$ and therefore is a submanifold of this space, of same dimension, n^2. The Lie algebra of $GL_n(\mathbb{R})$ is equal to $\mathcal{M}_n(\mathbb{R})$, and is generated by all $(\partial x_{ij}, i, j = 1, \ldots, n)$.

if $\xi \in \mathcal{M}_n(\mathbb{R})$, the associated right-invariant vector field is $X^\xi : g \mapsto \xi g$. The adjoint map is $\eta \mapsto g\eta g^{-1}$, and the Lie bracket is $[\xi, \eta] = \xi\eta - \eta\xi$. Finally, the exponential is the usual matrix exponential:

$$\exp(\xi) = \sum_{k=0}^{\infty} \frac{\xi^k}{k!}.$$

Special Linear Group

$SL_n(\mathbb{R})$ is the subgroup of $GL_n(\mathbb{R})$ containing matrices with determinant 1. The determinant is a C^∞ function. Its derivative at $g \in GL_n(\mathbb{R})$ is a linear map from $\mathcal{M}_n(\mathbb{R})$ to $\mathbb{R}$, given by $D\det(g).\xi = \det(g)\operatorname{trace}(g^{-1}\xi)$. Since this has rank one, Theorem B.12 implies that $SL_n(\mathbb{R})$ is a submanifold of $GL_n(\mathbb{R})$, of dimension $n^2 - 1$. The Lie algebra of $SL_n(\mathbb{R})$ is defined by $D\det(\operatorname{Id}_{\mathbb{R}^n})\xi = 0$, and therefore consists of matrices with vanishing trace.

Rotations

$O_n(\mathbb{R})$ is the group of matrices g such that $g^T g = \operatorname{Id}$. $SO_n(\mathbb{R})$ is the subgroup of $O_n(\mathbb{R})$ containing matrices of determinant 1. The map $\Phi : g \mapsto g^T g$ is C^∞, its differential is

$$D\Phi.(g)\xi = g^T \xi + \xi^T g.$$

The null space of $D\Phi(g)$ therefore contains matrices $\xi = g\eta$ such that η is skew symmetric, and has dimension $n(n-1)/2$. Thus, again by theorem B.12, $O_n(\mathbb{R})$ and $SO_n(\mathbb{R})$ are submanifolds of $\mathcal{M}_n(\mathbb{R})$, of dimension $n(n-1)/2$.

The Lie algebra of $O_n(\mathbb{R})$ is the space of skew-symmetric matrices of size, n.

Similitudes

$Sim_n(\mathbb{R})$ is the group of similitudes. It is composed with matrices g such that $g^T g = \lambda\operatorname{Id}$, for some $\lambda > 0$ in $\mathbb{R}$. In fact, one must have $\lambda = \det(g)^{2/n}$, so that $Sim_n(\mathbb{R})$ is the set of invertible matrices for which $\Phi(g) = 0$, with

$$\begin{aligned} \Phi : GL_n(\mathbb{R}) &\to \mathcal{M}_n(\mathbb{R}) \\ g &\mapsto g^T g - \det(g)^{2/n}\operatorname{Id} \end{aligned}.$$

One can check that Φ has constant rank and that $Sim_n(\mathbb{R})$ is a submanifold of $\mathcal{M}_n(\mathbb{R})$ of dimension $1 + n(n-1)/2$.

The Lie algebra of $Sim_n(\mathbb{R})$ contains matrices of the form $\alpha\operatorname{Id} + \xi$, with ξ skew-symmetric.

Affine Groups

Groups of affine transformations are obtained by combining the previous linear transformations with translations. Let G be one of the linear groups we have discussed. Assimilating a translation to a vector in $\mathbb{R}^n$, we can represent an affine transformation with linear part in G by a pair (g, a) with $g \in G$ and $a \in \mathbb{R}^n$. The set of such pairs is denoted $G \ltimes \mathbb{R}^n$. This notation indicates that we have a semi-direct product: if (g, a) and (g', a') belong in $G \ltimes \mathbb{R}^n$, their product must be defined (in order to be consistent with the composition of maps) by

$$(g, a)(g', a') = (gg', ga' + a)$$

(and not $(gg', a + a')$ which would correspond to a direct product).

One can also find alternative notation for some affine groups, like $GA_n(\mathbb{R})$ for $GL_n(\mathbb{R}) \ltimes \mathbb{R}^n$, or $SA_n(\mathbb{R})$ for $SL_n(\mathbb{R}) \ltimes \mathbb{R}^n$.

Affine groups of dimension n can also be represented as subgroups of $GL_{n+1}(\mathbb{R})$: to $(g, a) \in G \times \mathbb{R}^n$, one can associate

$$\Phi(g, a) = \begin{pmatrix} g & a \\ 0 & 1 \end{pmatrix}.$$

It is easy to check that this is a bijection and a group isomorphism, in the sense that

$$\Phi(g, a)\Phi(g', a') = \Phi((g, a)(g', a')).$$

This allows one to identify the Lie algebra of $G \ltimes \mathbb{R}^n$ to the set of matrices of the form

$$\begin{pmatrix} A & a \\ 0 & 0 \end{pmatrix}$$

where A belongs to the Lie algebra of the subgroup of $GL_n(\mathbb{R})$ on which the affine group is built, and $a \in \mathbb{R}^n$.

Projective Group

Definition B.17. *The set of real lines passing through the origin in an $(n+1)$-dimensional vector space is the n-dimensional projective space and is denoted $\mathbb{P}^n(\mathbb{R})$.*

An element of $\mathbb{P}^n(\mathbb{R})$ therefore is a collection of lines $m = \mathbb{R}.x = \{\lambda.x, \lambda \in \mathbb{R}\}$, x being a non-vanishing vector in $\mathbb{R}^3$. There is no loss of generality in assuming that x has norm 1, and since x and $-x$ provide the same point m, $\mathbb{P}^n(\mathbb{R})$ can be seen as the sphere S^n in which antipodal points are identified.

The set $\mathbb{P}^n(\mathbb{R})$ is a differential manifold of dimension n. It can be equipped with the quotient topology of the space $\mathbb{R}^{n+1}$ under the equivalence relation of being co-linear to the origin (we skip the details). One can define the local chart (U_i, Φ_i) by

$$U_i = \{m = \mathbb{R}.x : x = (x_1, \ldots, x_n), x_i \neq 0\}$$

and

$$\Phi_i : U_i \rightarrow \mathbb{R}^n$$
$$\mathbb{R}.x \mapsto \left(\frac{x_1}{x_i}, \ldots, \frac{x_{i-1}}{x_i}, \frac{x_{i+1}}{x_i} \ldots, \frac{x_n}{x_i}\right).$$

One can verify that these definitions do not depend on the choice of the representant $x \neq 0$ in the line $\mathbb{R}.x$. One can also easily check that the family (U_i, Φ_i) forms an atlas of $\mathbb{P}^n(\mathbb{R})$.

Definition B.18. *Since linear maps transform lines into lines, one can associate to each $g \in GL_{n+1}(\mathbb{R})$ an induced transformation of $\mathbb{P}^n(\mathbb{R})$, still denoted g, defined by*

$$g.(\mathbb{R}.x) = \mathbb{R}.(g.x).$$

The set of all such transformations $g : P^n(\mathbb{R}) \rightarrow P^n(\mathbb{R})$ is a group, called the n-dimensional projective group, and denoted $PGL_n(\mathbb{R})$. This group is a Lie group (we skip the demonstration). It can be identified to $GL_{n+1}(\mathbb{R})$ modulo the relation: $g \sim h$ if there exists $\lambda \neq 0$ such that $g = \lambda.h$. This group has dimension $(n+1)^2 - 1$. (In particular, the projective group $PGL_2(\mathbb{R})$, which corresponds to images, has dimension 8.)

The Lie algebra of $PGL_n(\mathbb{R})$ is the Lie algebra of $GL_{n+1}(\mathbb{R})$ in which two matrices are identified if they differ by a matrix of the form $\alpha\mathrm{Id}$. It can also be identified to the set of $(n+1)$ by $(n+1)$ matrices with zero trace.

A local chart for this group in a neighborhood of the identity is the mapping that associates to a matrix $g = (g_{ij}) \in GL_{n+1}(\mathbb{R})$ (or more precisely to the set $\{\lambda g, \lambda \in \mathbb{R}\}$), the $(n+1)^2 - 1$ coordinates $x_{ij} = g_{ij}/g_{n+1n+1}$, for $(i,j) \neq (n+1, n+1)$.

Remark: Projections of Three-Dimensional Motion

The projective group has a fundamental importance in computer vision [78, 132]. One can consider that a camera with focal point $C \in \mathbb{R}^3$ transforms a point $m = (x, y, z)$ in $\mathbb{R}^3$ into its perspective projection, defined as the intersection of the line (Cm) with a fixed plane, called the retinal plane of the camera. Using if needed a change of coordinates, let's assume that $C = (0, 0, 0)$, and that the retinal plane is $z = -f$. In this case, the projection has coordinates $-f.(x/z, y/z, 1)$. The observable quantities are the projected coordinates $(u, v) = (x/z, y/z)$.

Conversely, observing a point in the retinal plane may correspond to any point on the line generated by the vector $(u, v, 1)$ in $\mathbb{R}^3$. This implies that the observations provided by a camera are three-dimensional lines passing through the focal point C (cf. Figure B.1). This is the two-dimensional projective space.

It is important to analyze how three-dimensional motion projects with this model. We can first remark that an affine transformation $x \mapsto gx + a$

defined on $\mathbb{R}^{n+1}$ does not transform (unless $a = 0$) a line passing through the origin into another line passing through the origin. It therefore cannot induce a transformation of the projective space. In concrete terms, it is not possible to infer, from the simple observation of the set $\mathbb{R}.x$ (and not of the point x), what will be the set $\mathbb{R}.(gx + a)$. Returning to the perspective camera model, this implies that arbitrary displacements of objects in space cannot be modeled in general by a well-defined transformation of the retinal plane.

Some approximations can be made, however, in some situations. Assume that a three-dimensional object Ω is some subset of $\mathbb{R}^{n+1}$, and consider a displacement $x \mapsto gx + a$ in this space $\mathbb{R}^{n+1}$. We want to discuss whether there exists a transformation $\varphi : \mathbb{P}^n(\mathbb{R}) \rightarrow \mathbb{P}^n(\mathbb{R})$, such that *for* $x \in \Omega$, $\varphi(\mathbb{R}.x) = a(\mathbb{R}.x) + b$. Here, the question is whether considering only points in the object, Ω, can allow for a representation with a projective transformation.

It is clear that this problem has a solution in the following case: there exists a linear transformation $h \in GL_{n+1}(\mathbb{R})$ such that, for all $x \in \Omega$, one has $g.x + a = h.x$. If Ω contains $n + 2$ points which are affinely independent (for $n = 2$, this means that Ω is a 3D object that contains four points that do not belong to the same plane), then the affine transformation is uniquely specified by its restriction to Ω and therefore cannot coincide with a linear one unless $a = 0$. However, if Ω contains no affinely independent subset (in 3D, Ω is a flat object), such a representation is always possible. This is because there is at most $n + 1$ independent equations in the relations $hx = gx + a$, $x \in \Omega$, so that a solution h exists. We therefore have the following result.

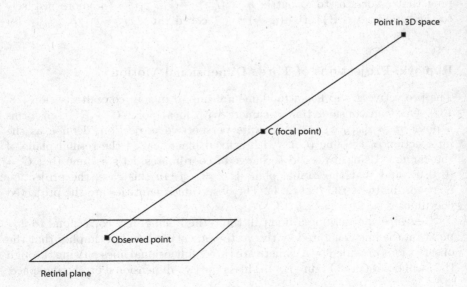

Fig. B.1. Perspective camera model.

Proposition B.19. *Let Ω be included in an affine submanifold of dimension at most n of $\mathbb{R}^{n+1}$, and $x \mapsto g.x + a$ an affine transformation in $\mathbb{R}^{n+1}$. There exists $h \in PGL_n(\mathbb{R})$ such that, for all $x \in \Omega$,*

$$h.(\mathbb{R}.x) = \mathbb{R}.(g.x + a).$$

Returning to the situation in which an object is observed with a camera, this proposition states that the displacement of a flat object in 3D induces a projective transformation of the coordinates. Such an assumption can be considered as a valid approximation in a large number of cases. This is because one generally observes opaque (non-transparent) objects, so that the assumption of flatness only has to apply to the visible part or the object. Moreover, even for non-flat surfaces, this is a reasonable approximation as soon as the thickness of the object is small compared to its distance to the focal point C.

Another, even stronger, approximation is often acceptable. Consider the 3D case, and let $(x, y, z) \in \mathbb{R}^3$. The projection of x on the retinal plane is (eventually after a change of coordinates) $(u, v) = (x/z, y/z)$. If an affine displacement transforms this point in another, say (x', y', z'), the new observation is $(u', v') = (x'/z', y'/z')$. If we assume that the depth z is almost constant on Ω, and make the same assumption for z', then we can consider that the new coordinates (u', v') can be deduced from (u, v) with an affine transformation.

We obtain the fact that the displacement of a 3D object in space can be modeled, with a small approximation, by a projective transformation on the retinal plane, and even, for objects that are sufficiently far away, by an affine transformation.

Of course, in this discussion, we did not take into account the possible appearance of previously hidden parts after displacement (of the occlusion of some previously visible part). This is a strong limitation, since such events typically happen as soon as the displacement is large enough, unless (i) the object is polyhedral, and only one face remains visible during the displacements, or (ii) the displacement is a rotation with axis Cz.

In this last case, the observed transformation is an arbitrary rotation in the retinal plane. This gives a particular importance to rotation invariance for shape recognition, since any 2D rotation of the image of an object is an equally valid image of the same object. For displacements other than these rotations, and smooth objects, apparitions and occlusions are frequent. This implies that, for such objects, the assumptions leading to modeling retinal variations by affine or projective transformation become invalid for large displacements.

B.5 Group Action

B.5.1 Definitions

One says that a group G acts (on the left) on a set M, if there exists a map, Φ from $G \times M$ to M which associates to a pair (g, p) the result of the action of g on p with the properties that $\Phi(g, \Phi(h, p)) = \Phi(gh, p)$ and $\Phi(id, p) = m$. The map Φ is a right action if the first property is replaced by $\Phi(g, \Phi(h, p)) = \Phi(hg, p)$. Left actions are usually denoted $(g, p) \mapsto g \cdot p$, and right actions $(g, p) \mapsto p \cdot g$, and the associativity property becomes $g \cdot (h \cdot p) = (gh) \cdot p$ in the first case and $(p \cdot h) \cdot g = p \cdot (hg)$ in the second case.

The orbit, or coset, of $p \in M$ under a left action is the set $G \cdot p = \{g \cdot p, g \in G\}$. Orbits either coincide or are disjoint, and they form a partition of M. We let $M/G = \{G \cdot p, p \in M\}$. A similar definition holds for right actions.

The action of G is transitive if there exists only one orbit, i.e., for all $p, p' \in M$, there exists $g \in G$ such that $g \cdot p = p'$.

The isotropy subgroup of a point $p \in M$ is the collection of elements $g \in G$ such that $g \cdot p = p$. It is denoted $\mathrm{Iso}_p(G)$, and forms a subgroup of G, (i.e., it is closed under the group product and under the group inverse). The isotropy subgroup of M is the intersection of all $\mathrm{Iso}_p(G)$, and denoted $\mathrm{Iso}_M(G)$.

When G is a Lie group and M is a manifold, one implicitly assumes that, in addition, the map $(g, p) \mapsto g \cdot p$ is C^∞.

B.5.2 Homogeneous Spaces

If H is a subgroup of G, the map $(h, g) \mapsto gh$ defines a right action of H on G. The coset space G/H is the set of orbits, $\{gH, g \in G\}$ for this action. When G is a Lie group and H a closed subgroup of G, G/H is called a homogeneous space. The differential structure of G can be projected onto G/H to provide this set with a structure of differential manifold.

The group G acts on the left on G/H through $g(g'H) = (gg')H$. This action is transitive and H is the isotropy subgroup of (idH). Conversely, the following is true.

Proposition B.20. *Let G be a group acting transitively on the left on a set M. Fix $p \in M$ and let $H = \mathrm{Iso}_p(G)$ be the isotropy subgroup of p. The map*

$$\Phi: \ G/H \to M$$
$$gH \mapsto g \cdot p$$

is a bijection which commutes with the actions of G on G/H and of G on M.

Proof. First note that this mapping is well-defined: if $[g]_H = [g']_H$, then $g^{-1}g' \in H$ so that $g^{-1} \cdot (g' \cdot p) = p$, which implies $g' \cdot p = g \cdot p$. It is onto because the action is transitive, and one-to-one because $g \cdot p = g' \cdot p$ if and only if $[g]_H = [g']_H$. The fact that Φ commutes with the actions means that $\Phi(g \cdot [g']_H) = g \cdot \Phi([g'_H])$ which is obvious. $\square$

When G is a Lie group and M a differential manifold, one shows [107] that, in addition, Φ is differentiable; this provides an identification between M and a homogeneous space.

B.5.3 Infinitesimal Action

Many Lie group concepts can be interpreted infinitesimally as essentially equivalent concepts on the group Lie algebra. This applies, in particular, to group actions.

We consider a left action, and focus on the mapping $\Phi_p : G \to M; g \mapsto g.p$, where G is a Lie group acting on a manifold M. Let ξ be an element of the Lie algebra of G, $\mathfrak{g}$ (i.e., a tangent vector to G at the identity id). For all $p \in M$, we let

$$\xi \cdot p = D\Phi_p(id)\xi.$$

The family $\rho(\xi) = (\xi \cdot p, p \in M)$ forms a vector field on the manifold M. This therefore provides a mapping $\rho : \mathfrak{g} \to \mathcal{X}(M)$, which is called the *infinitesimal action of G on M*.

The map ρ being linear, its range, $\rho(\mathfrak{g})$, forms a linear subspace of $\mathcal{X}(M)$. Its dimension is finite and must be smaller than or equal to the dimension of G. Its elements are called infinitesimal generators.

For the dimension of $\rho(\mathfrak{g})$ to be strictly smaller than the dimension of G, there must exists a non-vanishing ξ such that $\rho(\xi) = 0$. It is easy to see that this implies that $\exp(t\xi) \cdot p = p$ for all $p \in M$, which means that the isotropy group of G is non-trivial. Conversely, one can show that any element ξ of the Lie algebra of $\mathrm{Iso}_M(G)$ is such that $\rho(\xi) = 0$.

B.6 Riemannian Manifolds

B.6.1 Introduction

In this section, M is a differential manifold of dimension d.

Definition B.21. *A Riemannian structure on M is the definition of a C^∞ inner product between vector fields:*

$$(X, Y) \in \mathcal{X}(M) \times \mathcal{X}(M) \mapsto \langle X, Y \rangle \in C^\infty(M)$$

such that $\langle X, Y \rangle_M = \langle Y, X \rangle_M$, $\langle X, X \rangle \geq 0$, $\langle X, X \rangle = 0$ *if and only if* $X = 0$, *and for all* $\varphi, \psi \in C^\infty(M)$

$$\langle \varphi X + \psi X', Y \rangle = \varphi\langle X, Y \rangle + \psi\langle X', Y \rangle.$$

The value of $\langle X, Y \rangle$ at $p \in M$ will be denoted $\langle X, Y \rangle_p$, and it can be shown that it depends only on X_p and Y_p. An equivalent construction is to assume that, for all $p \in M$, an inner product denoted $\langle\ ,\ \rangle_p$ is given on T_pM, which is such that, if X and Y are vector fields, the function $p \mapsto \langle X_p, Y_p \rangle_p$ is C^∞. We shall use the notation

$$|\xi|_p = \sqrt{\langle \xi, \xi \rangle_p}.$$

In a local chart, $C = (U, \Phi)$, with coordinates $(x_1, \ldots, x_d)$, a tangent vector at $p \in U$ can be written as a linear combination of the $\partial_{x_i,p}$'s. From elementary linear algebra, there exists a positive definite symmetric matrix S_p, the coefficients of which being C^∞ functions of p, such that, if $\xi = \sum \lambda_i \partial_{x_i,p}$, $\eta = \sum \mu_i \partial_{x_i,p}$, then,

$$\langle \xi, \eta \rangle_p = \lambda^T S_p \mu.$$

The Riemannian structure permits, among other things, to measure lengths of displacements on the manifold. If $\mu : [0, T] \to M$ is continuous, piecewise differentiable, its length is defined by

$$L(\mu) = \int_0^T |\dot{\mu}_t|_{\mu(t)}\, dt.$$

In other terms, one defines infinitesimal length elements from the norms on the tangent spaces to M. Similarly, the energy of μ is defined by

$$E(\mu) = \frac{1}{2} \int_0^T |\dot{\mu}_t|_{\mu(t)}^2\, dt.$$

The extremal curves of the energy are called geodesics (one says that a curve is an extremal of a given variational problem if any first-order local perturbation of the curve has only second-order effects on the functional). In a chart where $\mu(t) = (y^{(1)}(t), \ldots, y^{(n)}(t))$, and where $S(y) = (s^{(ij)}(y))$ is the matrix associated to the inner product, we have

$$|\dot{\mu}_t|_{\mu(t)}^2 = \sum_{ij} s^{(ij)}(y(t)) \dot{y}_t^{(i)} \dot{y}_t^{(j)}.$$

Making a local variation $y^{(i)} \mapsto y^{(i)} + h^{(i)}$, extremals of the energy are characterized by: for all i

$$\int_0^T \sum_{lj} \dot{h}_t^{(l)} s^{(lj)}(y(t)) \dot{y}_t^{(j)} + \sum_{i,j,l} \dot{y}_t^{(i)} \dot{y}_t^{(j)} \partial_{x_l} s^{(ij)} h^{(l)} = 0,$$

which yields, after an integration by parts

$$-2 \int_0^T \left(\sum_{l,j} h^{(l)} \partial_t \left(s^{(lj)}(y(t)) \dot{y}_t^{(j)} \right) + \sum_{i,j,l} \dot{y}_t^{(i)}\, \dot{y}_t^{(j)} \partial_{x_l} s^{(ij)} h^{(l)} \right) dt = 0.$$

This relation being true for every h, we have, for all l,

$$-2 \sum_j s^{(lj)}(y) \ddot{y}_{tt}^{(j)} - 2 \sum_{ij} \partial_{x_i} s^{(lj)} \dot{y}_t^{(i)} \dot{y}_t^{(j)} + \sum_{i,j} \dot{y}_t^{(i)} \dot{y}_t^{(j)} \partial_{x_l} s^{(ij)} = 0.$$

Let $\tilde{s}^{(ij)}$ denote the coefficients of S^{-1}. The previous identities give (with a symmetrized second term)

$$-2\ddot{y}_{tt}^{(k)} = \sum_{ij} \dot{y}_t^{(i)} \dot{y}_t^{(j)} \sum_l \tilde{s}^{(kl)} (\partial_{x_i} s^{(lj)} + \partial_{x_j} s^{(li)} - \partial_{x_l} s^{(ij)}).$$

Denoting

$$\Gamma_{ij}^k = \frac{1}{2} \sum_l \tilde{s}^{(kl)} (\partial_{x_i} s^{(lj)} + \partial_{x_j} s^{(li)} - \partial_{x_l} s^{(ij)}), \qquad (B.4)$$

this is

$$\ddot{y}_{tt}^{(k)} + \sum_{i,j} \Gamma_{ij}^k \dot{y}_t^{(i)} \dot{y}_t^{(j)} = 0. \qquad (B.5)$$

The coefficients Γ_{ij}^k only depend on the Riemannian metric. They are called the Christoffel's symbols of the manifold at a given point. Therefore, geodesics (expressed in a local chart) are solutions of a second-order differential equation. This implies that they are uniquely specified by their value at time, say, $t = 0$ and their derivative $\dot{\mu}_t$ at $t = 0$. In particular, one defines the Riemannian exponential at $p \in M$ in the direction $v \in T_p M$ by

$$\mathrm{Exp}_p(tv) = \mu(t) \qquad (B.6)$$

where $\mu(.)$ is the geodesic with $\mu(0) = p$ and $\dot{\mu}_t(0) = v$. Such a geodesic exists, as a solution of a differential equation, at least for small times, so that the exponential is well-defined at least for small enough t. If this exponential exists at all p's for all times, M is said to be a complete manifold. The Riemannian exponential (defined on a neighborhood of 0 in $T_p M$) forms a local chart of the manifold.

B.6.2 Geodesic Distance

When M is a Riemannian manifold, one defines the distance between two points p and p' in M by the length of the shortest path which links them, setting

$$d(p, p') = \inf\{L(\mu) : \mu : [0, 1] \to M,$$

$$\mu \text{ continuous, piecewise differentiable }, \mu(0) = p, \mu(1) = p'\}.$$

The following theorem is standard, and may be proved as an exercise or read, for example, in [65]:

Theorem B.22. *The function d which is defined above is a distance M. Moreover,*

$$d(m, m') = \inf\{\sqrt{2E(\mu)} : \mu : [0, 1] \to M,$$

$$\mu \text{ continuous, piecewise differentiable }, \mu(0) = p, \mu(1) = p'\}.$$

B.6.3 Lie Groups with a Right-Invariant Metric

On Lie groups, Riemannian structures can be coupled with invariance constraints. As seen in Chapter 8 when considering groups of diffeomorphisms, the suitable way of "moving" within a group is by iterating small steps through the composition rule. For a curve $g(.)$ on the group, the length of a portion between $g(t)$ and $g(t + \varepsilon)$ should measure the increment $g(t + \varepsilon)g(t)^{-1}$. Fix t and let $u(\varepsilon) = g(t + \varepsilon)g(t)^{-1}$: one has $u(0) = 0$ and $g(t + \varepsilon) = u(\varepsilon)g(t)$. If there is a Riemannian structure on G, the length of the displacement from $g(t)$ to $g(t + \varepsilon)$ is

$$\int_t^{t+\varepsilon} |\dot{g}_s|_{g(s)} \, ds \simeq \varepsilon \, |\partial_\varepsilon g(t + \varepsilon)|_{g(t)}$$

where the last derivative is taken at $\varepsilon = 0$. The right-invariance constraint says that this length should in fact be only measured by the increment on the group and therefore be a function of $u(\varepsilon)$. But $u(.)$ is itself a curve on G, between id and $g(t + \varepsilon)g(t)^{-1}$ and its length is therefore essentially given by $\varepsilon \, |\dot{u}_\varepsilon(0)|_{id}$. Thus, with the invariance constraint, we should take

$$\left|\partial_\varepsilon g(t + \varepsilon)_{|_{\varepsilon=0}}\right|_{g(t)} = |\dot{u}_\varepsilon(0)|_{id} .$$

Introduce the right translation in the Lie group:

$$R_g : G \to G$$
$$h \mapsto hg$$

so that $g(t + \varepsilon) = R_{g(t)}(u(\varepsilon))$. We have, by the chain rule

$$\partial_\varepsilon g(t + \varepsilon)_{|_{\varepsilon=0}} = DR_{g(t)}(id)\dot{u}_\varepsilon(0).$$

This leads to the following definition:

Definition B.23. *A Riemannian metric on a Lie group G is said to be right invariant if and only if, for all $u \in \mathfrak{g} = T_{id}G$, for all $g \in G$*

$$|DR_g(id)u|_g = |u|_{id} .$$

Thus, the metric on any $T_g G$ may be obtained from the metric on $\mathfrak{g}$ by right translation.

The conservation of momentum (see section 11.5) is in fact true in any Lie group with a right-invariant metric. The proof is very similar to the one we have sketched for diffeomorphisms [11, 12, 136].

B.6.4 Covariant Derivatives

We briefly describe here how a Riemannian structure leads naturally to the notion of directional derivative of a vector field relative to an other. This can provide an introduction and a motivation for the more abstract theory of affine connections [65]. This will allow us to interpret a geodesic as a curve with vanishing acceleration, like straight lines in Euclidean spaces.

Let's work on the simpler case in which M is a submanifold of $\mathbb{R}^N$. In this case, the tangent spaces to M can be considered as affine subspaces of $\mathbb{R}^N$, which inherit its standard dot product, which is the Riemannian metric on M. A curve μ on M is also a curve in $\mathbb{R}^N$, and its energy is given by

$$E(\mu) = \frac{1}{2} \int_0^1 |\dot{\mu}_t|^2 \, dt$$

where the norm here simply is the Euclidean norm on $\mathbb{R}^N$. To compute the geodesics, we need the extremals of E, subject to the constraint that the curves must remain on M.

For such an extremal, μ, and a small perturbation h such that $h(t) \in T_{\mu(t)}M$ for all t, we have

$$E(\mu + h) \simeq E(\mu) + \int_0^1 \langle \dot{\mu}_t, \dot{h}_t \rangle dt + o(h) = E(\mu) - \int_0^1 \langle \ddot{\mu}_{tt}, h \rangle dt + o(h).$$

Thus, μ is an extremal if and only if

$$\langle \ddot{\mu}_{tt}, h \rangle = 0$$

for all $h \in T_{\mu(t)}M$, which is equivalent to, for all t

$$\Pi_{\mu(t)} (\ddot{\mu}_{tt}) = 0$$

where Π_p is the orthogonal projection of $\mathbb{R}^N$ on T_pM, with $p \in M$. This provides another characterization of geodesics (which does not require introducing local coordinates), in the particular case of a submanifold of $\mathbb{R}^N$.

Still restricting ourselves to this case, let's fix a curve μ on M, and define, for all vector fields Y on M, the derivative of Y along μ by

$$\frac{DY}{Dt}\Big|_{\mu(t)} = \Pi_{\mu(t)} \left(\frac{dY_{\mu(t)}}{dt} \right).$$

This is a vector field along μ, and the previous computation shows that a geodesic is characterized by the equation

$$\frac{D\dot{\mu}_t}{Dt} = 0.$$

One can show (we skip the details) that the expression in a local chart $C = (U, \Phi)$ of the derivative along μ of a vector field $Y = \sum_{i=1}^d \eta_i \partial_{x_i}$ is given by

$$\sum_{i=1}^d \rho_i \partial_{x_i, \mu(t)}$$

with

$$\rho_i = \partial_t(\eta_i \circ \mu(t)) + \sum_{j,k=1}^{d} \Gamma_{jk}^i(\mu(t))\eta_j(\mu(t))\eta_k(\mu(t)).$$

The Γ_{jk}^i's are the same Christoffel's symbols as defined in (B.4). Introducing $\lambda_1, \ldots, \lambda_d$ such that

$$\dot{\mu}_t(t) = \sum_{i=1}^{d} \lambda_i(t)\partial_{x_i}(\mu(t))$$

one can write

$$\rho_i = \sum_{j=1}^{d} \lambda_j(t)\partial_{x_j}\eta_i + \sum_{j,k=1}^{d} \Gamma_{jk}^i(\mu(t))\eta_j(\mu(t))\eta_k(\mu(t)).$$

This expression is intrinsic: it does not depend on the ambient space $\mathbb{R}^N$, but on quantities that are computed on the manifold. So, assuming now that M is a general Riemannian manifold, we can define, in a chart $C = (U, \Phi)$, and for two vector fields $X = \sum_{i=1}^{d} \xi_i \partial_{x_i}$ and $Y = \sum_{i=1}^{d} \eta_i \partial_{x_i}$, a third vector field, called the *covariant derivative* of Y along X by

$$\nabla_X (Y)_p = \sum_{i=1}^{d} \rho_i \partial_{x_i, p}$$

with

$$\rho_i = \sum_{j=1}^{d} \xi_j \partial_{x_j}\eta_i + \sum_{j,k=1}^{d} \Gamma_{jk}^i \eta_j \xi_k.$$

(Note that $\nabla_X (Y)_p$ only depends on the coordinates of X at p.)

From our definition of the Christoffel's symbols in the general case, we see that curves of minimal energy still satisfy

$$\frac{D\dot{\mu}_t}{Dt} := \nabla_{\dot{\mu}_t} (\dot{\mu}_t) = 0.$$

The equation $\nabla_{\dot{\mu}_t} (\dot{\mu}_t)$ is therefore called the geodesic equation for curves on M. Curves that satisfy it are called geodesics, even when they are not energy-minimizing (although they are always locally so [65]).

Covariant derivatives can be defined in more general contexts that on Riemannian manifolds [107, 65]. The one we have defined above is adapted to the Riemannian metric and called the *Levi–Civita connection*. It satisfies the two characteristic properties:

(1) $Z\langle X, Y \rangle = \langle \nabla_Z (X), Y \rangle + \langle X, \nabla_Z (Y) \rangle$
(2) $\nabla_X (Y) - \nabla_Y (X) = [X, Y]$ with $[X, Y] = XY - YX$.

B.6.5 Parallel Transport

Parallel transport (or translation) is the displacement of a vector along a curve with vanishing covariant derivative. It is the generalization of translation in space. Given a curve $t \mapsto \mu(t)$ on a Riemannian manifold M, a time-dependent tangent vector $t \mapsto X(t)$ with $X(t) \in T_{\mu(t)}M$ is said to be parallel on μ if its derivative along μ vanishes, namely

$$\nabla_{\dot{\mu}_t} X = \frac{DX}{Dt} = 0.$$

So, by definition, a geodesic is a curve with derivative moving parallel to itself. Parallel transport derives from a first-order differential equation (for X) along the curve, which, in a chart, is given by:

$$\partial_t(\rho_k \circ \mu) + \sum_{i,j=1}^{d} \Gamma_{ij}^k \rho_i \circ \mu \partial_t(\mu_j) = 0$$

with $X = \sum_{k=1}^{d} \rho_k \partial_{x_k}$. Since this is a first-order linear system of equations in $\rho \circ \mu$, it can be integrated uniquely as soon as an initial condition $\rho \circ \mu(0)$ is given. This leads to the following definition.

Definition B.24. *Let M be a Riemannian manifold, $p \in M$ and $\xi_0 \in T_pM$. Let μ be a curve on M with $\mu(0) = p$. The parallel transport of ξ_0 along μ is the time-dependent vector $\xi(t) \in T_{\mu(t)}M$ such that $\xi(0) = \xi_0$ and*

$$\nabla_{\dot{\mu}_t} \xi = 0.$$

It is important to remember that parallel transport is only defined *along* a curve. If $p, \tilde{p} \in M$, $\xi \in T_pM$, and μ and $\tilde{\mu}$ are two curves linking p to $\tilde{p}$, the results of the parallel transport of ξ along μ and $\tilde{\mu}$ are generally distinct.

B.6.6 A Hamiltonian Formulation

Geodesics in a chart have a Hamiltonian formulation that is sometimes convenient. Define the function (Hamiltonian)

$$H(p, a) = \frac{1}{2} a^T S(p)^{-1} a \tag{B.7}$$

where S is the metric, $p \in M$ and a is a d-dimensional vector. One can prove that a curve $t \mapsto m(t)$ on M is a geodesic if and only if it satisfies the system

$$\begin{cases} \dot{m}_t = S(\mu(t))^{-1} a(t) = \partial_a H(m(t), a(t)), \\ \dot{a}_t = -\partial_p H(m(t), a(t)). \end{cases}$$

Introducing the coefficients $\tilde{s}^{(ij)}$ of S^{-1}, this is

$$\begin{cases} \partial_t m_i = \sum_{j=1}^{d} \tilde{s}^{(ij)}(m) a_j, \\[2em] \partial_t a_i = \sum_{k,l=1}^{d} a_k a_l \partial_{x_i}(\tilde{s}^{(kl)})(m). \end{cases}$$

Note that the energy of the curve is (at time t)

$$\frac{1}{2}\langle \dot{m}_t, \dot{m}_t \rangle_M = \frac{1}{2}\dot{m}_t^T S(m) m = \frac{1}{2}a^T \dot{m}_t.$$

The vector $a(t)$ is *cotangent* to the motion (because it acts as a linear form on tangent vectors) and must be considered as an element of $T_p M^*$. It is the *momentum* of the motion.

The covariant derivative can also be written in terms of a. If $a = SX$, $b = SY$ and if we let

$$\nabla_a^* b = S(\nabla_X Y),$$

then, the kth coordinate of $\nabla_a^* b$ is

$$(\nabla_a^* b)_k = \sum_{j=1}^{d} \partial_{x_j}(b_k) X_j$$

$$+ \frac{1}{2}\left(\sum_{i,j=1}^{d} \partial_{x_k} \tilde{s}^{(ij)} a_i a_j + \sum_{i,j,l,q=1}^{d} s_{kl}\partial_{x_j} s^{(lq)}(b_q X_j - a_q Y_j) \right).$$

This expression also provides the equation for parallel translation of b along a curve m with $S\dot{m}_t = a$, namely

$$\partial_t b_k + \frac{1}{2}\left(\sum_{i,j=1}^{d} \partial_{x_k} \tilde{s}^{(ij)} a_i a_j + \sum_{i,j,l,q=1}^{d} s_{kl}\partial_{x_j} s^{(lq)}(b_q X_j - a_q Y_j) \right) = 0.$$

C

Ordinary Differential Equations

In the first part of this chapter, we review basic existence theorems and properties of ordinary differential equations (ODEs) on Banach spaces. The presentation will follow the ones provided in standard textbooks [49], although we will work with slightly relaxed regularity conditions. The second part will provide an overview of some of the most important numerical methods designed to solve ODEs.

We start with some definitions about differentials in infinite dimensions.

C.1 Differentials in Banach Space

A domain Ω in a Banach space $\mathbb{B}$ is open if, for every point $x \in \Omega$, there exists $\varepsilon(x) > 0$ such that the open ball $B(x, \varepsilon(x))$ is included in Ω, with

$$B(x, \varepsilon(x)) = \{y \in \mathbb{B} : \|y - x\|_{\mathbb{B}} < \varepsilon(x)\}.$$

A set F is closed in $\mathbb{B}$ is $\mathbb{B} - F$ is open. The closure of a set A (denoted $\operatorname{clos}(A)$ or $\bar{A}$) is the smallest closed set that contains it and its interior $(\operatorname{int}(A))$ is the largest open set included in it. Its boundary is $\partial A = \operatorname{clos}(A) \setminus \operatorname{int}(A)$.

A function f defined on a subset Ω' of a Banach set $\mathbb{B}'$ with values in $\mathbb{B}$ is continuous if $\|f(x_n) - f(x)\|_{\mathbb{B}} \to 0$ whenever (x_n) is a sequence in Ω' such that $\|x_n - x\|_{\mathbb{B}'} \to 0$ for some $x \in \Omega'$. The function f is Lipschitz if there exists a constant c such that

$$\|f(x) - f(y)\|_{\mathbb{B}} \leq c\|x - y\|_{\Omega'}$$

for all $x, y \in \Omega'$. The smallest c for which this is true is called the Lipschitz constant of f and denoted $\operatorname{Lip}(f)$.

The integral of a Banach-valued function $f : [a, b] \to \mathbb{B}$ can be defined, as in finite dimensions, by the limit of the integral of sums of simple functions (constant over a finite number of measurable subsets of $[a, b]$). This provides the Bochner integral [25, 62], with properties very similar to Lebesgue's integral on the line. An important characteristic property is that f is Bochner integrable if and only if $\|f\|_{\mathbb{B}}$ is Lebesgue integrable.

A function $f : \mathbb{R} \to \mathbb{B}$ is differentiable at $t \in \mathbb{R}$ if there exists an element of $\mathbb{B}$, denoted $\partial_t f(t)$ or $\dot{f}_t$ such that

$$\lim_{\varepsilon \to 0} \left\| \frac{1}{t}(f(t + \varepsilon) - f(t)) - \dot{f}_{t_{\mathbb{B}}} \right\| = 0.$$

If $\mathbb{B}$ and $\mathbb{B}'$ are Banach spaces and $f : \mathbb{B}' \to \mathbb{B}$, we say that f is *Gâteaux differentiable* at $x \in \mathbb{B}'$ is, for all $h \in \mathbb{B}'$, $t \mapsto f(x+th)$ is differentiable at $t = 0$ and there exists a continuous linear transformation, denoted $Df(x) : \mathbb{B}' \to \mathbb{B}$ such that, for all $h \in \mathbb{B}'$:

$$\partial_t f(x + th)(0) = Df(x)h.$$

$Df(x)$ is called the Gâteaux derivative of f at x.

We say that f is differentiable in the Fréchet sense, or *Fréchet differentiable* at $x \in \mathbb{B}'$ if there exists a linear transformation $Df(x) : \mathbb{B}' \to \mathbb{B}$ such that, for any $h \in \mathbb{B}'$,

$$\lim_{\|h\|_{\mathbb{B}'} \to 0} \frac{1}{\|h\|_{\mathbb{B}'}} \|f(x + h) - f(h) - Df(x)h\|_{\mathbb{B}} = 0;$$

$Df(x)$ is called the Fréchet derivative of f at x.

There is no ambiguity in the notation, because it is clear that if f is Fréchet differentiable, then it is also Gâteaux differentiable, and the two definitions for $Df(x)$ coincide.

Gâteaux differentiability implies Fréchet differentiability in the following case.

Proposition C.1. *Let $f : \Omega' \subset \mathbb{B}' \to \mathbb{B}$ be Gâteaux differentiable at all points $x \in \Omega'$, which is an open subset of the Banach space $\mathbb{B}'$. Assume that $Df(x)$ is continuous on Ω' in the sense that, if*

$$\omega(x, \delta, Df) =$$
$$\sup \left\{ \|(Df(y) - Df(x))h\|_{\mathbb{B}}, \ \|h\|_{\mathbb{B}'} = 1, \ x, y \in \Omega', \|x - y\|_{\mathbb{B}'} \leq \delta \right\},$$

then, for all $x \in \Omega'$,
$$\lim_{\delta \to 0} \omega(x, \delta, Df) \to 0.$$

Then f is also Fréchet differentiable.

(In other terms, if f is continuously Gâteaux differentiable, then it is Fréchet differentiable.)

Proof. Note that, for $x, h \in \Omega'$, such that the segment $[x, x + h]$ is included in Ω', the transformation $t \mapsto \partial_t f(x + th)(t) = Df(x + th) h$ is integrable (because $Df(x)$ is continuous) and

$$f(x + h) - f(x) = \int_0^1 \partial_t f(x + th) dt.$$

This implies

$$\begin{aligned}
f(x + h) - f(x) - Df(x)h &= \int_0^1 (Df(x + th) - Df(x))\, h dt \\
&\leq \omega(x, \|h\|_{\mathbb{B}'}, Df) \|h\|_{\mathbb{B}'} \\
&= o(\|h\|_{\mathbb{B}'}).
\end{aligned}$$

$\square$

A function $v : \mathbb{B} \to \mathbb{B}$ is called a vector field over $\mathbb{B}$ (or over $\Omega \subset \mathbb{B}$ if v is just defined on Ω). A time-dependent vector field is a function $v : [0, 1] \times \mathbb{B} \to \mathbb{B}$ (or $v : [0, 1] \times \Omega \to \mathbb{B}$).

If v is a time-dependent vector field on $\mathbb{B}$, we will say that a function $t \mapsto y(t)$, defined on some interval $[a, b]$ is a solution of the ordinary differential equation

$$\partial_t y = v(t, y) \tag{C.1}$$

with initial condition x at time t if $s \mapsto v(s, y(s))$ is integrable over $[a, b]$ and, for all $u \in [a, b]$,

$$y(u) = x + \int_t^u v(s, y(s)) ds.$$

If the vector field v does not depend on time, one says that (C.1) is *an autonomous equation*.

We recall the standard fixed-point theorem, which will be used later.

Theorem C.2 (Banach fixed-point theorem). *Let $\mathbb{B}$ be a Banach space. Let $U \subset \mathbb{B}$ be a closed set and Φ be a contraction of U, i.e., $\Phi : U \to U$ is such that there exists a constant $c \in [0, 1)$ satisfying, for all $x, y \in U$,*

$$\|\Phi(x) - \Phi(y)\|_{\mathbb{B}} \leq c \|x - y\|_{\mathbb{B}}.$$

Then, Φ has a unique fixed point in U, i.e., there exists a unique $x_0 \in U$ such that $\Phi(x_0) = x_0$.

C.2 A Class of Ordinary Differential Equations

Let Ω be an open subset of a Banach space $\mathbb{B}$. We define the class $\mathcal{L}(T, \Omega)$ of time-dependent vector fields $v : [0, T] \times \mathrm{clos}(\Omega) \to \mathbb{B}$ that vanish on $\partial \Omega$ and are Lipschitz with respect to the Banach space variable, with (for some fixed $x_0 \in \Omega$)

$$\mathrm{Lip}_T(v) := \int_0^T \|v(u, x_0)\|_\mathbb{B} dt + \int_0^T \mathrm{Lip}(v(t, .)) dt < \infty. \qquad (\mathrm{C.2})$$

Of course, the vanishing condition on $\partial\Omega$ becomes void if this boundary is empty.

C.2.1 Existence and Uniqueness

Theorem C.3. *Let $v \in \mathcal{L}(T, \Omega)$. For all $x \in \Omega$ and $t \in [0, T]$, there exists a unique solution on $[0, T]$ of the ordinary differential equation $\partial_t y = v(t, y)$ with initial condition $y(t) = x$. This solution is such that $y(s) \in \Omega$ for all $s \in [0, T]$.*

Proof. The proof slightly deviates from the standard Picard–Lindelöf theorem, which assumes that $\mathrm{Lip}(v(t, .))$ is uniformly bounded [49, 180]. We have here an integral condition instead, given by (C.2), so that the statement is slightly more general, although the proof, which follows [70, 202], is very similar.

We first prove the result for $\Omega = \mathbb{B}$. Fix $x \in \mathbb{B}$, $t \in [0, T]$ and $\delta > 0$. Let $I = I(t, \delta)$ denote the interval $[0, T] \cap [t - \delta, t + \delta]$. If φ is a continuous function from I to $\mathbb{B}$ such that $\varphi(t) = x$, we define the transformation $\Gamma(\varphi) : I \to \mathbb{B}$ by

$$\Gamma(\varphi)(s) = x + \int_t^s v(u, \varphi(u)) du.$$

The integral is well-defined because

$$\|v(u, \varphi(u))\|_\mathbb{B} \le \|v(u, x_0)\|_\mathbb{B} + \mathrm{Lip}(v(u, \cdot))\|\varphi(u) - x_0\|_\mathbb{B}$$

and $\varphi(u)$ is bounded since φ is continuous.

The function $s \mapsto \Gamma(\varphi)(s)$ is continuous and is such that $\Gamma(\varphi)(t) = x$. The set of continuous functions from the compact interval I to $\mathbb{B}$, equipped with the supremum norm, is a Banach space, and we now show that for δ small enough, Γ satisfies

$$\|\Gamma(\varphi) - \Gamma(\varphi')\|_\infty \le \gamma \|\varphi - \varphi'\|_\infty$$

with $\gamma < 1$. The fixed-point theorem will then imply that there is a unique function φ such that $\Gamma(\varphi) = \varphi$, and this is the definition of a solution of the ordinary differential equation on I.

Since $\Gamma(\varphi)(s) - \Gamma(\varphi')(s) = \int_t^s (v(u, \varphi(u)) - v(u, \varphi'(u))) du$ we have

$$\|\Gamma(\varphi) - \Gamma(\varphi')\|_\infty \le \|\varphi - \varphi'\|_\infty \int_I \mathrm{Lip}(v(u, .)) du.$$

But $\int_I \text{Lip}(v(u,.))du$ can be made arbitrarily small by reducing δ so that existence and uniqueness over I is proved. We can make the additional remark that δ can be taken independent of t. This is because the function $\alpha : s \mapsto \int_0^s \text{Lip}(v(u,.))du$ is continuous, hence uniformly continuous on the interval $[0, T]$, so that there exists a constant $\eta > 0$ such that $|s - s'| < \eta$ implies that $|\alpha(s) - \alpha(s')| < 1/2$, and it suffices to take $\delta < \eta/2$.

From this remark, we can conclude that a unique solution of the ordinary differential equation exists over all $[0, T]$, because it is now possible, starting from the interval $I(t, \delta)$ to extend the solution from both sides, by jumps of $\delta/2$ at least, until boundaries are reached.

This proves the result for $\Omega = \mathbb{B}$ and we now consider arbitrary open sets Ω. By extending $v(t)$ with 0 on Ω^c, the value of $\text{Lip}_T(v(t,.))$ remains unchanged and we can apply the result over $\mathbb{B}$ to ensure existence and uniqueness of the solution with a given initial condition. So, it only remains to show that solutions such that $y(t) \in \Omega$ for some t belong to Ω at all times. This is true because if there exists s such that $y(s) = x' \notin \Omega$, then the function $\tilde{y}(u) = x'$ for all u is a solution of the equation, since $v(u, x') = 0$ for all u. Uniqueness implies $\tilde{y} = y$ at all times, which is impossible. $\qquad\square$

Important special cases of this theorem are linear equations. The following corollary can be proved by taking $v(t, x) = A(t)x + b(t)$ in Theorem C.3.

Corollary C.4. *Assume that, for $t \in [0, T]$, $A(t) : \mathbb{B} \to \mathbb{B}$ is a continuous linear operator, $b(t) \in \mathbb{B}$ and that they are integrable and satisfy*

$$\int_0^T (\|A(t)\|_\mathbb{B} + \|b(t)\|_\mathbb{B})dt < \infty.$$

Then the equation

$$\partial_t y = A(t)y + b(t)$$

has a unique solution over $[0, T]$ for a given initial condition. (Here, $\|A(t)\|_\mathbb{B}$ is the operator norm of $A(t)$, namely $\|A(t)\|_\mathbb{B} = \sup\{\|A(t)h\|_\mathbb{B} : \|h\|_\mathbb{B}\} = 1\|.$)

C.2.2 Flow Associated to an ODE

Definition C.5. *Let $v \in \mathcal{L}(T, \Omega)$. We denote by $\varphi_{st}^v(x)$ the solution at time t of the equation $\partial_t y = v(t, y)$ with initial condition $y(s) = x$. The function $(t, x) \mapsto \varphi_{st}^v(x)$ is called the flow associated to v starting at s. It is defined on $[0, T] \times \Omega$ and takes values in Ω.*

From the definition, we have the following property.

Proposition C.6. *If $v \in \mathcal{L}(T, \Omega)$ and $s, r, t \in [0, T]$, then*

$$\varphi_{st}^v = \varphi_{rt}^v \circ \varphi_{sr}^v.$$

In particular, $\varphi_{st}^v \circ \varphi_{ts}^v = \text{id}$ and φ_{st}^v is invertible for all s and t.

Proof. If $x \in \Omega$, $\varphi_{st}^v(x)$ is the value at time t of the unique solution of $\partial_t y = v(t, y)$, which is equal to x at time s. It is equal to $x' = \varphi_{sr}^v(x)$ at time r, and thus also equal to $\varphi_{rt}^v(x')$ which is the statement of the proposition.   □

Theorem C.7. *Let $v \in \mathcal{L}(T, \Omega)$. The associated flow, φ_{st}^v, is at all times a homeomorphism of Ω.*

The proof of this result relies on Gronwall's lemma, that we first state and prove.

Theorem C.8 (Gronwall's lemma). *Consider two positive functions $\alpha(s)$ and $u(s)$, defined for $s \in I$ where I is an interval in $\mathbb{R}$ containing 0. Assume that u is bounded, and that, for some integrable function c, and for all $t \in I$,*

$$u(t) \leq c(t) + \left| \int_0^t \alpha(s)u(s)ds \right| . \tag{C.3}$$

Then, for all $t \in I$,

$$u(t) \leq c(t) + \left| \int_0^t c(s)\alpha(s)e^{\left| \int_0^s \alpha(u)du \right|} \right| . \tag{C.4}$$

When c is a constant, this upper bound becomes

$$u(t) \leq ce^{\left| \int_0^t \alpha(s)ds \right|} . \tag{C.5}$$

Proof. To address simultaneously the cases $t > 0$ and $t < 0$, we let $\varepsilon = 1$ in the first case and $\varepsilon = -1$ in the second case. Inequality (C.3) now becomes:

$$u(t) \leq c(t) + \varepsilon \int_0^t \alpha(s)u(s)ds.$$

Iterating this inequality once yields

$$u(t) \leq c(t) + \varepsilon \int_0^t c(s)\alpha(s)ds + \varepsilon^2 \int_0^t \int_0^{s_1} \alpha(s_1)\alpha(s_2)u(s_2)ds .$$

and iterating further, we get

$$u(t) \leq c(t) + \varepsilon \int_0^t c(s)\alpha(s)ds + \cdots$$

$$+ \varepsilon^n \int_{0 \leq s_1 \leq \cdots \leq s_n \leq t} c(s_n)\alpha(s_1)\ldots\alpha(s_n)ds_1\ldots ds_n$$

$$+ \varepsilon^{n+1} \int_{0 \leq s_1 \leq \cdots \leq s_n \leq t} \alpha(s_1)\ldots\alpha(s_{n+1})u(s_{n+1})ds_1\ldots ds_n ds_{n+1} .$$

Consider the integral

$$I_n = \int_{0 \le s_1 \le \cdots \le s_n} \alpha(s_1) \ldots \alpha(s_{n-1}) ds_1 \ldots ds_{n-1}.$$

Let σ be a permutation of $\{1, \ldots, n-1\}$: making the change of variable $s_i \to s_{\sigma_i}$ in I_n yields

$$I_n = \int_{0 \le s_{\sigma_1} \le \cdots \le s_{\sigma_{n-1}} \le s_n} \alpha(s_1) \ldots \alpha(s_{n-1}) ds_1 \ldots ds_{n-1}.$$

Obviously,

$$\sum_\sigma 1_{0 \le s_{\sigma_1} \le \cdots \le s_{\sigma_{n-1}} \le s_n} = 1$$

whenever $s_i < s_n$ and $s_i \ne s_j$ for $i \ne j \le n-1$. The $(n-1)$-tuples $s_1, \ldots, s_{n-1}$ for which $s_i = s_j$ for some j form a set of dimension $n-2$ which has no influence on the integral. Thus, summing over σ yields

$$(n-1)! I_n = \int_{[0,s_n]^{n-1}} \alpha(s_1) \ldots \alpha(s_{n-1}) ds_1 \ldots ds_{n-1} = \left(\int_0^{s_n} \alpha(s) ds \right)^{n-1}.$$

Therefore, using the fact that u is bounded, we have

$$u(t) \le c(t) + \sum_{k=1}^n \frac{\varepsilon^k}{(k-1)!} \int_0^t c(s_n) \alpha(s_n) \left(\int_0^{s_n} \alpha(s) ds \right)^{k-1} ds_n$$
$$+ \frac{\varepsilon^{n+1} \sup(u)}{(n+1)!} \left(\int_0^t \alpha(s) ds \right)^{n+1},$$

and passing to the limit yields (C.4). If $c = \text{cst}$, the previous upper bound can be written

$$u(t) \le c \sum_{k=0}^n \frac{\varepsilon^k}{k!} \left(\int_0^t \alpha(s) ds \right)^k + \frac{\varepsilon^{n+1} \sup(u)}{(n+1)!} \left(\int_0^t \alpha(s) ds \right)^{n+1},$$

which yields (C.5).   □

We now pass to the proof of Theorem C.7.

Proof (Theorem C.7). It suffices to show that φ_{st}^v is continuous, since Proposition C.6 implies that $(\varphi_{st}^v)^{-1} = \varphi_{ts}^v$. Take $x, y \in \Omega$. We have

$$|\varphi_{st}^v(x) - \varphi_{st}^v(y)| = \left| x - y + \int_s^t (v(s, \varphi_{sr}^v(x)) - v(s, \varphi_{sr}^v(y))) dr \right|$$
$$\le |x - y| + \int_s^t \text{Lip}(v(t, .)) |\varphi_{sr}^v(x) - \varphi_{sr}^v(y)| ds.$$

We apply Gronwall's lemma, equation (C.5), with $c = |x - y|$, $\alpha(r) = \text{Lip}(v(r, .))$ and $u(r) = |\varphi_{0r}^v(x) - \varphi_{0r}^v(y)|$, which is bounded since φ_{sr}^v is continuous in r. This yields

$$|\varphi^v_{sr}(x) - \varphi^v_{sr}(y)| \le |x - y| \exp\left(\int_s^t \mathrm{Lip}(v(r, .))dr\right), \tag{C.6}$$

which shows that φ^v_{st} is continuous on Ω, and even Lipschitz, with a Lipschitz constant smaller than $\exp\left(\mathrm{Lip}_t(v)\right)$. Since $(\varphi^v_{st})^{-1} = \varphi^v_{ts}$, the inverse is also continuous so that φ^v_{st} is a homeomorphism of Ω.   □

C.3 Numerical Integration of ODEs

We end this chapter with a few remarks on how ODEs can be solved numerically (in finite dimensions!). Of course, it cannot be our intention, in these few paragraphs, to include all the information that can be found in specialized text books (e.g., [195, 167, 196]) and we refer the reader to these references for a comprehensive presentation.

Consider the differential equation $\partial_t y = v(t, y)$ where $y \in \mathbb{R}^n$. Denote by $\varphi_{st}(z)$ the solution at time t of this equation with initial condition $y(s) = z$. One-step numerical schemes for this ODE implement iterations of the form

$$\begin{cases} z_{k+1} = \eta(t_k, z_k, h_k) \\ t_{k+1} = t_k + h_k \end{cases}$$

in which z_k is intended to provide a numerical approximation of $\varphi_{0t_k}(z_0)$. If $h_k = h$ is fixed, this is a fixed step method, and is h_k is optimized online, this is an adaptive step method.

The simplest such algorithm is the Euler method, for which

$$\eta(t, z, h) = z + hv(t, z).$$

This algorithm has the advantage of requiring only one evaluation of v per iteration, but it also has a limited accuracy, as measured by the difference

$$\delta(t, z, h) = \frac{1}{h}(\varphi_{t, t+h}(z) - \eta(t, z, h)).$$

For the Euler method, we have, making a first-order Taylor expansion of y around $y(t)$:

$$\delta(t, z, h) = \frac{h}{2}\ddot{y}_{tt}(t) + o(h)$$

with $\ddot{y}_{tt} = \dot{v}_t(t, y) + Dv(t, y)v(t, y)$ (therefore requiring that v is differentiable in time and space). A method is said to be accurate at order p if $\delta(t, z, h) = O(h^p)$. So the Euler method has order 1.

Slightly more complex is the modified Euler, or Euler midpoint method in which η is defined by the iteration

$$\begin{cases} u_1 = v(t, z), \\ \\ \eta(t, z, h) = z + hv(t, z + \frac{h}{2}u_1). \end{cases}$$

Using a second-order expansion, one easily proves that the Euler midpoint method has order 2, assuming this time two derivatives in time and space. A fourth-order method is provided by the Runge–Kutta iteration, defined by

$$\begin{cases} u_1 = v(t, z), \\[2mm] u_2 = v(t + \dfrac{h}{2}, z + \dfrac{h}{2} u_1), \\[2mm] u_3 = v(t + \dfrac{h}{2}, z + \dfrac{h}{2} u_2), \\[2mm] u_4 = v(t + h, z + h u_3), \\[2mm] \eta(t, z, h) = z + \dfrac{h}{6}(u_1 + 2u_2 + 2u_3 + u_4). \end{cases}$$

More generally, one defines Runge-Kutta methods with s stages using iterations of the form

$$\begin{cases} u_i = v\Big(t + c_i h, z + h \displaystyle\sum_{j=1}^{i-1} a_{ij} u_j\Big), \quad i = 1, \dots, s, \\[4mm] \eta(t, z, h) = z + h \displaystyle\sum_{i=1}^{s} b_i u_i \end{cases}$$

with $c_i = \sum_{j=1}^{i-1} a_{ij}$. This is therefore specified by the lower triangular s by s matrix (with null diagonal) (a_{ij}) and by the vector (b_i). Such an iteration requires s evaluations of the function v, but can reach higher orders of accuracy. Note that high orders of accuracy can only be achieved provided v has at least the same number of derivatives in time and space.

A lot can be gained in the accuracy vs. efficiency trade-off by using adaptive step size. This can be implemented by comparing two similar Runge–Kutta methods (using different coefficients in the expansion) and making sure that their difference is small. If it is not the step should be reduced; if it is too small (compared to a given tolerance level), it can be enlarged [66, 167].

The above methods are explicit, in the sense that z_{k+1} is explicitly given as a function of the current value z_k. Implicit methods relax this constraint, and require solving a nonlinear equation to compute $y(t_{k+1})$. In spite of this added complexity, such methods may exhibit improved performances, because they generally are more stable and allow for the use of larger steps h. The simplest example is the Euler implicit method, which iterates

$$z_{k+1} = z_k + h_k v(t_{k+1}, z_{k+1}).$$

To see why this can be interesting, consider the simple autonomous and linear case in which $v(t, y) = Ay$. The solution of the ODE is $y(t) = e^{tA}y(0)$. When A only has negative eigenvalues, the solution converges to 0 when t tends to infinity (one says that the resulting system is stable). An Euler discretization, say with constant step h gives $z_k = (\mathrm{Id} + hA)^k z_0$. Such an iteration will converge to 0 only if the eigenvalues of $(\mathrm{Id}+hA)$ are smaller than 1 in modulus, requiring h to be smaller than $2/|\lambda_{max}|$, where λ_{max} is the eigenvalue of A that has the largest absolute value. The implicit iteration yields $z_k = (\mathrm{Id}-hA)^{-k}z_0$ which always converges to 0.

More general one-step implicit schemes use iterations for which η is given by

$$\theta(t, z, \eta(t, z, h), h) = 0$$

for some function θ such that $\theta(t, z, z', h) - z'$ tends to 0 when h tends to 0. Since an analytical solution (expressing z' as a function of t, z and h) for

$$\theta(t, z, z', h) = 0$$

is almost never available, one typically uses fixed-point iterations. The sequence

$$\zeta_{q+1} = \zeta_q - \theta(t, z, \zeta_q, h)$$

will converge to the required z' for small enough h (ensuring that $\zeta \mapsto (\theta(t, z, \zeta, h) - \zeta)$ is contractant). Doing this may clearly require several evaluations of η (which may in turn require several evaluations of v), so that the gain obtained from being able to use larger steps may be lost if evaluating v is costly.

A natural situation in which implicit methods are needed is when one wants to solve an equation backwards in time, in a way which is consistent with the forward solution. Let's restrict ourselves to autonomous equations, so that v does not depend on t, assuming the same property at the discretization level (θ does not depend on t). The uniqueness theorem for ODEs implies that if y is a solution of $\partial_t y = v(y)$, then the solution of $\partial_t \tilde{y} = -v(\tilde{y})$ initialized with $\tilde{y}(0) = y(T)$ is $\tilde{y}(t) = y(T - t)$. This property is useful if one needs to recover the initial condition of an ODE given its state at time T. If one separately discretizes the forward and the backward equations, this property is generally not satisfied at the discrete level. One needs to make sure that the backward iteration is consistent with the forward one. This implies that, if the forward iteration is solved using (assuming constant steps)

$$\theta(z_k, z_{k+1}, h) = 0$$

then the backward scheme should use

$$\tilde{\theta}(\tilde{z}_k, \tilde{z}_{k+1}, h) = 0$$

with

$$\tilde{\theta}(\tilde{z}_k, \tilde{z}_{k+1}, h) = \theta(\tilde{z}_{k+1}, \tilde{z}_k, h)$$

and it is clear that θ and $\tilde{\theta}$ cannot be explicit together. Another interesting notion related to reversibility in time is the adjoint scheme

$$\theta^*(z, z', h) = \theta(z', z, -h),$$

which is such that $\tilde{\theta}(z, z', h) = \theta^*(z, z', -h)$. The adjoint scheme is also a forward scheme. For example, the adjoint of the Euler explicit scheme is the Euler implicit one.

One says that the scheme is symmetric is $\theta^* = \theta$. Symmetric schemes are interesting because they can be used to solve both the backward and the forward equation in a consistent way. One simple way to build a symmetric scheme from a non-symmetric one is to combine the latter with its adjoint, i.e., define z' from z by first solving $\theta^*(z, z'', h/2) = 0$, then $\theta(z'', z', h/2) = 0$. Starting with the Euler method, this yields the implicit midpoint method, which is defined by

$$y_{k+1} = y_k + \frac{h}{2}v(y')$$

where y' satisfies the equation $y' = y_k + (h/2)v(y')$.

Another important class of problems in which implicit methods are useful are for the solution of Hamiltonian systems, like in Section B.6.6, and like the EPDiff equation, which is extensively discussed in this book. Such systems are split over two variables, traditionally denoted p (momentum) and q (state) and involve a real-valued function $H(p, q)$ called the Hamiltonian.

The resulting Hamiltonian system is the ODE

$$\begin{cases} \partial_t q = \partial_p H, \\ \\ \partial_t p = -\partial_q H. \end{cases}$$

An important property of Hamiltonian systems is that their flow is symplectic. To define this notion, introduce the matrix

$$J = \begin{pmatrix} 0 & \mathrm{Id} \\ -\mathrm{Id} & 0 \end{pmatrix}.$$

With this notation, we can rewrite the Hamiltonian system in the form

$$\partial_t \begin{pmatrix} p \\ q \end{pmatrix} = J\nabla H(p, q)$$

or, letting Φ denote the flow associated to this equation:

$$\partial_t \Phi = J\,\nabla H \circ \Phi.$$

The differential of Φ satisfies

$$\partial_t D\Phi = J\left(D^2 H \circ \Phi\right)D\Phi$$

from which one easily deduces that $\partial_t(D\Phi^T J D\Phi) = 0$, which implies (since $\Phi = id$ at $t = 0$)

$$D\Phi^T J D\Phi = J.$$

Transformations Φ that satisfy this equation are called *symplectic*. In fact, such a property characterizes Hamiltonian systems, and a natural requirement is that it should be shared by discrete systems, too. This leads to the definition of symplectic integrator, which are integrators for which the function $z' \mapsto \eta(z, z', h)$ is symplectic for all h (there is no time variable because the Hamiltonian system is autonomous).

The simplest examples of symplectic integrators are the symplectic Euler methods. There are two of them, defined by

$$\begin{cases} p_{n+1} = p_n - h\partial_q H(p_{n+1}, q_n) \\ q_{n+1} = q_n + h\partial_p H(p_{n+1}, q_n) \end{cases} \text{ and } \begin{cases} p_{n+1} = p_n - h\partial_q H(p_n, q_{n+1}) \\ q_{n+1} = q_n + h\partial_p H(p_n, q_{n+1}) \end{cases}.$$

Note that the latter equation is the dual of the former. When composed, they form the Störmer–Verlet schemes (there are two of them, depending on which of the symplectic Euler schemes is applied first). Symplectic Euler is a first-order method, whereas Störmer–Verlet has order two. Another symplectic method of order two is the implicit midpoint, defined by

$$p_{n+1} = p_n - h\partial_q H(p'_n, q'_n)$$
$$q_{n+1} = q_n + h\partial_p H(p'_n, q'_n)$$

with $p'_n = (p_n + p_{n+1})/2$ and $q'_n = (q_n + q_{n+1})/2$. The proof of these statements a many more details can be found in [139, 73].

D

Optimization Algorithms

Like with ordinary differential equations (ODEs), we give in this appendix a very limited and very partial account of numerical optimization methods, restricting ourselves to the few concepts and algorithm that are used in this book, and inviting the reader to consult one of many available textbooks (e.g., [85, 158, 27]) for more information.

D.1 Directions of Descent and Line Search

Since the problems we consider in this book are nonlinear, and very often non-convex, we will consider the general problem of finding a (local) minimizer of a function $x \mapsto E(x)$, defined on a $\mathbb{R}^n$ and taking values in $\mathbb{R}$ without making any specific assumption on E except that it is sufficiently differentiable. We will discuss iterative methods that update a current value of x by first finding a good direction, $h \in \mathbb{R}^n$, then replacing x by $x + \varepsilon h$ for a suitably chosen ε. We will assume, in the following, that E has enough derivatives for the computations to make sense, without repeating the exact assumptions every time.

The minimal requirement for h is that it must be a direction of descent, i.e., there must exist $\varepsilon_0 > 0$ such that $E(x + \varepsilon h) < E(x)$ for $\varepsilon \in (0, \varepsilon_0)$. Once h is given (and most of our discussion will be on how to find a good h) determining ε is a one-dimensional operation which is usually referred to as a line search. If we expand $E(x + \varepsilon h)$ at first order in ε, we get

$$E(x + \varepsilon h) = E(x) + \varepsilon DE(x)h + o(\varepsilon)$$

and we see that a sufficient condition for h to be a direction of descent is that $DE(x)h < 0$. We will only consider directions h that satisfy this condition, and therefore always assume that it holds when speaking of directions of descent.

For such an h, we find that, for any $\gamma \in (0, 1)$, the expression

$$c(\varepsilon, x, h) = \frac{1}{\varepsilon}\big(E(x + \varepsilon h) - E(x) - \gamma \varepsilon DE(x)h\big)$$

converges to a non-vanishing negative number when $\varepsilon \to 0$. Given this, one devises a simple line search procedure as follows. Fix two constants $\alpha, \gamma \in (0, 1)$. Let x and a descent direction h be given. Start with some reasonably large value of ε, $\bar{\varepsilon}$, and replace ε by $\alpha\varepsilon$ iteratively, as long as $c(\varepsilon, x, h) > 0$. This is the backtracking line search technique.

The choice of $\bar{\varepsilon}$ is important because choosing it too large would require trying too many values of ε before finding a satisfactory one, and taking it too small may prevent the algorithm from making large steps. One possibility is to start a minimization procedure with some choice of $\bar{\varepsilon}_0$, and choose at step k of the procedure ε_k to be some multiple of ε_{k-1} by some factor larger than one. Let's summarize this in the following algorithm.

Algorithm 8 (Generic minimization with backtracking line search)
Start with an initial choice for x_0, $\bar{\varepsilon}_0$. Choose positive constants $\alpha, \gamma < 1$ and $\beta > 1$. Let x_k, $\bar{\varepsilon}_k$ be their current values at step k and obtain their values at the next step as follows.

1. *Compute a good direction of descent h_k.*
2. *Set $\varepsilon_k = \bar{\varepsilon}_k$. While*

$$E(x_k + \varepsilon_k h_k) - E(x_k) - \gamma\varepsilon_k DE(x_k)h_k > 0$$

   replace ε_k by $\alpha\varepsilon_k$.
3. *Set $x_{k+1} = x_k + \varepsilon_k h_k$ and $\bar{\varepsilon}_{k+1} = \beta\varepsilon_k$.*

One typically stops the algorithm if the decrease $(E(x_k + \varepsilon_k h_k) - E(x_k))/\varepsilon$ is smaller than a given threshold. Other, more elaborate line search methods can be devised, including a full optimization of the function $\varepsilon \mapsto E(x + \varepsilon h)$. One additional condition which is often imposed in the line search is the Wolfe condition, which ensures (taking $0 < \rho < 1$) that

$$DE(x_{k+1})h_k \geq \rho DE(x_k)h_k$$

or the strong Wolfe condition, which is

$$|DE(x_{k+1})h_k| \leq \rho|DE(x_k)h_k|,$$

which forces the steps to be larger as long as the slope of E along the direction of descent remains significantly negative. But one must not forget that the line search needs to be repeated many times during the procedure and should therefore not induce too much computation (the Wolfe condition may be impractical if the computation of the gradient is too costly).

The core of the algorithm is, of course, the choice of the direction of descent, which is now addressed, starting with the simplest one.

D.2 Gradient Descent

A direction of descent at x being characterized by $DE(x)h < 0$, a natural requirement is to try to find h such that this expression is as negative as possible. Of course, this requirement does not make sense unless some normalization is imposed on h, and this is based, for gradient descent, on the selection of a dot product that may depend on x.

So assume that, for all $x \in \mathbb{R}^n$, a dot product denoted $\langle \cdot, \cdot \rangle_x$ is selected (or, in other words, a Riemannian metric is chosen on $\mathbb{R}^n$; smoothness of the dot product as a function of x is not a requirement, but it is needed, for example, in error estimation formulae like (D.2) below). In finite dimensions, this is equivalent to associating to each x a symmetric, definite positive matrix $A(x)$ and to setting (denoting as usual $\langle \cdot, \cdot \rangle$ the standard dot product on $\mathbb{R}^n$)

$$\langle u, v \rangle_x = \langle u, A(x)v \rangle = u^T A(x)v.$$

Note that the choice of the metric can be inspired by infinite-dimensional representations of the problem, but it is important to make sure that it remains positive once discretized.

The gradient of E at x for this metric, denoted $\nabla^A E(x)$, is defined by

$$\forall h \in \mathbb{R}^d, \ DE(x) h = \left\langle \nabla^A E(x), h \right\rangle_x.$$

If we denote by $\nabla E(x)$ the column vector representation of $DE(x)$, i.e., the gradient for the usual dot product (with $A(x) = \mathrm{Id}$), this definition implies

$$\nabla^A E(x) = A(x)^{-1} \nabla E(x). \tag{D.1}$$

A gradient descent procedure for the Riemannian metric associated to A selects, at a given point x, the direction of descent h as a minimizer of $DE(x) h$ subject to the constraint $\langle h, A(x)h \rangle = 1$. The solution of this problem can readily be computed as

$$h = -\frac{\nabla^A E(x)}{|\nabla^A E(x)|}$$

(unless of course $\nabla^A E(x) = 0$, in which case no direction of descent exists).

Since directions of descent need only be defined up to a multiplicative factor, we may as well take $h = -\nabla^A E(x)$. This can be plugged into step 1 of Algorithm 8, which becomes: *Set* $h_k = -A(x_k)^{-1} \nabla E(x_k)$.

Gradient descent is, like all the methods considered in this discussion, only able to find local minima of the function E (one may even not be able to rule out the unlikely situation in which it gets trapped in a saddle point). The speed at which it finds it is linear [158], in the sense that, if x^* is the limit point, and if an exact line search algorithm is run, there exists $\rho \in (0, 1)$ such that, for large enough k,

$$(E(x_{k+1}) - E(x_*)) \leq \rho \left(E(x_k) - E(x_*) \right). \tag{D.2}$$

The smallest possible ρ in this formula depends on how "spherical" the function E is around its minimum. More precisely, if c^* is the ratio between the largest and the smallest eigenvalue (or condition number) of

$A_*^{-1/2}D^2E(x^*)A_*^{-1/2}$ where $A_* = A(x^*)$, then ρ must be larger than $(c^* - 1)^2/(c^* + 1)^2$.

The gradient descent algorithm can also be written in continuous time, namely

$$\partial_t x = -A(x)^{-1}\nabla E(x).$$

This is a convenient formulation, provided that one remembers that it is usually preferable to discretize it using a line search rather than using standard ODE methods (since the goal is not to solve the ODE, but to minimize E).

The time-continuous formulation can easily be extended to Riemannian manifolds. If M is such a manifold, and E is defined on M, the gradient descent algorithm runs

$$\partial_t x = -\nabla^M E(x).$$

When discretized, however, one must remember that additions do not make sense on nonlinear manifolds (even when they are submanifolds of Euclidean spaces). The Riemannian equivalent to moving along a straight line is to use geodesics, so that gradient descent should be discretized as

$$x_{k+1} = \mathrm{Exp}_{x_k}(-\varepsilon\nabla^M E(x_k))$$

where Exp is the Riemannian exponential (as defined in equation (B.6)).

D.3 Newton and Quasi-Newton Directions

If the condition number of $A_*^{-1/2}D^2E(x^*)A_*^{-1/2}$ is 1, which is equivalent to $A^* = D^2E(x^*)$, then ρ can be taken arbitrarily small in (D.2), which means that convergence becomes *superlinear*. This suggests using $A(x) = D^2E(x)$ in the metric, provided, of course, this matrix is positive (it is nonnegative at the minimum, but not necessarily everywhere if the function is not convex).

The direction $-D^2E(x)^{-1}\nabla E(x)$ is called the Newton direction. It is optimal up to a second-order approximation, as shown by the following computation. For a given x, we have

$$E(x + h) = E(x) + \langle \nabla E(x)\,, h \rangle + \frac{1}{2}h^T D^2E(x)h + o(h^2).$$

If we neglect the error, and if $D^2E(x)$ is a positive matrix, the minimum of the second-order approximation is indeed given by

$$\hat{h} = -D^2E(x)^{-1}\nabla E(x).$$

If $D^2E(x)$ is not positive, the Newton direction is not necessarily a direction of descent. It has to be modified, the simplest approach being to add a multiple of the identity matrix to the Hessian, i.e., to use $A(x) = D^2E(x) + \lambda(x)\mathrm{Id}$ for a large enough $\lambda(x)$. Given x, the choice of a suitable λ

can be based on ensuring that the Cholesky decomposition of $A(x)$ (i.e., the decomposition $A = LL^*$ where L is lower triangular and L^* the conjugate of the transpose of L [99]) only has real coefficients.

When the computation of the Hessian is too complex, or its inversion too costly, using a Newton direction is impractical. Alternative methods, called quasi-Newton, are available in that case.

One such method, called BFGS (based on the initials of its inventors), updates an approximation A_k of the Hessian at step k of the algorithm. Let x_k be the current variable at step k, and $h_k = -A_k^{-1}\nabla E(x_k)$ the associated direction of descent, computed using the current A_k. Then, the new value of x is $x_{k+1} = x_k + s_k$ with $s_k = \varepsilon_k h_k$. The BFGS method defines a matrix A_{k+1} for the next step as follows. Letting $y_k = \nabla E(x_{k+1}) - \nabla E(x_k)$ and $\gamma_k = \langle s_k, y_k \rangle$, define

$$A_{k+1} = \left(I - \frac{y_k s_k^T}{y_k^T s_k}\right) A_k \left(I - \frac{y_k s_k^T}{y_k^T s_k}\right) + \frac{y_k s_k^T}{y_k^T s_k}.$$

We refer the reader to [158] for a justification of this update rule. It has the important feature to ensure that A_{k+1} is positive as soon as A_k is, and $s_k^T y_k > 0$. This last condition can be written

$$\langle \nabla E(x_k + \varepsilon_k), h_k \rangle > \langle \nabla E(x_k), h_k \rangle$$

and one can always find ε_k such that this is satisfied, unless the minimum of E is $-\infty$. This is because the left-hand side is $\partial_\varepsilon E(x_k + \varepsilon h_k)(\varepsilon_k)$, and

$$\partial_\varepsilon E(x_k + \varepsilon h_k) \le \langle \nabla E(x_k), h_k \rangle < 0$$

for all $\varepsilon > 0$ implies that $\lim_{\varepsilon \to \infty} E(x_k + \varepsilon h_k) = -\infty$. The condition $s_k^T y_k > 0$ can therefore be added to the line search procedure. It is automatically satisfied if the Wolfe condition is ensured.

Equally important to the fact that A_k can be made to remain positive is the fact that an inverse to it can be computed iteratively too. If we let $B_k = A_k^{-1}$, then

$$B_{k+1} = B_k - \frac{B_k y_k y_k^T B_k}{y_k^T B_k y_k} + \frac{s_k s_k^T}{y_k^T y_k}, \tag{D.3}$$

which allows for an efficient computation of the direction of descent, $h_k = -B_k \nabla E(x_k)$.

A variation of the method offers an even more efficient update rule, namely

$$B_{k+1} = V_k^T B_k V_k + \rho_k s_k s_k^T, \tag{D.4}$$

with $\rho_k = 1/y_k^T s_k$ and $V_k = \mathrm{Id} - \rho_k s_k s_k^T$.

The BFGS method is not directly applicable for large n (the dimension), however, because the computation and storage of n by n matrices like B_k rapidly becomes impractical. One possibility is to use the iteration specified by (D.4) over a finite time interval only (say p iterations in the past), resetting the value of B_{k-p-1} to Id. The computation of $B_k h_k$ only requires storing the values of y_{k-j} and s_{k-j} for $j = 1, \ldots, p$, and can be done recursively using (D.4), with a storage and computation cost which is now linear in the dimension. This results in the limited-memory BFGS method.

D.4 Conjugate Gradient

Nonlinear conjugate gradient methods can be seen as intermediate in complexity and efficiency between basic gradient descent and quasi-Newton methods. They may provide an optimal choice for very large-scale methods, for which even limited-memory quasi-Newton methods may be too costly.

Since non-linear conjugate gradient derives from the linear one, and since linear conjugate gradient is a method of choice for solving large-scale linear systems and is needed in many places in this book, we start with describing the linear case.

D.4.1 Linear Conjugate Gradient

The goal of the conjugate gradient method is to invert a linear system

$$Ax = b$$

where A is an n by n symmetric, positive definite matrix. This problem is equivalent to minimizing the quadratic function

$$E(x) = \frac{1}{2}\langle x, Ax \rangle - \langle b, x \rangle.$$

Conjugate gradient works by generating a set of *conjugate directions*, $h_0, \ldots, h_p, \ldots$ that satisfy $\langle h_i, h_j \rangle = 0$ if $i \neq j$, and generate the sequence

$$x_{k+1} = x_k + \alpha_k h_k \tag{D.5}$$

in which α_k is the optimal choice for the minimization of $\alpha \mapsto E(x_k + \alpha h_k)$, namely

$$\alpha_k = -\frac{\langle Ax_k - b, h_k \rangle}{\langle h_k, Ah_k \rangle}.$$

It is easy to prove by induction that, if $h_1, \ldots, h_k$ are non-vanishing conjugate directions, then $Ax_{k+1} - b$ is orthogonal to span$(h_0, \ldots, h_k)$ and since this space has dimension $k + 1$, this implies that $Ax_n - b = 0$ so that the algorithm converges to a solution of the linear system in at most n steps. This also implies that

$$E(x_{k+1}) = \min \left\{ E(x), x = x_0 + t_1 h_1 + \cdots + t_k h_k, t_1, \ldots, t_k \in \mathbb{R} \right\}.$$

since $\langle Ax_{k+1} - b, h_j \rangle = \partial E/\partial t_j$.

So the core of the conjugate gradient algorithm is to specify a good sequence $h_1, \ldots, h_n$. Gram–Schmidt orthonormalization is a standard process to build a conjugate family of vectors starting from independent vector $r_0, \ldots, r_k$. It consists in setting $h_0 = r_0$ and, for $k \geq 1$,

$$h_k = r_k + \sum_{j=0}^{k-1} \mu_{kj} h_j$$

with

$$\mu_{kj} = -\frac{\langle r_k, Ah_j \rangle}{\langle h_j, Ah_j \rangle}.$$

The beautiful achievement made in conjugate gradient was to recognize that, if $r_k = Ax_k - b$ with x_k coming from (D.5) (which is algorithmically feasible), then $\mu_{kj} = 0$ except for $j = k - 1$. To prove this, one needs to first check, by induction, that

$$\mathrm{span}(h_0, \ldots, h_k) = \mathrm{span}(r_0, Ar_0, \ldots, A^k r_0),$$

the latter space being called the Krylov space of order k associated to A and r_0. This implies that, for $j \leq k - 1$,

$$Ah_j \in \mathrm{span}(r_0, Ar_0, \ldots, A^{j+1} r_0) = \mathrm{span}(r_0, \ldots, r_{j+1}) \subset \mathrm{span}(r_0, \ldots, r_k).$$

Since we know that r_k is perpendicular to this space, we have $\langle r_k, Ah_j \rangle = 0$ if $j \leq k - 1$.

Given these remarks, the following iteration provides a family of conjugate directions, starting with an initial x_0 and $h_0 = r_0 = Ax_0 - b$ (and letting $\beta_k = \mu_{k,k-1}$):

$$x_{k+1} = x_k + \alpha_k h_k \text{ with } \alpha_k = -\frac{\langle r_k, h_k \rangle}{\langle h_k, Ah_k \rangle},$$

$$r_{k+1} = r_k + \alpha_k Ah_k,$$

$$h_{k+1} = r_{k+1} + \beta_{k+1} h_k \text{ with } \beta_{k+1} = -\frac{\langle r_{k+1}, Ah_k \rangle}{\langle h_k, Ah_k \rangle}.$$

This is, in essence, the conjugate gradient algorithm. The computation can be made slightly more efficient by noting that $\langle r_k, h_k \rangle = \langle r_k, r_k \rangle$ and $\langle r_{k+1}, Ah_k \rangle = \langle r_{k+1}, r_{k+1} \rangle$ [99, 158].

The rate of convergence of conjugate gradient can be estimated from the eigenvalues of A. If $\lambda_1 \leq \cdots \leq \lambda_n$ are the ordered eigenvalues, and $\rho_k = \lambda_{n-k+1}/\lambda_1$, then,

$$|x_k - x_n| \leq \left(\frac{\rho_k - 1}{\rho_k + 1} \right) |x_0 - x_n| \tag{D.6}$$

(x_n being the final – and correct – state of the algorithm). So, if many eigenvalues of A are much larger than the smallest one, conjugate gradient will converge very slowly. One the other hand, if A has a few large eigenvalues, which then drop to being close to the smallest, then a few iterations will be needed to obtain a good approximation of the solution.

Note that we could have formulated the initial problem in terms of any dot product on $\mathbb{R}^n$. Let M be symmetric, and positive definite, and consider the dot product $\langle x, y \rangle_M = x^T M y$. Let $\tilde{A}$ be self-adjoint for this dot product, i.e.,

$$\forall x, y, \ \langle x, \tilde{A}y \rangle_M = \langle Ax, y \rangle_M \text{ or } M\tilde{A} = \tilde{A}^T M.$$

Then solving $\tilde{A}x = \tilde{b}$ is equivalent to minimizing

$$\tilde{E}(x) = \frac{1}{2}\langle x, \tilde{A}x \rangle_M - \langle \tilde{b}, x \rangle_M.$$

and the same argument we made for $M = \text{Id}$ leads to the algorithm (taking $\tilde{r}_0 = \tilde{A}x_0 - \tilde{b}$)

$$x_{k+1} = x_k + \alpha_k h_k \text{ with } \alpha_k = -\frac{\langle \tilde{r}_k, \tilde{r}_k \rangle_M}{\langle h_k, \tilde{A}h_k \rangle_M},$$

$$\tilde{r}_{k+1} = \tilde{r}_k + \alpha_k \tilde{A}h_k,$$

$$h_{k+1} = \tilde{r}_{k+1} + \beta_{k+1}h_k \text{ with } \beta_{k+1} = -\frac{\langle \tilde{r}_{k+1}, \tilde{r}_{k+1} \rangle_M}{\langle h_k, \tilde{A}h_k \rangle_M}.$$

Now, we can remark that given any symmetric matrix A, we get a self-adjoint matrix for the M dot product by letting $\tilde{A} = M^{-1}A$ and that the problem $Ax = b$ is equivalent to $\tilde{A}x = \tilde{b}$ with $\tilde{b} = M^{-1}b$, which can be solved using the M dot product. Doing this leads to the iterations (in which we set $\tilde{r}_k = M^{-1}r_k$): start with $r_0 = Ax_0 - b$ and iterate

$$x_{k+1} = x_k + \alpha_k h_k \text{ with } \alpha_k = -\frac{\langle r_k, M^{-1}r_k \rangle}{\langle h_k, Ah_k \rangle},$$

$$r_{k+1} = r_k + \alpha_k Ah_k,$$

$$h_{k+1} = M^{-1}r_{k+1} + \beta_{k+1}h_k \text{ with } \beta_{k+1} = -\frac{\langle r_{k+1}, M^{-1}r_{k+1} \rangle}{\langle h_k, Ah_k \rangle}.$$

This is preconditioned conjugate gradient. Its speed of convergence is now governed by the eigenvalues of $M^{-1}A$ (or, equivalently of $M^{-1/2}AM^{-1/2}$), and a lot of efficiency can be gained if most of these eigenvalues get clustered near the smallest one. Of course, for this to be feasible, the equation $M\tilde{r}_k = r_k$ has to be easy to solve. Preconditioning has most of the time to be designed specifically for a given problem, but it may result in a dramatic reduction of the computation time.

D.4.2 Nonlinear Conjugate Gradient

Now, assume that E is a nonlinear function. We can formally apply the conjugate gradient iterations by replacing r_k by $\nabla E(x_k)$, which yields, starting with $h_0 = -\nabla E(x_0)$,

$$x_{k+1} = x_k + \varepsilon_k h_k,$$
$$h_{k+1} = \nabla E(x_{k+1}) + \beta_{k+1} h_k$$
$$\text{with } \beta_{k+1} = -\frac{|\nabla E(x_{k+1})|^2}{|\nabla E(x_k)|^2}.$$

This is the Fletcher–Reeves algorithm. In this algorithm, ε_k should be determined using a line search. This algorithm can significantly accelerate the convergence of gradient descent methods, especially when closing up to a minimum, around which E will be roughly quadratic. However, h_k is not guaranteed to always provide a direction of descent (this is true when the line search is exact, or under the strong Wolfe condition; see [158]). It may sometimes be useful to reinitialize the iterations at regular intervals, setting $h_{k+1} = -\nabla E(x_{k+1})$ (or, equivalently, $\beta_{k+1} = 0$).

Variants of this algorithm use different formulae for β_{k+1}. One of them is the Polak–Ribière algorithm, which sets

$$\tilde{\beta}_{k+1} = \frac{\nabla E(x_{k+1})^T (\nabla E(x_{k+1}) - \nabla E(x_k))}{|\nabla E(x_k)|^2}$$

and $\beta_{k+1} = \max(\tilde{\beta}_{k+1}, 0)$.

E

Principal Component Analysis

E.1 General Setting

Assume that a random variable X takes values in a finite- or infinite-dimensional Hilbert space H (for example, a space of plane curves). Denote by $\langle \cdot, \cdot \rangle_H$ the inner product in this space. We assume that $x_1, \ldots, x_N$ are observed.

The goal of principal component analysis (PCA) is to provide, for each finite $p \leq \dim(H)$, an optimal representation of order p the form

$$x_k = \overline{x} + \sum_{i=1}^{p} \alpha_{ki} e_i + R_k, k = 1, \ldots, N$$

where $(e_1, \ldots, e_p)$ is an orthonormal family in H. The error terms, $R_1, \ldots, R_N$, should be as small as possible. More precisely, PCA minimizes the residual sum of squares

$$S = \sum_{k=1}^{N} \|R_k\|_H^2.$$

When $\overline{x}$ and $(e_1, \ldots, e_p)$ are fixed, $\sum_{i=1}^{p} \alpha_{ki} e_i$ must be the orthogonal projection of $x_k - \overline{x}$ on $\mathrm{Hilb}(e_1, \ldots, e_p)$, which implies that $\alpha_{ki} = \langle x_k - \overline{x}, e_i \rangle_H$.

Still assuming that $(e_1, \ldots, e_p)$ is fixed, it is easy to prove that the optimal choice for $\overline{x}$ is $\overline{x} = \frac{1}{N} \sum_{k=1}^{N} x_k$. For notational simplicity, we assume that $\overline{x} = 0$, which is equivalent to assuming that all x_k's have been replaced by $x_k - \overline{x}$.

From these remarks, $(e_1, \ldots, e_p)$ must minimize

$$S = \sum_{k=1}^{N} \left\| x_k - \sum_{i=1}^{p} \langle x_k, e_i \rangle e_i \right\|_H^2$$

$$= \sum_{k=1}^{N} \|x_k\|_H^2 - \sum_{i=1}^{p} \sum_{k=1}^{N} \langle x_k, e_i \rangle_H^2.$$

For $u, v \in H$, define

$$\langle u, v \rangle_T = \frac{1}{N} \sum_{k=1}^{N} \langle x_k, u \rangle_H \langle x_k, v \rangle_H$$

and $\|u\|_T = \langle u, u \rangle_T^{1/2}$ (the index T refers to the fact that this norm is associated to a training set). This provides a new quadratic form on H. The formula above shows that minimizing S is equivalent to maximizing

$$\sum_{i=1}^{p} \|e_i\|_T^2$$

subject to the constraint that $(e_1, \ldots, e_p)$ is orthonormal in H.

The newly introduced quadratic form can be orthogonalized in an orthonormal basis for H, i.e., there exists an orthonormal basis, $(f_1, f_2, \ldots)$ of H which is in addition orthogonal for $\langle , \rangle_T$. Its existence is a standard theorem in linear algebra. Letting $\lambda_n = \|f_n\|_T$, we assume that such a basis is ordered according to decreasing λ_n's (which vanish for $n > N$).

We have the following fact, which we state without proof:

Theorem E.1. *The optimal $(e_1, \ldots, e_p)$ must be such that $Hilb(e_1, \ldots, e_p) = Hilb(f_1, \ldots, f_p)$. In particular $f_1, \ldots, f_p$ provide a solution. They are called the p principal components of the observed data set. With this choice*

$$S = N \sum_{i > p} \lambda_i^2.$$

E.2 Computation of the Principal Components

E.2.1 Small Dimension

Assume that H has finite dimension, M, and that $x_1, \ldots, x_N \in \mathbb{R}^M$ are column vectors. Let the inner product on H be associated to a matrix A:

$$\langle u, v \rangle_H = u^T A v.$$

Then:

$$\langle u, v \rangle_T = \frac{1}{N} \sum_{i=1}^{N} (u^T A x_i)(x_i^T A v)$$

$$= u^T \left(\frac{1}{N} \sum_{i=1}^{N} A x_i x_i^T A \right) v$$

so that the training-set inner product is associated to

$$Q_T = \frac{1}{N} A \cdot \left(\sum_{k=1}^{N} x_k \, {}^t x_k \right) A.$$

To simultaneously orthogonalize Q_T and A, we must find a matrix F which is orthogonal for F (which means that $F^T A F = \mathrm{Id}$) and such that $F^T Q_T F$ is diagonal. Since the first condition is $F^T = F^{-1} A^{-1}$, the second can also be written: $F^{-1} A^{-1} Q_T F$ is diagonal. Thus, the columns of F are eigenvectors of $A^{-1} Q_T$. In other terms, the principal directions are solution of the so-called generalized eigenvalue problem

$$Q_T f = \lambda A f.$$

There are routines available in most numerical or statistical software packages to solve this problem. If they are not directly available, solutions of the generalized eigenvalue problem can be deduced from solution of the standard eigenvalue problem for symmetric matrices as follows. First compute $A^{-1/2}$, which is done by diagonalizing A as $A = PDP^T$ and letting $A^{-1/2} = PD^{-1/2}P^T$. Then compute the eigenvectors of $A^{-1/2} Q_T A^{-1/2}$, arranged in a matrix G. The solution is then given by $F = A^{-1/2} G$.

E.2.2 Large Dimension

It often happens that the dimension of H is much larger than the number of observations, N. In such a case, the previous approach is quite inefficient and one proceeds as follows.

The basic remark is that there are at most N principal components $f_1, \ldots, f_N$ with non-vanishing eigenvalues, and they must belong to the vector space generated by T. This implies that, for some $\alpha_{ik}, 1 \le i, k \le N$:

$$f_i = \sum_{k=1}^{N} \alpha_{ik} x_k.$$

With this notation, we have $\langle f_i, f_j \rangle_H = \sum_{k,l=1}^{N} \alpha_{ik} \alpha_{jl} \langle x_k, x_l \rangle_H$ and

$$\langle f_i, f_j \rangle_T = \sum_{l=1}^{N} \langle f_i, x_l \rangle_H \langle f_j, x_l \rangle_H = \sum_{k,k'=1}^{N} \alpha_{ik} \alpha_{jk'} \sum_{l=1}^{N} \langle x_k, x_l \rangle_H \langle x_{k'}, x_l \rangle_H.$$

Letting S be the matrix of the dot products $\langle x_k, x_l \rangle_H$, we have $\langle f_i, f_j \rangle_H = \alpha_i^T S \alpha_j$ and $\langle f_i, f_j \rangle_T = \alpha_i^T S^2 \alpha_j$. Thus, the previous simultaneous orthogonalization problem can be solved with respect to the α's by finding a basis simultaneously orthogonal for S and for S^2, which boils down to orthogonalizing S, and taking the first eigenvectors, normalized so that $\alpha_i^T S \alpha_i = 1$.

E.3 Statistical Interpretation and Probabilistic PCA

The statistical interpretation of linear PCA is quite simple: assume that X is a centered random vector with covariance matrix Σ. Consider the problem which consists in finding a decomposition $X = \sum_{i=1}^{p} \xi_i e_i + R$ where $(\xi_1, \ldots, \xi_p)$ forms a p-dimensional centered random vector, $e_1, \ldots, e_p$ is an orthonormal system, and R is a random vector, uncorrelated to the ξ_i's and as small as possible, in the sense that $E(|R|^2)$ is minimal. One can see that, in an optimal decomposition, one needs $\langle R, e_i \rangle = 0$ for all i, because one can always write

$$\sum_{i=1}^{p} \xi_i e_i + R = \sum_{i=1}^{p} (\xi_i + \langle R, e_i \rangle) e_i + R - \sum_{i=1}^{p} \langle R, e_i \rangle e_i$$

and $|R - \sum_{i=1}^{p} \langle R, e_i \rangle e_i| \leq |R|$. Also, one can always restrict oneself to uncorrelated $(\xi_1, \ldots, \xi_p)$ by a change of basis in $\mathrm{span}(e_1, \ldots, e_p)$.

Then, we can write

$$E(|X|^2) = \sum_{i=1}^{p} E(\xi_i^2) + E(|R|^2)$$

with $\xi_i = \langle e_i, X \rangle$. So, to minimize $E(|R|^2)$, one needs to maximize

$$\sum_{i=1}^{p} E(\langle e_i, X \rangle^2)$$

which is equal to

$$\sum_{i=1}^{p} e_i^T \Sigma e_i,$$

the solution for this problem being given by the first p eigenvectors of Σ. PCA exactly applies this procedure, with Σ replaced by the empirical covariance.

Probabilistic PCA is based on the statistical model in which it is assumed that X can be decomposed as $X = \sum_{i=1}^{p} \lambda_i \xi_i e_i + \sigma R$, where R is a d-dimensional standard Gaussian vector and $\xi = (\xi_1, \ldots, \xi_p)$ a p-dimensional standard Gaussian vector, independent of R. The parameters here are the coordinates of $e_1, \ldots, e_p$, the values of $\lambda_1, \ldots, \lambda_p$ and of σ. Introduce the $d \times p$ matrix W with columns given by $\lambda_1 e_1, \ldots, \lambda_p e_p$ to rewrite this model in the form

$$X = W\xi + \sigma^2 R$$

where the parameters are W and σ^2, with the constraint that $W^T W$ is diagonal. As a linear combination of independent Gaussian random variables, X is Gaussian with covariance matrix $WW^T + \sigma^2 I$. The log-likelihood of the observation $x_1, \ldots, x_N$ is

$$L(W, \sigma) = -\frac{N}{2}(d \log 2\pi + \log \det(WW^T + \sigma^2 I) + \mathrm{trace}((WW^T + \sigma^2 I)^{-1}\Sigma))$$

where Σ is the empirical covariance matrix of $x_1, \ldots, x_N$. This function can be maximized explicitly in W and σ. One can show that the solution can be obtained by taking $e_1, \ldots, e_p$ to be the first p eigenvectors of Σ, $\lambda_i = \sqrt{\delta_i^2 - \sigma^2}$, where δ_i^2 is the eigenvalue of Σ associated to e_i, and

$$\sigma^2 = \frac{1}{d-p} \sum_{i=p+1}^{d} \delta_i^2.$$

F

Dynamic Programming

This chapter describes a few basic dynamic programming algorithms.

F.1 Minimization of a Function on a Tree

We consider the issue of minimizing a function $E : x \mapsto E(x)$, defined for all x of the form $x = (x_s, s \in S)$, in the following context:

- S is a finite set which forms the vertices of an oriented tree. We will represent edges by relations $s \to t$, for some pairs (s, t) in S, assuming that there is no loop: there exists at most one path between s and s' (such that $s = s_0 \to s_1 \to \cdots \to s_N = s'$). For $s \in S$, we denote by $\mathcal{V}_s$ the set of all children of s, i.e., the set of all t such that $s \to t$.
- For all $s \in S$, x_s takes its values in a finite set A_s. We denote $A = \prod_{s \in S} A_s$.
- The function E takes the form $E(x) = \sum_{s,t \in S, s \to t} E_{st}(x_s, x_t)$.

We will consider the following partial order on S: $s < t$ if there exists a sequence $s = s_0 \to s_1 \to \cdots \to s_p = t$. For all $s \in S$, define

$$E_s^+(x) = \sum_{t \in \mathcal{V}_s} E_{st}(x_s, x_t) + \sum_{t > s, u > s, t \to u} E_{tu}(x_t, x_u).$$

Clearly, $E_s^+(x)$ only depends on x_s, and x_t for $t > s$. One can furthermore prove the relation

$$E_s^+(x) = \sum_{t \in \mathcal{V}_s} E_{st}(x_s, x_t) + \sum_{t \in \mathcal{V}_s} E_t^+(x). \tag{F.1}$$

Denote $F_s^+(x_s) = \min\{E_s^+(y), y \in A, y_s = x_s\}$. The following equation, which is a consequence of (F.1), is the core of the method.

$$F_s^+(x_s) = \min_{x_t, t \in \mathcal{V}_s} \left[\sum_{t \in \mathcal{V}_s} E_{st}(x_s, x_t) + \sum_{t \in \mathcal{V}_s} F_t^+(x_t) \right]. \qquad (\text{F.2})$$

This implies that, in order to compute the values of $F^+(x_s)$, it suffices to know $F^+(x_t)$ for all $t \in \mathcal{V}_s$, and all x_t. This yields an algorithm that successively minimizes the functions F_s^+, starting from the leaves of the tree (the minimal elements for our order) to the roots (the maximal elements). If $\mathcal{R}$ is the set of roots, we have

$$\min_{x \in A} E(x) = \sum_{s \in \mathcal{R}} \min_{x_s \in A_s} [F_s^+(x_s)]$$

so that this algorithm directly provides a minimum of E.

The practical limitations of this approach are as follows. First, the minimization involved in (F.2) should not be too difficult, because it has to be replicated at every s. This means that the product space of all A_t for $t \in \mathcal{V}_s$ must not be too large. Second, the memory load should remain tractable; for a given s, the value of $F_s^+(x_s)$ must be kept in memory for all $x_s \in A_s$, until the values of F_t^+ have been computed for the parent of s. One does not need to keep track of the configuration $x_t, t > s$ that achieves this minimum because the optimal configuration can be reconstituted if all the values of $F_s^+(x_s)$ have been saved. Indeed, if this is done, the optimal configuration can be reconstructed by starting with the roots of the tree, and keeping the best x_s at each step.

One can, however, devise upgraded versions of the algorithm for which the values and solutions of some of the $F^+(x_s)$ can be pruned a priori, based on lower bounds on the best value of E they can lead to.

F.2 Shortest Path on a Graph: a First Solution

The search for a shortest path on a graph is important for the methods discussed in this book, because it can be used to compute geodesics on discretized manifolds.

The set S now forms the vertices of a general oriented graph. A cost $\Gamma(s, t)$ is attributed to each edge $s \to t$; the global cost of a path $\mathbf{s} = (s_0 \to \cdots \to s_N)$ is defined by

$$E(\mathbf{s}) = \sum_{k=1}^N \Gamma(s, t).$$

Given s and t in S, we want to compute the shortest path (or path with lowest cost) $\mathbf{s} = (s = s_0 \to \cdots \to s_N = t)$. The variables therefore are $s_1, \ldots, s_{N-1}$, and the integer N. The optimal cost will be denoted $d(s, t)$. To avoid infinite minima, we will assume that there exists at least one path from s to t, for any $s, t \in S$.

We fix t and consider s as an additional variable. We denote by $\mathcal{V}_s$ the set of all $u \in S$ such that $s \to u$. We then have the formula, which is the analog of (F.2) in this case:

$$d(s,t) = \min_{u \in \mathcal{V}_s}[d(u,t) + \Gamma(s,u)]. \tag{F.3}$$

This equation provides an algorithm to compute $d(s,t)$ for all $s \in S$. Define, for all $N \geq 0$,

$$d_N(s,t) = \min\{E(\mathbf{s}) : \mathbf{s} \text{ path between } s \text{ and } t \text{ with at most } N \text{ points}\}.$$

We have

$$d_{N+1}(s,t) = \min_{u \in \mathcal{V}_s}[d_N(u,t) + \Gamma(s,u)]. \tag{F.4}$$

Let $d_0(s,t) = +\infty$ if $s \neq t$ and 0 if $s = t$. Equation (F.4) can be iterated to provide $d_N(u,t)$ for all u and arbitrary N. We have $d(s,t) = \lim_{N \to \infty} d_N(s,t)$, but it is clear that, if $d_N(s,t) = d_{N+1}(s,t)$ for all s, then $d_N(s,t) = d(s,t)$ and the computation is over.

This provides the distance. To also obtain the shortest path between s and t once all "distances" $d(u,t), u \in S$, have been computed, it suffices to start the path at s and iteratively choose the neighbor of the current point that is closest to t. More precisely, one lets $s_0 = s$, and, given s_k, take s_{k+1} such that $d(s_{k+1},t) = \min\{d(u,t), u \in \mathcal{V}_{s_k}\}$.

F.3 Dijkstra Algorithm

Dijkstra's algorithm provides an alternative, and significantly more efficient, method for the computation of the shortest path. This algorithm provides, for a given $s_0 \in S$, the shortest path between s_0 and any $t \in S$ with a number of operations of order $|S| \log |S|$.

To each step of the algorithm are associated two subsets of S; the first one is the set of all unresolved vertices, and will be denoted C; the second is the set of all vertices t for which $d(s_0,t)$ is known, and will be denoted D. Since $d(s_0,s_0) = 0$, the initial partition is $D = \{s_0\}$ and $C = S \setminus \{s_0\}$. We will also ensure that at each step of the algorithm

$$\max\{d(s_0,t) : t \in D\} \leq \min\{d(s_0,t), t \in C\}.$$

which is true at the beginning.

Introduce the function F defined on S by:

- $F(t) = d(s_0,t)$ if $t \in D$;
- $F(t) = \inf\{d(s,u) + \Gamma(u,t), u \in D\}$ if $t \in C$, with $\inf(\emptyset) = +\infty$.

We therefore start with $F(s_0) = 0$, $F(t) = \Gamma(s,t)$ for all $t \in \mathcal{V}_{s_0}$ and $F(t) = +\infty$ otherwise.

The basic remark is that, at each step, the vertex $t \in C$ for which $F(t)$ is minimal is such that $d(s_0, t) = F(t)$; indeed, we can already remark that

$$d(s_0, t) = \inf\{d(s_0, u) + \Gamma(u, t) : u \in S\} \le F(t).$$

Assume $d(s_0, t) < F(t)$. This implies that there exists $u \in C$ such that $d(s_0, u) + \Gamma(u, t) < F(t)$. Choose an optimal path between s_0 and u and let u' denote the first time that this path leaves D. We have $F(u') = d(s_0, u') \le d(s_0, u) < F(t)$, which is a contradiction to the fact that $F(t)$ is minimal. Moreover, for all $u \in C$, we have $d(s_0, u) \ge F(t)$: to prove this, it suffices to consider again the exit point from D of an optimal path between s_0 and u to have a contradiction.

We can therefore move t from C to D. It remains to update the function F accordingly. This function must be modified only for points $t' \in \mathcal{V}_t$, for which $F(t')$ must be replaced by

$$\min(F(t'), F(t) + \Gamma(t, t')).$$

The algorithm stops when $C = \emptyset$. Like in the previous section, the function F can be used to reconstruct the optimal paths.

F.4 Shortest Path in Continuous Space

We now provide a very elegant and efficient algorithm to compute a certain class of geodesic distances in $\mathbb{R}^d$ [185], which can be seen as a numerically consistent version of Dijkstra's algorithm. Although the algorithm generalizes to any dimension, we restrict ourselves here to $d = 2$ in order to simplify the presentation.

Let W be a positive function and consider the Riemanian metric on $\mathbb{R}^2$ given by $\|u\|_x = W(x)|u|$, for $u \in \mathbb{R}^2$. The length of a path $\gamma(.)$ defined over $[a, b]$ is therefore given by

$$L(\gamma) = \int_a^b |\dot{\gamma}(t)|\, W(\gamma(t))dt.$$

The geodesic distance is then given by the length of the shortest path between two points. Now let $f(x)$ denote the distance of x to a set $S_0 \subset \mathbb{R}^2$. The function f can be shown to satisfy the following *eikonal equation*

$$|\nabla f| = W \qquad\qquad (F.5)$$

and the algorithm consists in solving (F.5) given that $f = 0$ on S_0.

The principle is to progressively update f starting from S_0. The norm of the gradient is discretized according to the following formulae. Let

$$\partial_x^+ f(i,j) = (f_{i+1j} - f_{ij})/h,$$
$$\partial_x^- f(i,j) = (f_{ij} - f_{i-1j})/h,$$
$$\partial_y^+ f(i,j) = (f_{ij+1} - f_{ij})/h,$$
$$\partial_y^- f(i,j) = (f_{ij} - f_{ij-1})/h$$

and [175]

$$|\nabla_{ij} f| = \Big(\max \big(\max(\partial_x^- f(i,j), 0), - \min(\partial_x^+ f(i,j), 0) \big)^2$$
$$+ \max \big(\max(\partial_y^- f(i,j), 0), - \min(\partial^+ f_y(i,j), 0) \big)^2 \Big)^{1/2},$$

which can also be written

$$|\nabla_{ij} f| = \big((f_{ij} - \min(f_{ij}, U_{ij}))^2 + (f_{ij} - \min(f_{ij}, V_{ij}))^2 \big)^{1/2}$$

with $U_{ij} = \min(f_{i-1j}, f_{i+1j})$ and $V_{ij} = \min(f_{ij-1}, f_{ij+1})$.

Before proceeding further, let's try to understand why this is a well-chosen formula, starting with a one-dimensional example, and restricting ourselves to the particular case $W \equiv 1$, which corresponds to the Euclidean distance function. In one dimension, it suffices to consider the case in which the set with respect to which the distance is computed has two points, say 0 and 1. We also focus on the interval $[0, 1]$ in which interesting things happen. The distance function in this case is given by the first plot of Figure F.1. The following two plots in the same figure provide a discretization of the same function over respectively an odd and an even number of points. For such a function, it is easy to check that

$$|f_i - \min(f_i, f_{i-1}, f_{i+1})| = 1$$

for every point; this partially justifies the choice made for the discretization of the eikonal equation. Approximating the gradient by central differences (for example, by $(f_{i+1} - f_{i-1})/2$) would not satisfy the equation at the central points, since the result would be 0 in the odd case, and ± 0.5 in the even case. Note that the formula would have worked without including f_i in the minimum, but this is because, in one dimension, the situation in which f_i is smaller that its two neighbors does not happen, unless the distance is zero.

This can happen in two dimensions, however. Let's illustrate the formula in the simplest case when one computes the distance to a single point, say $(0,0)$; that is, we consider the function $f(x,y) = \sqrt{x^2 + y^2}$. Let this function be discretized over a grid $(ih, jh), i, j \in \mathbb{Z}$, h being the discretization step. When i and j are non-vanishing, a first-order expansion shows that the norm of the previous estimation of the gradient is equal to 1 at first order in h (we skip the computation). If, say, $i = 0$, the provided approximation formula works exactly, since in this case $\min(f_{1j}, f_{0j}, f_{-1j})$ is equal to f_{0j} so that the approximation of the first derivative is zero, while the approximation of the

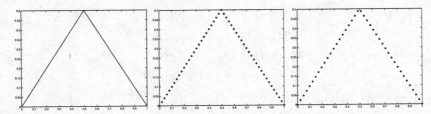

Fig. F.1. A typical distance map in 1D, followed by a discretization over an odd and even number of points.

other derivative is exactly 1. But this would not have worked if the term f_{ij} had been omitted in the minimum. The reader is referred to [183] for further justifications of this discretization.

We now solve the discretized eikonal equation,

$$(f_{ij} - \min(f_{ij}, U_{ij}))^2 + (f_{ij} - \min(f_{ij}, V_{ij}))^2 = W_{ij}^2,$$

with respect to f_{ij}. For this, it suffices to discuss the relative position of f_{ij}, U_{ij} and V_{ij}, and solve a second-degree polynomial equation in each case. The first remark is that, unless $W_{ij} = 0$, which is excluded, there is no solution, f_{ij}, which is smaller than both U_{ij} and V_{ij}. Let's first prove this when $U_{ij} \leq f_{ij} \leq V_{ij}$. In such a case the equation is

$$(f_{ij} - U_{ij})^2 = W_{ij}^2,$$

which yields $f_{ij} = W_{ij} + U_{ij}$. To satisfy the current assumption $f_{ij} \leq V_{ij}$, we need $W_{ij} \leq V_{ij} - U_{ij}$. We have the same conclusion inverting the roles of U_{ij} and V_{ij}, and both cases can be summarized in the unique statement:

$$f_{ij} = W_{ij} + \min(U_{ij}, V_{ij}) \text{ if } W_{ij} \leq |V_{ij} - U_{ij}|.$$

To have f_{ij} larger that both U_{ij} and V_{ij}, we must solve

$$(f_{ij} - U_{ij})^2 + (f_{ij} - V_{ij})^2 = W_{ij}^2,$$

which yields

$$f_{ij} = \frac{1}{2}\left(U_{ij} + V_{ij} + \sqrt{2W_{ij}^2 - (U_{ij} - V_{ij})^2}\right).$$

One can prove that f_{ij} is larger than $\max(U_{ij}, V_{ij})$ only if $W_{ij} \geq |U_{ij} - V_{ij}|$ which is complementary to the previous case. To summarize

$$\begin{cases} f_{ij} = W_{ij} + \min(U_{ij}, V_{ij}) \text{ if } W_{ij} \leq |U_{ij} - V_{ij}|, \\[2mm] f_{ij} = \left(\sqrt{2W_{ij}^2 - (U_{ij} - V_{ij})^2} + U_{ij} + V_{ij}\right)/2 \\[2mm] \quad\quad \text{if } W_{ij} \geq |U_{ij} - V_{ij}|. \end{cases} \quad\quad \text{(F.6)}$$

This will be called the *update formula*. From this formula, one can organize an iterative algorithm in which one progressively updates the f_{ij}'s one at a time according to this formula until stabilization [183]. Notice that, if, at some point, the values of U_{ij} and V_{ij} are exactly known, then the value of f_{ij} will also be exactly known after the application of the update formula.

The beautiful trick that has been introduced in [185] is that it is possible to organize the algorithm so that the correct values U_{ij} and V_{ij} are known at the time they are needed. This parallels the Dijkstra algorithm (Section F.3) and goes as follows.

Denote S the complete discretized grid. This set will be divided into two subsets which evolve during the algorithm. These subsets are denoted C and D, the former containing the pixels which have not been updated yet, and the latter those for which the values of f are already known. At the beginning of the algorithm, D contains the points in S_0 (after discretization), f being 0 on D, and C contains all the other points, on which we temporarily set $f = \infty$. The algorithm stops when $C = \emptyset$.

Preliminary step. Compute the value of f for all points m which are neighbors of D according to the update formula (F.6).

Then iterate the main loop until C is empty:

Main loop. Select a point t in C for which f is minimal, and add it to D. Then recompute f for the neighbors of t according to (F.6). (Provided of course these neighbors do not already belong to D.)

Figure F.2 provides an example of distance function to a plane curve, computed using this algorithm.

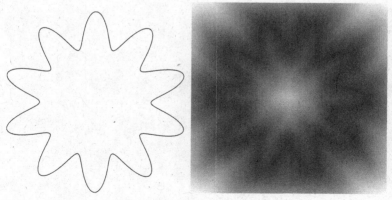

Fig. F.2. A curve and its distance map.

The first thing that can be proved about this algorithm is that, if t_n is the point added to D at step n, and f^n the current function f at the same time, the sequence $f^n_{t_n}$ is increasing. Indeed, consider t_{n+1}. If t_{n+1} is not a neighbor of t_n, then the value of f is not updated after step n and $f^{n+1}_{t_{n+1}} = f^n_{t_{n+1}} \geq f^n_{t_n}$ since t_n was the minimum. So, if $f^{n+1}_{t_{n+1}} < f^n_{t_n}$, then t_n and t_{n+1} are neighbors, and the value of f at t_{n+1} must have changed; this implies that the new value at t_n must have affected $U_{t_{n+1}}$ or $V_{t_{n+1}}$ depending on whether t_n and t_{n+1} are horizontal or vertical neighbors. Say it has affected $U_{t_{n+1}}$. The only way this can happen is when $U_{t_{n+1}} = f^n_{t_n}$. We are therefore in the situation when $f^{n+1}_{t_{n+1}} \leq U_{t_{n+1}}$, which implies $f^{n+1}_{t_{n+1}} = V_{t_{n+1}} + W_{t_{n+1}}$. But this value has not been affected by the previous step, since, for $V_{t_{n+1}}$ to be smaller than $f^n_{t_n}$, it must be equal to the value at a point which is already in D, which has not changed. This is a contradiction to $f^n_{t_{n+1}} < f^n_{t_n}$ and proves our initial statement.

Given this, any updated value after step n must be larger than the stored value at f_{t_n}. But this cannot affect the update formula, since this one only depends on the values of f at the neighbors which are smaller than f_{t_n}. So the update formula is satisfied, and remains so for all points added to D, which proves the convergence of the method.

References

1. R. Abraham and J. E. Marsden. *Foundations of Mechanics*. Perseus Publishing, 1994.
2. R. A. Adams. *Sobolev Spaces*. Academic Press, 1975.
3. Daniel C. Alexander, James C. Gee, and Ruzena Bajcsy. Strategies for data reorientation during non-rigid warps of diffusion tensor images. In *MICCAI*, pages 463–472, 1999.
4. L. Alvarez, P.-L. Lions, and J.-M. Morel. Image selective smoothing and edge detection by non-linear diffusion. *SIAM J. Numer. Anal.*, 29:845–866, 1992.
5. L. Alvarez, F. Guichard, P.-L. Lions, and J.-M. Morel. Axioms and fundamental equations of image processing. *Arch. Rational Mechanics*, 123(3):199–257, 1993.
6. N. Amenta and M. Bern. Surface reconstruction by Voronoi filtering. *Discrete and Computational Geometry*, 22:481–504, 1999.
7. Y. Amit, U. Grenander, and M. Piccioni. Structural image restoration through deformable templates. *JASA*, 86:376–387, 1989.
8. Y. Amit and P. Piccioni. A non-homogeneous markov process for the estimation of gausian random fields with non-linear observations. *Ann. of Probab.*, 19:1664–1678, 1991.
9. N. Arad, N. Dyn, D. Reisfeld, and Y. Yeshurun. Image warping by radial basis functions: application to facial expressions. *CVGIP: Graphical Models and Image Processing*, 56(2):161–172, 1994.
10. N. Arad and D. Reisfeld. Image warping using few anchor points and radial functions. *Computer Graphics Forum*, 14:35–46, 1995.
11. V. I. Arnold. Sur un principe variationnel pour les ecoulements stationnaires des liquides parfaits et ses applications aux problèmes de stanbilité non linéaires. *J. Mécanique*, 5:29–43, 1966.
12. V. I. Arnold. *Mathematical Methods of Classical Mechanics*. Springer, 1978. Second Edition: 1989.
13. N. Aronszajn. Theory of reproducing kernels. *Trans. Am. Math. Soc.*, 68:337–404, 1950.
14. F. Arrate, J. T. Ratnanather, and L. Younes. Diffeomorphic active contours. *SIAM J. Imaging Science*, 2010 (to appear).
15. J. Ashburner. A fast diffeomorphic image registration algorithm. *Neuroimage*, 38(1):95–113, 2007.

16. J. August, A. Tannebaum, and S. W. Zucker. On the evolution of the skeleton. In *Proceedings of ICCV*, pages 315–322. IEEE Computer Society, 1999.

17. R. Bajcsy and C. Broit. Matching of deformed images. In *The 6th International Conference in Pattern Recognition*, pages 351–353, 1982.

18. R. Bajcsy and S. Kovacic. Multiresolution elastic matching. *Comp. Vision, Graphics, and Image Proc.*, 46:1–21, 1989.

19. R. K. Beatson and L. Greengard. A short course on fast multipole methods. In M. Ainsworth, J. Levesley, Light W. A., and M. Marlette, editors, *Wavelets, Multilevel Methods and Elliptic PDEs*, pages 1–37, 1997.

20. M. F. Beg. *Variational and Computational Methods for Flows of Diffeomorphisms in Image Matching and Growth in Computational Anatomy*. PhD thesis, Johns Hopkins University, 2003.

21. M. F. Beg, M. I. Miller, A. Trouvé, and L. Younes. Computing large deformation metric mappings via geodesic flows of diffeomorphisms. *Int J. Comp. Vis.*, 61(2):139–157, 2005.

22. S. Belongie, J. Malik, and J. Puzicha. Shape matching and object recognition using shape contexts. *IEEE Trans. PAMI*, 24(24):509–522, 2002.

23. J.-D. Benamou and Y. Brénier. A computational fluid mechanics solution to the Monge-Kantorovich mass transfer problem. *Numer. Math.*, 84(3):375–393, 2000.

24. H. Blum. A transformation for extracting new descriptors of shape. In W Dunn, editor, *Proc. Symp. Models for the Perception of Speech and Visual Form*, pages 362–380. MIR press, 1967.

25. S. Bochner. Integration von funktionen, deren werte die elemente eines vectorraumes sind. *Fundamenta Mathmaticae*, 20:262–276, 1933.

26. J.-D. Boissonnat and M. Teillaud. *Effective Computational Geometry for Curves and Surfaces*. Springer, 2007.

27. J. F. Bonnans, J.-C. Gilbert, C. Lemaréchal, and C. A. Sagastizábla. *Numerical Optimization: Theoretical and Practical Aspects*. Springer, 2009. Second edition.

28. F. L. Bookstein. Principal warps: Thin plate splines and the decomposition of deformations. *IEEE Trans. PAMI*, 11(6):567–585, 1989.

29. F. L. Bookstein. *Morphometric Tools for Landmark Data; Geometry and Biology*. Cambridge University Press, 1991.

30. W. Boothby. *An Introduction to Differentiable Manifolds and Riemannian Geometry*. Academic Press, 2002. Original edition 1986.

31. Y. Brénier. Polar factorization and monotone rearrangement of vector-valued functions. *Comm. Pure and Applied Math.*, 44:375–417, 1991.

32. H. Brezis. *Analyse Fonctionnelle, Théorie et Applications*. Masson, Paris, 1983. English translation: Springer-Verlag.

33. A. M. Bruckstein, E. Rivlin, and I. Weiss. Scale space semi-local invariants. *Image and Vision Computing*, 15:335–344, 1997.

34. L. Caffarelli. The regularity of mappings with a convex potential. *J. Amer. Math. Soc.*, 5:99–104, 1992.

35. R. Camassa and D. D. Holm. An integrable shallow water equation with peaked solitons. *Phys. Rev. Lett.*, 71:1661–1664, 1993.

36. V. Camion and L. Younes. Geodesic interpolating splines. In M Figueiredo, J Zerubia, and K Jain, A, editors, *EMMCVPR 2001*, volume 2134 of *Lecture Notes in Computer Science*. Springer, 2001.

37. J. Canny. A computational approach to edge detection. *IEEE PAMI*, 6:679–698, 1986.

38. F. Cao. *Geometric Curve Evolution and Image Processing*. Springer, 2003.

39. F. Cao, J.-L. Lisani, J.-M. Morel, P. Musé, and F. Sur. *A Theory of Shape Identification*. Springer, 2008.

40. E. Cartan. *La Théorie des Groupes Finis et Continus et la Géométrie Différentielle Traitée par la Méthode du Repère Mobile*. Jacques Gabay, 1992. Edition originale Gauthiers-Villars 1937.

41. V. Caselles, R. Kimmel, and G. Sapiro. Geodesic active contours. In *Proceedings of the 1995 ECCV*, 1995.

42. S. L. Chan and E. O. Purisima. A new tetrahedral tesselation scheme for isosurface generation. *Computer and Graphics*, 22(1):83–90, 1998.

43. T. F. Chan and L. A. Vese. Active contours without edges. *IEEE Trans Image Proc.*, 10(2):266–277, 2001.

44. S. S. Chern. Curves and surfaces in euclidean space. In S. S. Chern, editor, *Studies in Global Geometry and Analysis*, 1967.

45. E. V. Chernyaev. Matching cubes 33: Construction of topologically correct isosurfaces. Technical report, CERN, 1995. Available at http://wwwinfo.cern.ch/asdoc/psdir/mc.ps.gz.

46. H.-I. Choi, S.-W. Choi, and H.-P. Moon. Mathematical theory of medial axis transform. *Pacific Journal of Mathematics*, 181(1):57–88, 1997.

47. K.-S. Chou and X.-P. Zhu. *The Curve Shortening Problem*. Chapman & Hall, 2001.

48. G. E. Christensen, R. D. Rabbitt, and M. I. Miller. Deformable templates using large deformation kinematics. *IEEE Trans. Image Proc.*, 1996.

49. E. A. Coddington and N. Levinson. *Theory of Ordinary Differential Equations*. McGraw-Hill, 1955.

50. L. D. Cohen. On active contours models and balloons. *Computer Vision, Graphics and Image Processing: Image Understanding*, 53(2):211–218, 1991.

51. D. Cohen-Steiner and J.-M. Morvan. Second fundamental measure of geometric sets and local approximation of curvature. *J. Differential Geometry*, 74:363–394, 2006.

52. H. Cohn. *Conformal Mapping in Riemann Surfaces*. Dover Publications, 1980.

53. J. L. Coolidge. *A Treatise on Algebraic Plane Curves*. Dover, 1959.

54. T. F. Cootes, C. J. Taylor, D. H. Cooper, and J. Graham. Active shape models: their training and application. *Comp. Vis. and Image Understanding*, 61(1):38–59, 1995.

55. R. Courant and D. Hilbert. *Methods of Mathematical Physics, I*. J. Wiley and Sons, 1955. Original edition: 1937.

56. T. M. Cover and J. A. Thomas. *Elements of Information Theory*. John Wiley and Sons, Inc., 1991.

57. J. Damon. Smoothness and geometry of boundaries associated to skeletal structures i: Sufficient conditions for smoothness. *Ann. Inst. Fourier*, 53:1941–1985, 2003.

58. J. Damon. Smoothness and geometry of boundaries associated to skeletal structures ii: Geometry in the blum case. *Compositio Mathematica*, 140(6):1657–1674, 2004.

59. J. Damon. Determining the geometry of boundaries of objects from medial data. *Int. J. Comp. Vis.*, 63(1):45–64, 2005.

60. M. C. Delfour and J.-P. Zolésio. *Shapes and Geometries. Analysis, Differential Calculus and Optimization.* SIAM, 2001.

61. F. Dibos, V. Caselles, T. Coll, and F. Catté. Automatic contours detection in image processing. In V Lakshmikantham, editor, *Proceedings of the First World Congress of Nonlinear Analysts,* pages 1911–1921. de Gruyter, 1996. .

62. J. Diestel and J. J. Uhl. *Vector Measures.* Mathematical Surveys and Monographs. American Mathematical Society, 1977.

63. Discrete differential geometry: An applied introduction, 2006. Course at SIGGRAPH'06.

64. M. P. Do Carmo. *Differential Geometry of Curves and Surfaces.* Prentice Hall, 1976.

65. M. P. Do Carmo. *Riemannian Geometry.* Birkaüser, 1992.

66. J. R. Dormand and P. J. Prince. A family of embedded runge-kutta formulae. *Journal of Computational and Applied Mathematics,* 6(1):19–26, 1980.

67. T. R. Downie, L. Shepstone, and B. W. Silverman. A wavelet approach to deformable templates. In V Mardia, K, A Gill, C, and L Dryden, I, editors, *Image Fusion and Shape Variability Techniques,* pages 163–169. Leeds University Press, 1996.

68. M. Droske and M. Rumpf. A variational approach to non-rigid morphological registration. *SIAM Appl. Math.,* 64(2):668–687, 2004.

69. I. L. Dryden and K. V. Mardia. *Statistical Shape Analysis.* John Wiley and Sons, 1998.

70. P. Dupuis, U. Grenander, and M. Miller. Variational problems on flows of diffeomorphisms for image matching. *Quarterly of Appl. Math.,* 1998.

71. N. Dyn. Interpolation and approximation by radial and related functions. In C. K. Chui, L. L. Shumaker, and J. D. Ward, editors, *Approximation Theory VI: vol. 1,* pages 211–234. Academic Press, 1989.

72. G. Dziuk, E. Kuwert, and R. Schäzle. Evolution of elastic curves in R^n: existence and computation. *SIAM J. Math. Anal.,* 33(5):1228–1245, 2002.

73. Hairer. E., C. Lubich, and G. Wanner. *Geometric Numerical Integration.* Springer, 2006. Second Edition.

74. H. Edelsbrunner. *Geometry and Topology for Mesh Generation.* Cambridge Monographs on Applied and Computational Mathematics. Cambridge Unversity Press, 2001.

75. C. L. Epstein and M. Gage. The curve shortening flow. In A. J. Chorin and A. J. Madja, editors, *Wave Motion.* Springer-Verlag, 1987.

76. L. C. Evans and J. Spruck. Motion of level sets by mean curvature I. *J. Diff. Geom.,* 33:635–681, 1991.

77. O. Faugeras. Cartan's moving frame method and its application to the geometry and evolution of curves in the euclidean, affine and projective plane. In J. L. Mundy, A. Zisserman, and D. Forsyth, editors, *Applications of Invariance in Computer Vision,* volume 825 of *Lecture notes in computer sciences,* pages 11–46. Springer Verlag, 1994.

78. O. Faugeras and Q.-T. Luong. *The Geometry of Multiple Images.* the MIT Press, 2001.

79. A. C. Faul, G. Goodsell, and M. J. D. Powell. A Krylov subspace algorithm for multiquadric interpolation in many dimensions. *IMA J. Num. Anal.,* 25:1–24, 2005.

80. H. Federer. Curvature measures. *Trans. Amer. Math. Soc.,* 93:418–491, 1959.

81. H. Federer. *Geometric Measure Theory.* Springer, 1969.

82. M. Fels and P. J. Olver. Moving coframes I. a practical algorithm. *Acta Appl. Math.*, 51:161–213, 1998.

83. M. Fels and P. J. Olver. Moving coframes II. regularization and theoretical foundations. *Acta Appl. Math.*, 55:127–208, 1999.

84. M. A. Fishler and R. A. Elschager. The representation and matching of pictorial structures. *IEEE Trans. Computers*, C-22:67–92, 73.

85. B. Fletcher. *Practical Methods of Optimization.* Wiley, 2000. Second edition.

86. C. Foias, D. D. Holm, and E. S. Titi. The Navier-Stokes-alpha model of fluid turbulence. *Physica D*, 152:505–519, 2001.

87. M. Gage. An isoperimetric inequality with applications to curve shortening. *Duke Math. J.*, 50(4):1225–1229, 1983.

88. M. Gage and R. S. Hamilton. The heat equation shrinking convex plane curves. *J. Differential Geometry*, 53:69–96, 1986.

89. W. Gangbo and R. McCann. The geometry of optimal transportation. *Acta Mathematica*, 177(113-161), 1996.

90. L. Garcin and L. Younes. Geodesic image matching: A wavelet based energy minimization scheme. In *EMMCVPR'05*, pages 349–364, 2005.

91. L. Garcin and L. Younes. Geodesic matching with free extremities. *J. Math. Imag. Vis.*, 25(3):329–340, 2006.

92. J. C. Gee and P. D. Peralta. Continuum models for Bayesian image matching. In K. H. Hanson and R. N. Silver, editors, *Maximum Entropy and Bayesian Methods.* Kluwer Academic, 1995.

93. P. J. Giblin and B. B. Kimia. On the local form and transitions of symmetry sets, and medial axes, and shocks in 2D. In *Proc. of ICCV*, pages 385–391, Greece, September 1999. IEEE Computer Society.

94. C. A. Glasbey and K. V. Mardia. A review of image-warping methods. *J. of Applied Stat.*, 25(2):155–171, 1998.

95. J. Glaunès. *Transport par difféomorphismes de points, de mesures et de courants pour la comparaison de formes et l'anatomie numérique.* PhD thesis, University Paris 13, 2005. in French.

96. J. Glaunès, A. Qiu, Miller M. I., and L. Younes. Large deformation diffeomorphic curve matching. *International Journal of Computer Vision*, 80(3):317–336, 2008.

97. J. Glaunès, A. Trouvé, and L. Younes. Diffeomorphic matching of distributions: A new approach for unlabelled point-sets and sub-manifolds matching. In *Proceedings of CVPR'04*, 2004.

98. J. Glaunès, M. Vaillant, and M. I. Miller. Landmark matching via large deformation diffeomorphisms on the sphere. *J. Math. Imag. Vis.*, 20:179–200, 2004.

99. G. H. Golub and C. F. Van Loan. *Matrix Computations.* Johns Hopkins University Press, 1996. Third Edition.

100. M. Grayson. The heat equation shrinks embedded plane curves to round points. *J. Differential Geometry*, 26:285–314, 1987.

101. U. Grenander. *Lectures in Pattern Theory*, volume 33. Applied Mathematical Sciences, 1981.

102. U. Grenander. *General Pattern Theory.* Oxford Science Publications, 1993.

103. U. Grenander, Y. Chow, and D. M. Keenan. *Hands: a Pattern Theoretic Study of Biological Shapes.* Springer-Verlag, 1991.

104. U. Grenander and M. I. Miller. *Pattern Theory: From Representation to Knowledge*. Oxford University Press, 2007.

105. N. A. Gumerov and R. Duraiswami. *Fast Multipole Methods for the Helmholtz Equation in Three Dimensions*. Elsevier, 2004.

106. N. A. Gumerov and R. Duraiswami. Fast radial basis function interpolation via preconditioned Krylov iteration. *SIAM J. Sci. Comp.*, 29(5):1876–1899, 2008.

107. S. Helgason. *Differential Geometry, Lie Groups and Symmetric Spaces*. Academic Press, 1978.

108. P. A. Helm, L. Younes, M. L. Beg, D. B. Ennis, C. Leclercq, O. P. Faris, E. McVeigh, K. Kass, M. I. Miller, and R. L. Winslow. Evidence of structural remodeling in the dyssynchronous failing heart. *Circulation Research*, 98(1):125–132, 2006.

109. G. Hermosillo. *Variational methods for multimodal matching*. PhD thesis, Université de Nice, 2002.

110. G. Hermosillo, C. Chefd'Hotel, and O. Faugeras. Variational methods for multimodal image matching. *Int. J. Comp. Vis.*, 50(3):329–343, 2002.

111. C. Hilditch. Linear skeleton from square cupboards. *Machine intelligence*, 6:403–409, 1969.

112. D. D. Holm. *Geometric Mechanics*. Imperial College Press, 2008.

113. D. D. Holm, J. E. Marsden, and T. S. Ratiu. The Euler–Poincaré equations and semidirect products with applications to continuum theories. *Adv. in Math.*, 137:1–81, 1998.

114. D. R. Holm, J. T. Ratnanather, A. Trouvé, and L. Younes. Soliton dynamics in computational anatomy. *Neuroimage*, 23:S170–S178, 2004.

115. D. R. Holm, A. Trouvé, and L. Younes. The euler poincaré theory of metamorphosis. *Quarterly of Applied Mathematics*, 2009. (to appear).

116. H. Hoppe, T. DeRose, T. Duchamp, J. McDonald, and W. Stuetzle. Surface reconstruction from unorganized points. In *ACM SIGGRAPH 1992*, 1992.

117. M. K. Hu. Visual pattern recognition by moment invariants. *IRE Trans. Inf. Theory*, IT-8:179–187, 1962.

118. S. Joshi. *Large Deformation Diffeomorphisms and Gaussian Random Fields for Statistical Characterization of Brain Sub-manifolds*. PhD thesis, Sever institute of technology, Washington University, 1997.

119. S. Joshi and M. Miller. Landmark matching via large deformation diffeomorphisms. *IEEE transactions in Image Processing*, 9(8):1357–1370, 2000.

120. S.H. Joshi, E. Klassen, A. Srivastava, and I. Jermyn. A novel representation for Riemannian analysis of elastic curves in r^n. In *Proceedings of CVPR'07*, 2007.

121. J. Jost. *Riemannian Geometry and Geometric Analysis*. Springer, 1998. 2nd edition.

122. M. Kass, A. Witkin, and D. Terzopoulos. Snakes: active contour models. *Int. J. of Comp. Vision*, 1988.

123. D. G. Kendall. Shape manifolds, Procrustean metrics and complex projective spaces. *Bull. London Math. Soc.*, 16:81–121, 1984.

124. D. G. Kendall, D. Barden, T. K. Carne, and H. Le. *Shape and Shape Theory*. Wiley, 1999.

125. D. Keren, D. B. Cooper, and J. Subrahmonia. Describing complicated objects by implicit polynomials. *PAMI*, 16(1):38–53, January 1994.

126. R. Kimmel, D. Shaked, N. Kiryati, and A. M. Bruckstein. Skeletonization via distance maps and level sets. *Computer Vision and Image Understanding*, 62(3):382–391, 1995.

127. E. Klassen, A. Srivastava, W. Mio, and S. H. Joshi. Analysis of planar shapes using geodesic paths on shape spaces. *IEEE Trans. Pattern Anal. Mach. Intell.*, 26(3):372–383, 2004.

128. N. Koiso. On the motion of a curve towards elastica. In *Actes de la Table Ronde de Géométrie Différentielle, Luminy 1992*, Sémin. Congr. 1. Soc. Math. France, 1996.

129. G. Letac. Mesures sur le cercle et convexe du plan. *Annales scientifiques de l'université de Clermont-Ferrand II*, 76:35–65, 1983.

130. W. E. Lorensen and H. E. Cline. Marching cubes: a high resolution 3d surface construction algorithm. In *Proceedings of SIGGRAPH 87*, 1987.

131. D. G. Luenberger. *Optimization by Vector Space Methods*. J. Wiley and Sons, 1969.

132. Y. Ma, S. Soatto, J. Kosecka, and S. S. Sastry. *An Invitation to 3-D Vision: From Images to Geometric Models*. Springer-Verlag, 2003.

133. F. Maes, A. Collignon, D. Vandermeulen, G. Marchal, and P. Suetens. Multimodality image registration by maximization of mutual information. *IEEE trans. Med. Imag.*, 16(2):187–198, 1997.

134. D. Marr and E. Hildreth. Theory of edge detection. *Proc. of the Royal Soc. of London, Series B*, 207(1167):187–217, 1980.

135. J. E. Marsden. *Lectures on Geometric Mechanics*. Cambridge University Press, 1992.

136. J. E. Marsden and T. S. Ratiu. *Introduction to Mechanics and Symmetry*. Springer, 1999.

137. S. Marsland and C. J. Twining. Clamped-plate splines and the optimal flow of bounded diffeomorphisms. In *In Statistics of Large Datasets, Proceedings of Leeds Annual Statistical Research Workshop*, pages 91–95, 2002.

138. R. McCann. Existence and uniqueness of monotone measure-preserving maps. *Duke Mathematical J.*, 80(2):309–323, 1995.

139. R. I. McLachlan and G. R. W. Quispel. Six lectures on geometric integration. In R. De Vore, A. Iserles, and E. Süli, editors, *Foundations of Computational Mathematics*, pages 155–210. Cambridge University Press, 2001.

140. J. Meinguet. Multivariate interpolation at arbitrary points made simple. *J. Applied Math. and Physics*, 30:292–304, 1979.

141. A. Mennucci and A. Yezzi. Metrics in the space of curves. Technical report, arXiv:math.DG/0412454 v2, 2005.

142. M. Meyer, M. Desbrun, P. Schroeder, and A. Barr. Discrete differential geometry operators for triangulated 2-manifolds. In *VisMath '02 Proceedings*, 2002.

143. P. W. Michor. *Topics in Differential Geometry*. Graduate Studies in Mathematics. American Mathematical Society, 2008.

144. P. W. Michor and D. Mumford. Vanishing geodesic distance on spaces of submanifolds and diffeomorphisms. *Documenta Mathematica*, 10:217–245, 2005.

145. P. W. Michor and D. Mumford. Riemannian geometries on spaces of plane curves. *J. Eur. Math. Soc.*, 8:1–48, 2006.

146. P. W. Michor and D. Mumford. An overview of the Riemannian metrics on spaces of curves using the Hamiltonian approach. *Applied and Computational Harmonic Analysis*, 23(1):74–113, 2007.

426   References

147. M. I. Miller, S. C. Joshi, and G. E. Christensen. Large deformation fluid diffeomorphisms for landmark and image matching. In A. Toga, editor, *Brain Warping*, pages 115–131. Academic Press, 1999.

148. M. I. Miller, A. Trouvé, and L. Younes. On the metrics and Euler-Lagrange equations of computational anatomy. *Annual Review of Biomedical Engineering*, 4:375–405, 2002.

149. M. I. Miller, A. Trouvé, and L. Younes. Geodesic shooting for computational anatomy. *J. Math. Image and Vision*, 24(2):209–228, 2006.

150. M. I. Miller and L. Younes. Group action, diffeomorphism and matching: a general framework. *Int. J. Comp. Vis*, 41:61–84, 2001. Originally published in electronic form in: Proceedings of SCTV 99, http://www.cis.ohio-state.edu/ szhu/SCTV99.html.

151. C. Montani, R. Scateni, and R. Scopigno. A modified look-up table for implicit disambiguation of marching cubes. *The Visual Computer*, 10:353–355, 1994.

152. F. Morgan. *Geometric Measure Theory: a Beginner's Guide*. Academic Press, 1988.

153. S. Mukhopadhyaya. New methods in the geometry of plane arc. *Bull. Calcutta Math. Soc.*, 1:31–37, 1909.

154. D. Mumford and J. Shah. Optimal approximation by piecewise smooth functions and associated variational problems. *Comm. Pure Appl. Math.*, 42:577–685, 1989.

155. B. K. Natarajan. On generating topologically consistent surfaces from uniform samples. *The Visual Computer*, 11(1):52–62, 1994.

156. G. M. Nielson. On marching cubes. *IEEE Transactions on Visualization and Computer Graphics*, 9(3):283–297, 2003.

157. G. M. Nielson and B. Hamann. The asymptotic decider: removing the ambiguity in marching cubes. In G. M. Nielson and L. J. Rosenblum, editors, *IEEE Visualization'91*, pages 83–91, 1991.

158. J. Nocedal and S. J. Wright. *Numerical Optimization*. Springer, 1999.

159. R. L. Ogniewicz. *Discrete Voronoï Skeletons*. Swiss federal institue of technology, 1992.

160. K. B. Oltham and J. Spanier. *An Atlas of Functions*. Springer, 1996.

161. P. J. Olver. *Equivalence, Invariants and Symmetry*. Cambridge University Press, 1995.

162. J. O'Rourke. *Computational Geometry in C*. Cambridge University Press, 2000.

163. S. Osher and R. Fedkiw. *Level Set Methods and Dynamic Implicit Surfaces*. Springer, 2003.

164. R. Osserman. The four of more vertex theorem. *The American Mathematical Monthly*, 92(5):332–337, 1985.

165. A. Pentland and S. Sclaroff. Closed-form solutions for physically-based shape modeling and recognition. *IEEE TPAMI*, 13(7):715–729, 1991.

166. F. P. Preparata and M. I. Shamos. *Computational Geometry: an Introduction*. Springer, 1985.

167. W. H Press, S. A. Teukolsky, W. T. Vetterling, and B. P. Flannery. *Numerical Recipes in C, The Art of Scientific Computing*. Cambridge University Press, second edition, 1992.

168. A. Qiu, L. Younes, L. Wang, J. T. Ratnanather, S. K. Gillepsie, K. Kaplan, J. Csernansky, and M. I. Miller. Combining anatomical manifold information

via diffeomorphic metric mappings for studying cortical thinning of the cingulate gyrus in schizophrenia. *NeuroImage*, 37(3):821–833, 2007.

169. R. D. Rabbitt, J. A. Weiss, G. E. Christensen, and M. I. Miller. Mapping of hyperelastic deformable templates using the finite element method. In *Proceedings of San Diego's SPIE Conference*, 1995.

170. S. T. Rachev. *Probability Metrics*. Wiley, 1991.

171. A. Rangarajan, H. Chui, and F. L. Bookstein. The softassign procrustes matching algorithm. In *Information Processing in Medical Imaging*, pages 29–42. Springer, 1997.

172. A Rangarajan, E. Mjolsness, S. Pappu, L. Davachi, P. S. Goldman-Rakic, and J. S. Duncan. A robust point matching algorithm for autoradiograph alignment. In K. H. Hohne and R. Kikinis, editors, *Visualization in Biomedical Computing (VBC)*, pages 277–286, 1996.

173. F. Richard. Résolution de problèmes hyperélastiques de recalage d'images. *C. R. Acad. Sci., serie I*, 335:295–299, 2002.

174. F. Riesz and B. S. Nagy. *Functional Analysis*. Ungar Pub. Co., 1955.

175. E. Rouy and A. Tourin. A viscosity solutions approach to shape from shading. *SIAM J. Numer. Analy.*, 29(3):867–884, 1992.

176. W. Rudin. *Real and Complex Analysis*. Tata McGraw Hill, 1966.

177. F. A. Sadjadi and E. L. Hall. Three-dimensional moment invariants. *IEEE TPAMI*, 2:127–136, 1980.

178. G. Sapiro. *Geometric Partial Differential Equations and Image Analysis*. Cambridge University Press, 2001.

179. I. J. Schoenberg. Metric spaces and completely monotone functions. *The Annals of Math.*, 39(4):811–841, 1938.

180. B. Schroder. *Mathematical Analysis: A Concise Introduction*. Wiley, 2007.

181. S. Sclaroff. *Modal Matching: a Method for Describing, Comparing and Manipulating Signals*. PhD thesis, MIT, 1995.

182. J. Serra. *Image Analysis and Mathematical Morphology vol 1 & 2*. Academic press, 1982, 88.

183. J. Sethian. *Level Set Methods and Fast Marching Methods: Evolving Interfaces in Computational Geometry, Fluid Mechanics, Computer Vision and Material Science*. Cambridge University Press, 1996. Second edition: 1999.

184. J. A. Sethian. A review of recent numerical algorithms for hypersurface moving for curvature dependent speed. *J. Differential Geometry*, 31:131–161, 1989.

185. J. A. Sethian. A fast marching level set method for monotonically advancing fronts. *Proc. Nat. Acad. Sci.*, 93(4):1591–1595, 1996.

186. R. Shaback. Error estimates and condition numbers for radial basis function interpolation. *Advances in Computational Mathematics*, 3:251–264, 1995.

187. J. Shah. H^0 type Riemannian metrics on the space of planar curves. *Quarterly of Applied Mathematics*, 66:123–137, 2008.

188. E. Sharon and D. Mumford. 2d-shape analysis using conformal mapping. In *Proceedings IEEE Conference on Computer Vision and Pattern Recognition*, 2004.

189. E. Sharon and D. Mumford. 2d-shape analysis using conformal mapping. *Int. J. Comput. Vision*, 70(1):55–75, 2006.

190. K. Siddiqi, B. Kimia, A. Tannenbaum, and A. Zucker. Shocks, shapes, and wiggles. *Image and Vision Computing*, 17(5-6):365–373, 1999.

191. K. Siddiqi, B. B. Kimia, and C.-W. Shu. Geometric shock-capturing ENO schemes for subpixel interpolation, computation and curve evolution. *GMIP*, 59(5):278–301, September 1997.

192. K. Siddiqi, K. J. Tresness, and B. B. Kimia. On the anatomy of visual form. In C. Arcelli, L.P. Cordella, and G. S. di Baja, editors, *Aspects of Visual Form Processing*, 2nd International Workshop on Visual Form, pages 507–521, Singapore, May-June 1994. IAPR, World Scientific. Workshop held in Capri, Italy.

193. C. Small. *The Statistical Theory of Shape*. Springer, 1996.

194. K. Stephenson. *Introduction to Circle Packing*. Cambridge University Press, 2005.

195. J. Stoer and R. Bulirsch. *Introduction to Numerical Analysis*. Springer-Verlag, 1992. Second Edition.

196. E. Süli and D. F. Mayers. *An Introduction to Numerical Analysis*. Cambridge University Press, 2003.

197. G. Sundaramoorthi, A. Yezzi, and A. Mennucci. Sobolev active contours. In *Variational, Geometric and Level Set Methods, VLSM'05*, volume 3752/2005 of *Lecture Notes in Computer Science*, pages 109–120. Springer, 2005.

198. G. Taubin. Estimating the tensor of curvatures of a surface from a polyhedral interpolation. In *Proceedings of ICCV'95*, 1995.

199. M. R. Teague. Image analysis via the general theory of moments. *J. Opt. Soc. Am*, 70(8):920–930, 1980.

200. H. Tek and B. B. Kimia. Curve evolution, wave propagation, and mathematical morphology. In Henk J. A. M. Heijmans and Jos B.T.M. Roerdink, editors, *Mathematical Morphology and its Applications to Image and Signal Processing*, volume 12 of *Computational Imaging and Vision*, pages 115–126. Kluwer Academic, Amsterdam, 1998.

201. J.-P. Thirion. Image matching as a diffusion process: an analogy with Maxwell's demons. *Medical Image Analysis*, 2(3):243–260, 1998.

202. A. Trouvé. Habilitation à diriger les recherches. Technical report, Université Paris XI, 1996.

203. A. Trouvé. Diffeomorphism groups and pattern matching in image analysis. *Int. J. of Comp. Vis.*, 28(3):213–221, 1998.

204. A. Trouvé and L. Younes. Diffeomorphic matching in 1d: designing and minimizing matching functionals. In D. Vernon, editor, *Proceedings of ECCV 2000*, 2000.

205. A. Trouvé and L. Younes. Local geometry of deformable templates. *SIAM J. Math. Anal.*, 37(1):17–59, 2005.

206. A. Trouvé and L. Younes. Metamorphoses through lie group action. *Found. Comp. Math.*, pages 173–198, 2005.

207. M. Vaillant and J. Glaunès. Surface matching via currents. In G. E. Christensen and M. Sonka, editors, *Proceedings of Information Processing in Medical Imaging (IPMI 2005)*, number 3565 in Lecture Notes in Computer Science. Springer, 2005.

208. M. Vaillant, M. I. Miller, A. Trouvé, and L. Younes. Statistics on diffeomorphisms via tangent space representations. *Neuroimage*, 23(S1):S161–S169, 2004.

209. A. Van Gelder and J. Wilhelms. Topological considerations in isosurface generation. *ACM Transactions on Graphics*, 13(4):337–375, 1994.

210. F.-X. Vialard. Hamiltonian approach to geodesic image matching. Technical report, ArXiv Mathematics, 2008.

211. F.-X. Vialard. *Hamiltonian Approach to Shape Spaces in a Diffeomorphic Framework: From the Discontinuous Image Matching Problem to a Stochastic Growth Model.* PhD thesis, Ecole Normale Supérieure de Cachan, 2009. http://tel.archives-ouvertes.fr/tel-00400379/fr/.

212. F.-X. Vialard and F. Santambrogio. Extension to bv functions of the large deformation diffeomorphisms matching approach. *Comptes Rendus Mathematiques*, 347(1-2):27–32, 2009.

213. C. Villani. *Topics in Optimal Transportation.* American Mathematical Society, 2003.

214. P. Viola and W. M. Wells III. Alignment by maximization of mutual information. *Int. J. Comp. Vis.*, 24(2):137–154, 1997.

215. G. Wahba. *Spline Models for Observational Data.* SIAM, 1990.

216. L. Wang, M. F. Beg, J. T. Ratnanather, C. Ceritoglu, L. Younes, J. Morris, J. Csernansky, and M. I. Miller. Large deformation diffeomorphism and momentum based hippocampal shape discrimination in dementia of the Alzheimer type. *IEEE Transactions on Medical Imaging*, 26:462–470, 2006.

217. L. N. Wasserstein. Markov processes over denumerable products of spaces describing large systems of automata. *Problems of Information Transmission*, 5:47–52, 1969.

218. T. Willmore. *Riemannian Geometry.* Clarendon Press, 1993.

219. C. Xu and J. Prince. Gradient vector flow: A new external force for snakes. In *Proceedings of the 1997 Conference on Computer Vision and Pattern Recognition (CVPR '97)*, 1997.

220. T. Yanao, W. S. Koon, and J. E. Marsden. Intramolecular energy transfer and the driving mechanisms for large-amplitude collective motions of clusters. *The Journal of Chemical Physics*, 130(14):144111, 2009.

221. K. Yosida. *Functional Analysis.* Springer, 1970.

222. L. Younes. Computable elastic distances between shapes. *SIAM J. Appl. Math*, 58(2):565–586, 1998.

223. L. Younes. Jacobi fields in groups of diffeomorphisms and applications. *Quart. Appl. Math.*, 65:113–134, 2007.

224. L. Younes, F. Arrate, and M. I. Miller. Evolutions equations in computational anatomy. *NeuroImage*, 45(1, Supplement 1):S40–S50, 2009. Mathematics in Brain Imaging.

225. L. Younes, P. Michor, J. Shah, and D. Mumford. A metric on shape spaces with explicit geodesics. *Rend. Lincei Mat. Appl.*, 9:25–57, 2008.

226. Laurent Younes. A distance for elastic matching in object recognition. *C. R. Acad. Sci. Paris Sér. I Math.*, 322(2):197–202, 1996.

227. M. Zähle. Integral and current representation of curvature measures. *Arch. Math.*, 46:557–567, 1986.

228. E. Zeidler. *Applied Functional Analysis. Applications to Mathematical Physics.* Springer, 1995.

229. H.-K. Zhao, T. Chan, B. Merriman, and S. Osher. A variational level set approach to multiphase motion. *J. Comput. Phys.*, 127:179–195, 1996.

Index